Defending the Master Race

UNIVERSITY OF VERMONT PRESS

Burlington, Vermont

Published by University Press of New England

Hanover and London

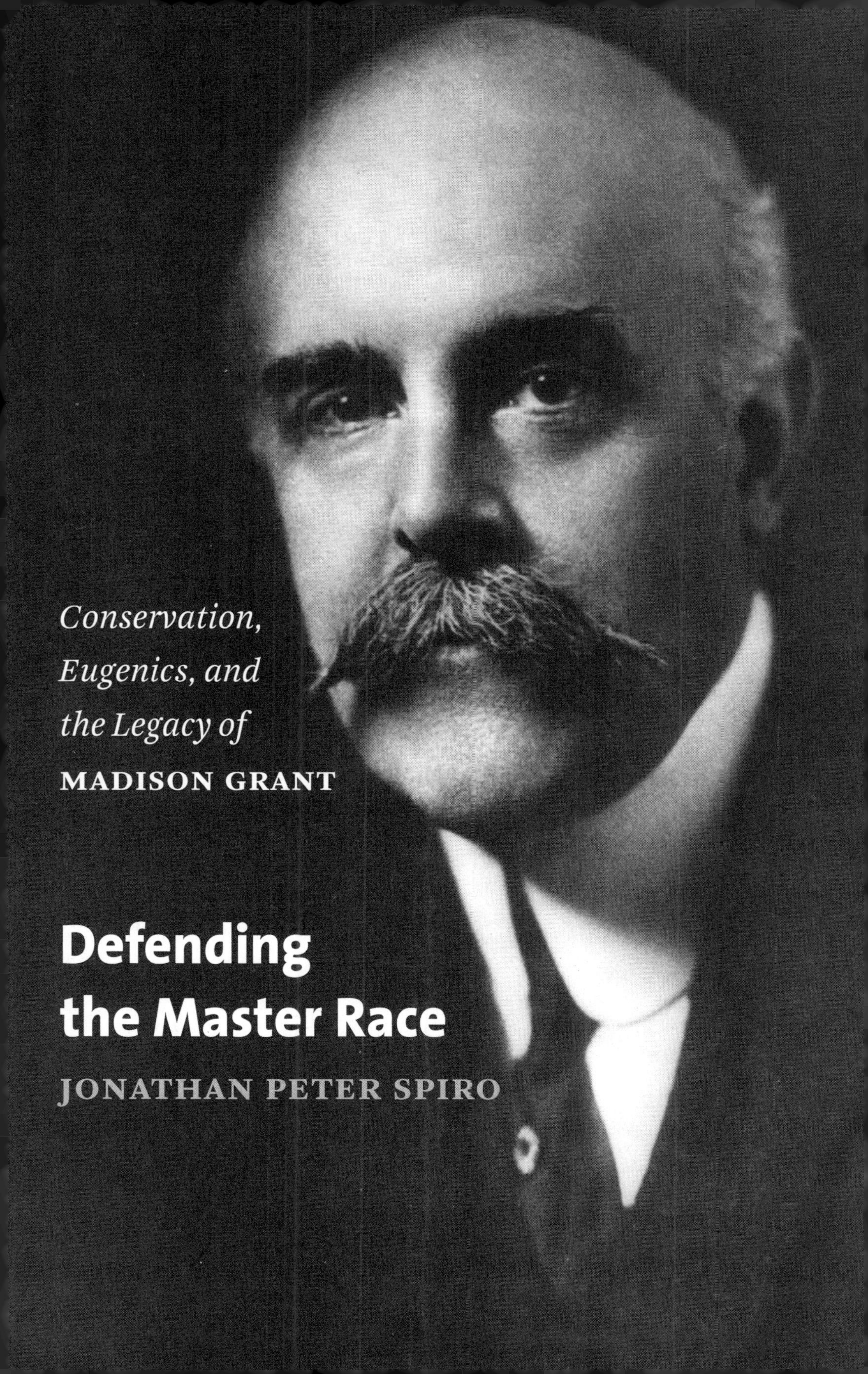

Conservation, Eugenics, and the Legacy of **MADISON GRANT**

Defending the Master Race

JONATHAN PETER SPIRO

UNIVERSITY OF VERMONT PRESS
Published by University Press of New England,
One Court Street, Lebanon, NH 03766
www.upne.com

Printed in the United States of America
5 4 3 2 1

Library of Congress Cataloging-in-Publication Data
Spiro, Jonathan Peter.
Defending the master race: conservation, eugenics, and the legacy of Madison Grant / Jonathan Peter Spiro.
p. cm.
Includes bibliographical references and index.
ISBN 978-1-58465-715-6 (cloth: alk. paper)
1. Grant, Madison, 1865–1937. 2. Grant, Madison, 1865–1937—Influence. 3. Conservationists—United States—Biography. 4. Hunters—United States—Biography. 5. Big game hunting—History. 6. Wildlife management—United States—History. 7. Zoologists—New York (State)—New York—Biography. 8. Eugenics—United States—History. 9. Racism—United States—History. 10. United States—Race relations. I. Title.
CT275.G677S65 2008
305.8'00973—dc22
[B] 2008038610

Dedicated with
immeasurable love and gratitude to
AUDREY AND MELFORD SPIRO—
the one instance where I hope that
Madison Grant was right when he
claimed that the sole determinants
of what we become are the genes
of our parents

Contents

Madison Grant

The Consensus

1940s
"The high priest of racialism in America."
Gunnar Myrdal

1950s
"Intellectually the most important nativist in recent American history."
John Higham

1960s
"The nation's most influential racist."
Mark Haller

1970s
"The dean of American racists."
Ethel W. Hedlin

1980s
"The most famous of the new scholars of race."
Page Smith

1990s
"One of the nation's foremost racists."
Steven Selden

2000s
"The great patriarch of scientific racism."
Matthew Guterl

Introduction

At the conclusion of World War II, the American Military Tribunal at Nuremberg indicted Major General Karl Brandt of the Waffen-SS for conspiracy to commit war crimes and crimes against humanity. Brandt had been Adolf Hitler's personal physician and the most important medical authority in the Third Reich. The specific crimes charged in the case of *United States of America v. Karl Brandt et al.* fell into three categories:

1. Implementing a "euthanasia" program in which the sick, the aged, the mentally ill, and the members of racial minorities were secretly executed in gas chambers.
2. Murdering concentration camp prisoners for the express purpose of collecting their skulls for research.
3. Performing medical experiments on defenseless death camp inmates against their will. These experiments involved sterilizing healthy men and women; forcing subjects to ingest lethal amounts of poison or seawater; performing mutilating and crippling bone, muscle, and nerve operations; and exposing inmates to typhus, malaria, yellow fever, mustard gas, smallpox, burning phosphorus, freezing temperature, high altitude, and epidemic jaundice.

In his defense, Brandt introduced into evidence a book published in Munich in 1925 that had vigorously advocated and justified the elimination of inferior peoples. Brandt highlighted for the court excerpts from the book that called on the state to destroy sickly infants and sterilize defective adults who were of no value to the community. Little wonder that upon reading the book, the Führer himself had announced: "This book is my Bible."

The American judges at Nuremberg were well aware that Brandt's defense exhibit was actually the German translation of a work originally published in the United States in 1916: *The Passing of the Great Race,* written by the prophet of scientific racism in America, Madison Grant. Grant's book held that mankind was divided into a series of hierarchically arranged subspecies, with the blond-haired, blue-eyed Nordics at the top of the ethnological pyramid and the other, less-worthy races falling into place beneath the master race. In the 1920s and 1930s, it had been quite common for congressmen to read aloud from Grant's book in the U.S. Capitol to argue for restricting the immigration of the "inferior" non-Nordic races and even to justify the lynching of African Americans. The Nuremberg judges therefore had to come to terms with the discomfiting irony that the Nazi doctor was tracing the roots of the Third Reich's eugenics program to a best-selling book by a recognized American scholar.

The tribunal nonetheless found Dr. Brandt guilty and sentenced him to death—and the world seemingly passed the same judgment on the philosophies espoused in *The Passing of the Great Race.* In fact, the very name of Madison Grant was consigned to the ash heap of history after World War II. But Grant and his ideas have been resurrected in the twenty-first century, where they simmer just below the surface of respectable society and inspire—and are promulgated on the websites of—various white-power groups and anti-immigration organizations.

There was a time, however, when Grant and his theories were accorded much greater respect. During the first four decades of the twentieth century, Grant was an important and admired figure who played a prominent role in several mainstream causes in the United States. Grant, for instance, was the leader of the eugenics movement, and in addition to convincing Congress to enact the immigration restriction legislation of the 1920s, his influence was crucial in the passage by a majority of the states of coercive sterilization statutes, by which tens of thousands of Americans deemed to be unworthy of procreation were sterilized from the 1930s to the 1970s. Grant also cooperated with southern white racists during this period to ban miscegenation, and he worked with northern black nationalists such as Marcus Garvey to repatriate America's Negroes back to Africa.

What is especially fascinating (or some might say distressing) is that even as Madison Grant sought to eliminate inferior races, he endeavored to preserve for posterity our nation's natural beauty, and along with his friend Theodore Roosevelt he became one of the founders of the conservation movement. Among his many accomplishments, Grant preserved the California redwoods, saved the American bison from extinction, founded the Bronx Zoo, fought for strict gun-control laws, built the Bronx River Parkway, helped to create Glacier and Denali National Parks, and worked tirelessly to protect the whales in the ocean, the bald eagles in the sky, and the pronghorn antelopes on the prairie.

In commemoration of his conservation efforts, the world's tallest tree, located in northern California, was dedicated to Madison Grant in 1931.

During the course of his life, Grant worked closely and became friends with a wide array of figures, including powerful politicians (e.g., Elihu Root, William Howard Taft, Franklin Delano Roosevelt), important naturalists (Gifford Pinchot, C. Hart Merriam, George Bird Grinnell), famous explorers (Carl Akeley, Lincoln Ellsworth, Admiral Peary), major philanthropists (Andrew Carnegie, George Eastman, John D. Rockefeller, Jr.), and leading scientists (Robert M. Yerkes, Edward L. Thorndike, and George Ellery Hale). And none of them thought that conservationism was incompatible with scientific racism. Grant dedicated his life to saving endangered fauna, flora, and natural resources; and it did not seem at all strange to his peers that he would also try to save his own endangered race. As Grant once explained to paleontologist Henry Fairfield Osborn: conservation and eugenics were two sides of the same coin, as both were "attempts to save as much as possible of the old America."[1]

It seems odd, at first glance, that a figure as diverse and influential as Madison Grant has not yet been the subject of a biography. Conservationists, of course, are more than a little reticent to acknowledge that one of our progenitors was a proto-Nazi. But even books on the history of the eugenics movement, nativism, immigration, or anti-Semitism—which almost always assert that Grant was one of the foremost racists in American history—usually devote just one or two paragraphs to his deeds. And it is always essentially the same one or two paragraphs: frankly, every scholar seems to be copying every previous scholar, in a scribal chain stretching back to the original obituary of Grant that appeared in the *New York Times* in 1937.

The main reason for the dearth of scholarship on Grant is that relatives destroyed his personal papers after his death in 1937. As this was a man who wrote hundreds of thousands of letters to scores of important persons during his lifetime, the loss to historians was immeasurable. It does not help that Grant shunned publicity and almost always refused requests from the press for interviews. Also, he never deigned to write his memoirs. When his friend William T. Hornaday urged him to write an autobiography, Grant declined on the grounds that "it is too much trouble and besides," he added mysteriously, "the things of real interest and importance would probably have to be omitted."[2]

Moreover, Grant seems particularly cursed by the gods of history. It is somewhat uncanny the number of fluke accidents that have befallen archival collections that we know at one time contained records relating to Grant. (One archive, for example, had a flood in which only the Grant documents, stored on the bottom shelves of the basement, suffered damage. In another archive, a well-meaning intern threw out a stack of letters from Grant that she mistakenly thought were copies of originals.) In addition, an inordinately large number of Grant's friends destroyed their personal papers. (Congressman Albert Johnson, for instance, who was the political leader of the immigration restriction movement and a close associate of Grant, burned his papers when he retired, thus eliminating a treasure trove of material on immigration restriction in general, and Madison Grant in particular.) Equally frustrating—and certainly more morally egregious—is the fact that Grant's correspondence with certain key figures who *did* save their papers has nonetheless "disappeared" from the archives. The boxes are there, but nothing is inside them. Whether this was effected by someone interested in protecting the reputation of Grant's friends, or by a scholar who intended to use the material to write a biography, we do not know; but the bleak result is the same.

Faced with this historiographical desert, historians have understandably given up trying to reconstruct the life—let alone the psychological motives or inner thoughts—of Madison Grant. Rushing in where angels fear to tread, I have scoured the newspapers of Grant's time and the memoirs of his peers, gleaning any and all mentions of Grant, and combed through the correspondence of his colleagues (dispersed in hundreds of archives throughout the country) attempting to decipher the occasional references to him. Thus, for example, we know that Grant was a vice president of the Immigration Restriction League. One of the founders of the league was Robert DeCourcy Ward, whose papers are in the Boston Public Library. And there, in a letter dated November 3, 1930, we find a person named Trevor saying the following to Ward: "Grant, of course, told me about Bradley's visit, but Bradley did not leave a copy of the letter to Hoover with him, so I only got a more or less garbled version. However, Bradley sent one to Johnson and I wired Johnson for permission to see it as Johnson sent me a copy of the letter which he wrote to Bradley about it. When I get it, I will send you a copy."[3] It is from such detritus that we are forced to unravel the facts of Madison Grant's life. (And, believe it or not, after immersing oneself in the archives for a number of years one can decipher a letter like this, which, it turns out, tells us a great deal about Grant's efforts in 1930 to lobby Congress to ban Mexican immigration to the United States.)

Still, in the absence of Grant's diaries, his letters, and his personal papers, there is no avoiding the fact that I cannot explain the most basic things about him. Why, for example, did this man—who expended a great deal of energy encouraging his fellow Nordics to produce as many children as possible—never marry? We will never know the answer to that question. Similarly (and more importantly), I have no idea what it was about Grant's upbringing or his intellectual training that influenced him to become a racist. To be sure, I hazard some educated guesses: as we shall see, I posit that a key event in Grant's philosophical development was his visit to the castle of Moritzburg in Saxony; similarly, I conjecture that the 1908 lecture of William Z. Ripley to the Half-Moon Club had a major effect on Grant. But these are suppositions only, and the reader will be perfectly justified in dismissing them as unsupported conjectures.

In full acknowledgment, therefore, of the sparseness of the documentary evidence, it is nonetheless the aim of this book to insert the once-famous (and now infamous) Grant back into the chronicle of twentieth-century America, and to explore how the founder of the Bronx Zoo wound up as Exhibit No. 51 at the Nuremberg Military Tribunal. The attentive reader will note that along the way I spend a great deal of time describing the activities and personalities of Grant's friends. This is done partly because they were an interesting bunch who happened to comprise the ruling class of the United States as it was heading toward the American Century, but also because in the absence of any hard data about Grant himself, all I can do is contextualize, on the presumption that in personality and outlook Grant was probably much like the men with whom he broke bread. Similarly, the reader will undoubtedly observe that this book is to a large extent a series of stories about the many organizations that Grant founded (see appendix B for some of the organizations run by Grant). Philosophically this reflects the fact that Grant, as an early twentieth-century progressive, had a veritable mania for forming organizations, and that he himself would have viewed his life as a series of organizational vignettes. But it also due to the fact that, unlike Grant, organizations tend to preserve their records. Again, the hypothesis is that Grant agreed with the positions and the actions of the organizations that he headed, and that in relating *their* history I am also relating *his* history. And, of course, Grant's friends and organizations led the United States as it was entering a period of unparalled growth and influence; thus we need to understand them if we are to understand ourselves and the nation we have become.

I knew at the outset that trying to cobble together a biography of Madison Grant would be a formidable challenge. But I have been sustained during this quixotic task by the spirits of a number of extraordinary figures, of whom I would like to single out four: Professor Robert Middlekauff, a thoroughly decent man who generously implanted in his graduate students the absurd idea that they could one day be as superb a scholar as he is; David Hollinger, whose calm support has been invaluable through the years; Mathew Guterl, a fellow sojourner through the historiographical swamps of racism, whose insights helped me to transform my 1,208-page dissertation into this book; and Gray Brechin: friend, scholar, and bon vivant, whose vast and intimate knowledge of world history never ceases to amaze me, and from whose learned lips I first heard the name "Madison Grant."

An earlier version of portions of chapter 12 appeared in Jonathan Spiro, "Nordic vs. anti-Nordic: The Galton Society and the American Anthropological Association," *Patterns of Prejudice* 36, no. 1 (January 2002), reprinted by permission of the publisher (Taylor & Francis Ltd, http://www.tandf.co.uk/journals).

Finally, a word on Grant's racism. Many of my students are aghast that I could spend four years living with the ghost of someone like Madison Grant. But Grant was not an evil man. He did not wake up in the morning and think to himself: "Hmm, I wonder what vile deeds I can commit today." To the contrary, he was by all accounts a sweet, considerate, erudite, and infinitely charming figure. By the standards of our own more enlightened time, of course, his racial views are abhorrent. But it would be more than a little arrogant to assume that just because Grant would not agree with *my* opinions (indeed, since I am Jewish, he would not even accede to my *existence*), he should therefore be dismissed as inhuman. I would only point out that, during the course of my research, if I told people that I was writing a biography of a leading conservationist, they would delightfully exclaim: "How wonderful!" On the other hand, if I told people that my subject was a leading eugenicist, they would invariably respond: "How dreadful!" It is instructive to remember that one hundred years ago, those reactions would have been reversed.

I

The Evolution of Scientific Racism

American of Americans,
with Heaven knew how many Puritans and Patriots
behind him. . . .
His world was dead.

Henry Adams

1 Big-Game Hunter

The vision of some of the most advanced thinkers is even yet obscured by the lingering cobwebs of the myths they absorbed in their youth.

Madison Grant

To Promote Manly Sport with the Rifle

In December of 1887, twenty-nine-year-old Theodore Roosevelt, back in his native Manhattan after a two-year stint playing cowboy in the Badlands, hosted a dinner party for ten of his closest chums. Among those in attendance were his dashing brother Elliott Roosevelt and the influential naturalist George Bird Grinnell (editor of the nation's foremost periodical for sportsmen, *Forest and Stream*). They were all men of wealth and prominence, and we can rest assured that there were no "mollycoddles" in the group: all were experienced in, and devoted to, the "manly outdoor sports," especially big-game hunting in the wilds of North America.[1]

After the plates were removed, the guests began spinning tales of their hunting exploits and frontier adventures. Teddy was enjoying it all immensely, and it occurred to him that it would be bully if the group could meet on a more regular and formal basis. He proposed that they form a club for big-game hunters who would gather to discuss matters of common interest and to share hunting lore. The club, according to Roosevelt, would be "emphatically an association of men who believe that the hardier and manlier the sport is, the more attractive it is, and who do not think that there is any place in the ranks of true sportsmen . . . for the man who wishes to . . . shirk rough hard work."[2]

This proposal was applauded by his guests, one of whom wryly suggested that the club be named "The Swappers," since they were obviously going to be spending the bulk of their time swapping stories, "true or otherwise," of their escapades. Roosevelt was not amused, and convinced the group to call their new association "The Boone and Crockett Club" in honor

Teddy in the 1880s, looking manly in custom-made buckskin and Tiffany-carved knife.

of "those two typical pioneer hunters Daniel Boone and Davey Crockett, the men who have served in a certain sense as the tutelary deities of American hunting lore."[3]

Roosevelt and company drew up a constitution declaring that the chief object of the Boone and Crockett Club was "To promote manly sport with the rifle." Membership was limited to an elite core of one hundred hunters who had killed large North American game animals of at least three different species (identified as bear, buffalo, caribou, cougar, deer, elk, moose, mountain sheep, musk ox, pronghorn antelope, white goat, and wolf). Though not explicitly stated in the original constitution, it was understood that said specimens must be full-grown adult males, as the killing of females or the young was considered beyond the pale. Furthermore, the trophies must have been killed "in fair chase," which meant that such unsportsmanlike practices as "crusting" (killing game rendered helpless in deep snow), "jacking" (shining lanterns into the darkness to hypnotize passing animals), and "hounding" (driving prey into a lake with dogs) were verboten.

Well-bred hunters like Theodore Roosevelt and George Bird Grinnell were outraged by such uncouth practices, which were "unworthy of gentlemen or of sportsmen." After all, anyone strong enough to pull a trigger could be a "hunter"; the true *sportsman* therefore had to find a way to set himself apart from the rude killers. This was accomplished via an aristocratic code of ethics that held that the hunter measured his success not by the quantity of game he killed but by the quality of the chase. The point was that a gentleman did not hunt for crass economic reasons; he hunted for sport—and an activity is not a sport unless there are challenges to be overcome and a clear set of rules about how to confront those challenges. Thus, for example, in addition to abjuring unsportsmanlike practices, the sport hunter willingly limited the technological sophistication of his weapon; he passed up the easy shot in favor of killing at the farthest possible range; he preferred the taking of a single fine specimen to the slaughter of a dozen inferior heads; and so forth. This was in direct contrast to the "market hunters," those commercial hunters (members of one of the oldest trades in America) who supplied the urban markets with game. Driven by the profit motive, the despicable market hunters utilized the most effective weaponry, actively sought the easy kill, and had no qualms about shooting young or even female animals. "It is becoming a recognized fact," huffed George Bird Grinnell's *Forest and Stream* in 1889, "that a man who wastefully destroys big game . . . has nothing of the true sportsman about him."[4]

The first official meeting of the Boone and Crockett Club took place in February 1888, and Theodore Roosevelt was elected president of the organization. Invitations to join the club were sent to a select number of candidates, and the membership roster eventually included some of the more influential citizens in the United States, such as Henry Cabot Lodge (the stalwart senator from Massachusetts), Gifford Pinchot (the conservationist), Albert Bierstadt (the landscape artist), T. S. Van Dyke (the most popular outdoor writer of his day), Clarence King (the Western explorer), Carl Schurz (the former secretary of the interior), Carl Akeley (the African explorer), Thomas B. Reed (the Speaker of the House), Lincoln Ellsworth (the explorer), Henry L. Stimson (secretary of war for both Taft and FDR, and secretary of state for Hoover), Henry Fairfield Osborn (America's leading paleontologist), John Hays Hammond (the international mining engineer), Francis G. Newlands (the powerful senator from Nevada), George Eastman (the founder of Eastman Kodak), Elihu Root (TR's secretary of state), Charles Curtis (the vice president of the United States), Owen Wister (the novelist whose best-selling *The Virginian* was dedicated to his Harvard classmate Theodore Roosevelt), and many others. They were all moneyed sportsmen "whose large wealth," noted George Bird Grinnell, enabled them to "indulge to the fullest extent their fondness for hunting." They were also the political and cultural leaders of the nation. Scions for the most part of venerable eastern families, and alumni of Ivy League schools, when they were not hunting together out west they were socializing together at the exclusive Century, Cosmos, Union, Metropolitan, and University Clubs of New York City and Washington, D.C. The members of the Boone and Crockett Club, explained *Forest and Stream,* were "men of social standing" whose opinion was "worth regarding" and whose influence was "widely felt in the best classes of society." They were, in short, the patricians of the United States of America.[5]

The Yale Man

In 1893, a key event in the history of the Boone and Crockett Club—and of the conservation movement in America—took place when debonair twenty-eight-year-old lawyer Madison Grant, himself of patrician stock, was admitted into the club.

Grant was born on November 19, 1865, in his grandfather's house in the posh Murray Hill area of Manhattan (three blocks south of the J. P. Morgan mansion and one block east of where the Empire State Building would be built). He was the heir of a rather distinguished American family. Madison Grant's mother, Caroline Manice, was a descendant of Jesse De Forest, the Walloon Huguenot who in 1623 recruited the first band of colonists to settle in the New Netherlands. After securing a number of land grants on Manhattan Island, De Forest's descendants prospered in the Dutch colony and played a prominent role in the social life of New Amsterdam (and then New York City).

On his father's side, Madison Grant's first American ancestor was Richard Treat, dean of Pitminster Church in England, who in 1630 was one of the first Puritan settlers of New England. Treat's descendants included Robert Treat (a colonial governor of Connecticut and founder of the city of Newark, New Jersey), Robert Treat Paine (a signer of the Declaration of Independence), Charles Grant (Madison Grant's grandfather, who served as an officer in the War of 1812), and Gabriel Grant (the father of Madison), a prominent physician and the health commissioner of Newark. When the Civil War broke out, Dr. Grant organized the Second New Jersey Volunteers and was eventually awarded the Congressional Medal of Honor, the highest military award for bravery that can be given to any citizen of the United States. The citation extolled in particular Grant's personal daring at the battle of Fair Oaks, where he engaged in actions far "beyond the call of duty, thus furnishing an example of most distinguished gallantry."[6] It was at Fair Oaks that Grant used his medical skills to save the life of General O. O. Howard, who went on after the war to found Howard University and head the Freedmen's Bureau. Dr. Grant and General Howard remained lifelong friends.

In sum, Madison Grant could proudly number among his ancestors Dutch grandees, Puritan divines, colonial magistrates, revolutionary patriots, and decorated soldiers. For centuries his antecedents had been accustomed to wealth, power, and deference, and in a country without a titled nobility they could lay as good a claim as any to being true American aristocrats.

Grant was the oldest brother among four siblings (DeForest was born in 1869, Kathrin in 1872, and Norman in 1877). The children's summers, and many of their weekends, were spent at Oatlands, the beautiful Long Island country estate built by their grandfather DeForest Manice in the 1830s. The turreted mansion at Oatlands, with its molded ceilings, rich oak paneling, and enormous stone fireplaces, was deemed by *Architecture* magazine to be "the best in design" of all American mansions of the Strawberry Hill style.[7] This impressive edifice was set amidst elaborate grounds and stables that, according to *Architecture,* had "the broad sweep and spacious dignity of the English park, together with the air of stability which belongs to the British country place." As a child, Madison Grant loved to roam the estate, which was famous, among other things, for its elegant flower gardens and tropical conservatory. In addition, Madison's grandfather had gone to great pains to plant all manner of unusual trees, including Chinese magnolias, Spanish chestnuts, a cedar of Lebanon brought from the Holy Land, and a European linden under which his guest the Comte de Paris (grandson of King Louis-Philippe and heir to the French throne) spent many summer afternoons during his post-1848 exile.[8]

Little wonder that Madison Grant was fascinated from an early age with natural history. As a result of his summertime forays through the Long Island countryside, the future founder of the Bronx Zoo amassed an extensive boyhood collection of rare reptiles and fishes, and he later confided to his friend Henry Fairfield Osborn that, from childhood, he had been interested in animals: "I began by collecting turtles as a boy and have never recovered from this predilection." Years later, after Grant had grown up and moved away, the Manice Woods on the north side of Oatlands were cut down to make room for the Belmont Park Race Track, and the mansion was transformed into the clubhouse of the Turf and Field Club (on whose governing board sat Madison Grant).[9]

As a member of the eastern patriciate, Madison Grant was educated by private tutors, though he obtained most of his worldly knowledge on trips abroad with his father. On one such excursion, they journeyed to the ruins of ancient Troy where, if the testimony of his friend H. E. Anthony is to be believed, young Madison "sat on the crumbling walls and chanted Homer's *Iliad.*"[10]

At the age of sixteen, Madison was sent to the German city of Dresden, where for the next four years European tutors provided him with the best possible classical education. During this time he managed to travel to every country in Europe (where he visited all the zoos and most of the natural history museums of the continent) and throughout North Africa and the Middle East as well. But his most significant visit was to Moritzburg, the baroque hunting lodge just outside Dresden, where my guess is that Grant found himself transfixed by the extensive collection of red deer antlers. The trophies—which had been collected three hundred years earlier—were impressively large, and the more the young student stared at them the more troubled he became. At some point, it occurred to him what was amiss: antlers of that size simply did not exist anymore on living European deer. Grant realized that, contrary to the Victorian understanding of evolutionary progress, the red deer had been getting smaller and smaller over the years. The species was actually *degenerating.*

Furthermore, Grant's naturalistically inclined mind apparently put together what he knew of the geographic range of the red deer, along with the sizes of the various specimens he had encountered in the wild, and he instantly envisioned a perfect continuum: At the far eastern edge of the red deer's range (in the Caucasus) the animal was almost as large as it had been in the sixteenth century. But toward the west (in the Carpathians) the deer began to diminish in size. Even farther west (in Saxony) the stags were smaller still, and at the far western limit of the animal's range (in Scotland) the red deer had shrunk to their smallest proportions.

Grant reasoned that this decline in size was indubitably the result of trophy hunting. Trophy hunters, of course, target the largest bulls with the finest antlers, which leaves the breeding to the inferior males. As one moves from east to west across Europe, the human population increases, as does the number of hunters, and the inevitable result is an ever-greater decline across space, and over time, in the size and vigor of the deer stock. In other words, as human civilization advanced, the deer declined. And Grant was struck by the fact that if the trend were to continue, the red deer would diminish in size and vitality to the point where ultimately the species would not be able to survive in the wild.[11]

After four years of study and travel abroad, Grant returned to the United States in 1884, and as a matter of genetic imperative applied to enter Yale University. Candidates for admission to Yale in the 1880s were examined over a three-day period in four subjects: mathematics, German, Greek grammar, and Latin grammar. It is a sobering thought that probably not a single American teenager is alive today who could have qualified for admission to Yale in 1884. Madison Grant, on the other hand, passed with flying colors; in fact, he was admitted as a sophomore after demonstrating his mastery of the freshman curriculum (including Socrates, Herodotus, Euripides, Livy, Horace, and trigonometry).

With the exception of the courses of Professor William Graham Sumner (who engaged the students in heated discussions that invariably concluded with the

Madison Grant, class of 1887. His classmates confessed that their chief hobbies at Yale were loafing, smoking, dear *hunting, swinging golf clubs, and "killing time and mosquitoes."*

professor assuring them that "as the rich grow richer, the poor grow richer also"), Grant's classmates did not find their Yale studies to be stimulating. "Most of our classrooms were dull," remembered William Lyon Phelps, "and the teaching purely mechanical; a curse hung over the Faculty, a blight on the art of teaching. Many professors . . . never showed any living interest, either in the studies or in the students." Certainly, Madison Grant seems not to have been wholly engaged with his studies. Not surprisingly to anyone who has read *The Passing of the Great Race,* Grant consistently earned among the higher scores in composition but ranked near the bottom of his class in logic. Within a few years, Grant's mind would possess a prodigious amount of knowledge about ethnology and natural history, but it would all be acquired by independent reading and experience subsequent to his graduation from Yale.[12]

But, of course, it was understood that the formative experiences at Yale took place not in the lecture halls but on the playing fields, at the eating clubs, and in the Greek-letter societies. And the end product of this New Haven–style socialization was what Santayana called "the Yale man," characterized by "trust in success, a ready jocoseness, a democratic amiability and a radiant conviction that there is nothing better than one's self." Strike "democratic" from that sentence, and you have a fairly good description of Madison Grant.[13]

Upon graduating from Yale, Grant returned to New York to attend Columbia Law School. He was admitted to the bar in 1890 and, after a brief stint with Seward, Guthrie & Morawetz, opened a law office of his own next to the New York Stock Exchange. But Grant had neither the financial need—nor the intellectual desire—to pursue seriously a legal career. Instead, for the first half of the Gay Nineties, the "breezy young New York lawyer" (as one friend described Grant in those days) devoted himself wholeheartedly to two endeavors: socializing and hunting. In rapid succession, he joined all the elite men's clubs of Manhattan, including the Union, Knickerbocker, University, Down Town, and Tuxedo Clubs, ensuring that every evening of the week could be spent hobnobbing with the *Herrenrasse* in a different salon. These clubs included many of the nation's wealthiest and most powerful figures, and Grant, according to his friend Henry Fairfield Osborn, "figured very prominently at the time and was regarded as a typical society and club man."[14]

The Society of Colonial Wars

In 1892, Madison and his brother DeForest (who had just graduated from Yale) helped to found a slightly different type of club: the Society of Colonial Wars, a fraternal organization with membership restricted to men "of good moral character and reputation" whose ancestors had attained distinction in the wars of the colonial period. The society was typical of the many hereditary patriotic societies springing up in the 1890s as a manifestation of uneasiness

The wretched refuse of Europe's teeming shore gathers in the Jewish section of the Lower East Side, 1912. Photography Collection, Miriam and Ira D. Wallach Division of Art, Prints and Photographs, The New York Public Library, Astor, Lenox and Tilden Foundations.

among old-stock Americans bewildered by—and antagonistic toward—rampant urbanization, industrialization, and immigration.[15]

For all of Madison Grant's life, a fairly steady average of 250,000 immigrants had entered the United States every year. But then in the early 1880s the rate had suddenly doubled, to well over 500,000 annually. The problem, as far as Grant was concerned, was not just the overwhelming number of newcomers but the alarming shift in their identity. Immigrants had heretofore come primarily from northwestern Europe (in particular the British Isles, Scandinavia, and Germany). But in the early 1880s an increasingly large number of immigrants began to arrive from southern and eastern Europe. These "New Immigrants" were often uneducated, unskilled, and illiterate peasants, who disconcertingly congregated in the large cities of the Northeast, especially New York.

Grant felt increasingly beleaguered by the waves of swarthy immigrants engulfing his city. They were filling up the almshouses, cluttering the streets, and turning Manhattan into a dirty, lawless, turbulent cacophony of foreign barbarians. Like Henry James, who observed the immigrants arriving in New York and was revolted by "the visible act of ingurgitation," Grant was disgusted by what he saw as he braved the congested sidewalks of his native city. He was repulsed by the bizarre customs, unintelligible languages, and peculiar religious habits of the foreigners. As he was jostled by Greek ragpickers, Armenian bootblacks, and Jewish carp vendors, it was distressingly obvious to him that the new arrivals did not know this nation's history or understand its republican form of government—indeed, upon landing at Castle Garden (just a few blocks from Grant's office), one of the first things the New Immigrants did was blithely sell their votes to New York's unscrupulous political bosses. Grant knew full well that classical Rome had fallen when she opened her gates to inferior races who "understood little and cared less for the institutions of the ancient Republic," and he feared for his country.[16]

In his *Education,* Henry Adams describes a visit he made to New York City at the turn of the twentieth century: "A traveller in the highways of history looked out of the club window on the turmoil of Fifth Avenue, and felt himself in Rome, under Diocletian, witnessing the anarchy."[17] Madison Grant gazed through that very same club window at all those Jewish and Catholic and Slavic peasants scurrying around on the pavement below him, and felt the exact same sense of aristocratic despair. Decades later, Grant recalled that when the New Immigrants began to arrive, "Americans were shocked to find what a subordinate place was occupied by the old American stock in the opinions of some aliens." Indeed, men like Grant and his brother DeForest, who were accustomed to striding the avenues of Manhattan the way that princes of the blood royal used to traverse the *tapis vert* at Versailles, were affronted that the newcomers did not recognize who they were, nor bother to show any deference to them. The name "Dr. Gabriel Grant" meant nothing to these helots; the fact that he had won the Congressional Medal of Honor carried no weight with them. They hardly cared that Robert Treat Paine had been a signer of the Declaration of Independence—they could barely *read* the Declaration of Independence. And so, as the number of ships that steamed through the Narrows from Naples and Hamburg increased year by year, and the streets of lower Manhattan grew ever more crowded, Madison Grant faced the painful realization that he was becoming a stranger in his own town. (As an editorial in the *Saturday Evening Post* expressed it a few years later: "We have not been assimilating our latter-day immigration; it has been assimilating us.")[18]

One of the ways that the Grants and other members of their class fought back was to form exclusive organizations that defensively touted their genealogical superiority to the newcomers and conspicuously asserted their patrician claims over the nation's heritage. The Society of Colonial Wars was one of the most active of these organizations.

The society held its first annual dinner at Delmonico's in December 1892, where the 250 members elected a governor and a nine-member council, on which Grant would serve for many years. At that dinner, the society's prized possession was unveiled: an elaborate, solid-silver punch bowl commissioned from Tiffany's. Etched upon the bowl in exquisite detail were two scenes from colonial history: one showed a three-masted vessel sailing into a harbor with a peaceful band of Indians waiting onshore to welcome the settlers; the other vignette depicted a group of Puritans waging bloody warfare against a tribe of Indians. The society's members were apparently oblivious to any irony implied by such a juxtaposition.

The stated objectives of the Society of Colonial Wars were threefold, and they reveal the anxiety these men felt about maintaining their status in the final years of the nineteenth century. The first objective was "To perpetuate the memory of . . . the men who . . . assisted in the establishment, defense, and preservation of the American Colonies, and were in truth the founders of this nation." Grant and his fellow patricians were deeply distressed that most Americans seemed to have forgotten—and the New Immigrants plainly had no interest in—the fact that the De Forests and Manices and Grants had been the bedrock upon which the nation had been built. They were upset that the veneration that was their due was no longer forthcoming from the masses, and hence they sought to foster "respect and reverence for those whose public services made our freedom and unity possible."[19]

The second objective of the Society of Colonial Wars was "To provide suitable commemorations or memorials relating to the American colonial period." As such, the society issued numerous publications dealing with events and personages of the colonial period, and erected markers on colonial battlefields all over the eastern seaboard. Madison Grant, for example, raised the funds for the memorial commemorating the 1745 capture of the fortress of Louisbourg; he also helped purchase the Oriskany battlefield for presentation to the state of New York.

The last goal of the Society of Colonial Wars was "To inspire in its members the fraternal and patriotic spirit of their forefathers," a spirit that seemed to have been on the wane for too long. The society employed a number of methods to inspire a sense of community among its members and to anoint them with the outward manifestations of distinction that would prove to hoi polloi that they were men deserving of honor and reverence. Each member, for example, received an elaborate diploma testifying to his qualifications for membership, and an intricate gold pendant hanging from a silk ribbon that he was to display on the left breast. In addition, the society adopted its own flag, great seal, and motto (*Fortiter Pro Patria*). It held regular pageantry-filled dinners, and it published an illustrated yearbook featuring a detailed genealogy of each member.[20]

Unlike organizations such as the Grand Army of the Republic, which admitted members on the basis of what they had *done,* and which wanted as *many* members as possible in order to enhance their political influence, the Society of Colonial Wars admitted members on the basis of who their *ancestors* had been, and consciously *limited* its size to assert its exclusivity. Hence it was, by any definition, an aristocratic institution. And to maintain it as such, Madison Grant, with his prodigious memory and interest in matters antiquarian, was called on over the years to scrutinize and verify hundreds of claims of colonial ancestry by prospective members.

Grant also joined at this time the Military Order of the Loyal Legion (open to Civil War officers and their eldest male descendants). His brother DeForest became a member of the Society of the War of 1812 and was elected as well to the St. Nicholas Society (whose exclusive membership was limited to 650 descendants of the original settlers of Manhattan). It was as if the Grant brothers and the other members of their class at the end of the nineteenth century—sensing that the manner of life they had always known was threatened by the New Immigration and the forces of modernization—turned away from the approaching twentieth century and sought refuge in the seventeenth and eighteenth centuries.

Grant, Grinnell, Roosevelt

When Madison Grant was not pursuing the life of a bon vivant in the clubs of Manhattan or paying obeisance to his respected ancestors, he was engaging in another sort of escape from modern life: hunting expeditions. Accompanied usually by his brother DeForest, Madison spent at least four months of every year tracking big game in far-off locations all over the North American continent. A friend remembered that Grant "had all the independence of a well-groomed musketeer—and more," and his excursions in those years took him at various times from Newfoundland to Alaska and most places in between.[21]

As he pursued wild game over the passes of the Rockies, down the tributaries of the Fraser, and up the fjords of the Kenai Peninsula, Grant began to realize that the large mammals of North America were dwindling, in terms of both sheer numbers and individual size. It was clear that the devastating predations of market hunters, along with the unsportsmanlike practices of amateur riflemen, were decimating the native fauna. He probably thought back to the castle of Moritzburg with its trophies of immense deer from a bygone age. And it struck him that he had been working via the Society of Colonial Wars to perpetuate the heritage of his forefathers, when right in front of his eyes the natural inheritance of the entire continent was being wiped out by hunters. "No more destructive animal has ever appeared on the face of the earth," he was forced to concede, "than the American back-woodsman with his axe and his rifle. Since the Civil War, we have plundered half a continent."[22]

Grant accepted that those in a position of power and prominence were obligated to husband the nation's wealth for the benefit of their less farsighted neighbors. And so, as he had done in founding the Society of Colonial Wars, Grant took up the patrician burden of stewardship over his native land. He decided that his role would be to alert his countrymen that "it is our duty as Americans to hand down to our posterity some portion of the heritage of wild life and of wild nature that was ours. In other words, to leave to them a country worth living in, with trees on the hillsides; with beasts in the forests; with fish in the streams; and with birds in the air." At some point in the early 1890s, Grant was transformed from a reckless rake known for his carousing and his shooting into a man committed to the cause of conservation (although that term would not be invented for another fifteen years). After Grant died, a colleague explained that Grant's "absorbing mission" in life had been to save "for posterity all the fine, noble and worthwhile things his generation had inherited . . . as if all these were his personal responsibility."[23]

As a result of his newfound interest in game conservation, Grant struck up what would become a lifelong friendship with a fellow member of the Society of Colonial Wars, George Bird Grinnell. A man of great dignity and quiet wit, Grinnell (1849–1938) is best remembered today for his monumental writings on the Plains Indians, but it was as a conservationist that he had his greatest impact on U.S. history (his *New York Times* obituary actually described Grinnell as the "father of American conservation").[24] Like Madison Grant, Theodore Roosevelt, and many other founders of the conservation movement, Grinnell was an enthusiastic big-game hunter who sprang from a wealthy New York family (his father was Cornelius Vanderbilt's broker) that numbered five colonial governors among its ancestors. After receiving his B.A. from Yale in 1870, Grinnell served as the naturalist on several journeys of exploration to the unmapped West, including George Armstrong Custer's excursion to the Black Hills in 1874 and Colonel Ludlow's reconnaissance of the Yellowstone in 1875. (Grinnell's duties at the Peabody Museum prevented him from accepting Custer's pressing invitation to accompany him to the Little Big Horn in 1876.) On these trips, Grinnell—who spoke several tribal dialects and was an adopted member of the Pawnees and an honorary chief of the Blackfeet—spent his spare time conversing with the Indian scouts and taking notes on their customs and folklore.

Grinnell went on to earn a Ph.D. from Yale in 1880, but he returned annually to the Great Plains to visit—and hunt bison with—his Indian friends. As the years proceeded, he became more and more concerned that the Indians, the bison, and the rest of the nation's wildlife were threatened with extinction by the forces of manifest destiny. In 1880, after four years as editor, Grinnell became owner of *Forest and Stream,* and from then until he stepped down in 1911 he made the magazine a vital advocate for conservation (and, as Grinnell him-

George Bird Grinnell: explorer, hunter, conservationist, ethnologist, editor of Forest and Stream, *and constant companion of Madison Grant.*

self admitted, the "mouthpiece of the Boone and Crockett Club").[25] As we have seen, he was a charter member of the Boone and Crockett Club, and served as the club's president from 1918 to 1927 (when he was succeeded by his best friend, Madison Grant).

George Bird Grinnell: famous explorer, accomplished naturalist, prototypical ethnologist, and ardent conservationist—it was inevitable that Grant would be entranced by him. Grinnell was also as inveterate an organizer as Grant, and would be involved in the creation of the American Ornithologists' Union, the American Game Protective Association, the Society of American Foresters, and the National Audubon Society. (He named the last-mentioned society for John James Audubon, previous owner of the home in which Grinnell had been raised. Lucy Audubon, the painter's widow, was Grinnell's teacher when he was a boy.) Madison Grant wrote admiringly in 1919 that "Grinnell, perhaps more than any other living man, represents the now disappearing class of educated easterners who went to the frontier in the buffalo and Indian days and devoted their lives to the welfare of the great West. . . . From the year 1870 [he] has freely given his time, his money, his scientific and literary attainments, and his talents to the cause of the preservation of the forests, the wild life of the country and, above all, the welfare of the Indians of the West."[26]

Grinnell and Grant spoke on the telephone daily and saw each other in person almost as often. They worked together for almost half a century to shape the agenda of the Boone and Crockett Club, the Society of Colonial Wars, the New York Zoological Society, the National Parks Association, and the American Society of Mammalogists. The word "love" appears exactly twice in Grant's surviving correspondence—once when Grinnell expresses his feelings for Grant, and once more when Grant expresses his feelings for Grinnell and his wife Elizabeth.[27] In addition to sharing the same background, interests, philosophy, and dry sense of humor, at least part of the attraction was that Grinnell, a Victorian family man who was Grant's senior by sixteen years, was positively enchanted by Grant's whirlwind bachelor's existence and his Beau Brummell flair. Once, when Grant (whom Grinnell referred to as "a lighthouse of fashion") was trying to decide what the topic of his next wildlife article should be, Grinnell suggested: "It is possible that your round of social and other gayeties undertaken last year might thrill us all." (It seems that Grant could never quite shake his reputation as a rascal. When he failed to appear at his office one morning, pleading a sore throat, colleague William White Niles immediately sent a note: "I trust the bad throat business is only a little bluff of yours to conceal some agreeable adventure with the fair sex which renders attendance at your office impracticable.")[28]

When George Bird Grinnell first met Madison Grant, he recognized the younger man as a fellow sportsman with grave concerns about the decreasing numbers of big-game animals. It did not take long for Grinnell to nominate Grant for membership in that most exclusive of men's associations, the Boone and Crockett Club.

Grant immediately became one of the most active members of the Boone and Crockett Club, and he and the club's president, Theodore Roosevelt, soon became good friends. Roosevelt admired Grant's devotion to the principles of sportsmanship, and—being TR—was greatly attracted to Grant's youthful energy. "I am inclined to think," Roosevelt confided to Grinnell, "that Madison Grant is a real acquisition; he strikes me as a good fellow."[29]

It would have been surprising had Roosevelt and Grant not hit it off. Both were the offspring of old Knickerbocker families and had been raised in Manhattan just eight blocks away from each other. Both boys, along with their three siblings, spent their summers at their families' respective estates on Long Island, when not traveling with their parents in Europe and the Middle East. While growing up, both maintained a collection of wild animals—of the live and the taxidermic variety—which carried over into an abiding interest in natural history. Teddy's best friend as a youth was Frederick Osborn ("a fine and manly young fellow"), while one of Madison's closest friends as an adult was Frederick's brother, Henry Fairfield Osborn.[30]

Grant and Roosevelt both possessed almost photographic memories, and continually astounded their friends and colleagues with their ability to recall the most arcane facts about geography, geology, and biology (and, in Grant's case, ethnology). Both teens were tutored in Dresden, attended an Ivy League school, and graduated from Columbia Law School; and though they were members of the New York bar, both became sidetracked by other ambitions and never formally practiced law.

Both men camped often in the Adirondacks and the Maine woods, and found their deepest solace tracking the large fauna of the West, where they earned reputations as fearless big-game hunters. They both served as president of the Boone and Crockett Club, and donated numerous specimens of animals they had killed to the American Museum of Natural History. In addition, both became obsessed with the possibility of race suicide and the dangers posed by excessive immigration, and were convinced of the need to implement eugenic measures to forestall the decline of the Anglo-Saxon race. During World War I, they would serve together as trustees of the American Defense Society, a rabidly pro-preparedness organization that helped to launch the postwar Red Scare.

Grant and Roosevelt were both wealthy conservatives and lifelong Republicans (with a famous lapse in 1912) who employed progressive means to attain their ends. There was, however, a difference in temperament between the two men. Grant had no need whatsoever to be in the public eye and, in fact, preferred to work behind the scenes; while Teddy, of course, "wanted to be the corpse at every funeral he attended and the groom at every wedding."[31] When Grant embarked on a project, it was less of a Rooseveltian crusade than a patient application of a reasoned plan of action. A quiet tête-à-tête over drinks at the Century Club, not a jeremiad from the bully pulpit, was Grant's preferred method of persuasion. Grant simply lacked TR's bellicosity, perhaps because he had less to prove (Grant's father, after all, was a genuine and certified Civil War hero, whereas Teddy always felt the need to atone for the fact that his father had avoided military service during the war by hiring a substitute).

With the support of Roosevelt (whose direct involvement with the Boone and Crockett Club had lessened after his move to Washington, D.C., to serve on the Civil Service Commission), Grant and his ally George Bird Grinnell set about transforming the club from a mere social lodge for wealthy hunters into the seminal conservation organization in America. As Grinnell put it: "However agreeable it may be for a number of hunters to dine together, and to exchange experiences and swap hunting stories, it must be acknowledged that the profit of such a meeting . . . is not great. . . . While the Boone and Crockett Club was perhaps established as a hunting club, and while its members do a great deal of hunting and enjoy it, it aims at something higher than being a mere social organization."[32]

The idea that a group of hunters might be interested in wildlife conservation was not entirely contradictory. To be sure, such men had a vested interest in maintaining conditions of game scarcity. After all, if the countryside were still teeming with game, as it had been when the settlers first came to the New World, hunting would provide little pleasure—and certainly no sense of achievement or honor—to these distinguished sportsmen. For them, just locating—let alone shooting—big-game animals at the end of the nineteenth century involved arduous excursions to the remoter parts of the continent. A great deal of prestige was attached to the man who could pit his skills and hardiness against the forces of the wilderness and return with a trophy of some rare species. Teddy Roosevelt, sounding like the loser of a faux-TR writing contest, drove home the connection between the effort involved in big-game hunting and the status that thereby accrued to the hunter: "Hunting big game in the wilderness is, above all things, a sport for a vigorous and masterful people. The rifle-bearing hunter . . . must be sound of body and firm of mind, and must possess energy, resolution, manliness, self-reliance, and capacity for hardy self-help. In short, the big-game hunter must possess qualities without which no race can do its life-work well."[33]

It was not unimportant that only the wealthy could afford to indulge in such a pastime. To arrange a western big-game hunt in the 1890s involved a tremendous amount of planning and a large investment of time and money. Given the difficulty of transportation, the antelope ranges of the Dakotas were much more remote for these eastern hunters than the veldts of Africa would be to the next generation of sportsmen. Conservationist Robert Sterling Yard remembered in 1928 that "to us in the East, it seemed more of an adventure to cross the Mississippi than it does now to circle the world."[34]

Thus, the fact that game was sparse was one of the factors that made sport hunting an honorable activity among the nation's elite. Scarcity meant that the pursuit was a true test of manly fiber; scarcity imparted value to the trophies; and scarcity ensured that only the wealthy could afford to engage in the performance. On the other hand, sport hunters certainly did not want the animals to be so scarce that they might actually die out. The extinction of the game would mean the demise of their sport and the disappearance of a valued source of status. Thus, while even the most sympathetic sportsmen did not want the game to flourish too much, even the most profligate hunters were in favor of at least some conservation of wildlife. To put it as straightforwardly as possible: Grant, Grinnell, and Roosevelt wanted to save America's animals in the present so that they could hunt them in the future.

But in the late nineteenth century, the desire to conserve wildlife was an indulgence of the idle rich. The average citizen could not be concerned with such a matter. In fact, it was in the immediate financial interest of most Americans—including farmers (who viewed game as varmints), tanners, milliners, and furriers—to kill as many wild animals as possible. And then there was the vast industry involved with transforming game into food. This included market hunters, railroad companies, cold-storage outfits, meat wholesalers, and restaurant owners. In those days, all manner of game mammals were commonly found on the American dinner table: skinks and squirrels, beavers and badgers, moose and mules, hares, raccoons, otters, muskrats, woodchucks, opossums, antelopes, porcupines, mountain goats, bighorn sheep, hams of bear, haunches of venison, saddles of elk, legs of caribou, tongues of deer, and so forth. In addition, virtually all species of birds, and their eggs, were available in the food markets of America's cities. Swans, geese, ducks, robins, grouse, coots, cranes, loons, blackbirds, sparrows, thrushes, warblers, vireos, woodpeckers, seagulls, goldfinches, prairie chickens, and passenger pigeons were all common parts of the American diet.

As a result of such liberal culinary tastes, the populations of many species of game were noticeably declining by the late 1800s. In the early nineteenth century, for example, there had been several billion (yes, *billion*) passenger pigeons in North America. They were almost certainly the most numerous avian of all time. Pioneers encountered individual flocks containing millions of birds each, which flowed all day long from horizon to horizon like aerial rivers. When the birds alighted, they gathered in Hitchcockian numbers that toppled trees. John James Audubon saw migrating flocks that "darkened the sun" for three days, and concluded that their numbers were "inexhaustible."[35]

But the passenger pigeons began to decline as more and more of their habitat was used for agricultural purposes and sporting clubs increasingly employed them for shooting contests. What finally spelled the passenger pigeon's doom was the murderous practices of the market hunters, who built smudge fires to suffocate the nestlings and drive the blinded adults into their nets. One group of hunters assaulted a colony of nesting birds near Petoskey, Michigan, in 1878, and within a few weeks had killed and sent to market 1,107,800,066 pigeons. The carcasses were sold to farmers (who turned them into hog feed), renderers (who turned them into soap), and cooks (who bought the birds for half a cent each and turned them into dinner).[36]

It was fairly clear to disinterested observers that the pigeon population was being reduced to the point where the species would not be able to recover. Yet in 1857 the Ohio legislature peremptorily *rejected* a bill to protect the animal with the observation that "the passenger pigeon needs no protection. Wonderfully prolific, having the vast forests of the North as its breeding grounds . . . it is here today and elsewhere tomorrow, and no ordinary destruction can lessen them." Ironically, it was in the state of Ohio that the last passenger pigeon in the world, named Martha, died at the age of twenty-nine in a cage in the Cincinnati Zoo on September 1, 1914.[37]

By that time, other North American game animals that had been exterminated, chiefly at the hands of the market hunters, were the heath hen, the

Labrador duck, the Eskimo curlew, the great auk, and the Carolina parakeet (our only native parrot, whose flocks used to add "splotches of tropic brilliance" to the American landscape).[38] And the outlook was bleak for a number of other species. Owing to human predation, the grizzly bear was almost extinct; the caribou was about to disappear from its last foothold in Maine; the pronghorn antelope was limited to a few secluded prairies; the moose and mule deer were rapidly slipping away; the wild turkey and the white-tailed deer had been eliminated nearly everywhere east of the Rockies; the elk, beaver, black bear, white goat, and mountain sheep had all been pushed back to a few wilderness pockets; and, most famously, the bison was quickly vanishing from the plains.

Thus, Madison Grant and George Bird Grinnell pointed out to their fellow Boone and Crocketteers that, as hunters, they really had very little choice but to embrace conservationism. Historian John F. Reiger is absolutely correct when he says, regarding the Boone and Crockett Club: "Though almost ignored by professional historians, it—and not the Sierra Club—was the first private organization to deal effectively with conservation issues of national scope." It is a fact that upper-class sportsmen were the progenitors of the nascent conservation movement in the United States (and it was for this reason that—until 1907, when the term "conservation" was originated—the movement to protect wildlife was called *game* protection). Conservationist William T. Hornaday had Madison Grant in mind when he wrote in 1913 that "Gentlemen sportsmen . . . are the very bone and sinew of wild life preservation. . . . These are the men who have done the most to put upon our statute books the laws that thus far have saved some of our American game from total annihilation."[39]

Banning Unsportsmanlike Hunting

A few months after he joined the Boone and Crockett Club, Madison Grant—at the encouragement of Theodore Roosevelt—produced his first published article, "The Vanishing Moose," for the January 1894 issue of *Century Magazine.* At the time, *Century* was at its height with a circulation of over two hundred thousand readers. It paid top dollar to the best writers in the United States who championed the latest progressive causes, including improving the tenements, ending the spoils system, and ridding the cities of boss rule. Grant's lengthy article fit into this reforming spirit, as it was a litany of the shameful "destruction in the flora and fauna of our land," and thus served as a wake up call for the nation's conservationists. Indeed, regarding most of the nation's trees and mammals, Grant stated unequivocally: "The end . . . is near."[40]

And thus, in his very first article, Grant introduced the pessimistic theme that would be a constant in his writings until the day he died: the frightening prospect that the spread of modern civilization was causing the demise of the native species of North America. It is of considerable significance that the second sentence of "The Vanishing Moose"—the second sentence that Grant ever published—stated: "The old order of things has largely passed away." It was an elegy that would issue time and again from his pen for the next forty-three years, whether the subject was the extermination of the bald eagles, the destruction of the redwoods, or the passing of the Nordics.[41]

Paradoxically, in the very same article in which Grant bewailed the dwindling numbers of moose, he also recounted the glorious moose hunts ("the noblest of American sports") in which he had participated—and he actually passed on to his readers detailed hunting tips. The seeming contradiction, of course, was not a contradiction to Grant: he genuinely mourned the passing of the moose, and he did want it to return—so that he could kill it. And so he now thought long and hard about the best means to preserve the moose—and indeed all of the continent's wildlife. He began with the unorthodox premise that the resources of a region do not belong to the local inhabitants but to the nation as a whole. "The interest," he later wrote, "of the entire people of the United States, and to some extent that of the civilized world, is centered in the continued existence of the forms of animal life which have come down to us from an immense antiquity through the slow process of evolution." Hence Grant concluded—in contradiction to the prevailing ethos that had elevated private property rights to the plane of metaphysical certitude—that "the game and forests belong to the nation and not to the individual." No private citizen, declared Grant, has "a divine commission to pollute the streams with sawdust, to destroy the forest by axe or fire, or to slaughter every living thing within reach of rifle, trap, or poisoned bait. The mere fact that he has the power to destroy . . . does not in itself confer a right."[42]

Accordingly, Grant decided that unsportsmanlike hunting practices (crusting, jacking, hounding, etc.) would have to end if big-game hunting was to survive beyond the nineteenth century. The matter was of particular urgency because in the early 1890s newly built railroads were spiriting an influx of hunters to upstate New York, and the animals of the Adirondacks were under siege by urban neophytes who rudely flouted the code of sportsmanship. An aghast Teddy Roosevelt told George Bird Grinnell, "I wish to see the Boone and Crockett Club *do something*"—and as always the phrase "do something" meant push for legislation. Roosevelt suggested that the new fellow, Madison Grant, though still in his twenties, be sent to Albany to secure a law outlawing unsportsmanlike hunting in the Adirondacks.[43]

Grant did just that, and brought to the legislature the Adirondack Deer Bill, banning in the state of New York all those practices that were so abhorrent to gentlemen hunters. But a coalition of market hunters, railroad companies, resort owners, and Adirondack guides came together to oppose the measure. They protested that an elite group of Manhattan amateurs was conspiring to deprive them of their liberty, not to mention their livelihood. Furthermore, they claimed that the wealthy Grant and his fellow Boone and Crocketteers did not truly care about the welfare of the deer: they were simply piqued that the Adirondacks, which had heretofore been their private hunting reserve, was now accessible to the general public thanks to the railroads. The opposition shrewdly insisted that the Deer Bill was the work of "monopolists" who did not want the "poor man" to have an opportunity to bag a deer.

There was some truth in that accusation. To be sure, as one historian points out, "No attack on environmentalism, no proposal to loot the country's natural assets, is complete without a phony-populist sneer at those who try to defend the environment."[44] But for many upstaters, Grant's interference in local affairs aroused a resentment that was quite genuine. Surely they, the inhabitants of the Adirondacks, were better qualified to run their own region than some pampered Ivy Leaguer.

We might bear in mind that the conservation "movement" at that time consisted of a few dozen persons—some big-game hunters, some professional foresters, a handful of naturalists, and perhaps a hydrologist or two. It was by definition an elitist movement, and most definitely did not represent the will of the people. The conservationists were few in number, and when a tiny minority of outsiders—even one that possesses a better education and a broader vision—tries to impose its views on a larger group of local residents, the situation poses a dilemma for a nation that worships democracy.

Of course, the one advantage Madison Grant possessed whenever the anti-conservation forces played the populist card was that he was not at all uncomfortable being labeled an elitist. Quite the contrary: he was proud to be a member of the haut monde, and indeed through his work with the Society of Colonial Wars he was making every effort to remind the public of his exalted status. America, after all, had been founded and built by his family. This was *his* country, and as its steward he would not sit idly by while an unholy alliance of market hunters, railroad corporations, and tourism magnates tried to despoil it. Grant and his fellow patricians from the city would save the democracy-loving residents of upstate New York from their own destructive inclinations, and devil take the hindmost.

To get the Adirondack Deer Law passed, Grant waged a three-year campaign in which he wrote letters to newspapers such as the *New York Times,* published articles in magazines such as *Harper's Weekly,* and carried on a series of face-to-face negotiations with key politicians in which he called on them to ignore their constituents and instead protect the game of New York from unregulated hunting.[45] As an American patrician and an 1887 graduate of Yale University, he never doubted that he would succeed.

And so it was that—with numerous revisions and amendments, and after a bitter floor fight—the Adirondack Deer Law, which banned unsportsmanlike hunting, was enacted. It took effect on June 1, 1897. Madison Grant had saved the wildlife of New York. An elated George Bird Grinnell congratulated Grant on the bill's passage: "Great credit is due you for having accomplished something that I thought two years ago quite impossible." Theodore Roosevelt told Grant that he had performed "wonders" and added: "I am extremely pleased with what you have accomplished."[46]

The Adirondack Deer Law, states Marvin Kranz, the historian of the New York conservation movement, "was of great significance." And it was also quite effective: not only did the deer of New York thrive, but it was not long before the beaver and the bear moved back to the Adirondacks, and in recent years even the moose (whose last-known Adirondack ancestor was killed in 1861) has returned. More importantly, with New York's Deer Law as the model, most of the other states in the Union soon adopted similar legislation. Throughout the country, practices like hounding, night hunting, and the killing of females all but disappeared. In a relatively short period of time, Grant had gone a long way toward ending unsportsmanlike hunting in the United States.[47]

Stopping the Market Hunters

Having ended unsportsmanlike hunting practices, Grant boldly decided that his next task would be putting an end to market hunting. Madison Grant was a big-game hunter, and he reveled in every aspect of the activity, from silently following the track of his quarry at sunrise to gloriously recounting every detail of the kill to his companions over the evening's campfire. But he could not understand how anyone could shoot a noble animal for monetary gain, and he was determined to put the despicable market hunters out of business. And it was in, of all places, the territory of Alaska that Grant began his anti-market-hunting campaign.

The unsullied territory of Alaska had a romantic mystique for Grant and his fellow Park Avenue hunters; it held the same allure for them as "darkest Africa" did for upper-class British sportsmen. The East was finished; the West was on the verge; but Alaska was still pristine. Grant appraised Alaska as "the finest hunting ground possessed by this nation," and he reminded his fellow citizens that "it is only in Alaska that we may be able to maintain primitive conditions approximating those of the whole country when first settled." Not only was Alaska's wildlife abundant; it was often quite unique. Witness the immense Kodiak bear ("the largest of living carnivores"), the awesome Alaskan wolf ("largest canine in the world"), and the giant moose of the Kenai ("the greatest among all living deer").[48]

Grant often hunted in Alaska and organized scientific expeditions to the region, as did such friends as George Bird Grinnell, E. H. Harriman, and Admiral Robert E. Peary. In 1897, one such expedition killed specimens of a very large caribou that Grant named *Rangifer osborni* in honor of his good friend Henry

This 1915 portrait of Rangifer granti *(by renowned wildlife artist Carl Rungius) hung for decades in the administration building of the Bronx Zoo. It is appropriate that a painting of the caribou named for Madison Grant is titled "Patriarch of the Herd."*

Fairfield Osborn. And in 1901, when Grant sent explorer Andrew J. Stone to reconnoiter the western Alaskan wilderness, Stone discovered another "magnificent" new species of caribou that was given the name *Rangifer granti* to honor the man who financed the expedition.

As an aristocratic hunter, Madison Grant had always identified closely with the large mammals of North America. The phenomenon of projection was even more intense now that one of the continent's grandest beasts bore his name, and he proudly hung in his residence the head of a Grant's caribou (next to his trophy of an Osborn's caribou). But it was with a deeply personal sense of sorrow that Grant lamented, of *Rangifer granti:* "Formerly they were exceedingly abundant [in Alaska], but of late they have been greatly reduced in numbers through the agency of the market hunters."[49] There is little doubt that, for Grant, the near extinction of *Rangifer granti* thanks to the butchery of the market hunters was not just representative of the fate of all North American wildlife but also a multivariate symbol of his own mortality, the decline of his class in America, and even the demise of the Nordic race in the world.

The caribou were declining because the discovery of gold in the Klondike in 1896 had brought destruction to the flora and fauna of Alaska by just the sort of greedy, grasping men that the Boone and Crocketteers had been raised to despise. In 1890 there had been only 4,300 whites in all of Alaska, two-thirds of them in the southeastern towns of Juneau and Sitka. But thanks to the Klondike gold rush, by the turn of the century there were tens of thousands of would-be millionaires feverishly prospecting throughout the territory. The miners and their hungry sled dogs all had to be fed, and opportunistic market hunters flocked to Alaska to supply the gold seekers with meat. What especially frightened Grant was that shipping companies were constructing huge cold-storage warehouses at Valdez and Skagway for the purpose of storing meat pending shipment to the United States.

Grant rightly feared for the future of Alaska's wildlife and warned that "the tragedy of the destruction of the American bison is being enacted over again in our Arctic province." In fact, he told his friend the congressman John F. Lacey, "in Alaska today tragedies of slaughter are being enacted far *worse* than the destruction of the buffalo, as they involve not one species, as in that case, but a dozen or more, some of them animals of great size, and exceptional beauty and interest."[50]

Grant decided to take action to prevent the extermination of Alaska's game. And—being a progressive—taking action meant pushing for federal legislation. Congress was a relatively small body of men gathered in a single location, and thus it was an institution amenable to being lobbied by a wealthy aristocrat like Grant. And Alaska's status as a territory meant that it was within reach of the eastern patrician via the long arm of the federal government. Accordingly, Grant drafted the highly progressive Alaska Game Bill, which attacked the problem on four fronts. Grant's bill (1) prohibited completely the commercial killing of all Alaskan game animals and wild birds; (2) banned the shipment from Alaska of wild meat; (3) established seasons for the private hunting of the major game species and limited the number of specimens that each hunter could kill; and (4) authorized the government to prohibit hunting entirely for species in danger of extinction.[51]

The residents of Alaska were furious about the bill, and projected their class and regional animosities onto Madison Grant. After all, what could be more elitist than a wealthy Manhattan sportsman, who shot animals for the sheer pleasure of it, using his influence and connections to prevent local market hunters from earning a living and providing sustenance to their clients? What kind of democracy was this, where the struggling pioneers of Alaska (who had no vote in Congress) were at the whim of an effete eastern conservationist in league with federal bureaucrats in Washington, D.C.?

Once again, however, if westerners thought they could cow Madison Grant by calling him antidemocratic or elitist, they were very mistaken. The aristocratic Grant cared as much about the views of local residents as he did about the opinions of the Anti-Defamation League, and he had no qualms whatsoever about delegating to himself the responsibility of deciding what was best for Alaska and Alaskans. "It should be clearly understood," wrote Grant, "that the game of Alaska, or of any other region, does not belong exclusively to the human inhabitants of that particular region, and that neither the white settlers nor the native inhabitants have any inherent right to the game." After all, "If control of the making and enforcement of the regulations be turned over to the residents without Federal control, it will be the death knell of many species of game." Conservationist William T. Hornaday concurred: "In Alaska we are now face to face with this question: Is it possible for the people of the United States to protect the game of Alaska against the residents of the territory who are determined to annihilate it?"[52]

Grant's friend Charles H. Sheldon, a leading member of the Boone and Crockett Club and an expert on Alaska, disagreed with this analysis. Sheldon thought it counterproductive to pass laws so unpopular with the locals that they were almost certain to be disobeyed. He felt that easterners could more wisely expend their efforts trying first to nurture local appreciation of the conservationist ethic and only afterward, when the residents were more amenable, turning to the federal government to enact legislation.

But Grant's position, if elitist, was valid. Aside from any sob-sister concern with the animals themselves, the long-term economic interests of Alaskans lay in preserving their abundant wildlife. If the game were to disappear, it assuredly would hurt the Alaskans more than Madison Grant. Wealthy eastern sportsmen could always afford to travel elsewhere to satisfy their recreational needs, but the residents of Alaska would indeed suffer without a steady supply of meat ensured by an intelligently managed wildlife population—not to mention the income derived from tourism. And if it took a patrician five thousand miles away to point this out to them, so be it. After all, wrote Grant, "This . . . is the history of the enforcement of game laws everywhere. The law itself must be in advance of public opinion."[53]

Grant, in short, was determined to save the Alaskans despite the fact that they had evinced no desire to be saved. Whether they appreciated it or not, the Alaska Game Bill, he wrote, had been drafted for their "true and permanent interest." (And recent indications are that westerners are finally beginning to understand that their "true and permanent interest" does indeed lie in preserving the natural wildlife and resources of their region. In the long term, there is far more money to be made—for far more people, and for a far longer period of time—in guiding, lodging, rafting, etc., than in market hunting, clear-cutting, and strip mining.)[54]

So Grant pressed ahead with his game bill, and with lobbying support from his colleagues in the Boone and Crockett Club, Congress passed the measure and President Roosevelt signed it on June 7, 1902—just six months after Grant had first conceived of the law. The Alaska Game Law of 1902 was one of the first important legislative triumphs for conservation on the national stage.[55]

Having put the market hunters out of business in Alaska, Grant turned his attention to the rest of the country, and within a few years he fatally crippled the market hunters in the Lower 48 by helping to promulgate (as I have recounted elsewhere) such seminal pieces of legislation as the No-Sale-of-Game Laws (in which legislatures from New York to California banned the sale of game) and the Migratory Bird Law of 1913 (which firmly established the principle of federal control over wildlife by authorizing the government to set bag limits, limit seasons, and prohibit spring shooting on a national scale). Furthermore, with Grant's active encouragement, Canadian authorities soon enacted similar game preservation laws for British Columbia and Newfoundland; and the success of these regulations then led British lawmakers to pass comparable legislation in His Majesty's colonies in East Africa.[56]

It was a wonderful time to be a progressive patrician ("progressive" in the sense that one expected the federal government to step in and implement legislative solutions to far-flung problems; "patrician" in the sense that one had sufficient power and influence to capture the ears of the appropriate government officials). A man like Madison Grant could see a problem, devise a solution, get Congress to approve it within a matter of months, have his friend the president of the United States sign it, and watch as authorities from Newfoundland to Uganda emulated it.

When Madison Grant began his fight to ban market hunting at the turn of the twentieth century, the effort was generally viewed as hopeless and absurd. But within a few years he had permanently enshrined the progressive idea that the federal government can and should regulate wildlife, and he had convinced the American public—in a major philosophical shift—to view game animals as recreational rather than commercial resources. And as a result, market hunting all but disappeared in the United States, which is why an admiring E. W. Nelson (head of the Biological Survey) declared in the early 1930s that Madison Grant had been "the godfather to . . . wild life conservation measures for the last 25 years."[57]

From the day Madison Grant joined the Boone and Crockett Club in 1893, the organization was made over in his likeness. Over the next five decades, Grant would serve as a member of the club's executive committee (from 1897 to 1903), secretary (1903–1913), vice president (1913–1923), and president (1928–1937). During that time, the Boone and Crockett Club emulated Grant's quiet, persistent, behind-the-scenes method of lobbying. Its members cultivated key legislators, entertained important newspaper editors, submitted articles to influential journals, and appeared frequently before congressional committees. The club was quite consciously and comfortably an exclusive association. While many modern conservation organizations spend a great deal of their time and resources appealing to the public in an effort to raise funds and increase their membership, the Boone and Crockett Club never asked the public for funds and purposely limited its membership to an elite core of one hundred. The very concept that a group that seeks to shape public policy would *limit* its membership may seem odd to us. How can you be influential in a democracy if you don't have hundreds of thousands of members backing you up? But when Madison Grant discussed, for instance, the plight of the pronghorn antelope with Senator Wadsworth over a drink at the University Club, he got far more accomplished than a million e-mail messages to the White House would accomplish today.

The Boone and Crockett Club, some might say, was the Trilateral Commission of its time. In fact, modern-day conspiracy theorists would have had a field day had they scanned the club's membership roster at the turn of the century. At the club's fifteenth annual meeting in 1902, for example, among the members present were influential conservationists (such as George Bird Grinnell and Madison Grant), famous explorers, important businessmen, powerful philanthropists, the leaders of several state agencies, the head of the Bureau of Biological Survey (forerunner of the U.S. Fish and Wildlife Service), the chief forester of the United States, numerous members of the House of Representatives, four U.S. senators, the secretary of war, a future secretary of state, and the president of the United States.

George Bird Grinnell admitted that the charter members of the Boone and Crockett Club had been more interested in conviviality than conservation. "Its original purpose," he confessed, "was to bring together a number of big-game hunters for social intercourse." By 1904, however, Grinnell could write that "a considerable proportion of its members now never hunt wild animals, but, looking to the future, are endeavoring to preserve for this country a reasonable stock of its indigenous wild creatures, which will be beautiful and historical objects for succeeding generations to admire."[58] And yet, despite the new dedication to conservation, the original constitutional provisions for membership in the club still remained in effect: candidates must have killed in fair chase at least three different species of American large game. It would be akin to the Sierra Club making it a requirement that prospective members must have personally slashed and burned one thousand acres of rain forest. Despite this anomaly, Secretary of the Interior Stewart Udall could state admiringly in 1963 that "the Boone and Crockett Club has made an outstanding contribution to our legacy of wild things." Of the club's members, he wrote: "Together they possessed a prestige and breadth of experience that gave them entrée to offices of influential men. They set the standards of sportsmanship for a generation and did much to save the big mammals of North America. No conservation organization in our history has had more political know-how."[59]

The wider public may never have heard of the Boone and Crockett Club, but on issues relating to conservation, it became a force—*the* force—to be reckoned with in the halls of Congress, the agencies of the federal government, and the legislatures of the various states. The club owed its remarkable success to the three patricians who most influenced its methods and its goals: Theodore Roosevelt, George Bird Grinnell, and—most crucially—Madison Grant. After the death of Grant in 1937, conservationist H. E. Anthony accurately evaluated his influence on the conservation movement: "Because Grant had such a wide range of contacts and enjoyed the confidence of such a broad circle of friends, [he] exerted an influence for conservation that has probably been exceeded by no other individual in private life."[60]

2 The Bronx Zoo

Order is Heaven's first law, and must be ours, also. The warfare against dirt and disorder must be constant.

Rule No. 1 for Employees of the Bronx Zoo

As Madison Grant hunted the ever-dwindling mammals of North America in the 1890s, he began to dream of creating a zoological park in which the continent's endangered species could be preserved "in surroundings as nearly as possible similar to those of their native habitat." "No civilized nation," he stated, "should allow its wild animals to be exterminated without at least making an attempt to preserve living representatives of all species that can be kept alive in confinement."[1]

But Grant had little desire to create an Old World–style game preserve encompassing thousands of acres of fenced-in land on a nobleman's estate, where species were indeed protected from poachers but where the average citizen could not see the creatures. Grant wanted to locate his zoo in the midst of the nation's metropolis, New York City, for it was his belief that an American game sanctuary should provide access to the urban public, who would thereby become educated about—and be alerted to the beleaguered status of—their country's native fauna.

On the other hand, Grant had no interest in building a typical nineteenth-century urban zoological garden. The leading European zoos of the time (e.g., London, Paris, and Antwerp) and their North American emulators (e.g., Philadelphia) measured only about thirty acres in size. Grant was disgusted by these cramped institutions where all the species, irrespective of their particular needs or habits, were locked up like dangerous prisoners in bare, solitary cells of uniform size and shape, lined with tile or cement and fronted with thick iron bars.

Grant was determined to create a zoo "entirely divergent" from the established institutions. He envisioned a zoological park about three hundred acres in size (which would make it five times the size of the largest zoo in Europe, the sixty-three-acre Berlin zoological garden). Rather than solitary confinement, the animals would live in groups as in the wild; and instead of cramped, sterile enclosures, they would roam in large, realistic habitats. In this manner, they would receive stimulation from interacting with other creatures and with their environment, and visitors would be able to view and study these healthy animals in beautiful, natural settings.[2]

Zoological experts of the day thought the idea extreme and wrongheaded, and argued that the public preferred small cages, with one species per exhibit, where the animals could be seen up close and easily identified. But Grant's novel scheme did resonate amongst at least some segments of a society increasingly aware that the frontier no longer existed and that the nation's flora and fauna were now completely surrounded—and threatened—by the encroaching forces of modern civilization. Further, many people were entranced by Grant's vision of the zoological park as a patch of nature in the urban wasteland and a place where the masses could get in touch with the natural setting in which America's values had first flourished. At a time when cities were dirty, violent, unsanitary breeding places of crime and disease, lorded over by corrupt bosses and overridden by foreigners, the creation of the Bronx Zoo was part of a conscious turn to nature among old-stock Americans, who were at the same time landscaping parks, creating urban playgrounds, reading the stories of Owen Wister, enrolling their sons in the Boy Scouts, and nodding affirmatively as Frederick Jackson Turner explained that it was not the overcrowded city but the wide-open frontier that had forged the American character.

Still, as a dyed-in-the-wool Republican, Madison Grant simply did not possess the requisite contacts within Tammany Hall to have his concept even considered by the municipality of New York. But then, fortuitously, developments in 1894 provided him the opportunity to obtain the political influence necessary to implement his project.

Reforming New York City

New York City, observed Grant's friend Rudyard Kipling, is "a despotism of the alien, by the alien, for the alien, tempered with occasional insurrections of decent folk." The year 1894 witnessed one of those insurrections, and Grant and his brother DeForest were among the young reformers who helped to lead it. In that year, a number of organizations standing for civic reform created the nonpartisan "Committee of Seventy" for the purpose of exposing the corruption permeating the municipal government and proposing concrete programs of urban improvement. Among other remedies, the Committee of Seventy (which included a number of Grant's friends from the Society of Colonial Wars) called on the municipality to expand the civil service, improve the public schools, reform tenement housing, extend transit facilities, ensure clean water, and provide more and better street cleaning, garbage disposal, public lavatories, and parks. Then, after considering a number of individuals (including Theodore Roosevelt and Seth Low), the reformers chose respected businessman William L. Strong to be their nonpartisan candidate for mayor in the 1894 election.[3]

The New York municipal reform movement of 1894 was a proto-progressive crusade, and its adherents—the Grant brothers included—were the type of men we typically associate with late nineteenth-century status anxiety. One scholar notes that the Committee of Seventy comprised men "who by ancestry and political leadership had deep roots locally, who belonged to the best clubs, but who now found themselves surpassed by the leaders of the new industrial order."[4] But as much as the reformers feared the rising power of the new plutocracy, it was their concern about the detrimental effects of the New Immigration that spurred them into action in 1894. To most of these patricians, Tammany Hall was a threat to the social order not because of its ties to the robber barons but because of its corrupt relationship with the newly landed foreigners.

In the twenty-nine years that Madison Grant had been alive, the population of New York City had increased 250 percent. And the main source of this increase had been immigration. More than two-thirds of the nation's immigrants were entering through the port of New York, and nearly half of them were settling permanently in the city. By 1894, New York had over 1.8 million residents, of whom an astounding 1.4 million had been born abroad or had at least one foreign-born parent. In other words, foreigners and their children now outnumbered old-stock Americans three to one in the city. New York City had more Italians than Rome; it had twice as many Irish as Dublin; and it would soon be home to the largest Jewish community in history. Grant's friend George Horace Lorimer (editor of the *Saturday Evening Post*) looked out on this "Babel of aliens" and realized with consternation that "New York is now a great foreign city with an American quarter."[5]

The influx of all these foreigners was rendering the city all but unrecognizable to Grant. Large swaths of his hometown were now characterized by teeming slums, filthy sweatshops, rowdy saloons, noise and drunkenness, crime and disorder, poverty and disease. Grant was repulsed by the chaos, the squalor, and the stench emanating from the immigrant neighborhoods of Manhattan's East Side, where the streets were often filled with dead horses, mounds of offal, and spilled barrels of fish. But even more than the anarchy and the vice, it was the political corruption associated with the immigrants that disturbed Grant. For the immigrants were bound indissolubly to Tammany Hall, and the congenially venal operatives of that graft-ridden institution were too busy enriching themselves at the public trough to address the fact that the city of New York, with its exploding population, deteriorating infrastructure, and social pathologies, had become all but unlivable.

Madison Grant and his brother DeForest were indignant at what was happening to their native city, and they enthusiastically assumed a leading role in the reform movement of 1894. They were extremely active in the mayoral campaign of William L. Strong and applauded when the Reverend Charles H. Parkhurst (pastor of the Madison Square Presbyterian Church and an important supporter of Strong) denounced "the damnable pack of administrative bloodhounds . . . that, under pretense of governing this city, are feeding day and night on its quivering vitals. They are," thundered Parkhurst while pointing toward Tammany, "a lying, perjured, rum-soaked, and libidinous lot." Come Election Day, Strong ran well in all areas of the city—except, of course, the East Side—and defeated the Tammany candidate.[6]

Upon assuming office in January 1895, Mayor Strong proved to be a staunch supporter of civil service reforms, public health measures, and increased expenditures for the public schools ("Many future generations of school children will rise up and call him blessed," stated Nicholas Murray Butler).[7] Strong also made excellent appointments, including Theodore Roosevelt as police commissioner and noted expert George E. Waring as sanitation commissioner.

For the first time in many years, New York City was governed honestly and efficiently. And that, of course, proved to be the downfall of William L. Strong. For, once he was inaugurated, some of the elements that had been "outraged" by Tammany's spoils system and had worked to bring about Strong's election began to demand what was rightfully theirs. Many of them were disappointed to discover that they had put into office a genuinely honest, nonpartisan man, who ungratefully refused to make appointments on the basis of political contributions. To make matters worse, the expanded municipal services provided by the Strong administration, along with its various public works projects, required increased taxes. This, combined with Commissioner Roosevelt's policy of enforcing the law against drinking on the Sabbath, led to a general backlash against "reform" in New York. Even if he had wanted a second term (which apparently he did not), the political climate made it impossible for Strong to seek reelection in 1896. In that year, Tammany easily elected its candidate (Robert Van Wyck) and celebrated its triumphant return to City Hall with the gleeful chant:

> Well, well, well.
> Reform has gone to Hell![8]

The Triumvirate Builds the Zoo

The ephemeral Strong administration proved to be a boon for the ambitions of young Madison Grant. His campaign activities in 1894 had given him a certain amount of political visibility, and as soon as Strong was elected, Theodore

Roosevelt dined with Grant and advised him that if he was ever going to turn his dream of a zoological park into a reality, now was the moment. Accordingly, Grant drafted and introduced into the state legislature at Albany a bill creating the New York Zoological Society, a corporation that would be empowered to establish a zoological park somewhere in New York City.[9]

But the measure was immediately attacked by a coalition of politicians identified by Grant as "certain East Side representatives," who instinctively opposed any proposal coming from a supporter of Mayor Strong. The opponents of Grant's bill hit upon an effective strategy when they charged that the zoo was a scheme by wealthy Fifth Avenue property holders to do away with the Central Park Menagerie. The menagerie was a tiny (nine acres) and olfactorily offensive collection of mangy goats and diseased camels discarded from passing carnivals. It did not quite approach Grant's vision of a world-class zoological park: "A more wretched exhibition of ill-kept specimens," he sniffed, "cannot be found in any large city in the world." But the menagerie was a favorite weekend destination for what the *New York Times* referred to as "the teeming thousands of the East Side." And it thus received protection from the New York City bosses, who existed in a symbiotic relationship with the tenement dwellers of lower Manhattan. To allow the construction of a new zoo in some other part of the city, most likely beyond the range of a five-cent streetcar fare, would have been politically harmful to the minions of Tammany, who accused Grant of plotting to abolish the Central Park Menagerie in order to please "the rich people on Fifth Avenue who live near it, and object to it."[10]

In truth, when the Zoological Society was in the midst of a fund-raising drive a few years later, Grant did appeal confidentially but explicitly to the wealthy residents of Fifth Avenue on the grounds that their property "would be greatly benefited by the opening of our zoo and the consequent [demise] of the existing menagerie." However, the elimination of the "disreputable" menagerie was not Grant's primary goal in 1895 (though the idea that he could have his new zoo and at the same time thumb his nose at Tammany—and the unwashed *Untermenschen* who supported it—must have had its appeal).[11]

Early in 1895, Grant traveled to Albany to counteract the opposition of the East Side politicians. As "the bold young Yale graduate" prowled the halls of the state capitol he impressed upon the legislators "what a mission and opportunity" they had "in these closing days of the century . . . to preserve to future generations some remnant of the heritage which was our fathers'." To bolster his offensive, Grant called on his fellow Boone and Crocketteers to utilize their rather extensive influence with certain key legislators; for his part, it did not hurt that Grant was good friends with Hamilton Fish, the Speaker of the Assembly. After a prolonged battle, the "Act to Incorporate the New York Zoological Society and to Provide for the Establishment of a Zoological Garden in the City of New York" was passed in April 1895. Mayor Strong approved the bill at once, and the governor's signature was procured without difficulty. Theodore Roosevelt, for one, was delighted: "I congratulate you with all my heart upon your success with the Zoo bill," he wrote to Grant. "Really, you have done more than I hoped. I always count myself lucky if I get one out of three or four measures through."[12]

Grant then ensured his permanent control over the New York Zoological Society by stocking the board with numerous members of the Boone and Crockett Club and other friends connected to him through his social clubs and/or the campaign to elect Mayor Strong. To fill the office of president of the Society, Grant chose a very prestigious figure: Levi P. Morton, onetime vice president of the United States (under Benjamin Harrison), a former minister to France (who had accepted the Statue of Liberty for the United States on July 4, 1884), and the governor of New York who had endorsed Grant's Adirondack Deer Law.

The duties of secretary of the Zoological Society were split between George Bird Grinnell and Madison Grant. It was entirely characteristic of Grant to shun the limelight by taking on the more lowly position of secretary. Also, we should bear in mind that since he was still less than thirty years old ("a young aristocrat," in the words of a friend, "with the ink on his Yale diploma only half-way dry"), it would have been impolitic for him to accept the position of president or even vice president.[13] But everyone knew that Grant had conceived and brought into existence the New York Zoological Society and was the actual locus of power in the organization. Grant would serve on the executive committee from the day the society was formed in 1895 until his death in 1937. During those years he gradually assumed ever greater (and overlapping) official duties, beginning as secretary (1895–1925), then chairman of the executive committee (1909–1937), second vice president (1915–1916), first vice president (1916–1925), and finally president (1925–1937).

Next to Grant, the other major force within the Zoological Society was his closest friend (besides George Bird Grinnell), world-famous paleontologist Henry Fairfield Osborn. Like Grant, Osborn (1857–1935) was born into a wealthy and prominent New York family headquartered at Castle Rock, their palatial estate on the Hudson. His father was a railroad tycoon who associated with the likes of J. P. Morgan, Cornelius Vanderbilt, and Morris K. Jesup. As with Grant, however, it was not business but the natural sciences that attracted Henry Fairfield Osborn as a youth. He earned his A.B. from Princeton in 1877 and eventually became head of the Department of Vertebrate Paleontology at the American Museum of Natural History, where he trained numerous students, many of whom became distinguished zoologists and paleontologists.

Like Grant, Osborn was an active member of the Century, Metropolitan, and University Clubs and, of course, the Boone and Crockett Club. And thus, regardless of what other names appeared on the letterhead, the New York Zoological Society was, and for decades would be, the bailiwick of Madison Grant and

Henry Fairfield Osborn. In the ensuing years, Osborn would serve as vice president, chairman of the executive committee, and then president (1909–1923) of the society.

To serve as director of the Zoological Park, Grant and Osborn selected forty-one-year-old William Temple Hornaday (1854–1937), former director of the National Zoo in Washington, D.C. Hornaday had been a big-game hunter who abandoned that avocation and devoted the rest of his life to the tireless protection of wildlife. He helped to found many conservation organizations (e.g., the Camp Fire Club, the America Bison Society, and the National Audubon Society) and was an active member of many others. While Hornaday's biographer exaggerates when he says that Hornaday was "undoubtedly the greatest wildlife conservationist in the history of the United States," it is true that his dignified appearance, commanding personality, and jeremiads about the decline of the country's fauna ensured that he received more publicity than any other preservationist of his time. "It is ironic," William Bridges claims, "that Hornaday, whose name is almost synonymous with the conservation of wildlife in the first third of this century, should have spent so many years in killing animals." Ironic, to be sure, but hardly surprising; as we have seen, many of the early American conservationists started as hunters.[14]

For the next three decades, the New York Zoological Society would be ruled by Grant and Osborn, and the Bronx Zoo would be run by a triumvirate consisting of Grant, Osborn, and William T. Hornaday. And every day during those three decades, Hornaday called Madison Grant on the telephone at 11:30 A.M. sharp. Grant's first order to Hornaday was to select a site for the zoo. He sought an area that was secluded yet accessible, and that had a varied topography to accommodate the needs of what was certain to be an eclectic collection of species. After surveying all the undeveloped lands of New York City, Hornaday—with Grant's approval—decided that Bronx Park would make an ideal site for the zoo. "At that time," remembered Hornaday, Bronx Park "was an unbroken wilderness, to the eye almost as wild and unkempt as the heart of the Adirondacks." His first sensation at seeing the park was one "of almost paralyzing astonishment and profound gratitude. It seemed incredible that such virgin forest . . . had been spared in the City of New York until 1896! But there [it was]—waiting for us. The magnificent possibilities of the place as an ideal home for wild animals in comfortable captivity—freedom in security—unrolled before me in panoramic clarity."[15] After a series of protracted and contentious negotiations, Grant convinced the city to hand over to the Zoological Society all of Bronx Park south of Pelham Avenue—a tract of 261 acres.

In order to begin constructing the zoo, Grant needed to raise $250,000. He was fully aware that a number of well-established cultural institutions had first call on the funds of wealthy New Yorkers and confessed that he felt "Like young Lochinvar at the wedding." But all those years wastreling in the elite clubs of

Manhattan now paid off for Grant: "The suave young aristocrat" (as Hornaday referred to him) drew up a list of the wealthy men with whom he had savored a fresh shipment of Macanudos at one time or another and called on them to ask for contributions. Grant also enlisted the well-connected Henry Fairfield Osborn in the fund-raising effort, sending him off with "best wishes for a successful raid on the finances of your friends."[16]

Their fund-raising was abetted by the spirit of the age. Within two years, the Spanish-American War would prove to the world that the United States deserved to be ranked with the great *military* powers; and in the meantime, the leaders of New York desperately wanted Europe to take notice of the fact that their city deserved to be ranked with the great *cultural* centers. The Metropolitan Museum of Art was gobbling up almost any art object that came on the market, the New York Public Library was collecting virtually everything that was being published, and the cultural arbiters of New York were determined to amass a world-class zoological collection for their soon-to-be world-class city. Accordingly, among those who responded to Grant's financial appeal were such luminaries as Andrew Carnegie, Collis P. Huntington, John D. Rockefeller, Cornelius Vanderbilt, William C. Whitney, and J. P. Morgan. All were well known philanthropists with a history of supporting the cultural institutions of the city, many were members of the Society of Colonial Wars, and several had been on the Committee of Seventy that had elected Mayor Strong. Their pockets—and Madison Grant's charm—were such that he easily raised the quarter of a million dollars, and in the summer of 1898 hundreds of workmen from thirty-one different contractors invaded Bronx Park and commenced construction of the largest zoo on earth.

Several times a week Grant jumped onto his bicycle (bicycle riding was a veritable craze among fashionable New Yorkers in the Gay Nineties) and pedaled up to the Bronx to inspect the progress. Zoo employees knew to be on their best behavior when the founder of the Zoological Society was around; news that "Mr. Grant has just come in the Boston Road Gate" would spread anxiously from exhibit to exhibit, and woe be to any worker he found with jacket unbuttoned or cap askew. As he traversed the ground—avoiding the streams of supply trucks, the piles of construction materials, and the scores of curious onlookers who milled about—the thirty-two-year-old Grant could see his vision of a zoo being created before his eyes: here was the Alligator Pool and the Beaver Pond, there the Antelope Range and the Bear Den, yonder the Pheasants' Aviary and the Bison Prairie. Like a seigneur surveying his estate, Grant proceeded to name the various features of the park (using the surnames of famous American naturalists): Lake Agassiz, Baird Court, Merriam's Hill, Goode's Rock, Cope Lake, Audubon Court, Osborn's Walk, and so on. Later, there would also be a Grant's Walk.

As construction progressed, the triumvirate of Grant, Osborn, and Hornaday hired curators and staff members, designed uniforms and badges, composed rules and regulations, devised feeding procedures, procured supplies, and purchased animals. Decisions were not always made quickly or smoothly; all three were very opinionated and prideful gentlemen. Although Hornaday had been a nationally known naturalist for years, and was older than the other two men, he always felt like a social inferior in the company of Grant and Osborn and suspected that in their eyes he was just "a Philistine from the jungles of Buffalo."[17] He and Osborn had a number of hotly worded collisions in those early days. Hornaday, to be sure, was irrationally sensitive to the most unintended of slights, but it did not help that the aristocratic Osborn—who could be insufferably haughty even to those who were his superiors—was condescending at best, and impatiently dismissive at worst, to those he deemed beneath him.

While they had fewer blowups, a discernible note of resentment also crept into Hornaday's relations with Grant, who, after all, was eleven years younger than his subordinate and not yet the world-famous figure that Osborn or Hornaday was. But even if Hornaday had not minded taking orders from the green society man, his dealings with Grant were bound to be testy, as there was a large ideological gap between the two. Hornaday, the onetime hunter, had utter contempt for so-called "conservationists" like Grant who continued to hunt. He considered that to be the worst form of hypocrisy, and he could never excuse the fact that sportsmen like Grant were primarily interested in conservation in order to ensure a future supply of game. This was in spite of the fact that—or maybe because—Hornaday himself, in his early years, had killed as wide a variety of endangered species as any other American. (Stephen Fox correctly characterized Hornaday as "a classic example of the repentant hunter" who "with all the zeal of the convert, set about atoning for his early sins.")[18] As the years passed, however, Hornaday developed a deep respect for Grant, and in his autobiography he declared: "In Mr. Madison Grant, the world beholds a unique, picturesque and powerful Coadjutor of Science. . . . He was, and yet is, a man of delightful personality, broad vision, high ambition and indomitable spirit. . . . Warm-hearted, sympathetic and helpful, Mr. Grant was born in 1865 for the very evident purpose of originating the New York Zoological Society in 1895, and taking a tremendously important part in founding, designing and developing a zoological institution for the millions."[19]

As opening day for the Bronx Zoo drew near, offers of animals poured in from all over the world, including from Madison Grant's explorer friends (the expeditions of men such as Admiral Peary and Colonel Roosevelt, for example, resulted in numerous accessions to the park). As each animal arrived at the zoo, great thought was given to the labeling of its enclosure, and signs were posted conveying the animal's natural history. (This was in contrast to most zoos, which offered no information at the cages as a ploy to force visitors to purchase the guidebook.) Grant felt that a judicious use of maps and charts to disseminate information regarding the diminishing numbers and dwindling habitats of endangered species was a painless method to introduce the public to the importance of conservation. In addition, the act of labeling symbolized for Grant his patrician appropriation of these animals. By capturing and displaying as many species as possible he was demonstrating his mastery over the animal kingdom, and by naming and describing each species in a label he was imposing a sense of order onto the natural world. Further, having paid for the collection themselves, Grant and the other members of the Zoological Society felt a proprietary interest over their animals. They did not view themselves as the caretakers of these wild creatures but rather as their *owners*. And they enjoyed the fact that by owning these beasts—preferably the largest and fiercest fauna available—a great deal of status accrued to them.

On a wider level, all residents of New York and indeed of the nation took pride in the zoo's growing collection, for the importing of large animals from all parts of the globe served to showcase the country's newfound power. William T. Hornaday once averred that the triumvirate "sought only to build a humane and beautiful home for wild animals, and to amass collections that would be worthy of the United States of America—worthy of the metropolis of the western hemisphere." But it is probably fair to surmise that a collection of, say, dung beetles, no matter how extensive, would not have been deemed worthy of "the metropolis of the western hemisphere." No, the nation that had just defeated Spain in a "splendid little war" and acquired a worldwide empire deserved to show off the world's biggest and strongest animals in the world's biggest and most magnificent zoo.[20]

By May of 1899, six months prior to opening day, all manner of animals were arriving daily in the Bronx. Unfortunately, not all of their quarters had been completed; thus, for example, a group of rather sullen orangutans had to be put up in the Hornadays' living room (where the ever-patient Mrs. Hornaday fed them "a teaspoonful of castoria daily in order to get [their] digestive apparatus in good working order"). In such disordered conditions, escapes were common: the sea lions fled their pool for the dubious safety of the Bronx River, and the zoo's first snake, a "very vigorous and vicious python about 16 feet long," slithered to freedom through a hole in its enclosure on its very first night in the park. When Hornaday arrived at work the next day and was informed of the disaster, he blurted out: "Is *this* the way to start a new Zoological Park?"[21]

Not all the escape attempts were successful. One of the European wolves wiggled out of her temporary den and raced from the premises, but zoo officials finally recaptured the poor animal, which, after a week loose on the streets of New York City, was undoubtedly grateful to be returned to the friendly confines of the zoo. And then there was the Bornean sun bear who climbed over its fence and commenced to lumber away toward freedom; an infuriated Hornaday strode right after the animal, grabbed him by the scruff of the neck, and started dragging him kicking and growling back to the zoo. When the unhappy bear bit Hornaday on the hand, Hornaday (according to one witness) ordered: "Shoot the bear and give every man a steak!"[22]

The poor Bornean bear aside, the (unintended) death rate among the newly arrived animals was extremely high in those early months. In part, this was because proper accommodations had not yet been built. In part, it was because the zoo bravely (or foolishly) accepted foreign species that had never before been kept in North America (many of the tropical birds, for instance, and one of the pythons died when a cold wave hit New York in October). But the main problem was that veterinary skills and knowledge of nutritional needs still had a long way to go in the 1890s, and thus the mortality rate in the zoo's first year was an appalling 30 percent. It seems to have been accepted as a matter of course that a given animal would probably survive for only a few months in captivity and would then have to be replaced.

Nevertheless, by the fall of 1899 the zoo had on hand 843 animals of 157 species, and the New York Zoological Park, known colloquially as the Bronx Zoo, opened to the public on November 8. Levi P. Morton, president of the Zoological Society, gave a speech welcoming the two thousand politicians, philanthropists, scientists, and celebrities on hand. Madison Grant, as usual, declined to speak in public, but Henry Fairfield Osborn—to no one's surprise—opted to address the crowd. As a true progressive, he promised that the zoo had not just re-created nature but actually improved on it: "All the animals of North America . . . will be seen just as they live in the woods—happier perhaps because safe from the rifle of the hunter, free from the keen struggle for existence, generously quartered and fed." And he stressed the educational mission of the zoo: "What our museums are doing for art and natural science, this Park . . . will do for Nature, by bringing its wonders and beauties within the reach of thousands and millions of all classes who cannot travel or explore. [This zoo has been] designed in every detail for the benefit and enjoyment of the people, and as a supplement to the work of our schools and colleges."[23]

When Osborn finished speaking, the dignitaries took a tour of the zoo. They were enchanted by the animals and amazed that so much had been accomplished in such a short time. (It had been only four years since Grant received the legislature's approval, and just fifteen months since the initial groundbreaking.) An English ornithologist who visited the zoo shortly after it opened declared: "I feel as if I have seen the world's wildest of animals in their native haunts." As far as he was concerned, "not even the happy hunting grounds of East Africa" could compare to the animal enclosures of the Bronx Zoo.[24]

When the last opening-day visitor finally straggled out of the park at sunset, Grant congratulated the staff and immediately departed on a moose-hunting expedition to Canada. Grant loved to collect moose trophies (although he regretted that the animal's nose displayed "Jewish characteristics").[25] And while it may seem paradoxical that the leader of an organization dedicated to saving animals would celebrate by killing them, we must again emphasize that for a Boone and Crocketteer, the whole point of the Bronx Zoo was to conserve species so they could be hunted.

From the beginning the Bronx Zoo was an immensely popular attraction. In 1900, during its first full year, more than half a million people visited the zoo despite the lack of public transportation (the subway would not reach the zoo for another six years). Within three years, annual attendance had doubled to one million visitors; and by 1914 attendance had doubled again to two million (the same level as today).

The number of animals in the zoo's collection rose along with attendance, reaching an impressive five thousand creatures in 1909. Thus, within ten short years, the Bronx Zoo was not only hailed as the most beautiful zoo in the world but was also the largest, in terms of both geographical size and number of specimens.

In 1902, less than three years after the Bronx Zoo opened to the public, the Zoological Society assumed control of the New York Aquarium at Battery Park, at the southern tip of Manhattan. (Known today as Castle Clinton National Monument, it is the circular stone building where tourists embark for the ferry ride to the Statue of Liberty.) For decades the aquarium had been a gloomy, odoriferous, poorly managed institution, with a shady reputation as a Tammany Hall patronage scheme. The aquarium's directors were always political hacks who were more often found dining on fish at fancy restaurants than caring for the ones unhappily swimming in the aquarium's polluted tanks (most specimens never survived more than one season at the aquarium). But Madison Grant convinced the state legislature that the New York Zoological Society could take over management of the aquarium and bring to the Battery the same scientific and nonpartisan expertise that it had brought to the Bronx.

As director of the aquarium, Grant and Osborn selected Charles H. Townsend, one of the world's leading experts on marine animals and a devoted conservationist. Townsend would be involved with Grant in a number of preservation battles over the years, including successful efforts to win protection for the Alaskan fur seals, the Galapagos tortoises, and the world's endangered whales.

With Grant's enthusiastic support, Townsend removed the Tammany-appointed staff of the aquarium, doubled the size of the collection, and installed new exhibition tanks featuring the most natural habitats possible. The aquarium soon boasted of the largest and most aesthetically pleasing collection of marine fauna in the world, and it immediately became New York's most popular attraction. In 1903, its first full year of operation, the aquarium had 1.5 million visitors; by 1909 an astonishing 3.8 million visitors toured the aquarium. By 1910, Grant could proclaim with complete accuracy that the aquarium "is now the foremost institution of its kind in the world. . . . And it has an atten-

The most popular museum on earth: the New York Aquarium in the 1930s.
© Wildlife Conservation Society.

dance that is greater than that of all the other public museums of New York together."[26] For many years, in fact, the New York Aquarium was the most popular museum on the planet. And for decades to follow, no aquarium anywhere was built without direct input from the staff of the New York Aquarium, which developed state-of-the-art methods of collecting, displaying, feeding, breeding, and rearing fishes.[27]

Man in the Zoo

Madison Grant had conceived of the Bronx Zoo as a place where the public could come to be enlightened about conservation, but he had never really paused to consider that the public was made up of actual human beings, who were not going to maintain in the zoo the standards of decorum that obtained in the Knickerbocker Club. Particularly after the subway line was extended to the

Bronx, and the great unwashed began bypassing the Central Park Menagerie in favor of the Zoological Park, the triumvirate began to dread (as Hornaday put it) "the tramp of the feet of our annual millions."[28]

The fact is that all throughout the Grant-Osborn-Hornaday reign there was a tension between, on the one hand, their genuine desire to uplift and educate the masses and, on the other, their lingering sense that the zoo was their own personal fiefdom and that the patrons were therefore to be treated as annoyances at best and interlopers at worst. The crowds that streamed up from the East Side on weekends were especially obnoxious to the men who ran the zoo, and the three patricians openly paraded their contempt for the "degenerates" who babbled in Yiddish and Sicilian, napped on the lawns, and littered the grounds of their park. A frustrated Hornaday charged that most of the zoo's visitors were "low-lived beasts who appreciate nothing, and love filth and disorder." He patronizingly admitted, "It is difficult to understand the workings of the minds of persons who can blithely and cheerfully scatter rubbish on beautifully kept lawns that have been created for *their* benefit."[29]

The attitude of the triumvirate reflects the historical truth that zoos are the modern descendants of aristocratic hunting preserves, updated and popularized for the age of democracy. This comes through in Henry Fairfield Osborn's assurance to the public that, in designing the zoo, he and his colleagues had given "as close personal attention . . . as they would have given had they been laying out *one of their own country places.*" The Bronx Zoo, in the minds of its founders, was another of their country estates, like Grant's Oatlands or Osborn's Castle Rock, and the citizens of New York who came to gawk at the animals were so many trespassers.[30]

One of the most infamous episodes of gawking took place in 1906 when the zoo put Ota Benga, an African Pygmy, on display in the Primate House. The roots of this tragic episode lay in the St. Louis World's Fair of 1904, held to commemorate the centenary of the Louisiana Purchase. The organizers of the fair wanted to create an anthropology exhibit that would feature living representatives of all the world's races, from the Ainu of Japan to the Zulu of South Africa. As with the animals of the Bronx Zoo, these peoples would be displayed in simulations of their own natural habitats. Late in 1903, the fair employed noted explorer/missionary Samuel P. Verner to sail to Africa and procure a group of Pygmies. A few months later, in a slave market in the Congo, Verner purchased—for a pound of salt and a bolt of cloth—a Pygmy named Ota Benga, whose village had been destroyed and wife and children killed in a massacre perpetrated by the Belgian army. Verner was especially delighted with Benga's teeth, which had been filed to sharp points and would amuse visitors to the fair. He brought Benga back to St. Louis with him, where Benga and a group of other Pygmies helped build an "authentic" African village. When the fair closed in December, Verner was awarded the Grand Prize for his successful exhibit.[31]

Ota Benga, looking none too happy in his role as ambassador to the Bronx from "the savage African jungle." © Wildlife Conservation Society.

In 1905, Verner and Benga returned to Africa and spent many months collecting artifacts and searching for diamonds. When Verner informed Benga that he was going back to the United States and would not return to Africa for at least a year, Benga asked if he could come along. Verner assented, with the thought that Benga might prove useful on the lecture circuit. They landed in New York in August 1906, and Verner went straight to the zoo to see if Madison Grant and William T. Hornaday wanted to purchase a chimpanzee he had brought with him. They agreed to buy the chimp for $275 and also to take care of Benga for a few weeks while Verner tended to personal affairs in North Carolina. In later years, Hornaday was always careful to say that Benga had been "employed" at the zoo while Verner was away. And Verner always concurred that the zoo "simply gave [Benga] temporary employment . . . and a safe and comfortable home for a short time." And, in fact, for the first two weeks Benga did help the keepers at the zoo with their chores.[32]

But on Saturday, September 8, Benga was encouraged to enter an empty cage in the Primate House "so that," explained Hornaday, "he might show visitors how they did things in Africa." A crowd of several hundred people soon gathered, and keepers occasionally urged Benga to charge the bars, teeth bared, to give the throng a thrill.[33]

The spectacle of a man in an animal cage immediately drew the attention of the press. The headline the next day in the *New York Times* announced: "BUSH-

MAN SHARES A CAGE WITH BRONX PARK APES," and that afternoon several thousand persons took the subway to the zoo to ogle "the little black man . . . in the monkey cage." The keepers placed an orangutan and a parrot in the enclosure and scattered bones about the floor to evoke the savagery of darkest Africa. They also posted a label on the cage:

> The African Pygmy, "Ota Benga."
> Age, 23 years. Height, 4 feet 11 inches.
> Weight 103 pounds. Brought from the Kasai River,
> Congo Free State, South Central Africa,
> by Dr. Samuel P. Verner.
> Exhibited each afternoon during September

The front-page story in the *Times* the next day reported that "there was always a crowd before the cage, most of the time roaring with laughter." Indeed, "over and over again the crowd laughed at" Ota Benga, and the children, especially, "laughed uproariously." One of the most amusing things, said the *Times,* was that "the pygmy was not much taller than the orang-outang, and one had a good opportunity to study their points of resemblance. Their heads are much alike, and both grin in the same way when pleased." So similar were the two that "many of those in the crowd who watched Benga's antics doubted if he was a human being."[34]

The exhibit proved extremely popular, and Hornaday announced that "the little savage" would be on display until late in the fall. He explained that Benga was "in no sense a prisoner"—he was free to come and go, except that every afternoon at two o'clock he had to be in the enclosure. When a reporter asked Hornaday whether there was not something unsavory about placing a man in a monkey cage, the director explained that everything was being done with the full approval of Madison Grant, and he assured the reporter that "the little black man is really very comfortable there." After all, he told the *New York Globe,* "The little fellow . . . has one of the best rooms in the primate house."[35]

A stormy debate erupted among the public over the propriety of making an exhibit out of Ota Benga. Headlines blared, experts argued, editors pontificated, and letters flooded in. A committee of clergymen from the Colored Baptist Ministers' Conference indignantly declared the dehumanizing exhibition "an outrage" and vowed to free Benga from his cage. But when the ministers went to City Hall to ask Mayor George B. McClellan to halt the "degrading exhibition," the mayor refused to take any action (which will come as no surprise to those familiar with his father's inaction as head of the Army of the Potomac). The ministers went straight from City Hall to Madison Grant's law office and demanded custody of "the dwarf." Grant "was pleasant enough," according to Reverend James H. Gordon, but refused to hand over the African, as he was holding him in trust for Dr. Verner. The committee concluded that this was "no very satisfactory reply" and threatened legal action to free the pygmy.[36]

Grant saw little to be gained from a courtroom showdown, and the next day he ordered that Benga no longer be officially displayed in the Primate House. But Benga still resided in the park (and returned to the Primate House to sleep at night), awaiting Verner's return. And the following Sunday over forty thousand New Yorkers jammed into the zoo to catch a glimpse of what the *Times* called "the star attraction in the park—the wild man of Africa. They chased him about the grounds all day, howling, jeering and yelling. Some of them poked him in the ribs, others tripped him up, all laughed at him." Understandably, Benga did not react well to being pestered in this manner. He threw a tantrum after being denied a soda, he threatened to bite his keepers, and at one point he nearly caused a riot when he got hold of a knife and brandished it about the park. The Zoological Society pleaded with Samuel Verner to return to New York as soon as possible and reclaim him. Hornaday explained to Verner that since Benga was so much in the public eye, he could not be reprimanded when he misbehaved: "Finding himself immune from punishment, the boy" (Benga was a grown man) "does quite as he pleases, and it is utterly impossible to control him."[37]

Verner finally arrived at the end of September and arranged to transfer Benga to the Colored Orphan Asylum in Brooklyn, where Reverend Gordon was superintendent. According to the *Times,* Gordon's phone began ringing off the hook as anxious relatives of his charges called to protest. "For the lands' sake, Mr. Gordon," one woman reportedly cried, "I read in the papers this morning that you are going to take the wild man over to your place. Why, he'll eat my Matilda alive!" Nonetheless, Benga was placed in the care of the reverend, and the zoo's involvement with Ota Benga came to an end.[38]

But the month-long controversy over Ota Benga was more complicated than it appeared. To be sure, the main reason for the anger of the ministers was that keeping a black man in a monkey cage had humiliating connotations. "Our race, we think, is depressed enough," explained Reverend Gordon, "without exhibiting one of us with the apes. We think we are worthy of being considered human beings." But almost as important for the clergymen was the implication that Negroes were the missing link between apes and humans. The ministers were angered that the display was a direct endorsement of evolution, and Reverend Gordon denounced it on the grounds that "the exhibition evidently aims to be a demonstration of the Darwinian theory of evolution. The Darwinian theory is absolutely opposed to Christianity, and a public demonstration in its favor should not be permitted." In other words, it was not just that the zoo was claiming that *Negroes* were akin to apes but that *humans* were akin to apes. The scandal was almost as much a conflict of religion versus science as it was one of black versus white.[39]

There is no question that Benga was placed in the cage to reaffirm for the white viewers that they were superior to the half-naked Pygmy, but he was also put there to show the audience, in those pre-Scopes days, that humans—all humans—are primates. William T. Hornaday, declaring "I am a believer in the Darwinian theory," explained that he was "giving the exhibitions purely as an ethnological exhibit." The *New York Times* supported this position: "The reverend colored brother should be told that evolution, in one form or another, is now taught in the text books of all the schools, and that it is no more debatable than the multiplication table." And a letter to the editor the following day praised "the scientific character" of the display and expressed the hope that the zoo's actions would "help our clergymen to familiarize themselves with the scientific point of view so absolutely foreign to many of them."[40]

Madison Grant had little patience for those who rejected science in favor of religious superstition. (A few years later, when the Church of St. Anne in New York displayed the forearm of St. Anne, Grant called in Henry Fairfield Osborn, and the two concurred that the forearm was actually a chicken femur. They challenged the church to submit the bone for examination, whereupon the hiearchy quietly withdrew the relic from exhibition.) In the Ota Benga case, the scientists were definitely on one side and the clerics were on the other. Of course, we need to remember that "science" in 1906 was thoroughly racist. Hence Hornaday's prediction that Benga would never be able to "bridge the gap" between the "savage African jungle" and "the civilized elysian fields." And hence the explanation by the *New York Times* that while "to the average non-scientific person in the crowd of sightseers there was something about the display that was unpleasant," the fact is that Benga "doesn't think very deeply," for he belongs to a race that "scientists do not rate high in the human scale." A later *Times* editorial claimed: "We do not quite understand all the emotion which others are expressing in the matter. . . . It is absurd to make moan over the imagined humiliation and degradation Benga is suffering. The pygmies . . . are very low in the human scale, and the suggestion that Benga should be in a school instead of a cage ignores the high probability that school would be a place. . . from which he could draw no advantage whatever. The idea that men are all much alike except as they have had or lacked opportunities for getting an education out of books is now far out of date."[41]

In 1907, Samuel P. Verner made another expedition to the Congo, but Benga declined the opportunity to go with him. Both men assumed that Verner would be making many other trips to Africa and that Benga would be able to return whenever he wished. In the meantime, Benga's English improved, he converted to Christianity, and even his teeth were capped to help him look more "normal." In 1910, the Colored Orphan Asylum sent him to Lynchburg, Virginia, to attend elementary school at the Baptist Seminary. A few years later, Benga decided that the time had come to return to Africa. Unfortunately, the Great War had broken out, and travel to the Congo was under interdict. Benga was heartbroken, but he secured part-time work in a tobacco factory for the duration. However, as the war dragged on and the prospects of getting back to Africa grew ever more remote, Benga sank deeper and deeper into depression. On March 20, 1916, he stole a revolver, sneaked inside a cow stable, and killed himself.[42]

Exactly five days after Ota Benga died, on the other side of the country, the Yahi Indian Ishi—the "Last of His Tribe"—passed away. Ishi had lived for five years in the Anthropology Museum of the University of California under the care of anthropologist A. L. Kroeber. When Ishi died, his brain was removed and shipped to the Smithsonian, where for decades it floated in a jar of formaldehyde in a storage area—despite Ishi's wishes that he not be autopsied. Kroeber's treatment of Ishi elicited none of the criticism that was heaped on Madison Grant for his display of Ota Benga.[43]

Nor did anthropologist Franz Boas (Kroeber's mentor and Madison Grant's long-standing foe on matters of race) receive any negative press when he put a group of Eskimos on display at the American Museum of Natural History. In 1897, at the request of Boas, Arctic explorer Robert E. Peary had returned from Greenland with six Eskimos, including a girl of eleven and a boy of six. Twenty thousand people visited Peary's ship when it docked in Manhattan, paying twenty-five cents each to see the exotic strangers. Peary eventually deposited his northern cargo in the American Museum of Natural History, where visitors could view them in their living quarters in the basement. Unfortunately, the Eskimos developed tuberculosis, and within eight months four of them were dead. Boas had the flesh stripped from their skeletons, which were bleached and added to the museum's collection.[44]

Several years later, a man wrote to the Bronx Zoo and offered to put *himself* on display in the Ape House as an educational exhibit. Director Hornaday responded: "Your application for a position in the Zoological Park as an exhibit of the genus *Homo* has caused me to turn Ruminant, and chew the cud of Bitter Reflection. Your perfectly legitimate aspiration recalls the ghost of vanished Ota Benga, pygmy negro of the Congo, who was our first offense in the display of Man as a Primate." Declaring that once was enough, Hornaday declined the offer to exhibit a man in a cage. Of course, the man in this case was a Caucasian.[45]

Madison Grant was extremely busy in those early years of the Bronx Zoo. While he continued to go daily to his law office (now located at 11 Wall Street), all his time and attention were devoted to overseeing the zoo and the aquarium and ensuring the solvency of the New York Zoological Society. Thanks to his efforts, membership in the society climbed 1,000 percent in five years (from 118 in 1897 to 1,182 in 1902), leading a chic magazine to pronounce "Not to be in the New York Zoological Society is not to be in society." And the annual members' dinner held every January at the Waldorf-Astoria—an event known as "the penguin show"—was an important date on Manhattan's social calendar.[46] (In those days before air-conditioning, January and February were the height of the social season in New York.)

Still, raising funds to run the zoo and aquarium was a never-ending proposition. Grant once compared his situation—unfavorably—to that of the head of the Metropolitan Museum of Art: "The value [of animals] is highly ephemeral, and tends to decrease rather than to increase. A collection of pictures or other museum material is constantly growing in value with the increasing rarity of the specimens, but animals are not immortal and require constant replacement."[47] In other words, the Met could raise money to purchase a Renoir and then sit back and grin contentedly as the painting appreciated in value. But when Grant bought a cheetah (in the days when captive breeding programs were still far in the future), he knew that the animal, in addition to consuming enormous amounts of costly food, would depreciate in value every day until it died, at which point it would be worthless; and Grant would then have to go out and raise the money to replace it.

It did not help that—in keeping with their mission to bring conservation to the masses—admission to the zoo and aquarium was free most days of the week. For a number of years Madison Grant himself gave thousands of dollars to keep the Zoological Society solvent. But it was clear that deeper pockets than his were going to have to be tapped if the zoo was to survive. One of Grant's most gratefully remembered fund-raising efforts took place in 1914, when the Zoological Society was desperately trying to create a pension fund for the 171 workers in its employ. Grant called on Andrew Carnegie and asked him to give the staggering amount of $100,000 as principal for the fund. Carnegie showed little interest in the project but finally, and reluctantly, agreed to contribute $10,000 (which was still a substantial sum). Without so much as blinking, Grant "politely and sorrowfully" declined the offer, turned around, and walked out of the room. At which point a shocked Carnegie, whose largesse had never before been declined, eagerly yelled after Grant that he would gladly provide the entire $100,000. The employees of the society later gave "heaven-sent Madison Grant" a silver loving cup in token of his care for them.[48]

Eventually, even larger donations were secured. John D. Rockefeller, Jr., for example, announced that he was "deeply impressed with the work of the New York Zoological Society and the place which it occupies in the life of the community," and contributed $1 million.[49] Anna M. Harkness also donated $1 million to the society, and other large sums came from J. P. Morgan, Margaret Olivia Sage, George F. Baker, and Jacob Schiff. (The generosity of Schiff was particularly extraordinary, as he was well aware of Grant's anti-Semitism and of his lifelong efforts to keep Jews off the board of the Zoological Society.)

Thanks to Grant's labors, the Zoological Society was put on a secure footing, and it thrives to this day. Now known as the Wildlife Conservation Society, it runs—in addition to the Bronx Zoo and the New York Aquarium (which has since moved to a fifteen-acre site on Coney Island)—the Queens Zoo, the Prospect Park Zoo, and the Central Park Zoo. Grant would be very pleased that the Wildlife Conservation Society remains an international leader in exhibit design, scientific research, and conservation. He would be less happy that the society has removed from its website and historical literature any mention of the now anathematized Madison Grant.

3

From Conservation to Preservation

No greater conservationist than Madison Grant ever lived!

Horace M. Albright, director of the National Park Service

Grant the Naturalist

By the turn of the twentieth century, with the Bronx Zoo and the New York Aquarium firmly established as the largest and most popular institutions of their kind in the world, Madison Grant decided he could devote more time to his studies of North American mammals. He had not written about natural history since "The Vanishing Moose" in 1894 (the year before he founded the Zoological Society). But beginning in 1901, and for the next four years, he produced a series of erudite monographs on the large fauna of North America, including "Moose," "The Caribou," "The Rocky Mountain Goat," "Notes on Adirondack Mammals," and "The Origin and Relationship of the Large Mammals of North America."

Grant's articles on mammals covered in detail their evolutionary history, physical characteristics, geographic distribution, and habits in the wild. The articles were scholarly in tone and thoroughly accurate, and they earned Grant a high reputation among naturalists. Without ever formally studying biology or paleontology, Grant could plausibly discuss anything from the effects of glaciation to the evolution of palmated antlers. And while he may not have been the scientist he or his friends thought he was, Grant was an excellent popularizer. In particular, he had a marvelous talent for explaining Darwinism to the sophisticated layman. When telling the story of mammalian evolution, he could weave together in a comprehensible fashion arcane facts about geology, speciation, parallelism, and prehistoric climatic changes—with a few hunting anecdotes and quotes from the *Nibelungenlied* thrown in for good measure.

Three themes continually reappeared in Grant's natural history writings: *typology* (the concept that for each genus there was a classic "type"—a sort of platonic ideal—which was invariably the largest and "handsomest" version of the animal); *deterioration* (the claim that even the "types" were degenerating as a result of trophy hunters killing the largest bulls); and *invasive species* (the idea that Americans must prevent the introduction of Old World animals, as they could mingle with native animals and form "a mongrel race," or even completely drive out the native species).

Grant's writings from 1901 to 1905 made him a recognized expert on the fauna of North America.[1] One decade later, guided by the principle that "the laws which govern the distribution of the various races of man and their evolution through selection are substantially the same as those controlling the evolution and distribution of the larger mammals," Grant would apply to *Homo sapiens* his ideas about typology, deterioration, and foreign species, and the effort would earn him a new reputation as an expert on anthropology and the prophet of scientific racism.[2]

Game Refuges

In the meantime, as Grant continued to study the fauna of North America, he recognized far in advance of most of his compatriots that habitat destruction would pose a mortal danger to the continued viability of wildlife during the twentieth century. Grant saw that the continent's wetlands were being drained by agriculturists, its forests ravaged by lumbermen, its soil depleted by homesteaders, its native plants despoiled by ranchers, its riparian areas poisoned by miners, and hitherto untouched areas of the continent invaded by railroads. He realized that the forces of development threatened to destroy within a matter of years what remained of the American wilderness, and thereby do far more harm to the native fauna than armies of market hunters had ever done. The word "melancholy" appears very often in Grant's nature writings, and it was with melancholy that he warned that "sooner or later the development and population of the country at large will reach a point when there will be no room for the larger forms of mammalian life."[3]

By the first decade of the twentieth century, having ended unsportsmanlike and market hunting, Grant decided that the time had come to promulgate a new initiative: the creation of game refuges. These would be inviolate sanctuaries set aside by the federal government in discrete locations throughout the country within which animals could roam and breed undisturbed by hunters or settlers. "However efficient the game laws may be," explained Grant, "the only permanently effective way to continue in abundance and in individual vigor any species of game is to establish proper sanctuaries." The Bronx Zoo and the New York Aquarium were in essence game refuges—they just happened to be located in the midst of the continent's largest city. But Grant wanted to create other, bigger, more natural refuges in the West, where the animals on the verge of extinction could be preserved and perhaps even flourish in their native habitats.[4]

The problem was that the federal government was not normally in the habit of fencing off segments of the continent and telling land-hungry Americans that they had to keep out so a bunch of animals could forage in peace. To the contrary, from its inception the government had pursued a vigorous policy of transferring into private hands the vast quantities of land (known as the "public domain") it had obtained as a result of the nation's westward expansion. A variety of disposal laws encouraged land speculators, railroad magnates, cattle kings, mining barons, timber moguls, and others to exploit "usable" areas of the public domain.

Toward the end of the nineteenth century, however, groups such as the Boone and Crockett Club had begun urging the government to "withdraw" certain lands from the operation of the disposal laws. They understood that the only way to permanently protect areas of exceptional beauty or scientific interest was to keep those lands under the control of the government. The creation of Yellowstone National Park (when Congress underwent a temporary fit of sanity and created the country's first national park in 1872) provided the model for the practice of withdrawing discrete areas from the public domain in order to preserve their unique qualities.[5] And the next major step in this process had been the creation of the forest reserves in 1891.

Largely through the influence of Boone and Crockett Club member William Hallett Phillips (a prominent Washington, D.C., attorney and confidant of Henry Adams), a few lines were innocuously inserted into the Civil Service Bill of 1891 authorizing the president to withdraw from the public domain areas that were "wholly or in part covered with timber." The creation by executive order of such "forest reserves" (now known as the "national forests") would put those lands beyond the reach of the lumber companies that were rapaciously decimating the country's forests. The Boone and Crockett Club supported the creation of such reserves because in addition to preserving trees they would provide protection for fauna and thereby ensure the continuance of big-game hunting in the future. Without being debated on the floor—without, in fact, even being considered in committee—the Forest Reservation Act (as it came to be known) slipped through Congress in the waning hours of the session on March 3, 1891. It was a typically elitist maneuver by the Boone and Crockett Club. The bill had not been read by most legislators present, and George Bird Grinnell later admitted that of those who did peruse it "probably no one . . . understood what the section meant." And yet the Forest Reservation Act, in the estimation of historian Charles A. Beard, was "one of the most noteworthy measures ever passed in the history of the nation."[6]

Secretary of the Interior John W. Noble (another member of the Boone and Crockett Club, and a friend of William Hallett Phillips) induced President Benjamin Harrison to sign the bill. Noble then selected fifteen areas, totaling some seventeen million acres, for the president to set aside as forest reserves. To avoid facing the wrath of westerners, who vigorously objected to having their lands removed from the public domain, Harrison waited until just before leaving office to create most of the reserves. His successor, President Grover Cleveland, was spurred to set aside another thirteen reserves embracing twenty-two million acres of forest land—though he too waited until he was safely a lame duck and issued the orders ten days before leaving office. (Nonetheless, there was talk of impeachment, and angry western congressmen managed to push through in one day a bill restoring the forests to the public domain. As it was late in the session, however, Cleveland was able to kill the bill with a pocket veto.)

The forest reserves, as noted, were extremely unpopular in the West, where the average citizen, remarked Theodore Roosevelt, had "but one thought about a tree, and that was to cut it down."[7] Westerners bitterly resented the federal "lockup" of public lands and grew increasingly angry over the magnitude of presidential withdrawals. They believed that the executive branch, influenced by an elite cabal of eastern conservationists, was depriving them of their inalienable right to purchase and utilize land in the public domain. Politicians and presidential advisers noted the level of the westerners' enmity and began listening attentively to their demand that the forest reserves be restored to public sale.

But on September 14, 1901, Leon Czolgosz's bullet put the founder of the Boone and Crockett Club (or, as Mark Hanna put it, "that damned cowboy") in the White House. And Theodore Roosevelt was fully supportive of the actions of his predecessors in creating and expanding the forest reserves. "When Theodore Roosevelt became President," wrote Stewart Udall, "The Boone and Crockett wildlife creed . . . became national policy."[8]

The new president promptly put another prominent member of the Boone and Crockett Club, Gifford Pinchot, in full charge of the forest affairs of the nation. Born the same year as Madison Grant, Pinchot was a fellow graduate of Yale, a man of considerable wealth, and an accomplished big-game hunter who—after killing a bear, a bighorn ram, and a number of deer—had been voted into the Boone and Crockett Club in 1897. He quickly joined Madison Grant on the club's executive committee, where—despite differences in temperament and ideology—the two became fairly close friends. (Pinchot was also a member of the New York Zoological Society, and when the Bronx Zoo was under construction Grant had not hesitated to ask him for advice on landscaping matters.) Pinchot was even closer friends with Theodore Roosevelt. They first met in 1899 when the forester was taken by Grant's friend C. Grant La Farge (one of the zoo's architects and at that time secretary of the Boone and Crockett Club) to visit the roughriding governor at the Executive Mansion in Albany. Pinchot later remembered that after discussing forestry policy, "TR and I did a little wrestling, at which he beat me; and some boxing, during which I had the honor of knocking the future President of the United States off his very solid pins." At which point, of course, Teddy fairly fell in love with the man, and they became immediate and lifelong chums.[9]

Now that Pinchot was the nation's Chief Forester, he created many new forest reserves, occasionally soliciting Madison Grant's opinion as to the best locations for such endeavors. But Pinchot realized that if the reserves were to be protected in perpetuity, the opposition of the West would have to be taken into consideration. He traveled all over the region and explained to suspicious westerners that the Roosevelt administration had no intention of "locking up" the forests forever. Rather, he and TR simply sought to replace wasteful, short-term exploitation by selfish lumber barons with efficient, long-term management by the federal government. "Land," declared Pinchot, "is to be devoted to its most productive use for the permanent good of the whole people, and not for the temporary benefits of individuals or companies." In Pinchot's theory, forests—if protected properly and harvested judiciously—could be renewable resources that would last forever. Just as the Boone and Crockett Club wanted to save animals now so they could be hunted later, Pinchot's Bureau of Forestry wanted to conserve trees now so they could be harvested later.[10]

Pinchot assured his relieved auditors out west that he wanted to conserve forests not because they were aesthetically pleasing but because, if irresponsible businessmen continued their wasteful practices, no timber would be left for future generations. Furthermore, he pointed out that forests removed from private exploitation not only would yield timber forever but, if wisely managed, would accommodate land for grazing and also help to conserve the water supply, which in turn would provide irrigation for agriculture. Thus, forest reserves would benefit local (i.e., western) residents most of all and were not just a pet cause of effete eastern tree lovers. To drive home the point, Pinchot changed the name of the forest reserves to "national forests," as the former term implied that the trees were being reserved *from* the nation's use, while the latter implied they were being conserved *for* the nation's use. "The object of our forest policy," repeated Pinchot, "is not to preserve the forests because they are beautiful. . . . The forests are to be used by man. Every other consideration comes secondary."[11]

In other words, Pinchot's defense of the national forests provided the manifesto of the nascent conservation movement, which sought to "conserve" the resources of the nation in the present so there would still be a supply of them in the future. The conservation movement—whose very name was introduced into the English language by Pinchot in 1907—mirrored the progressives' enthrallment with "scientific management," rational use of resources, and large-scale and long-term planning. The Pinchotian conservationists did not have much patience for those who wished to leave the forests in their natural state, untouched by lumbermen, stockmen, or hydrologists. For Pinchot, "the first great fact about conservation is that it stands for development," and when he gazed upon a forest he saw "a manufacturing plant for the production of wood." It was not at all illogical, therefore, that in 1905 (as a result of a bill introduced by Boone and Crocketteer John F. Lacey) Congress removed the national forests from the jurisdiction of the Interior Department and placed them under the control of the Department of Agriculture. "Forestry," explained Pinchot, "is tree farming."[12]

By the time President Roosevelt left office in 1909, he and Pinchot had enlarged the extent of the national forests from 45 million to 151 million acres, for a total of 159 forests (covering an area the size of Maine, New Hampshire, Massachusetts, Rhode Island, Connecticut, Vermont, New York, New Jersey, Pennsylvania, Delaware, Maryland, Virginia, and West Virginia combined).

As the federal government created more and more national forests, Madison Grant realized that they could serve as perfect locations for his proposed game refuges. He and George Bird Grinnell now tried to convince the membership of the Boone and Crockett Club to support the idea of transforming the national forests into wildlife sanctuaries. The problem was that such a policy would severely limit the opportunities for big-game hunting, since the majority of America's big-game animals (with the exception of deer and antelope) were by this time to be found only in the national forests. If Grant's scheme were enacted, most big game would be placed forever beyond the reach of hunters. Grant, in other words, was asking his fellow sportsmen to severely curtail their hunting opportunities by selflessly supporting game refuges for no other reason than that preserving the nation's wildlife was a decent thing to do.

Up until now, Madison Grant's conservation activities had all been undertaken for reasons that a man like Gifford Pinchot could readily approve. The creation of the Bronx Zoo and New York Aquarium, the Adirondack Deer Law, the Alaska Game Law, the No-Sale-of-Game Laws, the Migratory Bird Law—all had been brought about to conserve resources in the present and thereby ensure their utilization in the future. But Grant's advocacy of game refuges in the national forests indicated that an important transformation had occurred in his thinking. He was now working to *preserve* North America's animals, not because of any selfish desire on his part to one day hunt them, but because he had come to the conclusion that they had an inalienable right to exist.

This metamorphosis in Grant's philosophy highlights the fact that the growing conservation movement was actually fed by two different streams at the beginning of the twentieth century. On one side were the utilitarian conservationists (epitomized by Gifford Pinchot), who were interested in conserving the nation's resources so they could continue to be utilized by future generations. Congressman John F. Lacey summed up their philosophy when he proudly told Grant that every piece of legislation he authored had favored "conservation for use, not conservation *from* use."[13] On the other side were the aesthetic preservationists (whose spiritual leader was John Muir), who were interested in preserving the nation's resources so they could withstand forever the encroachments of modern civilization. Preservationists like Madison Grant wanted to protect forests not for the sake of the next generation but for the sake of the trees. As far as the preservationists were concerned, the only way that the nation's forests were going to be exploited by humans would be as sites for contemplation and spiritual uplift. In many ways, the preservationists were part of a long American tradition that continues to this day, wherein citizens respond to the ills of urbanization with a romantic yearning to "get back to nature." And at the turn of the twentieth century, the aesthetic charms of the outdoors were ever more inviting to those disgusted by the congestion of cities, the corruption of politics, and the intrusion of immigrants.

While both the conservationists and the preservationists were viewed, in the popular mind, as part of "the conservation movement," there was a large gulf between those who looked at a forest and saw a farm for lumber, and those who looked at a forest and saw an inviolate temple of nature. When Madison Grant joined the Boone and Crockett Club in 1893, he had been a utilitarian conservationist (whose philosophy was: conserve game now so it can be hunted later). By the beginning of the new century he had evolved into an aesthetic preservationist (whose philosophy was: preserve game now so it can be protected always). And just like John Muir, who had once been rather friendly with the Chief Forester, Grant—as soon as he moved into the preservationist camp—found his relations with Gifford Pinchot beginning to cool.

Pinchot, emphatically, viewed preservationism as folly. The nation's resources were eventually going to be used by someone: the forces for development were never going to permit the land and its riches to be locked up forever. The only question worth asking, therefore, was whether the natural assets of the country were going to be used wastefully or wisely. Would they be extracted for short-term profit by the plutocrats or conserved for long-term use by the masses? "The natural resources," entreated Pinchot, "must be developed and preserved for the benefit of the many, and not merely for the profit of a few. Conservation," he declared, "means the greatest good to the greatest number for the longest time."[14]

It should be conceded that there is a possibility here for confusion. The preservationist wing of the conservation movement eventually developed into "environmentalism"—the post-*Silent Spring* mass movement that arose in the 1960s when conservationists suddenly realized that *Homo sapiens* was itself an endangered species. These modern-day environmentalists are normally thought of as being on the left of the political spectrum. But their progenitors (the early-day preservationists) were often to the *right* of the utilitarian conservationists on issues unrelated to nature. Some of the very wealthiest American industrialists, for example, were avid supporters of the preservationists. (Men such as Andrew Carnegie, Cornelius Vanderbilt, Collis P. Huntington, J. P. Morgan and the Rockefellers donated vast sums to Madison Grant's preservationist projects.) Thus, we should not think that Grant's move from conservationism to preservationism indicated that he was moving to the political left—to the contrary, while he was embracing John Muir with one arm, he was embracing fascism with the other.

Grant successfully convinced the majority of the Boone and Crockett Club to endorse aesthetic preservationism, and to support the creation throughout the country of game refuges in the national forests. Congressman John F. Lacey was one of the converts. At the Boone and Crockett Club's fifteenth annual dinner, he likened the club to "Saul of Tarsus, who, having seen the light, became a defender of the faith." The club, "which had been inaugurated to encourage killing," had now become "an association of game protectors."[15]

Foiled by Pinchot

After consulting closely with Madison Grant, John F. Lacey introduced in Congress a bill authorizing the president to declare that the national forests would henceforth be inviolate game refuges as well. Congressman Lacey is one of the great—and inexplicably ignored—pioneers of conservation in America. (So is Madison Grant. But in Grant's case, at least, the reason for the neglect is clear: like the vanishing commissar, his untenable political views made it necessary to remove his portrait from the history of conservation.) As chairman of the House Committee on Public Lands for most of the years from 1889 to 1907, Lacey (R-Iowa) worked often with Grant and other members of the Boone and Crockett Club to craft some of the most consequential legislation in the field of conservation.

Grant and Lacey spent a lot of time perfecting the wording of their Game Refuge Bill to ensure it would receive the broadest possible base of support. But Gifford Pinchot was afraid the bill would raise the hackles of the westerners he had so recently brought into the pro-forestry fold. It was only with great effort and patience that the Chief Forester had managed to convince recalcitrant westerners that conservationists did not want to "lock up" the forests forever—they simply wanted to forestall their wasteful destruction. But now the preservationist wing of the conservation movement was putting forth this sentimental game refuge scheme that would actually prevent the residents of the West from hunting and fishing in their own backyards. Pinchot was extremely concerned that Grant's bill would give westerners an excuse to relinquish their tentative support for the entire concept of national forests.

Grant understood and appreciated Pinchot's concern. He realized that to garner the support of westerners and their congressmen he would have to come up with an argument that spoke to the enlightened self-interest of the populace. He accordingly stressed that if wildlife refuges were established in the national forests, they would serve as breeding grounds that would create a never-ending "overflow" of game that could be hunted outside the refuges by local residents. Grant shrewdly appealed to all three constituencies—preservationists, conservationists, and even unreformed hunters—when he said: "No American, who has at heart the preservation of the remnant of our wild life, who desires to increase the supply of meat, or who desires to utilize at their highest efficiency lands belonging to the government and not suitable for other purposes, can fail to appreciate the great step forward in genuine civilization that would be accomplished by the proposed legislation."[16]

The Boone and Crockett Club endorsed Grant's strategy of "demonstrating . . . to the people of this country that it is for their benefit—economically, financially, and even from a purely selfish standpoint—that these Refuges shall be established." While the club admitted that "the real and intimate, the vital and esoteric reason which lies at the bottom of *our* interest is that we wish to see the game undiminished in numbers," it understood that "our argument to the *public* must be on purely *economic* and *utilitarian* lines."[17]

But Gifford Pinchot was unmoved. He was certain that westerners would never be reconciled to having their hunting privileges taken away by the federal government, and he concluded that the refuge bill would have to be defeated if conservationists were not to lose all credibility. Therefore, via a clever parliamentary maneuver (attaching to Grant and Lacey's bill a rider that he knew would be defeated), Pinchot was able to kill the bill.

In retrospect, the main reason for the defeat of the Game Refuge Bill was the inability of the two wings of the conservation movement to join in a common effort. Grant placed a great deal of the blame on Gifford Pinchot, from whom he was moving further and further apart philosophically. To Grant, forests could be many things—places of beauty, subjects for scientific inquiry, and even symbols of traditional Nordic values—but they most certainly were not living warehouses for the construction industry. To Pinchot, on the other hand, mere "scenery" would always be subordinate to the economic needs of the public, and Grant's position on preserving forests was "sentimental nonsense." After the defeat of the Game Refuge Bill, and for the next thirty years, the two men politely but studiously avoided each other, though they were members of the same clubs, organizations, and societies.[18]

There was one temporary thaw in their normally icy relations. It came during the infamous Ballinger-Pinchot controversy of 1909 (which arose when Pinchot protested the attempt by President Taft's Secretary of the Interior to reopen public lands in the West that had been withdrawn by TR). Whatever Grant's personal feelings toward Pinchot, he interpreted the actions of Interior Secretary Ballinger as hostile toward the entire conservation movement and concluded that Pinchot must be defended. He offered to see President Taft on Pinchot's behalf, and even lunched with Pinchot to devise a strategy to bring together the various sportsmen's organizations to "take a decided stand" in defense of the Chief Forester. But before they could implement their plan, Taft ended the Ballinger-Pinchot quarrel by summarily removing Pinchot from office.[19]

And then, inevitably, Grant and Pinchot split again—this time on the question of whether San Francisco should be allowed to dam the Tuolumne River and thereby flood the Hetch Hetchy Valley in Yosemite National Park. The protracted battle over Hetch Hetchy symbolized in many ways the struggle between utilitarian conservationists and aesthetic preservationists. Pinchot, who supported the utilization of resources in the national parks, spoke out in favor of building the dam, which would create a water reservoir for San Francisco. He did so both as a conservationist and as a progressive advocate of public utilities. He was joined by progressive Congressman William Kent of northern California (who had been a Yale classmate and long-standing friend of Madison Grant). The dam, after all, was desired by Mayor James D. Phelan, the reform mayor of San Francisco who advocated municipally run utilities and wanted to protect his constituents from the monopolistic practices of the privately owned Spring Valley Water Company, which specialized in poor service, high prices, and unsafe water. Furthermore the citizens of San Francisco, in a 1908 referendum (itself a progressive innovation), had voted for the dam by an impressive seven-to-one margin.

But John Muir and his preservationist allies—including Madison Grant, George Bird Grinnell, Henry Fairfield Osborn, William T. Hornaday, Charles Eliot, Ernest Thompson Seton, and Frederick Law Olmsted, Jr.—adamantly opposed the project. They were incredulous that anyone could even think of destroying the spectacular beauty of the priceless Hetch Hetchy Valley, and fought for years to prevent construction of the dam.

The difference between the two sides was summarized by Mayor Phelan when he accused John Muir of engaging in "aesthetic quibbling" while "the 400,000 people of San Francisco are suffering from bad water." The preservationists, in short, were once again painted as opposing the legitimate needs of the public, and Congressman William Kent lashed out at them for having no "social sense."[20]

John Muir and the preservationists eventually lost the battle over Hetch Hetchy, and the valley was flooded. Grant sadly declared that building the dam in Yosemite National Park was a "fatal precedent," and angrily held Gifford Pinchot responsible for "sacrificing the Hetch Hetchy Valley to commercial exploitation."[21] The two men never spoke to each other again.

Preserving the Bison

After Gifford Pinchot put the kibosh on Madison Grant's Game Refuge Bill in 1903, Grant put on hold his campaign to have all the national forests declared game refuges at one stroke. He now understood that if he was ever going to see federally sponsored game sanctuaries in his lifetime, they were going to have to be created one refuge at a time. He would have to pick a specific national forest—preferably one with as few nearby settlers as possible so as to lessen political opposition—and convince Congress to pass a special act authorizing the president to set aside a particular area of that forest as a game refuge. And each time he wanted to create a new refuge, the process would have to be repeated.[22]

Given that this promised to be an enormously lengthy undertaking, it was chillingly clear to Grant that the first refuge would have to be devoted specifically to the American bison, which in 1903 was in imminent danger of extinction. The astounding fact was that within Grant's lifetime the bison had been America's most abundant big-game animal: at one time at least thirty million bison had roamed the North American continent. Colonel Richard Dodge, a member of the Boone and Crockett Club, remembered the time he was patrolling the Arkansas River in 1871 and encountered a single bison herd that was fifty miles long and twenty-five miles wide, and contained at least four million head. George Bird Grinnell, who had hunted his share of bison during the 1870s, sadly remembered that "it was believed that the buffalo never could be exterminated."[23]

But construction of the transcontinental railroad in the 1860s had triggered the demise of the species. First of all, thousands of bison were butchered to feed the huge work crews. Also, as the railroad tracks advanced they split the herds, and then split them again and again as new lines branched off through the prairies. Then the railroads began to bring out sportsmen on excursion trains who sat in comfort and shot the bison through the windows of the rail cars. (A conductor on the Santa Fe Railroad remarked that one could walk for a hundred miles along the railroad's right-of-way without stepping off the carcasses of dead bison.)[24] And finally, the railroads made possible the shipment of unspoiled bison products to the meat markets of the East. After the Civil War, market hunters headed out from the rail centers of Leavenworth, Cheyenne, and Dodge City by the thousands and made fortunes by stripping the hides from the slaughtered bison (which sold as robes for $1.25) and cutting out the tongues (which were considered a delicacy and fetched twenty-five cents apiece). The remainders of the half-ton carcasses were left to rot on the plains.

"Nowhere is the blind, senseless human appetite for carnage, for destruction, more strikingly, more lamentably evinced," wrote Horace Greeley after his tour of the West, "than in the rapidly proceeding extermination of the buffalo." In the late 1860s and on into the 1870s, market hunters were slaughtering a staggering two to five million bison every year. The bloodbath was exacerbated by pioneers who coveted the range for homesteads, and military authorities who saw the destruction of the herds as essential to keeping the Plains Indians on their reservations. (General Philip Sheridan, an honorary member of the

Boone and Crockett Club, stated that the extermination of the bison would do more to solve the Indian problem than the army had done in thirty years.)[25]

Modern evolutionary biologists often cite the 50/500 rule, which states that "a species must have at least fifty adult members to survive the short run, and five hundred adults for the long run"—otherwise the lack of genetic diversity will spell ruin in only a few generations. In 1886, William T. Hornaday, then chief taxidermist of the Smithsonian (and a man who had himself hunted bison in his earlier days), traveled throughout the United Sates to inventory the surviving bison herds. According to his meticulous count, the tens of millions of bison that had existed before the Civil War had been reduced to a pathetic total of just 541 animals—of which only 85 still existed in the wild. Total extinction was now a distinct possibility, either at the hand of poachers or perhaps from accident, disease, or inbreeding.[26]

Such a prospect was quite disturbing to Madison Grant and his friends, who well remembered the West that was. George Bird Grinnell movingly admitted that "often, late at night, when the house is quiet, I sit before the fire, and muse and dream of the old days [and lament that] of the millions of buffalo which even in our own time ranged the plains in freedom, none now remain. From the prairies which they used to darken, the wild herds, down to the last straggling bull, have disappeared." Indeed, the prairies that the bison used to "darken" were now *whitened* in all directions by the presence of three hundred thousand tons of bones (which were eventually scavenged and ground up for fertilizer, save for the tips of the horns, which were sold to the umbrella trade at thirty dollars a ton). To Grant, the vanishing bison exemplified the plight of all American big game. (It was not an accident that a portrait of a bison appeared on all the early publications of the Boone and Crockett Club, as well as many publications of the New York Zoological Society.) If the bison were allowed to disappear, thought Grant, it would be "a standing reproach to the civilization of our country."[27]

Some wealthy individuals had attempted to preserve captive herds on their private estates, but most had allowed their bison to breed with their cattle (resulting in "catalos"). Madison Grant was aghast at this "contamination of the pure-breeds," and declared—in typical Grantian fashion—that "a half breed is an abomination." He insisted that "it is of the utmost importance to preserve all remnants of the American bison *without any cross-breeding*."[28]

It therefore seemed clear to Grant that the only way to ensure the perpetuation of the bison was to secure a refuge in some national forest located in the bison's former range in the West and to stock it with specimens of "absolutely pure full blooded stock" selected from the Bronx Zoo and other captive herds.[29] He hoped to eventually create at least four such refuges, purposefully separated to ensure that no one calamity or disease would endanger the entire species. After scanning the roster of existing national forests, Grant and Hornaday de-

Bison skulls waiting to be ground into fertilizer in the 1880s. By the time this photo was taken, the wild bison population—which a few decades earlier had stood at 30 million—had been reduced to eighty-five. Burton Historical Collection, Detroit Public Library.

cided that the Wichita Mountains Forest Reserve in southwest Oklahoma would be a good candidate for the first bison refuge. The Wichita Forest Reserve was a panorama of thousands of acres of rolling grassland embracing some of the best grazing grounds of what had been the great southern herd of North American bison. Grant surmised there would be little political opposition to declaring it a game refuge, as the area was sparsely settled and there were few politicians to object (Oklahoma was not yet a state). Since there was no game on the proposed wildlife refuge (hunters having long ago cleaned out the area), the project was not going to deprive any westerners of their hunting, and Grant therefore anticipated, correctly, that Gifford Pinchot—who was keenly interested in the fate of the bison—would not oppose the plan.

Accordingly, at the instigation of the New York Zoological Society and the Boone and Crockett Club, the U.S. Congress authorized the president to set aside a portion of the Wichita Forest Reserve as a game refuge. President Roosevelt quickly exercised his newfound power, and on June 2, 1905, he created the Wichita Mountains National Wildlife Refuge—the nation's first big-game refuge.

In the fall of 1907, William T. Hornaday inspected the bison herd at the Bronx Zoo, selected six bulls and nine cows (representing four distinct strains of blood), and coaxed the nervous beasts onto an Oklahoma-bound train at Grand Central Station. Western newspapers gave extensive coverage to the bison transfer, and the passage of the train through the various communities en route was attended with considerable interest and excitement. Elwin Sanborn, who accompanied the animals for the Zoological Society, was moved by the number of Americans—including aged Indians—who "pined to see the bison" and who applauded as the train pulled into each town. The same scene was repeated at every stop as "throngs of men, women and children rushed up to get a glimpse of the famous animals." They "crowded the cars on both sides . . . and the people only departed when they were forced out by the speed of the train."[30]

After their weeklong journey, the bison arrived in Oklahoma and were released into their new home. By the end of the first year two calves had been born (one of which was named "Hornaday"), and the Wichita herd went on to prosper after that. Madison Grant could not help but remark on the somewhat amazing irony that "the restocking of the West with this typical American animal is being carried out largely with bison bred in the City of New York."[31]

Today, the Wichita Mountains Wildlife Refuge covers over fifty-nine thousand acres (ninety-two square miles), and is one of the outstanding showplaces of the National Wildlife Refuge System. Nature writer Robert Murphy says that "it is one of the very few places left that shows what the old West was like, how it looked before the white man changed it." The tall native grasses, intermixed with brilliant wildflowers, have all come back and flourished, and a thousand descendants of that original herd of fifteen bison from the Bronx Zoo roam through them. A breathtaking variety of other animals have been successfully introduced (or migrated into the refuge on their own), including red wolves, gray foxes, wild turkeys, great horned owls, armadillos, bobcats, mountain lions, elk, and mink. Bald and golden eagles spend the winter in the refuge, and numerous ducks fill up its lakes during their spring and fall migrations. Cottonwoods lining the streams provide nesting sites for over two hundred species of birds, from white-breasted nuthatches to red-bellied woodpeckers. And all these animals are observed by the one million enchanted members of the *Homo sapiens* species who visit the refuge annually.[32]

The Wichita Mountains Wildlife Refuge was an unqualified success. The long and tragic decline of the American bison population had been reversed, and the precedent for creating refuges to protect the country's wildlife had been set: the Roosevelt administration went on to create over fifty national wildlife refuges (and today the National Wildlife Refuge System has grown to over five hundred refuges encompassing ninety-three million acres).

But Madison Grant felt that at least three more bison herds should be established in various parts of the United States to ensure the animal's continuance as a viable species. Since the Boone and Crockett Club and the New York Zoological Society were already involved in so many other conservation activities, Grant and Hornaday decided to spin off a specialty organization that would concentrate specifically on this one goal. To that end, the American Bison Society was organized at the end of 1905 with the stated purpose of "promoting measures calculated to preserve the American Bison from ultimate extinction."[33] William T. Hornaday, who was the leading figure in the American Bison Society in the early years, was chosen as its first president, and Theodore Roosevelt was named honorary president. Madison Grant, as always, played a less visible but controlling role by serving on the board of managers, the executive committee, and the advisory board.

The American Bison Society sponsored lectures and distributed numerous newspaper and magazine articles in the United States and Canada extolling the cause of bison preservation. Thanks to the society's propaganda campaign and its dramatic logo (in which Maxfield Parrish depicted a magnificent bull standing on a rocky mound), the public became very aware of the plight of the bison, donations poured in to the ABS, and membership in the society tripled between 1907 and 1914.

The American Bison Society sought long and hard to find suitable locations on federal lands on which to situate three more herds to supplement the one in Oklahoma. Many congressmen, particularly those from the West, were not eager to tie up lands that could be used more profitably by farmers and ranchers. However, Congress had witnessed the success of the Wichita Wildlife Refuge, and the nation's representatives were lobbied at key moments by Grant, Hornaday, and other leaders of the American Bison Society, the New York Zoological Society, and the Boone and Crockett Club. (This was the same Boone and Crockett Club—or perhaps we should say that this was *hardly* the same Boone and Crockett Club—that twenty years earlier had bestowed an honorary life membership on General Phil Sheridan, chief advocate of annihilating the bison as a means of pacifying the Indians.) Accordingly, by the time the Great War broke out, Congress had established three more bison refuges: the beautiful Montana National Bison Range in northwestern Montana, the Wind Cave National Game Preserve in the Black Hills of South Dakota, and the lush Niobrara National Wildlife Refuge in northern Nebraska.

The bison refuges were—and are—national treasures. With thriving herds in Oklahoma, Montana, South Dakota, and Nebraska, Madison Grant's original goal of four national bison herds at varied latitudes of the Great Plains was now complete. Working with Grant's two other preservation organizations (the Zoological Society and the Boone and Crockett Club), the American Bison Society had shown excellent judgment in the identification of ranges and in the selection of animals from particular strains to stock those ranges. And in just a few years surplus animals from those herds were being sent to zoos and preserves throughout the country and indeed throughout the world as the refuge movement spread outside the United States to Canada and Europe. Amazingly, as early as the year 1914 William T. Hornaday could state that "the future of the Bison as a species is now absolutely secure."[34]

And so it was that Madison Grant and his friends preserved the bison. When they first started working to rescue the animal, there were only some five hundred pure-blooded bison in the United States. By the time of Grant's death in 1937 there were well over twenty-five thousand bison (the number today is over two hundred thousand), and the American Bison Society had disbanded.

The project to save the bison had been carried out for purely preservationist motives; no one involved in the effort had any thought of increasing the bison's numbers so they could one day be hunted for either meat or sport. Furthermore, in saving the bison, the American Bison Society pioneered the notion of reintroducing endangered species into the wild from captive stock. Looking back on the effort in 1933, Grant wrote that the recovery of the bison was "the most dramatic incident of the restoration of a vanishing species in America, and is an example of what can be done with other large mammals." Henry Fairfield Osborn could not help but agree, and he was not guilty of exaggeration (for once) when he declared that the American Bison Society was "one of the most active and effective societies ever organized in the United States for purposes of conservation."[35]

A few months after Grant died, Martin Garretson (former secretary of the American Bison Society) published *The American Bison,* a history of the decline and resurrection of the species. He dedicated the book "to the memory of Madison Grant, late President of the New York Zoological Society, who for forty years was an active defender and preserver of the big game and other wild life of North America." Horace Albright, former director of the National Park Service, stated in his review of the book: "It is most appropriate that this volume is dedicated to Madison Grant [who] to the end of his long and useful life devoted his time and energy to American wild life, forest and scenery preservation."[36] As Garretson and Albright well knew, the bison was just one of many species that Madison Grant saved from extinction. I have elsewhere traced the inspiring saga of how Grant used his enormous organizational skills and political influence to preserve numerous other North American animals, including the bald eagle, the pronghorn antelope, the Alaskan bears, and the fur seals.[37] Which is why A. Brazier Howell was expressing the feelings of all conservationists when he stated in 1931 that he was proud to have served in the preservation movement with Madison Grant, "whose name commands the greatest respect." Howell told Grant, "I feel that no one will contradict me when I say that you are the dean of mammal conservationists in this country."[38]

Preserving Scenery

Madison Grant served a key role in the creation of several national parks, including Denali National Park in Alaska, Everglades National Park in Florida, and Olympic National Park in Washington. I have recounted the history of those campaigns elsewhere.[39] For now, I will merely touch on one of his more important efforts: the successful endeavor to create Glacier National Park.

It began in 1885, when Grant's best friend, George Bird Grinnell, went on a hunting expedition to the lustrous forests and lofty mountains of northern Montana. Other than an occasional mountain man or trapper, Grinnell was one of the first white men to see the remote area and its sixty living glaciers. He was staggered by the majestic beauty of the towering peaks and sculptured valleys, breathless at the most exquisite wildflower displays in North America, and overwhelmed by the thousand waterfalls that plunged from glacial snow masses into sparkling, jade-colored lakes. Boone and Crocketteer Raphael Pumpelly may well have been right when he proclaimed it in 1925 "the grandest scenery in the United States."[40]

Grinnell was equally impressed by the varied and copious wildlife of the area. As he hiked through the alpine meadows and dense forests, he could see that bighorn sheep, mountain goats, wapiti, black bears, and white-tailed deer were still flourishing in the lush environment. He noted that the large carnivores—timber wolves, grizzly bears, and mountain lions—were thriving in the exceptionally productive conditions (Glacier is the only location in the forty-eight contiguous states where these three major predator populations still occur naturally). He saw that the streams teemed with trout and whitefish and provided habitat for beavers and river otters. And when he looked overhead, he could count more than two hundred locally prevalent bird species, including osprey, ptarmigan, and golden eagles. "At a time when game nearly everywhere else was on the decline," writes James Trefethen, "Grinnell must have felt that he had found the sportsman's Valhalla."[41]

Grinnell returned to the region each summer to hunt, fish, explore, and commune with the Blackfeet Indians. The names he gave to the mountains, valleys, lakes, and glaciers of the region still persist, including a number of features named for fellow Boone and Crockett Club members: Mount Stimson (for Henry L. Stimson), Mount James (for Walter B. James), Mount Gould (for George H. Gould), Pumpelly Glacier (for Raphael Pumpelly), and Mount Pinchot (for Gifford Pinchot). As Trefethen has noted, "Portions of the map of northwestern Montana resemble the roster of a Boone and Crockett Club committee meeting."[42]

Its geographical isolation and diverse flora and fauna had combined to make the Glacier region one of the largest and most intact ecosystems in North America. And so, at the beginning of the 1890s, with white settlers threatening to invade the area and the tracks of the Great Northern Railroad approaching ever closer, Grinnell first conceived the idea of somehow preserving the region, perhaps by having it declared a national park.

At the time, however, there was only one national park: Yellowstone. (Just as World War I was referred to as "The Great War" until 1939, Yellowstone in those days was called "The National Park"—there being no thought that there would ever be another one.) Furthermore, mining interests viewed the Glacier area as potentially rich in gold, silver, and copper, and had no intention of allowing Congress to withdraw the land from private exploitation. Indeed, in 1898 the area was thrown open by the government, and a frenzied incursion of miners proceeded to scour every inch of the territory in search of valuable minerals. For the most part, their avaricious dreams were not realized. After a few years, most of the miners gave up hope of striking a bonanza and moved on—having cut down much timber and killed much game.

With the retreat of the initial wave of speculative excitement, Grinnell decided after the turn of the century that the time was propitious to advance his plan of turning Glacier into a national park. He asked Madison Grant and the Boone and Crockett Club for assistance. We have noted the ever more ethereal path taken by Grant within the conservation movement, as his interests evolved from regulating hunting to preserving wildlife. But now, with the campaign to turn Glacier into a national park, it was not even animals, necessarily, that he was trying to preserve—it was scenery: impractical, intangible, nonutilitarian scenery. This, then, was Muirian preservationism at its purest.

For a number of reasons, Grinnell and Grant had high hopes that the traditional western opposition to withdrawing land would not materialize in the upcoming battle over Glacier: it was obvious by now that there were not enough minerals in the area to justify mining; also, most of the timber was inaccessible, and the grazing land was poor. (There had been a short-lived flurry of interest in the area's supposed oil reserves, but it finally came to nothing, having been spurred by an oil company that secretly brought in casks of petroleum, poured them into its test well during the night, and then pumped out the oil during the day to impress investors.) Furthermore, in the years since Grinnell had first put forth the idea of declaring the area a national park, Congress had created nine other such parks (General Grant, Sequoia, Yosemite, Mount Rainier, Crater Lake, Wind Cave, Platt, Sullys Hill, and Mesa Verde). If 768-acre Sullys Hill deserved to be a national park, then surely the Glacier area deserved that status as well.

Still, Grant and Grinnell were men with keen political instincts. They themselves may have long since diverged from Gifford Pinchot and crossed over to the ranks of the preservationists, but they knew that the public had little reason for wanting to save Glacier's waving fields of blue heather or its majestic stands of red cedar. No, the public and its political representatives needed a more tangible motive than "scenery" if they were going to support a lockup of thousands of acres of western land. Thus Grant, Grinnell, and the Boone and Crockett Club decided to employ conservationist means to achieve preservationist ends, and they began publicizing the value of Glacier's forests as a potential watershed for the semiarid West.

In addition, Grant shrewdly began, as he put it, "carefully to cultivate the acquaintance" of prominent Montanans. By including them in the process, by judiciously bestowing memberships in the Boone and Crockett Club on key political figures, and by cleverly arguing that the Glacier region possessed "enormous value as a source of water," which could "be used to irrigate the dry plains," he convinced Montana's congressional delegation to support the park plan. Senator Thomas B. Carter (R-Montana) enthusiastically introduced a bill (written by Grinnell) setting aside one million acres (1,300 square miles) of northern Montana as Glacier National Park. After many revisions the bill passed the Senate but was defeated in the House. When the bill was later reintroduced, it again passed the Senate and again failed in the House.[43]

Paradoxically, the railroad—whose approach had first catalyzed a frightened Grinnell into launching the park effort—now came to the rescue. Louis W. Hill of the Great Northern Railroad, who, as Grant put it, "saw important material advantage" in the park scheme, influenced the House delegation from Montana to support the bill. Hill foresaw that Glacier National Park would be a desirable tourist destination and that all but the hardiest visitors would have to ride on his trains to get there. With support from Hill, the Glacier bill finally passed both houses of Congress and was signed by President Taft on May 11, 1910—exactly a quarter of a century since Grinnell had first seen the region.[44]

The Great Northern Railroad now expended thousands of dollars building trails, campsites, and eventually hotels in Glacier National Park. It ran advertisements in popular magazines touting Glacier as "your vacation land supreme—wondrous lakes, glistening glaciers, mighty peaks and trout-filled streams." Readers were also encouraged to come see the Blackfeet Indians, "specimens of a Great Race soon to disappear." And the Great Northern helpfully reminded Americans that Glacier "is your only national park on the main line of a transcontinental railroad."[45]

The railroad's campaign reveals that the business community was finally beginning to comprehend that scenery could actually spell profits for those prepared to accommodate the tourist trade (as a glance at any recent issue of the Sierra Club's magazine reveals, the majority of the advertisements are from companies trying to lure readers to spend their time and money at some scenic destination). The realization that "nature" could be exploited for something other than extractive resources was a very important development for the preservationist wing of the conservation movement, which could now tout tourism as an economic rationale for wilderness preservation. (The fact that the wilderness had not originally included the luxury lodges, four-star restaurants, and golf courses that had to be built to attract those tourists was a paradox not yet addressed by the preservationists. Nor were they certain how to respond when park rangers began poisoning the "bad" coyotes, wolves, mountain lions, and eagles in order to increase the number of "good" deer, mountain goats, and elk that tourists expected to see.)

Madison Grant would eventually revolt against the conservation movement's use of economic subterfuges. Toward the end of his life he began to formulate concepts that would later form the heart of the wilderness movement. For not only did Grant insist that there was no place for hunters, miners, or lumbermen in the national parks; he decided in the end that there was no place for *any* kind of person whatsoever. He concluded that "nature itself has some rights," and that one of those rights was freedom from all human contact. And so he insisted that some portions of the continent, at least, must be left in their primordial state, absolutely free of any development, "so as to preserve as far as possible their pristine condition of wildness."[46]

But in the meantime, thanks to Madison Grant, George Bird Grinnell, and the railroad, the Glacier region had been saved for all time. In 1995, UNESCO declared Glacier to be a World Heritage Site, meaning that the international community recognizes it as an area deemed especially worthy of preserving for future generations—which was something that George Bird Grinnell had known the instant he laid eyes on the area in 1885.

Madison Grant was a classic American big-game hunter. But in the 1890s, in order to ensure continued hunting, he became a conservationist. He banned unsportsmanlike practices, put market hunting on the road to extinction, and educated the public about wildlife conservation by building the Bronx Zoo and operating the New York Aquarium.

After ten years as a conservationist, Grant moved into the preservationist camp when he realized that the forces of modernization were so strong—and habitat destruction was proceeding so rapidly—that merely regulating hunting was not the answer. The only way that endangered species were going to survive was by creating inviolate wildlife sanctuaries. So he set up game refuges and helped to create national parks where fauna, flora, and scenery would be preserved in their pristine state. Grant was enormously successful in these efforts. "It is safe to say," wrote W. Redmond Cross in 1937, "that no other individual has obtained greater results in [preservationism] during the last forty years."[47]

In a letter Grant sent to the *Herald Tribune* one month before he died, he was able to state with satisfaction that "the tide of wildlife destruction is ebbing and the tide of conservation is coming in."[48] In fact, at the time of his death, despite rampant population growth and extensive urbanization, many kinds of wild animals were far more numerous than they had been at the turn of the century. The American bison and the pronghorn antelope, which had been thoroughly swept from the plains, now numbered in the tens of thousands. Songbirds and waterfowl, which had been shot from the sky, and the fur seals and whales, which had been dragged from the sea, were protected by international treaties. The Rocky Mountain goat, the bighorn sheep, the moose, grizzlies, the caribou, and the elk, all of which had been clinging to precarious existences in a few scattered bands, were spectacularly restored.

Furthermore, when Grant began his career, the only part of the public domain that had been set aside for preservation was Yellowstone National Park. At the time of his death there was a vast system of twenty-six national parks, comprising over eight million acres (and providing refuge to some eighty-six thousand large game animals). There were over two hundred million acres of national forests (containing 1.5 million head of big game). There were seventy-five national monuments covering ten thousand square miles. And there was a national wildlife refuge system that embraced over eleven million acres and was well on its way to providing in its natural habitats at least one sanctuary for each North American big-game mammal.

It is clear that these were noble—and necessary—achievements. What is less clear is what any of it has to do with Dr. Karl Brandt and the war crimes trials at Nuremberg. But in the next chapter we will see that, ten years after joining the preservationists, Grant took one further step: rather than simply placing animals behind fences and passively hoping for the best, he began to actively manage wildlife populations to ensure their healthy survival. In other words, Grant's initial activities as a conservationist and then a preservationist had been essentially defensive. The fauna of the country were being decimated, and he instinctively tried to place them beyond the encroachments of modern civilization. But by the mid-1910s he understood that a more dynamic approach was called for, and so he formulated the tenets of wildlife management. And it is those tenets that will take us from Denali to Dachau.

4

Wildlife Management

On this generation rests the responsibility of saying what forms of life shall be preserved in what localities, and on what terms.

Madison Grant

The Interlocking Directorate

At the time of its founding in 1888, and for many years thereafter, the Boone and Crockett Club stood alone as the only national organization devoted to protecting America's big game. By the second decade of the twentieth century there were many such organizations. Some of the more important were the New York Zoological Society, the American Bison Society, the National Audubon Society, the American Game Protective Association, the American Ornithologists' Union, the Camp Fire Club, and the American Society of Mammalogists.

Like the New York Zoological Society and the American Bison Society, many of them were spin-offs of the Boone and Crockett Club. Robert Sterling Yard of the National Park Service wrote that during this time numerous "popular organizations to conserve forests, wild life, scenery, and natural resources of many kinds, sprang into existence in every corner of the country," but they all "followed the leadership of the Boone and Crockett Club, the pathfinder and pioneer."[1] However, unlike the Boone and Crockett Club, most of the new organizations had full-time, salaried staffs. This transition from amateurs to professionals was an intrinsic feature of the progressive period, and a recurring motif in Madison Grant's life. (In the conservation movement, however, the new professionals were Grant's intellectual heirs, not his enemies. Such would not be the case, as we shall see, in the field of anthropology.)

As time went on, these professional wildlife organizations collaborated, fought, federated, dissolved, reorganized, merged, divided, and subdivided into an enormously complex welter of specialized groups. Some were guided by a philosophy of utilitarian conservationism, others by aesthetic preservationism; some concentrated on big game, others on birds; some emphasized public education, others legislative solutions; and so forth. The growth in the number of organizations, and their success in securing their agendas, should not mislead us as to the popularity of the conservation movement in the first few decades of the century. In general, the "movement" was not a universal uprising on the part of the people but rather a narrow-gauged effort that succeeded precisely because its core consisted of a small but well-connected elite. Madison Grant could—and did—attend a meeting of the New York Zoological Society in the morning, a banquet of the American Bison Society in the afternoon, a dinner of the Boone and Crockett Club in the evening, and a fund-raiser for the American Museum of Natural History in the after hours—and he would see the same faces at all four functions.

Indeed, a quick perusal of the letterheads of these different groups confirms that many of the conservation movement's leaders—almost all of whom lived in the East, and most of whom had known each other for years—served on the board of more than one organization, and lent each other assistance when the situation called for it. Madison Grant knew that if the Boone and Crockett Club faced a legislative crisis in Congress, he could rely on the fact that supporting resolutions, lobbying assistance, editorial support, expert testimony, and even financial aid would soon be forthcoming from the New York Zoological Society, the Camp Fire Club, the American Ornithologists' Union, and others. (Often, when the Zoological Society passed a resolution endorsing some conservation measure, Grant's secretary would simply cross out "New York Zoological Society" on the carbon copy and type in "Boone and Crockett Club," on the assumption that whatever position the one group supported, the other would support as well.) And so we can say that the eastern leaders of these organizations comprised, in essence, an "interlocking directorate" that controlled the wildlife conservation movement.

We can get a sense of this network from appendix B, which lists some of Grant's colleagues in the early movement and a few of the organizations to which they belonged. As the years wore on, and the conservation movement dealt with problems of increasing magnitude, it was often Madison Grant who mediated between the various factions of the interlocking directorate. Grant, stated conservationist H. E. Anthony, "was the nestor of American conservationists, the wise counselor to whom one turned first when a conservation crisis impended."[2]

William T. Hornaday, the director of the Bronx Zoo, was irked that Madison Grant, who was by now a full-fledged preservationist, had no qualms about working side by side with members of the interlocking directorate who did not share the preservationist ethos. Hornaday could never grasp the political truth that if preservationists were to succeed in North America, they would have to form coalitions with utilitarian conservationists and even with hard-core hunters. During the campaign to pass the Migratory Bird Law, for example,

Grant carefully cobbled together an eclectic coalition that included a number of arms manufacturers (who had learned from the conservationists that if the game were not saved today, there would be nothing left to shoot tomorrow, and hence little need for anyone to purchase weapons and bullets). Hornaday refused to cooperate with these groups, and his contemptuous attitude threatened on numerous occasions to rip apart Grant's fragile entente. Grant was a preservationist, but he was also a realist, and he pointed out to Hornaday that sportsmen were just about the only natural constituents that wildlife had at the time.

In his work for preservation, Hornaday had always manifested a certain streak of righteous indignation, and he grew increasingly pugnacious as he got older. Unlike Grant, who always displayed patience and a willingness to compromise, Hornaday personified impatience and an absolute refusal to yield an inch. His motto was: "No compromises with the enemy. Never!" He thrived on fights, and the spirit of his hero Theodore Roosevelt can be seen in Hornaday's call for a conservation movement that would be "virile, militant, uncompromising." In 1920, at the age of sixty-six, he confided to his friend Edmund Seymour: "If I were ten years younger, I would join in fighting the enemies of [preservationism] with a rifle in one hand and a meat axe in the other."[3]

Hornaday was in a continual state of agitation. He detected slights in the most innocent of remarks and saw snubs in the most trivial of incidents. And with Hornaday, every disagreement, no matter how innocuous, turned into a lifelong vendetta. Scores of letters issued forth from his vitriolic pen every day, threatening to prosecute a perceived enemy, fire an unfortunate subordinate, attack some imagined transgressor, or work nevermore with a "disloyal" colleague. If nothing else, one must admire Hornaday's tremendous energy: he was able to churn out thousands of such letters (and countless editorials, pamphlets, articles, and books) all while running the Bronx Zoo—an extremely complex, million-dollar operation—with great distinction and competence.

W. Reid Blair, Hornaday's successor as director of the Bronx Zoo, wrote in Hornaday's obituary that "he had an uncommon faculty of making enemies." Indeed, Hornaday *reveled* in antagonizing others. Historian James Trefethen contrasted Hornaday to Madison Grant: "Always the gentleman, both in appearance and in fact, Grant was a clean boxer who fought only with gloved fists. Hornaday was a brawler who used knees, boots, and brass-knuckles, and his attacks often left scars." The problem was that Hornaday inflicted those scars not just on the lumber executives and land developers but also on any conservationists he judged too willing to engage in compromise. Anyone within the interlocking directorate who happened legitimately to disagree with Hornaday on the suitable size of a wildlife sanctuary, the proper length of a closed season, or the need to regulate a particular weapon, was classified in his mind as one of the "Benedict Arnolds of conservation."[4]

Whenever conservationists opened a letter in the first four decades of the

William T. Hornaday at his desk in 1910. Note the collection of rifles and the trophies of animals he killed before converting to preservationism.

twentieth century, one glance at the underlined oaths, declaratory sentences, and uppercase vows to "FIGHT TO THE DEATH!" immediately told them that the author was their former friend William T. Hornaday. His published writings employed a similarly hysterical tone. Scholars who have read only one or two books by Hornaday tend to be impressed by his "combative spirit." Viewing him as a forerunner of the environmental movement, they admire his passion and extol his militancy in the cause of preservationism. Indeed, his constant calls for "drastic action," "severe measures," and "stern reprisals" sound admirable and stirring—until one realizes that they arose as much from his emotional imbalance as from his devotion to wildlife.

To Hornaday's credit, he had principles, and was unwilling to yield them. "To me," he wrote in his eightieth year, "the saving of wild life always was more important than 'harmony' with its destroyers." The problem was that if a man like Madison Grant had followed Hornaday's practice and refused to court "harmony" with his philosophical opponents, there would be today no bison or redwoods, no Adirondack Deer Law or Alaska Game Law, and no Glacier or Denali National Parks.[5]

Two "Success" Stories

The old problems facing Grant and the interlocking directorate had been relatively straightforward: the bison of the Great Plains were imperiled, the grizzlies of Glacier were endangered, the bighorn sheep of Denali were threatened. The solutions were equally straightforward: gather up the few survivors, put them behind a fence, and let them graze—and hopefully breed—in peace. And most of Grant's various wildlife preservation efforts along those lines had succeeded. The American bison were secure, the grizzlies were thriving, the bighorn sheep were flourishing.

But the success of preservationism had created a new challenge, and this one was much more complex. As early as 1910, there were hints that Grant's projects may have succeeded *too* well. This is an amazing statement to make, since only one decade earlier the preservationists had been a minuscule and not very hopeful band of dreamers. Nonetheless, the new problem was real, and it consisted of the fact that those species that had been saved were expanding so rapidly in population that they were threatening their own habitats. Two examples will suffice.

In 1894, the year after Grant joined the Boone and Crockett Club, the club succeeded in convincing Congress to pass the Yellowstone Park Protective Act, which made it illegal to hunt wildlife in Yellowstone. As a result, the elk herd within the national park flourished. In fact, under the ideal conditions of Yellowstone the prolific elk increased so rapidly that they began to venture outside the boundaries of the park in search of food. But whenever that happened, they were met by a hail of lead from hunters. The elk quickly learned that the only safe place was within the park. There they remained, secure and protected, their numbers continuing to build season after season. By 1910, unfortunately, the Yellowstone herd had increased to the point where it was exhausting its natural food supplies in the park (as well as the forage of several other species, including bighorn sheep and white-tailed deer). Rangers were forced to purchase $20,000 worth of feed from outside the park to prevent the elk from starving during the winter. Madison Grant made sure that the interlocking directorate contributed financially to this effort.

But it did not take long for the Yellowstone elk to become dependent on the handouts. Indeed, the more the animals were fed, the more their numbers increased—and the more they therefore had to be fed. It was a vicious cycle, and yet the feeding program could not be terminated without threatening the existence of the now totally dependent animals. Furthermore, as the elk crowded expectantly into the feeding area every winter, they denuded land, eroded trails, and became subject to parasites and crowd diseases.

These were problems that none of the early preservationists had foreseen. Looking back from 1943, Ira Gabrielson, director of the U.S. Fish and Wildlife Service (and member of the Boone and Crockett Club), wrote: "Realization was beginning to be forced upon those responsible for refuge administration that creating reservations and leaving them without management does not always produce the desired results." In fact, "the practice can often do as much harm as good to the species for which protection is sought."[6]

A number of steps were taken in an attempt to forestall disaster in Yellowstone. Additional lands for winter grazing grounds were acquired, many animals were transferred to other wilderness areas, and some were shipped off to zoos. Again, Madison Grant and other members of the Boone and Crockett Club helped pay for these efforts. But such measures accounted for only a fraction of the elk's population increase, and preservationists were at a loss as to how to resolve what they themselves had wrought in the national park.

A few years after the problems in Yellowstone manifested themselves, similar troubles appeared in the Grand Canyon National Game Preserve on Arizona's Kaibab Plateau. President Roosevelt had created the million-acre refuge in 1906 to protect the three thousand endangered Rocky Mountain mule deer of the area, noted for their large size and spectacular antler development. Hunting was prohibited in the refuge, except by agents of the Forest Service who went after the deer's main predators (wolves, mountain lions, bobcats, and coyotes) with a vengeance. No new predators ever migrated in to replace them, as the Kaibab Plateau was surrounded by deep canyons and inhospitable deserts and was thus a biological island. Within a few years the protected mule deer had managed to double their numbers, and the Grand Canyon National Game Preserve was hailed as a great success. Arizonans proudly proclaimed that the plateau now protected "the biggest deer herd in the world."[7]

But with no natural enemies, the Kaibab deer herd kept right on growing. Between 1906 and 1924 the herd increased from three thousand to perhaps as many as one hundred thousand animals, far more than could be supported by available forage. Having depleted their natural food supply, the hungry deer proceeded to chew most of the remaining plants down to the coarse stems, at which point malnutrition, disease, and starvation became real possibilities.

Madison Grant's friend George Shiras III, a former congressman and a leading member of the Boone and Crockett Club, visited the Kaibab in 1923 and was alarmed at the "deplorable" situation on the range. He reported that thousands of scrawny deer were "on the verge of starvation" and urged the government to take action.[8] Grant himself toured Arizona the following year and was most distressed at what he saw. Accordingly, Secretary of Agriculture Henry C. Wallace appointed a committee of experts from the Boone and Crockett Club and other conservation organizations to investigate developments on the Kaibab. They visited the area in 1924 and regretfully recommended that to avoid disaster the deer population should be cut immediately by 50 percent. The Forest Service accepted the advice and announced that hunters would be permitted to enter the refuge to reduce the herd.

But having devoted years to convincing the public and Congress of the need for inviolate wildlife refuges, preservationists now found it extremely difficult to sell the idea that protected animals should be subject to "population control." In fact, the "uninformed and uninformable sentimentalists" (as one government official called them) were outraged at the decision to allow federally sanctioned hunting on the Kaibab Plateau. Steven Mather of the National Park Service, for example, announced that "there is no such thing as too many deer," and stood up to oppose the plan.[9] Novelist Zane Grey, who occasionally socialized with Madison Grant and whose *Last of the Plainsmen* had been set on the Kaibab Plateau, thought he could preclude the need for any shooting by driving the excess deer from the overstocked area. In December 1924 he gathered a mounted force of forty cowboys and seventy Navajo Indians, lined them up side by side across a broad front, and with a great deal of yelling and whooping invaded the refuge. But the elusive deer, rather than being driven ahead of the horses, simply slipped between them, and the roundup failed.

When the secretary of agriculture directed the deer hunt to proceed, the governor of Arizona, supported by antihunting forces, invoked states' rights and sued to stop the action. The hunt was put on hold pending the decision of the courts. And until then, the governor threatened to call out the Arizona militia to protect the deer and arrest any hunter who entered the refuge. The court case dragged on for years as state and federal authorities wrangled over who had jurisdiction over the Arizona mule deer. But in the meantime, the emaciated deer began dying by the tens of thousands. From its peak of one hundred thousand, the starved herd had plummeted to fifteen thousand gaunt animals by the time the Supreme Court finally ruled in *Hunt v. United States* (1928) that the federal government did indeed have the authority to regulate wildlife in the refuge. Licensed hunters immediately commenced to cull the herd. But the range had deteriorated so much by then that the herd actually had to be cut far below even its original numbers in order to give the vegetation an opportunity to recover. Years were to pass before the Kaibab regained enough forage to maintain even a modest herd of deer.

Despite the clear lessons to be drawn from the fate of the elk in Yellowstone and the deer in Arizona, other "Kaibabs" soon developed, largely due to the unwillingness of pure preservationists, along with some states' righters, to admit that protected populations had increased beyond the capacity of their habitats. (The white-tailed deer population of Pennsylvania, for instance, had stood at one thousand in 1905. But thanks to the hunting regulations and gun-control laws promulgated by the Boone and Crockett Club, there were nearly one million deer in the state by 1925. At that point, overpopulation took its toll: the animals commenced to die of starvation by the thousands, and their emaciated carcasses began showing up everywhere.) As the President's Committee on Outdoor Recreation explained in 1927: "Over-protection, paradoxical as it may seem, defeats its end, and under its stimulus certain types of game animals multiply beyond their means of subsistence and cruel starvation ensues."[10]

The Genius of Aldo Leopold

Professional forester Aldo Leopold (alumnus of Yale and devotee of George Bird Grinnell) meticulously examined the problems of the Kaibab deer. His resultant conclusions, when elucidated for naturalists in 1925 in an article titled "Ten New Developments in Game Management," officially gave birth to the field of wildlife management. In his article, Leopold advanced the theory that each unit of habitat can support only a certain number of wild animals of a given species. If and when a fully stocked population is allowed to expand beyond that limit, the surplus will be eliminated by disease, starvation, and predation. But Leopold posited that wildlife officials do not have to stand by and let that happen. For one thing, they can try to increase the number and size of natural habitats, as Madison Grant was doing with his game refuge efforts. For another, they can maintain the wildlife population at its optimum level through selective castration, controlled breeding, and regulated hunting.[11]

According to the devotees of wildlife management, preservationists need to understand and accept that when a game population produces a surplus (i.e., more animals than are needed to maintain adequate stocks), the excess can and should be removed by hunters. It is true that the redundant population, if not culled by humans, will die of other causes anyway (e.g., the aforementioned disease, starvation, and predation), but not before endangering the breeding stock by spreading illness, denuding habitat, and enticing predators into the range.

Aldo Leopold understood that many members of the public were upset that hunters had been allowed to shoot the "excess" mule deer on the Kaibab Plateau. But according to the theories of wildlife management, understandable but misplaced sympathy for the fate of an individual animal must not be allowed to override concern for the survival of the herd as a whole. To oppose regulated hunting out of inappropriate sentimentalism is a dereliction of duty that can only result in such tragedies as had occurred in Yellowstone and on the Kaibab Plateau. Just as we cut down diseased trees that threaten the overall health of the forest, so we should cull individual animals that endanger the survival of the herd. And the most important part of any herd—the absolutely essential part—is the breeding stock. It is the fundamental task of wildlife officials to ensure that the fittest members of the herd survive to propagate the species. In an ideal world, of course, all members of the herd would be allowed to survive, but we must accept the fact that some elements are, biologically speaking, superfluous. And if these undesirable elements have to be eliminated for the greater good of the community, it is not at all helpful to have well-meaning but unsophisticated preservationists standing in the way.

For sentimentalists to decry the hunting of excess animals, thought Leopold, was as absurd as trying to "save" a crop of corn by leaving it unharvested in the field. Indeed, Leopold often compared wildlife management to farming, and advanced the idea that wildlife populations produce "crops" of game that can be "harvested" just like agricultural products. "We have learned," he wrote, "that game, to be successfully conserved, must be *positively produced,* rather than merely negatively protected."[12] (It is not a coincidence that the journal in which Leopold's 1925 article appeared, the *Bulletin of the American Game Protective Association,* had at one time featured on its masthead the epigram: "We must sow and tend the game crop, if we would reap it.")

The tragedy on the Kaibab Plateau not only alerted Leopold to the danger of mindless preservationism but also taught him to appreciate the beneficial role that predators play in maintaining the balance and health of ecosystems. He asked preservationists to acknowledge "the inter-relationships of living populations," and to understand that to single out one species for protection at the expense of others was to disrupt a natural equilibrium that may have been eons in the making. After Leopold, wildlife officials began to see that it was absurd to categorize some species as "good" and others as "bad," and they slowly began to appreciate those animals that the public viewed as "varmints," if not for their intrinsic value then at least for their role in managing populations and ecosystems.[13] (In the past few years, ecologists have debated whether wildlife officials in the 1920s overestimated the effect on the Kaibab deer of removing their predators; but what is of interest to us is not what actually happened to the deer herd but what conservationists like Leopold *thought* happened.)

Aldo Leopold called for a new generation of scientifically trained professionals, conversant in population dynamics and the operation of food chains, to become involved in wildlife protection. He noted that "agriculture and forestry began to apply science to their crops decades ago. Game management must now do so, or fail." He even called for those concerned with wildlife to apply to its management "the same principles of sound organization which we apply to our factories and offices." Leopold's progressive interest in *management* and *sound organization* seems very much in the utilitarian tradition of the conservation movement. And, too, his frequent use of the terms *crop* and *harvest* would seem to bring him awfully close to Gifford Pinchot, who, we recall, saw forests as farms for the production of lumber. Indeed, it would be surprising if Leopold, having graduated from the Yale Forestry School (established by Pinchot's family) in 1909, had not absorbed the lessons and the terminology of the master. But we need to remember that the path from conservationism to wildlife management had gone through preservationism, and therein lays the difference between the two. Unlike conservationism, wildlife management conserves the game not for *man's* benefit but for the *game's* benefit; in other words, it is motivated not so much by a desire for eventual utilization as by a concern for permanent and sustainable preservation.[14]

In 1933, Aldo Leopold became the nation's first professor of wildlife management, at the University of Wisconsin. He was enormously influential at the time, and remains so to this day. To be sure, the sophisticated (and somewhat troubling) precepts of wildlife management were much more difficult for the public to understand than the simplistic (but emotionally satisfying) preaching of a rabid preservationist like William T. Hornaday, who kept insisting that the only way to save wildlife was to kill hunters. Nevertheless, Leopold's seminal *Game Management* has been continuously in print since 1933 and is still regarded as a basic statement of the techniques of wildlife management. And Leopold's brilliant and moving *Sand County Almanac,* published posthumously in 1949, joined the works of Henry David Thoreau, John Muir, and Rachel Carson as the founding texts of the environmental movement of the 1960s. Years before Carson posited the idea that nature exists in, and depends on, a delicate balance, Leopold was teaching that a region's flora and fauna survive in an intricate web of interdependencies. His theories prefigured the modern science of ecology (defined as "the study of the interrelationships of organisms to one another and to the environment"), and his words are echoed today by proponents of the "Gaia hypothesis," for Leopold thought that humans had the capacity intuitively to "realize the indivisibility of the earth—its soil, mountains, rivers, forests, climate, plants and animals, and respect it collectively not only as a useful servant but as a living being."[15]

For the Good of the Species

Aldo Leopold was an enthusiastic and accomplished hunter who believed that "the instinct that finds delight in the sight and pursuit of game is bred into the very fiber of the race."[16] It is almost needless to say that he was a longtime member of the Boone and Crockett Club. And it is interesting that in the early 1910s, more than two decades before Leopold's *Game Management* was published, Madison Grant and Leopold's progenitors in the Boone and Crockett Club were already formulating a set of ideas that sound very much like the tenets of the future science of wildlife management and, indeed, of ecology.

In 1913, Grant went on a hunting excursion to Montana with his brother DeForest, after which he toured Yellowstone National Park to witness firsthand the deteriorating elk situation. Theodore Roosevelt, in *The Outlook,* had already informed the public that the elk had been so thoroughly "but unintelligently" protected "that their increase has outstripped the food supply, and in consequence multitudes now perish in the most miserable way by starvation." Grant's trip confirmed this, and upon his return to New York he began discussing with his wildlife colleagues possible solutions to the elk problem. In particular, Grant promulgated the idea that—even though the Yellowstone elk were protected—surplus animals could and should be eliminated for the good of the herd. The fate of the species, in Grant's mind, now overrode the fate of any individual animal, especially if that individual was not making a positive contribution to the breeding stock. And so Grant boldly recommended that the Boone and Crockett Club consider a more active program of wildlife management to replace the defensive policy of knee-jerk preservationism.

Not all Boone and Crocketteers were enthusiastic about Grant's proposal. The rank and file, after all, had been put through a rather wrenching experience during the previous few years. They had been asked by Grant to progress from big-game hunting in the 1880s, to utilitarian conservationism in the nineties, to aesthetic preservationism in the aughts, and now to this newfangled concept of wildlife management in the teens. But in 1915, the Boone and Crockett Club's Game Preservation Committee, under the chairmanship of Grant's close friend Charles Sheldon (and including such Grantian stalwarts as W. Redmond Cross, George Bird Grinnell, E. W. Nelson, and Charles H. Townsend), issued a momentous report courageously admitting that a simple philosophy of "preservationism" was no longer the solution to the problem of endangered species. Citing the situation in Yellowstone, the committee explained that a naïve policy of absolute protection had led inevitably to too many animals for the range. What the elk herd needed most was "scientific management," and the committee revealed the hard truth that this would necessitate the "elimination" of several thousand elk every year. Grant's colleagues, in short, endorsed the killing of individuals for the good of the species. We might note the year the report was issued and conjecture that the sacrificial atmosphere suffusing a world war was perhaps conducive to such thinking. We might also note that the committee researched and wrote its report at the same time that Grant was researching and writing *The Passing of the Great Race*.[17]

Although overlooked by historians, the 1915 report of the Boone and Crockett Club was the first public pronouncement of the theory of wildlife management. Theodore Roosevelt declared that he fully concurred with the findings of the committee and admitted that he "was regarded as rather hardhearted for saying so." But he explained that since the population of the protected elk was exploding, the animals "must be kept down by disease or starvation, *or else by shooting*. It is a mere question of mathematics, to show that if protected as they have been in the Park, they would, inside of a century, fill the whole United States; so that they would *then* die of starvation."[18]

In sum, long before Aldo Leopold addressed the issue, the leadership of the Boone and Crockett Club was beginning to understand that granting protection to a favored species destroys the balance of nature, and that the resultant population boom inevitably ends in disease and starvation unless professionals cull the herd.

Madison Grant's other refuge-creating organization, the American Bison Society, embraced wildlife management soon after the Boone and Crockett Club. In 1925, the Bison Society adopted a policy that all "undesirable" bison on its refuges were to be "eliminated and disposed of." In the category of "undesirable bison" the society bluntly included "surplus bulls," "barren cows," "deformed and maimed animals," and "all such as are not up to a fair standard."[19]

But it took a long time for the interlocking directorate to sell the public on the idea of killing animals to save them. ("America cannot see the herd for the deer," was one conservationist's explanation for the public's resistance to wildlife management.)[20] For years, the National Park Service chose to ignore the advice of the Boone and Crockett Club and the American Bison Society, and rather than take steps to control the elk population in Yellowstone, it did everything possible to *increase* the number of animals by continuing to administer feeding programs. Madison Grant and his colleagues were forced to watch with increasing distress as the park's environment continued to deteriorate, the feeding yards became virtual deserts, and hundreds of animals succumbed to crowd diseases. It was not until 1943, three decades after the Game Preservation Committee of the Boone and Crockett Club had issued its groundbreaking report, that the Park Service, supported by most of the national conservation organizations, finally overrode public sentiment and began killing a certain number of elk every year to bring the population within the carrying capacity of the range. (In what might appear to be an unrelated development, 1943 was also the year that the Nazi death camps started operating at full capacity.)

Defending Predators

With regard to predation—and wildlife management's acceptance of its beneficial role—Madison Grant was also ahead of his time. As early as 1911, for example, he had tried to halt the game warden at New York's Tuxedo Park from destroying what were officially classified as "noxious animals." Grant explained that "The old fashioned method of removing so-called vermin has been found to disturb the balance of nature, and frequently ends in most unexpected and undesirable results." It was Grant's contention that "vermin" (e.g., wolves and coyotes) target chiefly the "surplus" game animals—the weak, the old, and the diseased. And he concluded that this actually benefits the preyed-upon species in the long run, as the fleet and the strong are left to reproduce. If the predators are destroyed, the "inferior" animals will survive and propagate, with dangerous consequences for the evolutionary fitness of the species. The preyed-upon species, in other words, *need* the predator species.[21]

Furthermore, Grant felt that the very term "predator" is deceiving. Every animal—indeed, every living thing—is a predator. It is simply a matter of perspective. As soon as sentimental humans choose to bestow their sympathy on a particular species (often one with big eyes and/or soft fur), the animals above that species in the food chain are labeled vicious "predators," while those organisms below it in the food chain are viewed as necessary "food supply."

Madison Grant was putting forth the view that predators are crucial to the health of the ecological community at a time when Aldo Leopold was still advocating the extermination of wolves, coyotes, mountain lions, lynxes, eagles, and other "vermin" as the most efficacious means of protecting valued game species. Indeed, the federal government, encouraged by agricultural, ranching, shepherding, and hunting interests, officially declared war on the nation's "varmints" in 1915. The Biological Survey, after receiving a congressional appropriation of $125,000, announced that predators "no longer have a place in our advancing civilization" and immediately commenced an extensive and long-term program of "repressing" (i.e., shooting, trapping, poisoning, and fumigating) "undesirable" forms of wildlife. The goal was to cleanse the continent of wolves, cougars, foxes, bobcats, lynxes, coyotes, prairie dogs, fishers, martens, otters, pelicans, eagles, and bears (who, one official explained, "can only be shunted from their evil ways by complete elimination").[22]

William T. Hornaday, the great protector of wildlife, applauded these efforts. He thought that when it came to "wild-animal pests," all methods—"firearms, dogs, traps and strychnine"—were "thoroughly legitimate weapons of destruction. For such animals, no half-way measures will suffice." The peregrine falcon, according to Hornaday, was just one example of an animal that deserved to be shot on sight. Nature writer John Burroughs agreed with Hornaday that varmints "certainly needed killing." The "fewer of these there are, the better for the useful and beautiful game." And so America had reached the state of affairs where preservationists like Hornaday and Burroughs fully encouraged the killing of predators (e.g., wolves) but were aghast at the killing of selected members of favored species (e.g., elk), while Madison Grant—having embraced the philosophy of wildlife management—opposed the killing of predators but supported the killing of surplus members of favored species.[23]

By the end of the 1920s, the Bureau of the Biological Survey, urged on by an unlikely alliance of farmers, stockmen, and preservationists, had strewn millions of traps across the landscape and could triumphantly declare that "the end of the wolf is in sight." The government was annually destroying one hundred thousand coyotes (an animal of "truly satanic cunning"), and as for the cougar, the lynx, and the bobcat, the Survey promised: "Their depredations have been controlled, and their ultimate elimination . . . is only a matter of time." By that point, two-thirds of the Biological Survey's budget was devoted to predator elimination (and the portion would soon rise to three-fourths). In fact, since there was a direct relationship between the numbers of dead predators and the level of funding Congress appropriated to the Survey, the agency paid bounties to professional hunters to needlessly kill predators even in areas remote from human activity where the animals could not possibly pose a threat. The cost to the taxpayers to control predators rose to a hundred times the value of the losses caused by predators to stockmen.[24]

For years, Madison Grant was practically a lone voice in trying to stop the eradication programs of the federal government. In the face of congressmen from rural districts who kept increasing appropriations to the antipredator effort, Grant insisted to friends like Senator Frederic C. Walcott (R-Connecticut) that the government must take into consideration "the effects of disturbing the balance of nature by the undue destruction of *so-called* predatory animals."[25]

By the late 1920s Grant had finally convinced the interlocking directorate to take up the cause of the predators:

- In 1928, the New York Zoological Society passed a resolution condemning "the indiscriminate killing of predatory animals" (a practice it labeled "unscientific") and called on the government to suspend its policies pending further study.
- One year later, the Boone and Crockett Club (in yet another example of the total transformation the club had undergone) adopted a resolution *opposing* National Park Service practices that favored game animals over predators. The club urged the Park Service to adopt a policy "whereby all predatory animals in these parks be allowed to retain their primitive relations to the rest of the fauna."
- And one year after that, the American Society of Mammalogists challenged the Biological Survey's use of poison against predatory animals. The society was especially disturbed by the spread of thallium, a very dangerous, accumulative poison. Even C. Hart Merriam, the father of the Biological Survey (who in 1908 had urged Congress to destroy the predators since "the great bulk of [these] mammals are pests [and] of no service to man"), now joined with Madison Grant in opposing the government's eradication policy.[26]

The government dismissed the criticism of Grant and the interlocking directorate as the unfounded sobbing of "a few cranks and conservation fanatics." The chief of the Biological Survey defended the use of strychnine and thallium on the dubious grounds that millions of acres of public lands were "so densely populated" with ground squirrels and prairie dogs that "in many places erosion is rapidly developing." At the behest of agricultural and ranching interests, the Survey and other government agencies, including the National Park Service and the U.S. Forest Service, continued for decades to murder what it viewed as "pests" (and what are now classified as "endangered species"). It was not until the early 1970s that the government completely accepted Madison Grant's understanding of the role of predators in healthy ecosystems and modified its pest policy. The decision was aided by environmentalist concerns about the effects of poisons (first voiced by the interlocking directorate back in the 1920s) and by changing attitudes toward wildlife: in a dramatic turnaround, the public decided that despised "vermin" like wolves and hawks were now noble and even lovable creatures.[27]

* * *

We can see that well before wildlife management became fashionable among professional conservationists, Madison Grant and his associates were espousing many of its theories. They had a good understanding both of positive management techniques (such as preserving reproductive stocks and maintaining the food-population equilibrium) and of negative techniques (such as culling herds to eliminate the surplus and allowing predation to maintain the balance of nature).

One of the reasons that Grant and the interlocking directorate accepted the philosophy of wildlife management so quickly and enthusiastically was that wildlife management fit the tenor of the times. In the 1910s the progressives were actively trying to regulate not only big business but also municipal governments, public utilities, workplace conditions, public health, and personal habits such as consumption of alcohol—and now even the wild animals of the forests were going to be managed scientifically. Through expert analysis and intelligent planning, the most fundamental processes of nature were going to be controlled.

Progressives like Grant accepted that just as unregulated capitalism contains the seeds of its own destruction, so unrestricted preservationism eventually results in disasters like the runaway mule deer population on the Kaibab Plateau. They were prepared to regulate the economy to ensure competition, and they were equally willing to practice wildlife management to maintain the balance of nature. Wildlife management, in sum, was the penultimate progressive idea.

5

From Mammals to Man

You interest me very much, Mr. Holmes. I had hardly expected so dolichocephalic a skull. . . . A cast of your skull, sir, until the original is available, would be an ornament to any anthropological museum.

Arthur Conan Doyle

The Museum

Madison Grant and Henry Fairfield Osborn were close friends and colleagues for forty years, from the formation of the New York Zoological Society in 1895 until the death of Osborn in 1935 (two years before Grant). They socialized in the same clubs, served on the boards of the same organizations, and fought for the same causes. The two friends communicated daily, by either telephone or letter, and dined together about once a week. In addition, Osborn rendezvoused with Grant every Saturday at the Bronx Zoo, whence the two would survey their domain.

Grant and Osborn first met when the latter was appointed head of the Department of Vertebrate Paleontology at the American Museum of Natural History. The museum had been founded and incorporated in 1869 by some of New York City's richest men, including Robert Colgate, A. G. Phelps Dodge, Morris K. Jesup, J. P. Morgan, Levi P. Morton, Henry Parish, and Theodore Roosevelt, Sr. They hoped that a museum dedicated to natural history would raise the intelligence of the general community and educate the working classes about the laws of nature (a goal that was somewhat stymied for the first two decades by the refusal of the largely Presbyterian trustees to open the building on the Sabbath, thus preventing the city's working people from ever seeing the inside of the museum). They eventually situated the museum on a twenty-three-acre site on Central Park West, and it soon became the most extensive natural history repository in the Western Hemisphere. The museum had millions of artifacts and thousands of displays (the first "habitat group" in the museum was a group of orangutans killed and mounted by the young taxidermist William T. Hornaday in 1880), and by the beginning of the twentieth century it was the largest building in New York. "It was inevitable," remarked Henry Fairfield Osborn many years later, "that the Museum should become a World Museum, as New York has become a World City and as the United States has become a World Power."[1]

The American Museum of Natural History also became—and remained until the Great Depression—one of the pet philanthropies of the New York patriciate. The museum had the same number of members (approximately two thousand) as the New York Zoological Society, and many of the same benefactors (including Hugh Auchincloss, George F. Baker, George Eastman, Levi P. Morton, Percy Pyne, Margaret Olivia Sage, Jacob Schiff, William Sloane, the Dodges, the Harknesses, the Huntingtons, the Jesups, the Morgans, the Rockefellers, the Thornes, the Vanderbilts, and the Whitneys). Ronald Rainger points out that they were all "part of a closely connected socioeconomic network whose families intermarried and whose members held similar religious, social, and often political commitments."[2]

In the 1890s, Grant and Osborn were on parallel professional paths. Grant had chosen to be an expert on zoology, and Osborn had decided to be an expert on paleontology (which, he was fond of saying, is simply "the zoology of the past"). Grant had founded the Bronx Zoo in 1895 to present specimens of large North American mammals, and Osborn had joined the museum's Department of Vertebrate Paleontology in 1891 (when it was the Department of *Mammalian* Paleontology) to display fossils of large North American mammals. Both men had rapidly achieved their goals: Grant accumulated in the zoo the world's finest collection of living animals, and Osborn acquired for the museum the most extensive collection of fossilized animals. Both men explicitly hoped their institutions would teach the citizenry about nature, and therefore serve as centers for promulgating the conservationist ethos. And just as Grant was the first to display such animals in their native habitats, Osborn was the first paleontologist to place his specimens in large, realistic exhibits. Osborn always made sure that the museum showcased the largest possible dinosaurs and mastodons to attract the largest possible attendance. He sponsored a number of paleontological expeditions to find such specimens, including the 1905 Montana excursion that discovered the dinosaur he named *Tyrannosaurus rex*. More than anyone else, Osborn popularized paleontology in North America and made "dinosaur" a household word.[3]

Grant and Osborn were socially and ideologically inseparable as they advanced in the world. William T. Hornaday remembered the time he first encountered these "two young men, of fine education, excellent social standing, and high ambitions for public service. . . . They became welded into that one purpose, . . . and for [forty] long and arduous years pulled together like the best team of horses that my field battery ever knew. Their fine example set the pace

The always self-satisfied Henry Fairfield Osborn in the 1920s. He named Tyrannosaurus rex, *and then ruled the American Museum of Natural History as a tyrant king.*

for literally all others who were in a position to promote the general progress." By the beginning of the twentieth century, both men had become respected and well-known figures in their chosen fields. Unfortunately, as Osborn acquired more fame and power, his personality—which even at its most modest moments had been somewhat vainglorious—ballooned into all but insufferable pomposity. While he could be charming to those in a position to help him expand his sphere of influence, Osborn was usually pretentious around his professional colleagues and condescending to his assistants, who almost universally disliked him. In 1930 Osborn published *Fifty-two Years of Research, Observation, and Publication,* which he described as "an auto-bibliography." It listed for public edification his memberships in over 150 societies, and contained the titles of an incredible 904 works written by Osborn. A close look, however, reveals that some works were listed under more than one category and that the total was padded with such "published" ephemera as speeches at class reunions and letters to his college newspaper. In addition, the list included earlier Osborn bibliographies and, in the bibliographic version of a Möbius strip, even counted *Fifty-two Years* itself as one of his publications.[4]

What makes Osborn's conceit in this instance all the more outrageous is that much of his work was apparently ghostwritten by underlings. Osborn himself admitted that he was a "synthesizer" whose assistants performed much of the detailed work for him. Harry L. Shapiro of the museum's Anthropology Department testified that Osborn used to get staff members to help him write his books: "'I want a section on thus-and-so,' he would say," and the section would then appear unchanged as a chapter in one of his books. Similarly, Osborn routinely asked Madison Grant to research various topics for him and often "borrowed" Grant's notes and papers, which were then turned into speeches and articles signed by Osborn.[5]

Still, Madison Grant lionized the older (by eight years) and wealthier (by millions) Osborn. After all, any man who could put up with William T. Hornaday's paranoiac ravings on a daily basis could certainly abide a little Osbornian vanity. Relations between the two men were undeniably smoothed by the fact that Grant had almost as high a regard for Osborn's abilities as did Osborn himself. Where others saw arrogance, Grant perceived only patrician confidence. Where others suspected plagiarism, Grant noticed only wide-ranging erudition. In 1910, Osborn's *Age of Mammals* was published (with the usual borrowings from Grant's monographs on North American mammals), and Grant gushed that it was "the most monumental work of the kind that I have ever opened. I think it is epoch making." Grant's glowing review in *Forest and Stream* led George Bird Grinnell to timidly hint: "Is it not a *little* strong to say that it is one of the most notable books on evolution since the appearance of *The Origin of Species?*"[6]

In 1908, two years after turning down the secretaryship of the Smithsonian Institution, Osborn succeeded Morris K. Jesup to become the fourth and most important president of the American Museum of Natural History. Osborn was a tremendous organizer and outstanding fund-raiser, and during his twenty-five-year reign as president the museum became one of the major scientific institutions of the world. Its space more than doubled; the scientific staff trebled; the membership quadrupled; the city's annual appropriation increased from \$160,000 to \$500,000; and the endowment skyrocketed from \$2 million to over \$14 million.

President Osborn appointed a number of friends from his generation to the museum's board, including J. P. Morgan, Jr., and Madison Grant, who would be his closest confidant among the trustees. Grant also chaired or served on a number of the museum's organizing committees, including the Conservation Committee, the Anthropology and Archaeology Committee, the Mammals of the World Committee, the Comparative Anatomy Committee, the Wild Life Protection Committee, and the Nominating Committee. It was the task of the latter group to nominate new trustees for the museum, and the fact that Grant was the committee's perennial chair is one reason that the museum's board resembled so closely those of the New York Zoological Society and the Boone and Crockett Club. Prominent trustees from the interlocking directorate included George F. Baker, Jr., Suydam Cutting, Lincoln Ellsworth, William Averell Harriman, Walter B. James, Ogden Mills, George D. Pratt, and Percy R. Pyne. And since Grant, as he did with all his organizations, followed the aristocratic practice of replacing deceased trustees with their sons, the board was also replete with second- and third-generation Davisons, Dodges, Fricks, Morgans, Rockefellers, Roosevelts, Warburgs, and Whitneys.

In the same year that Osborn became president of the museum, a crucial transformation occurred in the intellectual life of Madison Grant. Where his scientific work had heretofore focused exclusively on mammals, in 1908 he began to take a passionate interest in the field of physical anthropology. The catalyst for this major change, I suspect, was economist William Z. Ripley, and the venue was the Half-Moon Club.

Ripley at the Half-Moon

One of the many joint endeavors of "those two sporty and reckless men" (as Hornaday referred to Grant and Osborn) was the Half-Moon Club.[7] This was an exclusive men's club they created in 1906 for New York aristocrats who had an interest in science and adventure. The idea was to periodically hold a dinner at which a famous explorer or scholar would give a talk about his latest achievements. It would be America's version of the Royal Geographical Society.

In keeping with Grant's interest in matters colonial, the Half-Moon Club was named for the ship upon which the original explorer of New York, Henry Hudson, sailed up the eponymous river in 1609. Hudson had claimed the area for the Dutch and opened the land for the settlers that followed, including Grant's Huguenot ancestors.

The "crew" of the Half-Moon Club, formally attired in tails and white tie, held their "cruises" approximately twice a year. Grant sent out the elegant invitations depicting Hudson's ship sailing to the New World under a half moon. In elaborate silver calligraphy, the members were requested to "reserve a cabin" for the upcoming "voyage," which usually "set sail" at seven bells in the University Club. Each dinner, with approximately twenty members in attendance, was hosted by a Master Mariner (often Grant) who invited and introduced that evening's Pilot who spoke about his recent adventures. Thus, for example, the "launch and first voyage of the Half Moon" was held on December 6, 1906, with Osborn presiding as the Master Mariner and Howard Crosby Butler as the invited Pilot, speaking on "Explorations in Northern Syria." The "grog" (gin and vermouth) was provided by Grant. Subsequent speakers included some of the leading archeologists, astronomers, architects, and adventurers of the day.[8]

The fourteen charter members, known officially as "The Gentlemen Adventurers of the Half Moon," included (in addition to Grant and Osborn) Thomas Hastings (designer of the New York Public Library, the Frick house, and the congressional office buildings in Washington, D.C.), William Milligan Sloane (president of the National Academy of Arts and Letters, and founder with Baron Pierre de Coubertin of the modern Olympic Games), Michael I. Pupin (the physicist whose inventions made possible long-distance telephone communication and X-ray photography), Charles F. McKim (head of the most successful and influential architectural firm of its time, responsible for the Boston Public Library, the Columbia University Library, the Metropolitan Club, the University Club, the Morgan Library, and the original Penn Station; six months before the formation of the Half-Moon Club, McKim's partner Stanford White had been shot and killed on the roof of one of their creations, Madison Square Garden, by the jealous husband of actress Evelyn Nesbit), and such mainstays of Grant's social world as John L. Cadwalader, Cleveland H. Dodge, Moses Taylor Pyne, William A. Wadsworth, and J. P. Morgan Jr.

Later members of the Half-Moon Club included Nicholas Murray Butler (president of Columbia University), Ralph Adams Cram (the foremost Gothic Revival architect in the United States), Daniel Chester French (who sculpted the statue of Abraham Lincoln for the Lincoln Memorial, of John Harvard at Harvard College, and *The Minute Man* at Concord), Charles Dana Gibson (whose Gibson girl drawings delineated the American ideal of femininity at the turn of the century), George Ellery Hale (the famed astronomer), Percival Lowell (the equally famed astronomer), John Russell Pope (designer of the National Gallery of Art and the Roosevelt Memorial that comprises the eastern part of the American Museum of Natural History), Frederic C. Walcott (the senator from Connecticut), Edmund Beecher Wilson (the prominent biologist), and Leonard Wood (co-organizer of the Rough Riders and chief of staff of the U.S. Army). It was quite an impressive little group, and they continued to meet through 1934, when the declining health of Grant and Osborn finally brought an end to the voyages of the Half Moon.

On February 6, 1908, the brilliant economist William Z. Ripley was the Pilot for the fifth voyage of the Half Moon, and his lecture that evening was titled "The Migration of Races." At the time of his appearance before the Half-Moon Club, Ripley was a professor of economics at M.I.T., and he was considered the country's greatest authority on railroad corporations. But he also dabbled in anthropology, and in 1899 he had written *The Races of Europe*, an extensively researched and ingeniously reasoned 624-page tome (supplemented by a separately published 160-page bibliography) analyzing the population of Europe from an anthropological point of view.[9]

As with all its proceedings, no record was kept of Ripley's lecture to the crew of the Half Moon, but based on his published writings and the text of the Huxley Memorial Lecture he gave later on in 1908, we can surmise the major points he covered. Ripley was interested, first of all, in classifying the European population from the standpoint of anatomy. He hoped that by quantifying such features as stature, eye color, and skull shape, he could discover the ideal European racial "type" that underlay the historical veneers of environment, ethnicity, and nationality. "Race," he declared, "denotes what man *is*; all these other details of social life represent what man *does*." Ripley, in other words, sought to analyze the human population as a mammalogist would analyze, say, the caribou species.[10]

And Ripley's investigations showed that Europe was peopled not by one race (usually called the "Caucasian" race) but by three races. Northern Europe, according to Ripley, was home to the tall, long-headed "Teutons," with their flaxen hair, blue eyes, and narrow noses. Central Europe was peopled by the stocky, round-headed "Alpines," characterized by brown hair, grayish hazel eyes, and broad noses. And southern Europe was home to the medium-statured, long-headed "Mediterraneans," with their dark hair, dark eyes, and broad noses.

This was of immediate interest to Ripley's American audience because every year, in ever-increasing numbers, a "horde" of Alpines and Mediterraneans was entering the United States. In 1907, the year before Ripley spoke to the Half-Moon Club, immigration to the United States had reached its record peak of 1.28 million newcomers in a single year. Between 1900 and 1908, over six million immigrants had landed on the nation's shores (one-quarter of the total immigration since the founding of the republic). Six million immigrants, Ripley pointed out, were enough people to repopulate all of New England (with types of people who would hardly be mistaken for native New Englanders). Ripley was not nearly as obsessed with the New Immigration as Madison Grant would be. Still, he could not help but notice that "wave has followed wave, each higher than the last," and that the influx of foreigners was causing "violent and volcanic dislocation" and threatening American society with disease, overcrowding, low wages, moral chaos, and political corruption. "The tide will rise higher," predicted Ripley (employing the cataclysmic metaphor so often employed by his generation of frightened Americans) until we are "inundated by the engulfing flood."[11]

In addition to the disruptive effects caused by the sheer numbers of immigrants, Ripley had serious concerns about the effect of the newcomers on the nation's gene pool. Madison Grant had always argued, regarding mammals, that the importation of nonnative species would either lead to mongrelization or displacement of the native species. And now Ripley conjectured the same of the New Immigrants, who in 1896 for the first time had outnumbered the traditional Teutonic immigrants. "We have tapped the political sinks of Europe," lamented Ripley, and it was distressingly obvious that no people were "too mean or lowly to seek an asylum on our shores." The net result was the existence in America of "a congeries of human beings, unparalleled for ethnic diversity anywhere else on the face of the earth." And the question was: what would happen if the incoming Alpines and Mediterraneans began to intermarry with the already-established Teutons? For an answer, Ripley turned to the new field of botany and borrowed the principle of reversion, which stated that the crossing of two varieties of domesticated plants sometimes produced an offspring with the traits of an ancient wild variety. To Ripley, this implied that if a "dark Italian type" should emigrate to the United States and mate with a native blond Teuton, the hybridized children could exhibit the traits of some primitive ancestor of man.[12]

And finally, as if all this was not frightening enough, Ripley reminded his listeners of something they hardly needed to be reminded of: that one of the European groups emigrating to the United States in ever larger numbers was the Jews. While Ripley was fairly certain that, technically speaking, the Jews were not a separate race, he pointed out that they did have their own peculiar physiognomy that rendered them instantly recognizable. There was, first of all, the Jewish nose, with its exaggerated degree of "hook," its marked "convexity," and its "nostrility." Jewish eyebrows were closer together than normal. Their eyelids were "rather full," revealing the "suppressed cunning" of the people, and there was "a peculiar separation of the teeth." "Quite persistent" also was "a fullness of the lips, often amounting in the lower one almost to a pout." And finally, Jews were "prone to nervous and mental disorders; insanity is fearfully prevalent among them."[13]

Ripley, like Madison Grant and many conservationists, was convinced of "the unfavorable influence of city life," and held it to be a general rule that "the urban type is physically degenerate." And who was the city dweller par excellence? The Jew. To congregate in commercial centers was "an unalterable characteristic of this peculiar people" because Jews had an inherent dislike for outdoor labor and physical exercise. It was the migration of the Jews into an urban environment, according to Ripley, that accounted for the fact that they were "one of the most stunted peoples in Europe" with their "narrow chests," "defective stature," and "deficient lung capacity." (On hearing this, Grant the mammalogist almost certainly concluded that the Jews were a degenerate form of the human "type.") In fact, claimed Ripley, the "unhappy country" of Poland was so "saturated with Jews" that Germany, to the west of Poland, "shudders at the dark and threatening cloud of population of the most ignorant and wretched description which overhangs her eastern frontier."[14]

But Germany was not the only country that should fear the Jews. For as a result of the lax immigration standards of the United States, "this great Polish swamp of miserable human beings, terrific in its proportions, threatens to drain itself off into our country as well, unless we restrict its ingress."[15] At this point, Grant assuredly reflected on the danger to his native city, where the influx of European Jews—desperately escaping epidemics, famines, political repression, and pogroms—had continued to increase with each passing year. From 1880 to 1914, a staggering one-third of eastern Europe's Jews migrated to New York City. The Jewish population of the city, which had numbered some 80,000 in 1880 (a tiny 4 percent of the city's population), was up to 1,250,000 by 1910 (more than 25 percent of the population). This was a number greater than the total population of the city just twenty-five years earlier. By the time Ripley spoke to the Half-Moon Club, New York had become—by far—the largest Jewish city in the history of the world. And every year, another 100,000 additional Jews poured in. Fifty-five percent of the city's cigar makers and tinsmiths were Jews. Sixty percent of the watchmakers and milliners, 68 percent of the tailors and bookbinders, and 80 percent of the hatmakers and wholesale butchers were Jews. Half a million Jews (perhaps one-tenth of the world's total) were packed into the 1.5 square miles of the Lower East Side, also known as "Jewtown." It was the most densely crowded quarter of the city, with 1⁄82 of the city's land area but 1⁄6 of its population. The Tenth Ward, with 700 people per acre, was more crowded than Bombay, India; it had, in fact, the highest population density in the world.[16]

According to Ripley, what made the Jews' presence in America particularly insidious was that they seemed to thrive here. Although unquestionably inferior physically to the Teutons, the Jews displayed "an absolutely unprecedented tenacity of life." Ripley cited statistics showing that "despite the unsanitary tenements, the overcrowding, [and] the long hours in sweat shops," the death rate among Jews due to disease or accident was half that of native-born Americans. (Ripley's numbers were accurate. The Lower East Side was one of the healthiest sections of the city.) Furthermore, Ripley pointed to the fact that suicide was "extraordinarily rare among Jews"—possibly due to innate cowardice.[17] (Ripley himself would commit suicide in 1941.)

In sum, we can assume that William Z. Ripley's lecture to the crew of the Half Moon Club described a situation in which hordes of Jews and other inferior types were descending upon the cities of America, where they flourished and propagated in increasing numbers and posed an evolutionary danger to the native Teutons via intermarriage and the consequent threat of reversion. "Is it any wonder," asked Ripley in words that must have made Grant's conservationist pulse race, "that serious students contemplate the racial future of Anglo-Saxon America with some concern? They have witnessed the passing of the American Indian and the buffalo. And now they query as to how long the Anglo-Saxon may be able to survive."[18]

Race Suicide

Madison Grant no doubt felt a myriad of emotions that February evening in 1908. On the one hand, he assuredly found it exciting to discover in William Z. Ripley a kindred scientific spirit. Working at the same time as Grant, Ripley had done for the white races what Grant had done for the large mammals. Grant's articles on mammals had described their evolutionary history, geographical distribution, and physical characteristics; and that is exactly what Ripley had done for the races of Europe. And all the themes that characterized Grant's natural history writings—typology, deterioration, and the need to prevent the introduction of foreign species—were there in the work of Ripley as well.

On the other hand, Ripley's speculations about the biological danger posed by the influx of degenerate, urbanized Jews must have filled Grant with despair. He now had to accept the fact that yet another mammal—the blue-eyed, long-headed Teutons—needed to be added to the list of endangered North American species. After all, it is one thing to learn that the bison are headed for extinction; it is quite another to learn that you yourself are similarly doomed.

Grant, of course, had suspected as much for some time. When he went to his Wall Street office in the morning, old Jews wearing phylacteries and prayer shawls scurried past him, headed for one of the 326 synagogues in Jewtown. When he ventured out to lunch, peddlers in caftans and untrimmed beards jostled him on the sidewalks. And on the way home in the late afternoon his ears were accosted by the newsboys on the corner, hawking the city's five Yiddish dailies in that unintelligible gibberish. As a bitter William T. Hornaday put it, New York was "an alien city" that had been overridden by "Jews from the slums of Riga." Or, as Grant told William Howard Taft, if he doubted that the natives were an endangered species, all he had to do was "walk down Fifth Avenue during the noon hour, as far as Washington Square," and judge for himself.[19]

The Jews that Grant knew personally—the Schiffs, the Warburgs, the Schemerhorns, the Guggenheims, the Seligmans—were all uptown German Jews who, frankly, were as revolted as Grant by the squalor of their fresh-off-the-boat eastern European cousins, and were condescending at best and hostile at worst toward their bizarre cultural practices. Even Emma Lazarus, after all, had referred to them as "wretched refuse." Grant must have thought the world had gone mad when, a few months after William Z. Ripley spoke to the Half-Moon Club, a character in a new play by Israel Zangwill actually *celebrated* the influx of New Immigrants—and coined a moniker heard round the world—when he rhapsodized that the United States was "the great *Melting Pot* where all the races of Europe are melting and re-forming!"[20]

The idea that the Teutons in America were endangered had not originated with William Z. Ripley. It dated back to the early 1890s and General Francis A. Walker. Like Ripley, Walker was a respected economist (he had been a professor of political economy at the Sheffield Scientific School of Yale and was the first president of the American Economic Association, a position subsequently held by Ripley). Walker was also the longtime president of M.I.T., where Ripley had studied and served on the faculty. After distinguishing himself as a general in the Civil War, Walker had been appointed chief of the Bureau of Statistics and superintendent of the census, and in that capacity he noticed a puzzling demographic trend. From 1790 to 1820, the population of the United States went up 35 percent per decade, the highest increase in human history. This had been accomplished, as he indelicately put it, "wholly out of the loins of our own people" (in other words, during a period of minimal immigration to the country). But beginning in the late 1830s, even though the number of immigrants to the United States increased substantially, the rate of population growth began to level off. This anomaly reached a climax in the 1880s, when immigration climbed to "the monstrous total" of 5.25 million, yet the total population of the United States increased at the lowest rate ever. "In other words," wrote Walker, "as the foreigners began to come in large numbers, the native population more and more withheld their own increase." As a result, immigration amounted "not to a re-enforcement of our population, but to a *replacement* of native by foreign stock." Such a process had been one of Madison Grant's chief fears regarding the importation of nonnative mammals.[21]

Walker had a social and an economic explanation for the low U.S. birthrate. First, native-born Americans did not want to bring children into a society degraded by filthy, ignorant foreigners who were "unfit to be members of any decent community"; and second, natives were afraid that their children would be obliged to compete in the marketplace with the "vast hordes of foreign immigrants" who were all too willing to work for inhuman wages.[22]

Looking to the future, Walker could not help but be pessimistic. In a very influential argument (that would echo all the way down to *The Bell Curve*), he posited that European railways and steamships had "reduced almost to a minimum the energy, courage, intelligence, and pecuniary means required for immigration; a result which is tending to bring to us no longer the more alert and enterprising members of their respective communities, but rather the unlucky, the thriftless, the worthless." Modern transportation, in short, was facilitating the immigration—and the survival—of the unfittest. "They are," he explained, in a phrase repeated continually in the ensuing decades, "beaten men from beaten races; representing the worst failures in the struggle for existence."[23]

It is important to note that Walker did not claim that the New Immigrants were a *biological* menace to the nation. Writing in the midst of the depression of the 1890s, he emphasized that they were dangerous because, as ignorant peasants, they posed a social, political, and above all an economic threat. But they were not a genetic hazard, and he admitted that with time they were fully capable of being civilized. Nevertheless, Walker was certain that no one could "be enough of an optimist to contemplate without dread the fast rising flood of immigration now setting in upon our shores." And he suggested that the time had arrived to place restrictions on immigration. "For one, I believe that the United States have, by a whole century of unrestricted hospitality, and especially by taking in five and a quarter millions of foreigners during the past ten years, fully earned the right to say to all the world, 'Give us a rest.'"[24]

If Walker felt this way in the 1890s, we can only imagine how he would have reacted the following decade, when the rate of immigration doubled. Luckily for Walker, he died in 1897. But four years later, progressive sociologist Edward A. Ross took up Walker's theme and coined the term "race suicide" to describe the declining birthrate of old-stock Americans in the face of increased immigration. It was bad enough, declaimed Ross in an address before the American Academy of Political and Social Science, that "our average energy and character is lowered by the presence in the South of several millions of an inferior race," but now, on top of the Negro problem, "the last twenty years have diluted us with masses of fecund but beaten humanity from the hovels of far Lombardy and Galicia." Ross was reminded—as were so many observers at the time—of the Roman Empire, when "the Latin husbandman vanished before the endless stream of slaves poured into Italy by her triumphant generals." "For a case like this," concluded Ross, "I can find no words so apt as 'race suicide.'"[25]

The phrase was soon adopted by his friend president Theodore Roosevelt, who in a series of speeches and articles chastised the middle and upper classes for their low birthrate, a phenomenon he felt was "the capital sin, the cardinal sin, against the race and against civilization." TR accused American parents of willfully limiting the size of their families in order to maintain their standard of living. He was livid at the "selfishness" of couples who cared "only for ease and gross material pleasures" and shrank from "the most elemental duties of manhood and womanhood." Students have often noted that Roosevelt, by trumpeting the duty of the family to the state, sounded like a protofascist. But in denouncing those who dared to place individual aspirations over their obligations to the nation, he also sounded like a wildlife manager upholding the good of the herd over the well-being of the individual.[26]

Given the infant mortality rate, and the number of Americans who could not have children, Roosevelt reckoned that patriotic couples would have to produce at least four children each to maintain the population of the United States. Upon these men and women "the whole future of the Nation, the whole future of civilization," rested. The president threatened that unless there was a turnaround, he would have to tax the unmarried, for otherwise "the race will in a short time vanish." As a frustrated Roosevelt told his friend Owen Wister: "They seem unable to see that it's simply a question of the multiplication table. If all our nice friends in Beacon Street, and Newport, and Fifth Avenue, and Philadelphia, have one child, or no child at all, while all the Finnegans, Hooligans, Antonios, Mandelbaums and Rabinskis have eight, or nine, or ten—it's simply a question of the multiplication table. How are you going to get away from it?"[27]

As for Madison Grant, he was a prime example of race suicide. A Teuton through and through, and as native an American as any white man on the continent, his loins produced exactly zero children during his lifetime. His sister Kathrin, who never married, also had no children. His younger brother DeForest did finally marry, at the age of forty-seven, but he too produced no children. His youngest brother Robert was the sole Grant sibling to procreate, and even he had only two children. The three brothers and one sister, who by President Roosevelt's reckoning should have produced sixteen children if civilization were to be saved, begat only two native-born Americans to compete with the peasant hordes from Europe. Grant's classmates from Yale, incidentally, were similarly infecund, producing only 1.7 children each.[28]

Grant's Taxonomy

Whatever his personal responsibility for the demographic disaster befalling the United States, Grant could hardly be accused of ignoring the issue. And whether or not my speculation is correct that William Z. Ripley's appearance at the Half-Moon Club was the catalyst, there is no denying that after 1908 Grant never again pursued research in natural history. Instead, he switched his scholarly focus from mammalogy to anthropology and resolved to do for the imperiled Teutons what he had done for so many other endangered species. Grant never felt that his background in zoology disqualified him from taking up the study of man. To the contrary, he declared that "man is an animal differing from his fellow inhabitants of the globe, not in kind but only in degree of development," and therefore "an intelligent study of the human species must be preceded by an extended knowledge of other mammals."[29]

In *The Races of Europe,* Ripley had expressed the hope that his book would render his anthropological findings "accessible to future investigators." Madison Grant was that future investigator. He evidently read and reread Ripley's book, and then used the two thousand titles in Ripley's supplementary bibliography to familiarize himself with the findings of the most prominent European anthropologists. He read John Beddoe and A. C. Haddon in English, Paul Topinard and René Collignon in French, Otto Ammon and Rudolf Virchow in German, Giuseppe Sergi and Ridolfo Livi in Italian, and countless other European anthropologists, and with his photographic memory made himself in a very short time an expert on physical anthropology. Grant was already familiar with some of this material, as America's leading anthropologist, Daniel Garrison Brinton, happened to be the father-in-law of his brother, DeForest Grant. Brinton held that the human species is divided into a graded series of races, with the simian-like Negro race at the bottom and the superior white race—which is the ideal "type" of the genus *Homo*—at the top. Whereas the myth of Adam and Eve had condemned previous generations to a blind acceptance of the brotherhood of mankind, a modern scientist like Brinton was no longer bound by such strictures and was free to proclaim what was obvious to anyone with eyes to see:

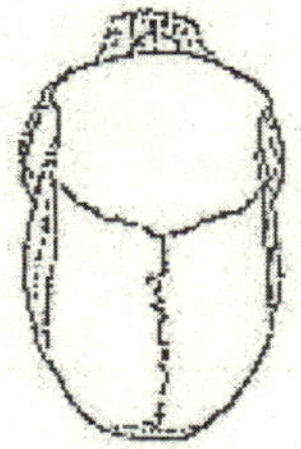

Dolichocephalic
(long headed)

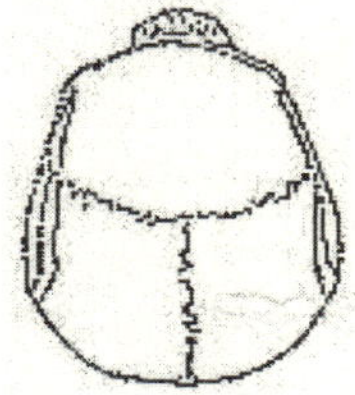

Brachycephalic
(round headed)

we are not related to each other, and we are not equal. According to Brinton, the races possess permanent physical and mental differences, and he warned that miscegenation between higher and lower peoples does not produce a blend of the two races but rather a specimen that reverts to a mentally inferior—and probably infertile—type.[30]

In an effort to categorize and rank the races of man, anthropologists had systematically and inexhaustibly measured all the parts of the human body (a practice known as anthropometry). But Grant soon discovered that they disagreed over which parts of the body were the most significant in identifying racial "types." Different scientists looked at different features, including skin color, hair color, hair texture, eye color, eye shape, ear shape, nose shape, nipple shape, body shape, body odor, body temperature, body stature, body lice, ratio of radius to humerus, distance between navel and penis, size of internal organs, rate of digestion, rate of pulse, strength of menstruation, strength of libido, size of genitals, size of brain, convolutions of the brain, and dimensions of the skull ("craniometry").

Craniometry was popular among anthropologists, as it was easy to perform (unless, like A. von Török, you insisted that five thousand separate measurements were necessary to classify properly a single skull); it could be applied to dead as well as living bodies; it was not as subjective as trying to categorize skin color; and the dimensions of the skull were thought to be resistant to environmental influences and hence reliable indicators of a subject's "true" race. One of the most prevalent craniometric measurements—and the one that Grant thought the most relevant—was the cephalic index, which is the ratio of the maximum width of the head to its maximum length. Hence, as the width of the head increases (that is, as the head grows rounder as viewed from above) the cephalic index increases. An index above 80 was classified as "brachycephalic" (roundheaded); below 75 was deemed "dolichocephalic" (longheaded).[31]

Assigning meaning to the cephalic index was not as easy as computing it. Scientists differed over which populations were roundheaded and which longheaded, and they argued over whether brachycephalic or dolichocephalic heads were the more intelligent. (Not surprisingly, there seemed to be a rather direct relationship between the shape of each scientist's skull and the particular shape he thought was superior—an example, for once, of correlation most definitely implying causation.)

And even though some twenty-five million Europeans had undergone anthropometric measurement by the mid–1890s, researchers could not agree on how many races there were, nor on what the "type" for each race should look like. (When William Z. Ripley asked German anthropologist Otto Ammon to supply a photograph of a "pure Alpine type," Ammon, who had personally measured tens of thousands of subjects, had to admit that he had never encountered one. "All his round-headed men," muttered a bewildered Ripley, "were either blond, or tall, or narrow-nosed, or something else that they ought not to be.")[32] Anthropologists, truth be told, could not even resolve on a *definition* of "race." Each scholar was therefore free to devise his own taxonomy of the human species. There is no need to peruse here each of these schemes in detail. (A mere enumeration of the number of races different scholars thought existed will provide a glimpse of the classificatory swamps into which such a discussion would lead us, perhaps never to escape: Cuvier claimed there were three human races, Saint-Hilaire thought there were four, Quatrefages five, Virey six, Peschel seven, Agassiz eight, Flower eleven, Mueller twelve, Saint-Vincent fifteen, Desmoulins sixteen, Topinard nineteen, Morton twenty-two, Broca twenty-seven, Deniker twenty-nine, Haeckel thirty-four, Crawford sixty, Burke sixty-three—and so on.)

To Madison Grant, with his love of measuring and compulsion for classifying, the findings of the physical anthropologists were a never-ending source of pleasure. Their inability to agree on a taxonomy of mankind did not overly perturb him. In fact, it allowed him, with all the misplaced confidence of the dilettante, to boldly propose his own scheme. In doing so, he relied mainly on six characteristics: stature, eye color, hair color, skin color, nose shape, and the cephalic index. He also took into account lesser traits such as texture of hair, length of limbs, shape of lips, shape and position of eyes, shape and proportion of jaw and chin, abundance of body hair, musculature, and instep. Using these indices, Grant concluded that the genus of mankind consisted of three species (which he sometimes called "subgenera"): the Caucasians, the Mongoloids, and the Negroids. These three species were in turn subdivided into numerous subspecies (or "races"). The Caucasians, for example, consisted of three subspecies: the Nordics, the Alpines, and the Mediterraneans. Grant borrowed the terms "Alpine" and "Mediterranean," of course, from Ripley. But he was not comfortable with the term "Teutons," which he felt had been unfairly expropriated by the nationalists of the Second Reich. So he adopted from Deniker the appellation *la race nordique,* anglicizing it to "Nordic," and after he employed it in *The Passing of the Great Race,* the term—along with Grant's entire taxonomy—was immediately adopted worldwide.

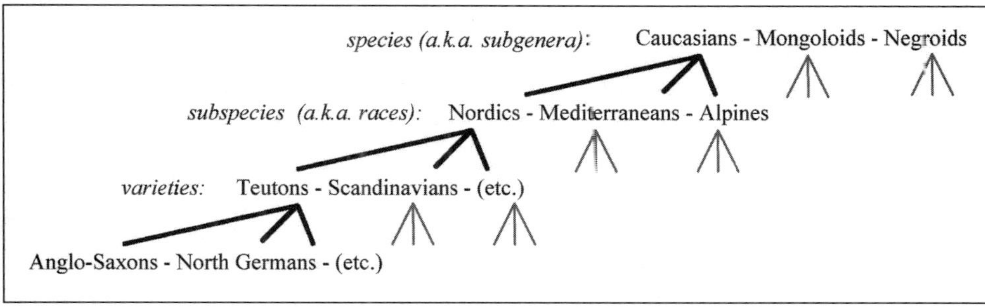

Grant's taxonomy of mankind.

At the same time that Madison Grant made the transition from mammalogy to anthropology, his friend Henry Fairfield Osborn, on a parallel track as always, made the shift—at Grant's urging—from ancient mammals to ancient men. Beginning in 1908, Osborn pronounced that paleoanthropology would be his "top priority" at the American Museum of Natural History, and thus it was that the museum became a center for researching and promulgating Grant's and Osborn's views on anthropology and evolution.

Aryanism

As Grant excitedly pursued his anthropological research, one of the ideas he found most congenial was the notion that the Nordics (who were called "Teutons" by Ripley and the Germans but "Aryans" by most Europeans) were superior to all other races. The Aryanists who had the most influence on Grant were three of the most intriguing figures in the Western canon: Arthur de Gobineau, Houston Stewart Chamberlain, and Georges Vacher de Lapouge.

Gobineau

Joseph Arthur, Comte de Gobineau (1816–1882) was a French diplomat and writer who established the stereotype of the superior Aryan race and was accordingly anointed by the Nazis as the founding father of racism. Gobineau was born (ironically enough, given his animosity toward the French Revolution) on the *quatorze Juillet,* into a bourgeois but royalist family in Ville d'Avray. Gobineau harbored a deep disdain for democracy (the family rumor was that Gobineau's mother was the daughter of the bastard son of King Louis XV), and his writings display a romantic longing to return to the hierarchy, tradition, and stability of the ancien régime. For Gobineau, urbanization, industrialization, miscegenation, and the June Days of 1848 were all proof that society—with its love of materialism and worship of equality—had deteriorated beyond repair.

With his personal charm and proficiency in languages, Gobineau was tabbed to be secretary to Alexis de Tocqueville during the latter's brief term as foreign minister in 1849 and, during the Second Empire, Gobineau held a number of important diplomatic positions abroad. In 1853, after familiarizing himself with the anthropological literature, Gobineau published his most influential work: the bitter, pessimistic, four-volume *Essai sur l'inégalité des races humaines* (published in English as *The Inequality of Human Races*).

According to the *Essai*, there are three human races: the white, the black, and the yellow. As the title suggests, they are far from equal. The black race is the lowest of the three, a fact that can be verified at a glance, as the black man is repellently ugly. (Truth be told, the majority of mankind is "a pestilent congregation of ugliness," but the Negro takes the palm in this department.) Commensurate with his physical repulsiveness, the black man behaves like a beast. "The animal character, that appears in the shape of the pelvis, is stamped on the negro from birth, and foreshadows his destiny. . . . What he desires is to eat, to eat furiously, and to excess; no carrion is too revolting to be swallowed by him." Furthermore, "his mental faculties," like those of animals, "are dull or even non-existent," and as a result the Negro is incapable of making "distinctions of good and evil." In fact, "He kills willingly, for the sake of killing."[33]

The yellow man is superior to the black, but no civilized society could ever be created by this race, for "the yellow man has little physical energy, and is inclined to apathy. . . . He tends to mediocrity in everything. . . . His whole desire is to live in the easiest and most comfortable way possible [and he has] a general proneness to obesity." Indeed, "the brutish hordes of the yellow race seem to be dominated by the needs of the body." In sum, when conceiving the yellow race, "clearly the Creator was only making a sketch."[34]

The white race is far and away the superior race, possessing a "monopoly of beauty, intelligence, and strength." Furthermore, whites are driven by a sense of honor (a word that, "together with all the civilizing influences connoted by it, is unknown to both the yellow and the black man"). Among other noble traits, whites are courageous, idealistic, and "have a remarkable, and even extreme, love of liberty."[35]

Because of miscegenation over the millennia, the three races no longer exist in their pure form. For Gobineau, therefore, the black, the yellow, and the white races are actually anthropological "types" to which current peoples correspond more or less closely depending on the amount of intermixture they have undergone. However, one branch of the white race—the Aryan—comes closer to the original type than any other. To the Aryans we owe everything "great, noble, and fruitful in the works of man on this earth." Thus, the great civilizations of India, Egypt, Greece, Rome, Germany, and even China (founded by an Aryan colony from India) are attributable to the blood of the Aryans, who formed the ruling classes of all these societies. "Such," declares Gobineau in his most quoted assertion, "is the lesson of history. It shows us that all civilizations derive from the white race, that none can exist without its help, and that a society is great and brilliant only so far as it preserves the blood of the noble group that created it."[36]

Disconcertingly, history also teaches us that no society has ever managed to preserve unsullied the racial purity of its ruling group. Like Madison Grant, Gobineau was obsessed with degeneration, which recurs in his *Essai* as a tragic paradox of history that cannot be avoided. For whenever a race like the Aryans becomes strong enough, it naturally subjugates the inferior races surrounding it; but it soon engages in miscegenation and begins to decline. And so we see time and again in history the weak races being conquered from without by the arms of the strong, and then the strong being conquered from within by the genes of the weak. The Roman Empire, for example, fell because the inferior races managed to sully the purity of the blood of the patriciate. And behind all the other significant events of history lurks a similar racial explanation. What the class struggle is to Marx, racial conflict is to Gobineau; for him, the story of man is the story of race, and "the racial question overshadows all other problems of history."[37]

Gobineau's oft-expressed fear of being "swamped by the influx of foreign elements" explains why he was opposed to the imperialism of his day, and why he would have been horrified by the American melting pot of Grant's time. As it was, Gobineau was sufficiently disgusted by Jacksonian America, with its democratic ideals and mongrel population composed, in his words, of "the detritus of all the ages." As for his own countrymen, Gobineau observed with incredulity that the French continued to cling "to the liberal dogma of human brotherhood" and that they actually believed that inferior races could somehow be elevated through education and other environmental reforms. He was dismayed by the gullibility of those who heard of Negroes that knew "how to read, write, count, dance, and speak like white men" and therefore concluded that "the negro is capable of everything!" He reminded his readers that the ability to imitate civilization is quite different from the ability to *create* it. If all races were equally capable, then why "in the course of ages, has [the American Indian] not invented printing or steam power?" No, it was Gobineau's task to teach the world that inferior racial traits were permanent and incapable of amelioration: "The savage races of today have always been savage, and we are right in concluding . . . that they will continue to be so, until the day when they disappear."[38]

By "savages," Gobineau did not just mean the natives of darkest Africa or the American West. Take, for example, the "brutal and ignorant" peasants of France: "Different governments have made the most praiseworthy attempts to raise the peasants from their ignorance. . . . Yet the agricultural classes learn nothing whatever." In fact, Gobineau's analysis of history showed that the French people were anthropologically two different groups: the nobles, who were descended from the Aryans, and the rabble, who were of Gallo-Roman stock. In the French Revolution, the inferior rabble overran the superior Aryans, with predictable results for the course of the nation. The lesson is a universal one: whenever the Aryan aristocracy is submerged by "inferior stocks," the nation is doomed.[39]

If only the primordial white race had never intermarried, then "the yellow and black varieties would have crawled for ever at the feet of the lowest of the whites." But this was not to be, and it was certainly too late to do anything about the Aryans' decline at this juncture. Indeed, history was already nearing the end, and in another three or four millennia "the white species will disappear henceforth from the face of the earth," at which point "the lifeless earth will continue, without us, to describe its apathetic orbits in space."[40]

It is safe to assert that the count was not a hopeful man. And in a century that was increasingly democratic and optimistic, Gobineau's aristocratic pessimism was not well received. The *Essai sur l'inégalité* was initially all but ignored in France. In 1843, in the first letter de Tocqueville ever wrote to Gobineau, he confessed to his young protégé that "one is fascinated both by what you could be and by what one fears you may become." Ten years later, after reading the *Essai,* Tocqueville's fears were realized: "I believe that [your doctrines] are probably quite false; I know that they are certainly very pernicious." Gobineau replied that his book presented anthropological truth, and that any sentimentalist who rejected it lacked the learning and the courage to deal with the cold facts of science.[41]

Not unexpectedly, Gobineau did achieve some success in the antebellum South, where a truncated translation of his book appeared in 1856 specifically to bolster the pro-slavery argument. After the Civil War, however, Gobineau was forgotten in the United States, until Madison Grant revived his theories. Grant appreciated Gobineau for obvious reasons. To be sure, Gobineau's epistemology was pre-Darwinian (and when *The Origin of Species* appeared in 1859, Gobineau, as a good Catholic, rejected it: "Nous ne descendons pas du singe," he is reported to have said, "mais nous y allons").[42] Ordinarily, this would have made him anathema to Grant, the stalwart Darwinian. But Gobineau's identification of race as the key to history, his ascription of different mental traits and abilities to different races, his espousal of nature over nurture, his defense of aristocracy, and his application to mankind of zoological concepts (including typology, degeneration, and the evils of crossbreeding) all resonated with Grant. And Grant, not incidentally, admired the verve of Gobineau's prose (which would also earn the admiration of critics such as Gide, Proust, and Cocteau).[43]

Aside from Madison Grant, it was among nationalists in Germany—whose Aryan civilization was deemed by Gobineau to be "one of the most glorious monuments ever erected by the genius of man"—that the count had the greatest influence. Both Renan and de Tocqueville had predicted to Gobineau that this would be the case, as they observed that the Germans seemed to be more obsessed than the French with racial matters. Princess Carolyn von Sayn Wittgenstein confirmed the German view of Gobineau: "You consider yourself a man of the Past. I am firmly convinced that you are the man of the Future." And yet, the German embrace of Gobineau is not as obvious as it might seem. For one thing, Gobineau glorified the Germans as a racial rather than a national category, and he did so not for the further glory of Prussia but in order—à la Tacitus—to hold up to his own degenerate countrymen an idealistic standard of virtue. Further, Gobineau contended that thanks to miscegenation even fewer Aryans remained in modern Germany than in modern France. And, most importantly, Gobineau's profoundly pessimistic philosophy explicitly denied the efficacy of political action. The forces of history, intoned Gobineau, are "lead[ing] societies down to the abyss of nothingness whence no power on earth can rescue them." He had no program to offer, for there was nothing that anyone could do to halt the process of racial degeneration; and were it not so tragic, we might appreciate the irony that a man with *no* solution would come to be hailed as the messenger of the *final* solution.[44]

And yet, rational or not, a number of Germans were entranced by Gobineau. Schopenhauer was of that company, as was Nietzsche, whose sister read the *Essai* aloud to him (whereupon Zarathustra addressed his disciples: "Tell me brothers: what do we account bad and the worst of all? Is it not *degeneration*?"). Richard Wagner enthusiastically hosted Gobineau at Wahnfried in 1876, and again for lengthy visits in 1881 and 1882. "My husband," Cosima Wagner informed Gobineau, "is quite at your service, always reading *The Inequality of Human Races* when he is not at work with the staging." In return, a grateful Gobineau assured Wagner that *The Ring* was the musical embodiment of his racial theories. In 1894, Wagner's followers founded the Gobineau Vereinigung to translate the count's works and introduce his theories to the German public. One receptive reader was Adolf Hitler, whose debt to Gobineau is evident in *Mein Kampf*:

> History . . . shows, with a startling clarity, that whenever Aryans have mingled their blood with that of an inferior race the result has been the downfall of the people who were the standard-bearers of a higher culture. . . . All the great civilizations of the past became decadent because the originally creative race died out, as a result of contamination of the blood. . . . Every manifestation of human culture, every product of art, science and technical skill, which we see before our eyes today, is almost exclusively the product of the Aryan creative power. . . . Should he be forced to disappear, a profound darkness will descend on the earth; within a few thousand years human culture will vanish and the world will become a desert.

Gobineau was subsequently pointed to by all the Nazi race experts as their original inspiration. In 1939, the Third Reich brought out a new edition of the *Essai,* and extracts were assigned in German schools and reprinted in popular anthologies.[45]

Chamberlain

Until Hitler, Gobineau's most important follower was Houston Stewart Chamberlain (1855–1927), who added anti-Semitism and a patina of biology to Aryanism, and thereby earned a place next to Gobineau in the Nazi pantheon. Born into an aristocratic English family, Chamberlain's father was an admiral in the British navy, and all four of his uncles were army officers of high rank. In his early teens, Chamberlain suffered a breakdown at school and was sent to the Continent to recover. He never again lived in England. Declaring himself a disciple of "an incomparable master, Charles Darwin," he studied botany at the university in Geneva but had to give up his dream of becoming a scientist after suffering a second breakdown at the age of thirty. Fortunately, he received financial support from his family and had no need to work.[46]

Over the years, Chamberlain felt increasingly drawn to the history and culture of Germany. He married a German woman, and in 1885 they moved to Dresden, arriving just a few months after young Madison Grant had finished his studies there and returned to the United States to attend Yale. Like Grant, Chamberlain would spend four years in Dresden, where he became absorbed in the music and philosophy of Richard Wagner, who had served there as royal kapellmeister. Chamberlain published numerous books in German on the composer and his compositions, and became the chief popularizer of the Bayreuth cult.

Although Jews comprised just 0.27 percent of the population of Saxony, the area at that time was a hotbed of anti-Semitism. Chamberlain's circle of friends included a number of angry conservatives who accused the cosmopolitan Jews of being behind all manner of threats to the area's traditional patterns of life—threats that included the rise of modern industrialism, bourgeois materialism, scientific socialism, and liberal parliamentarism. Chamberlain's budding anti-Semitism was confirmed when he moved to Vienna in 1889. Though he opted to live there for the next twenty years, Chamberlain, like Adolf Hitler, was horrified by the polyglot city, which was experiencing a massive influx of Jews from the east. These Israelites, with their strange costumes and bizarre customs, were dirty and disgusting to Chamberlain, and he began to see them as human vermin who threatened the health of the native Aryans. The Jews were like an agricultural "pest," warned Chamberlain, and their immigration must be cut off lest they "spread like a poison over the whole earth."[47]

Like so many anti-Semites and eugenicists in both Germany and the United States, when Chamberlain warned of the danger posed by Jewish immigrants, his terminology revealed that the underlying fear was that the Jews were a threat not just to the body politic but to the actual human body. The influx of unclean Jewish immigrants was often likened to an invasion of germs, and men like Chamberlain, obsessed with the health and cleanliness of their own body, seemed to project all their fear of pollution and contamination onto the poisonous Jews. Hence the constant refrain that society must prevent miscegenation in order to maintain the "purity" of the blood. Chamberlain's idol (and fellow vegetarian) Richard Wagner, for instance, who referred to the Jews as a people mired in the "ferment of decomposition," insisted that humanity's only salvation was to "cleanse the blood of the human race from all impurities." Chamberlain, taking up Wagner's obsession with purity, accused the Jews of encouraging their daughters to marry outside the race as part of a plot to "infect the Indo-Europeans with Jewish blood." He called on his fellow Teutons to purge their blood of Jewish "poison" lest the white race turn into "a herd of pseudo-Hebraic mestizos, a people beyond all doubt degenerate physically, mentally and morally."[48]

At the suggestion of Cosima Wagner, who more or less adopted Chamberlain after her husband's death, the expatriate Englishman studied Gobineau's *Inequality of Human Races,* which he assessed as "brilliant." Chamberlain now realized that Vienna, like imperial Rome, was "a union of mongrels" that was going to destroy itself through miscegenation. He proceeded to peruse the books of the most important contemporary anthropologists, and in 1899 he published his two-volume, 1,500-page opus, *Die Grundlagen des Neunzehnten Jahrhunderts* (published in English as *The Foundations of the Nineteenth Century*), which an enchanted Kaiser Wilhelm II immediately pronounced a marvelous "Hymn to Germanism" (and a dissenting Ashley Montagu some years later referred to as "that stupendous miracle of nonsense").[49]

The Foundations of the Nineteenth Century is a sweeping analysis of European culture showing that the wellspring of all history is race. Verily, the idea that *race is all* reaches its apotheosis in Chamberlain's dictum: "There are no good and bad men. . . . On the other hand there are certainly good and bad races." Like Gobineau, Chamberlain asserts that anything good or important in modern civilization was created by the Teutonic branch of the Aryans, and that "the importance of each nation as a living power today is dependent upon the proportion of genuinely Teutonic blood in its population." The Teutons are for the most part tall, fair, and dolichocephalic; but above and beyond any particular physiognomy, Teutons are imbued with an idealistic spirit, a virile sense of loyalty, and an enduring love of freedom.[50]

The Teutons are "by right . . . the lords of the world," and yet ever since the thirteenth century the creative and regenerative spirit of the Teutonic race has been locked in a dialectic "struggle for life and death" with the negative and destructive force of the Jews. History teaches us that the Teutons and the Jews stand "always as alien forces face to face" separated by a "deep abyss." Jews are the bearers of a "peculiar and absolutely un-Aryan spirit"; they are the personification of "negation"; they are the antithesis. Jews are "the enemy, open or secret, of every other human being, and a danger to every culture." Their goal is to "put [their] foot upon the neck of all the nations of the world and be Lord and possessor of the whole earth."[51]

Chamberlain denied any personal animus against individual Jews, and in his introduction he attacks the "perfectly ridiculous and revolting tendency to make the Jew the general scapegoat for all the vices of our time"—and he then spends some five hundred pages doing exactly that. According to Chamberlain, the Jews have made no discoveries in science or exploration, and they have contributed nothing to art or philosophy. As to the Ten Commandments, they were actually of Egyptian origin. (Also, Saint Paul was almost certainly not a pure Jew, a fact clear to anyone who studies the "deepest inner nature" of the apostle. King David was at least three-quarters Aryan, as proven by his biblical description as "ruddy and of a fair countenance," plus the fact that "his daring [and] his spirit of adventure" are hardly Jewish traits. And Christ himself, who manifested all the noble traits of a typical Aryan, "had not a drop of genuinely Jewish blood in his veins.") The Jews do possess "an abnormally developed" understanding of high finance and thus play a "dominant" role in modern society; nevertheless, the "parasitic" Jews are and always will be an "alien element . . . in our midst."[52]

Chamberlain was deeply fearful that Teutonic blood was deteriorating because of interbreeding with the Jews. In his book he warns that miscegenation is "an incestuous crime against nature; it can only be followed by a miserable or a tragical fate." One only has to look at the pitiful mestizo states of South America—whose "mongrel" populations are on the verge of falling into "bestial barbarism"—to see what happens when two races engage in "unnatural incest." (This observation, coming just one year after the United States seized suzerainty over such peoples in the Spanish-American War, gave Madison Grant pause.) But the foremost example of miscegenation, of course, is ancient Rome, which succumbed to racial degeneracy when Caracalla extended the franchise throughout the empire. "Like a cataract the stream of strange blood overflooded" the eternal city, transforming it "into the *cloaca gentium,* the trysting-place of all the mongrels of the world." Rome soon became "a chaos of unindividualised, speciesless human agglomerates"—in a word, a *völkerchaos.*[53]

And there was every indication that nineteenth-century Germany was itself slipping into "chaotic mongreldom." Recent anthropological investigations had shown that "the short round-skulls are constantly increasing in numbers and so gradually superseding the narrow dolichocephali." But unlike Gobineau, Chamberlain was not pessimistic. For one thing, by positing the Jews as an "other" against which the Teutons could struggle, Chamberlain introduced action and even optimism into Gobineau's defeatist schema. For another, he thought that the precepts of biology could save the Teutons. Having studied intently the works of Charles Darwin, Chamberlain understood that man is simply an animal among other animals, and thus reasoned that the same practices employed in animal husbandry could be used in the propagation of human beings. In his book he points out that thoroughbred race horses and champion hunting dogs are not "produced by chance and promiscuity" but by "artificial selection and strict maintenance of the purity of the race." Hence, by scientifically planning our matings, and by outlawing interracial couplings, the degeneration of the Teutons can be brought to a halt. And as for any purebred Germanic babies who nonetheless exhibit signs of weakness or inferiority—well, Chamberlain reminds his readers that killing sickly babies by exposing them to freezing temperatures was one of "the most beneficial" customs of the ancient world.[54]

According to Chamberlain, the science of anthropology has just as much to teach us as biology, for the one great achievement of modern anthropology is its conclusive proof of the inequality of mankind. Craniometry has shown that the shape of the skull (which is inherited "with ineradicable persistency") differs from race to race, and that certain races are mentally superior to others. The notion of "equality" is a temporary and unfortunate product of Enlightenment sentimentality, a "foolish humanitarian day-dream." It reached its apotheosis during the catastrophe known as the French Revolution. "Let us hope the day may come when every sensible person will know the proper place for such things as the Declaration of the Rights of Man, namely, the waste-paper basket."[55]

Yet for all his admiration for science, Chamberlain's concept of race is at its core more mystical than physiological. After all, he assures us, any true German can *intuit* that Luther was a fellow Teuton—not by the shape of his skull, but because he can "picture this man fifteen hundred years ago, on horseback, swinging his battle-axe to protect his beloved northern home." Similarly, irrespective of what science may tell us, it is clear that Francis of Assisi was a Teuton, as was Dante, a fact "so clear from his personality and his work that proof of it is absolutely superfluous." No, we do not need calipers and skin-color charts to tell us about race. "It frequently happens," Chamberlain confidently informs us, "that children who have no conception of what 'Jew' means, or that there is any such thing in the world, begin to cry as soon as a genuine Jew or Jewess comes near them!"[56]

In sum, *The Foundations of the Nineteenth Century* was a reheating of Gobineau's racist stew, with a pinch of animal husbandry, a dash of craniometry, and a large dose of romantic anti-Semitism. Like Madison Grant, Chamberlain was a popular synthesizer whose work was a patchwork of random borrowings from scholarly sources. The end result was vague, confused, and contradictory, but (as would be the case with the anthropological books of Madison Grant) few members of the reading public were qualified to refute the seemingly profound and erudite assertions of Herr Chamberlain. *The Foundations* was a huge success, not just with *völkisch* and Pan-German groups but also with the public at large. The book attracted admirers from Albert Schweitzer to George Bernard Shaw (who called it "a masterpiece of really scientific history"). It was an especial favorite of Wilhelm II, who summoned Chamberlain to a private audience to thank him "for what you have done for Germany." According to the histrionic kaiser, Chamberlain's insights "bring order into the confusion, light into the darkness . . . and reveal the paths which must be followed for the salvation of the Germans and thus the salvation of mankind! . . . God sent your book to the German people and you personally to me." He ordered *The Foundations* to be read aloud at court, and arranged for free copies to be distributed to officers of the army and displayed in schools and libraries throughout the Reich. When it became known that Chamberlain was "the kaiser's anthropologist," a familiarity with *The Foundations* became de rigueur among the Prussian elite. The book went through three editions in the first year alone, and by World War I had sold one hundred thousand copies and been translated into English, French, and Czech.[57]

American author Carey McWilliams was one of many observers who correctly deemed Madison Grant to be Chamberlain's "most influential disciple in this country." Although Grant, as a rational American, rejected Chamberlain's quasi-mystical *völkisch* nationalism, he did learn much about anti-Semitism from *The Foundations,* and many of Chamberlain's ideas found their way into *The Passing of the Great Race.* Several other Americans were impressed by *The Foundations,* including Ellery Sedgwick (editor of *Atlantic Monthly*), Senator Albert T. Beveridge (who visited Chamberlain in Bayreuth in 1915), political scientist John Burgess of Columbia (who was sent a copy of *Foundations* by the kaiser), Theodore Roosevelt (who urged that Chamberlain's statement of the inequality of races was a useful counterweight to the "thoroughly pernicious doctrines taught by well-meaning and feeble-minded sentimentalists"), and historian Carl Becker (who raved about Chamberlain's "intellectual mastery" and "brilliant originality," and predicted that "among historical works" it was "likely to rank with the most significant of the nineteenth century").[58]

As for the now-famous Chamberlain, he was fed up with Vienna, and moved to Bayreuth in 1909 to be closer to Cosima and the Wagner circle. In short order he divorced his first wife and married Wagner's daughter (and Franz Liszt's granddaughter), Eva. Despite two marriages, and his assertion that "to beget sons, sons of the right kind, is without question the most sacred duty of the individual towards society," Chamberlain produced no children.

When the Great War broke out, Chamberlain explained to the kaiser that the chief cause of the conflict was the unquenchable ambition of "*Judentum* and its near relative Americanism for the control of the world." He renounced his British citizenship and spent the war years churning out extremist propaganda contrasting the freedom-loving Germans to the Semitic-loving Allies, who had fallen "totally into the hands of the Jews." In the middle of the war, the Reich created a special civilian version of the Iron Cross to reward "the sage of Bayreuth" for his services to the Hohenzollern dynasty.[59]

Chamberlain was deeply depressed by the defeat of 1918 and deteriorated rapidly. Like Madison Grant, for the last ten years of his life Chamberlain was

Houston Stewart Chamberlain, like Madison Grant, was confined to his bed for the last decade of his life. Adolf Hitler admired the writings of both men.

largely bedridden (with a painful form of multiple sclerosis or Parkinson's disease, which partly paralyzed his limbs). He continued to correspond with the exiled kaiser about the destructive poison of *Judentum,* and his spirits were buoyed in the early 1920s by the rise of the Teutonic savior, Adolf Hitler. Geoffrey Field reminds us, however, that the courtly, refined Chamberlain never explicitly advocated violence against Jews. To be sure, he favored legislation restricting both their immigration and their rights, but his anti-Semitism was largely academic, not activist. The goal of his rhetoric, like that of Gobineau, was more to inspire his (adopted) countrymen than to bring about the destruction of the Jews. Still, Chamberlain was the first prominent figure to endorse the Nazi movement, and the Führer never forgot that after the failure of the Beer Hall Putsch the ailing Chamberlain continued to send him encouraging letters in Landsberg.

Houston Stewart Chamberlain, who was born ten years before Madison Grant was born, died ten years before Grant died, in 1927, in Bayreuth. The funeral was attended by numerous notables, including a somber Adolf Hitler and his lieutenants.

Lapouge

Other than Gobineau and Chamberlain, the foremost exponent of Aryanism in Europe was the anthropologist Georges Vacher, Comte de Lapouge (1854–1936). A disciple of both Gobineau and the German zoologist Ernst Haeckel, Lapouge was the founder of the French school of *anthroposociologie,* an ultraconservative movement that interpreted all history in terms of racial struggle and sought to organize society along zoological principles. Lapouge sought to replace "the fictions of Liberty, Equality, and Fraternity" with "the reality of Forces, Laws, Races, Evolution." Indeed, as Lapouge told Madison Grant: "I have commenced to conceive a religion out of evolution." By applying the laws of biology, he hoped that his anthroposociology would forestall "the decadence of the nation and the deterioration of the racial stock."[60]

Lapouge was (and remains) virtually unknown in the United States, but his work was esteemed by the Nazis and the Vichy regime, and he exerted a great deal of influence on Madison Grant. Like Grant, Lapouge—whose family had comprised the nobility in Poitiers for four centuries—had been a childhood naturalist who had originally trained as a zoologist. The two became friends and corresponded regularly for decades, until Lapouge passed away one year before Grant. Lapouge was far from an original thinker, and added little to the theories of Gobineau and Chamberlain, but Grant respected him because he was widely conversant in history, sociology, biology, and anatomy, and he was fluent in Latin, Greek, Hebrew, Chinese, and Japanese. Grant also appreciated Lapouge because his anti-Semitism, while fairly crude, exhibited none of the mysticism of Houston Stewart Chamberlain, but instead seemed to be grounded in firm zoological principles. Among other things, Lapouge was renowned for his anthropometric skills ("Good observers," bragged Lapouge, "can recognize the race of a woman merely by examining her genital organs"—and the amount of energy that Lapouge devoted to discussing, in detail, the female genitalia certainly indicates that he was as conscientious an observer as any). Henry Fairfield Osborn considered Lapouge "the leading authority on racial anthropology," and the kaiser (that unerring judge of character) hailed Lapouge as "the only great Frenchman."[61]

In his main work, *L'Aryen: son rôle social* (1899), Lapouge agrees with Gobineau that France is divided into two antagonistic races. Unlike Gobineau, he claims he has the anthropometric measurements to prove it. The aristocrats of France, according to Lapouge's data, are tall, blond, dolichocephalic Aryans, whereas the peasants are short, brown, brachycephalic Alpines. The two races differ not just physically but mentally as well. The intelligent, Protestant Aryans are by far the superior race, with many eminent Frenchmen to their credit. Ambition, courage, the ability to reason, an aptitude for command—all these are inherited traits peculiar to the Aryans. In truth, all that is good in the world can be attributed to this innately superior race. And it is certainly no accident that the most commercially enterprising countries (England, Holland, and the United States) are also predominantly Aryan. (Lapouge was a sort of Max Weber with calipers.)

Opposed to the Aryans are the brachycephalic Alpines, "a cautious, unprogressive race." One would be hard-pressed to name a single Alpine who has achieved distinction in science, literature, or politics. If the Aryans are the brains of civilization, the Alpines are the muscles. Slow, servile, dull, the Catholic Alpines are incapable of comprehending the spirit of liberty that was invented by the Aryans. "The ancestors of the Aryans," claims Lapouge, "were already cultivating wheat when those of the brachycephalics were still living like monkeys." And make no mistake: the inferiority of the Alpines is genetic. It is nature, not nurture, that creates the man, and the sentimental belief in the equipotentiality of races is a relic.[62]

According to Lapouge, the Aryans and the Alpines have always been, and still are, locked in a bitter conflict. (The French Revolution, for example, "was above all a replacement of the dolicho-blonds by the brachycephalics.") Thus, what the proletarians, egged on by their academic tribunes, mistakenly perceive as a class struggle is actually, subconsciously, a race struggle. And Lapouge is not optimistic about the outcome of the contest, for the superior Aryans are being rapidly outbred by the inferior Alpines. As Lapouge told Grant, if present demographic trends continue much longer, "France will cease to be France." The problem is that the superior "dolicho-blonds" unwittingly sow the seeds of their own destruction, not, as Gobineau thought, by interbreeding, but—in a Gallic version of "race suicide"—by *refusing* to breed.

Lapouge is at great pains to remind us that in evolution there is a difference between the *best* and the *fittest.* The Aryans are unquestionably the best race, but they are not as fitted to the conditions of modern civilization as are the roundheads. "Dark, short, heavy, the brachycephalic reigns today from the Atlantic to the Black Sea. As bad money chases good, so the brachycephalic has replaced the superior race. He is inert, he is mediocre, but he multiplies" (a process encapsulated by the "Law of Lapouge," which states that "bad blood drives out good"). Furthermore, the brachycephalics have an ally: the parasitic Jew. To be sure, the morphology of the Jew reveals that he has no creative power and no capacity for original thought. But the Jew from birth is a sly speculator and a greedy swindler, and when his financial ability, so well suited to modern plutocratic society, is combined with the breeding ability of the brachycephalics, it is not difficult to see why the Aryans have chosen to withdraw from the demographic contest.[63]

The Alpines and the Jews sense that the Aryans are declining, and cunningly maintain their power by trumpeting the virtues of democracy—a game that the numerically inferior Aryans can never win. Universal suffrage, Lapouge told Grant, means only one thing: "Power passes into degenerate hands." To make matters even worse, France in the twentieth century was being inundated by its colonial subjects. In an example of what Lapouge called "colonization in reverse," Moroccans, Senegalese, and Indo-Chinese were migrating to France, exercising the franchise, and (even worse) intermarrying with French women—which was precisely why Gobineau had warned against imperialism.[64]

Lapouge is not certain what can be done to save society, though he does allude to the Spartan practice of exposing sickly infants. In any event, if emergency measures are not taken, the Comte de Lapouge will have to concur with the prediction of his colleague Otto Ammon that Western civilization will soon witness an *Aryanendämmerung*.[65]

The works on physical anthropology and Aryanism that Madison Grant was reading as he moved from mammals to man share a number of properties. First, the stipulation that man is part of the natural world and that it is proper to apply to *Homo sapiens* the zoological concepts of typology, degeneration, and the dangers of crossbreeding. Second, the claim that the key to history is not class struggle but race struggle. Third, the ascription of different mental traits and abilities to different races, and the consigning of the Jews to a subordinate place in the hierarchy of mankind. Fourth, the assurance that environmental reforms can have but little effect on the inherent nature of man. And fifth, the realization that the conditions of modern civilization are permitting the inferior races to outbreed the superior races.

In the early 1910s, Grant was assimilating these principles, and thinking about their relevance to the future of the United States. As for Count Lapouge, he was fully cognizant of where all this was headed: "The conflict of races is now about to start," he correctly predicted in 1887. "I am convinced that in the next century people will slaughter each other by the million because of a difference of a degree or two in the cephalic index."[66]

6

The Eugenics Creed

A good tree cannot bring forth evil fruit, neither can a corrupt tree bring forth a good fruit.

Matthew 7:17, 18

One year after William Z. Ripley's lecture at the Half-Moon Club confirmed for Madison Grant the dangers that immigration posed to the Nordic race, Grant had to face another kind of crisis. On the morning of Sunday, June 13, 1909, his sister Kathrin Manice Grant died suddenly, at the age of just thirty-six. It was, of course, a shocking personal blow to Grant, and he spent the entire summer in mourning. Madison was the firstborn, and Kathrin the only daughter; hence it had always fallen to those two to care for their aging parents and two brothers. Neither he nor Kathrin had ever married; both had grown up and lived in the same house together all their lives. (In fact, for decades all the Grant siblings, their parents, and their wives shared the same residence on East Forty-ninth Street, one block east of where Rockefeller Center is today. It was not until 1926 that Madison, at the age of sixty-one, finally moved into a place of his own—though he did not stray far: his new residence on Park Avenue, across the street from the Waldorf-Astoria Hotel, was just around the corner from the family home, and his brother DeForest moved in next door to Madison.)

In addition to the disturbing speculations of Ripley, Gobineau, et alia about the impending demise of the Nordic race, Grant now had to deal with an actual death within his own family. The dying of his younger sibling must have given the forty-four-year-old patrician, possibly for the first time, intimations of his own mortality.

Stunningly, just five months later, on November 8, Grant's father, Dr. Gabriel Grant, also passed away. It was obviously a time of great emotional turmoil for Madison. To the eldest son, the death of his father—the paterfamilias and heroic recipient of the Medal of Honor—was devastating. Grant went into seclusion for some months, and we can guess that during this time his thoughts turned to the subjects of death, impermanence, and extinction. And the passing of the sister and the passing of the father may well have become intertwined with thoughts of the passing of the race.

I do not think it an accident that it was around this time that Grant, searching (I believe) for personal and philosophical solace, embraced eugenics. As someone who lacked religious convictions, and who liked to think of himself as a scientist, eugenics offered to Grant—in a time of personal anguish—a body of thought that provided meaning to his existence and direction to his life.

A Faith Is Born

Eugenics was founded by the remarkable Sir Francis Galton (1822–1911), who believed that personality characteristics—including intelligence—are just as heritable as physical characteristics. Born into a wealthy family in Birmingham, England, Galton trained as a physician at the order of his father. But, like Houston Stewart Chamberlain, he was plagued at school (and for the rest of his life) by constant headaches and the occasional nervous breakdown. Luckily, like Chamberlain, he was financially independent and never had to work for a living. He left Cambridge without completing his degree and spent the rest of his existence, like Madison Grant, exploring, hunting, and enjoying the life of a gentleman scientist.

Galton was obsessed with numbers and was one of the founders of the field of statistics; he devised a number of important tools used to this day (including the correlation coefficient and the law of regression). He also satisfied his predilection for quantification by engaging in less academic pursuits, such as using a sextant to covertly measure the buttocks of Hottentot women, tabulating the number of fidgets in an audience to determine its level of boredom, and constructing a "beauty map" of Great Britain by assigning a rating of "attractive," "indifferent," or "repellent" to the women he passed on the street (it turned out that London had the most beautiful women, and Aberdeen the least). And, being British, he also collected reams of data on all the variables (temperature of the water, amount of water, amount of tea, etc.) that went into making the perfect cup of tea. Galton seemed to be interested in anything and everything, as long as numbers and measurements were involved. During his fascinating career, he led expeditions to the tropics to determine the proper longitude and latitude of British possessions, established fingerprinting as a method of criminal identification, pioneered in the use of twins for genetic research, connected points of equal temperature and pressure and thereby invented weather maps, and actually published a scientific article discussing the optimum ways of cutting a cake to preserve its freshness.

After the 1859 publication of *The Origin of Species* by his cousin Charles Dar-

Sir Francis Galton: Darwin's cousin and the childless founder of eugenics.

win, Galton became interested in evolution and the rules of heredity. He conducted experiments with sweet-pea plants to gather statistical data on the subject. Unbeknownst to Galton, Gregor Mendel (who was born in the same year as the Englishman) was concurrently engaged in the same course of action in the garden of the Augustinian monastery at Brünn.

In 1865, Galton focused his enormous energies on the subject of human heredity and published "Hereditary Talent and Character," the founding document of the eugenics movement. Eighteen sixty-five was coincidentally the year that Mendel announced to an unreceptive world *his* findings on heredity. (Of slightly less import, it was also the year that Madison Grant was born.) In "Hereditary Talent and Character" (expanded four years later into his most famous book, *Hereditary Genius*), Galton used biographical dictionaries and memoirs to analyze the pedigrees of hundreds of prominent men, and discovered that intelligence (which he called "talent") seemed to run in families. According to his computations, one-half of the men who had superior talent had at least one near relative who was equally accomplished in some field. Furthermore, specific abilities seemed to run in families. By Galton's reckoning, for example, one out of every ten musicians had a near relative who was also a famous musician, one in six famous writers had a distinguished literary relative, and so forth. Given that great ability is a rare commodity (the distribution of talent among the general population following the dictates of a bell-shaped curve), these frequencies were much higher than would be statistically expected. Indeed, whether he looked at the families of judges, statesmen, military commanders, scientists, artists, or clergymen, the facts bore out Galton's contention that "talent is transmitted by inheritance in a very remarkable degree."[1]

Galton felt he had thereby proved that "intellectual capacity" was inherited in much the same manner as physical features. He granted that one's childhood environment and family connections could have an effect on the position one eventually attained in society; but favorable nurture could hardly account for the fact that even in the "open" fields of science, literature, and the law—where, claimed Galton, ability, not connections, was the determinant of success—a high proportion of eminent practitioners had brothers and sons who were also famous scientists, writers, and lawyers. "Everywhere," concluded Galton, "is the enormous power of hereditary influence forced on our attention," a fact he summed up most famously in his declaration that "nature" is more important than "nurture" in determining man's character.[2]

Given that intelligence and ability were primarily hereditary, Galton propounded that the human race could be drastically improved in just a few generations if marriages were planned properly. "If talented men were mated with talented women . . . generation after generation, we might produce a highly-bred human race." Conversely, if alcoholics, paupers, and criminals were *prevented* from marrying, their unfortunate traits could be bred out of the race. Like Darwin and Houston Stewart Chamberlain, Galton argued that just as superior breeds of domesticated animals can be created by "preventing the more faulty members of the flock from breeding, so a race of gifted men might be obtained, under exactly similar conditions." Little could he foresee the consequences that would flow from that seemingly inoffensive phrase: *preventing the more faulty members of the flock from breeding.*[3]

In 1883, Galton coined the term "eugenics" to describe his program of improving the human race through controlled breeding. To be successful, eugenics would have to employ a two-pronged strategy: "positive eugenics" would encourage the fittest members of society to have more children, while "negative eugenics" would discourage the propagation of the unfit. (Eugenicists of lesser refinement than Sir Francis would later refer to the positive and negative aspects of their program as "breeding and weeding.") As examples of positive measures, Galton suggested that "worthy" young men and women could be issued certificates verifying their eugenic fitness. Philanthropists could then dispense bonuses to eugenically suitable couples that agreed to marry, and the state could sponsor public marriages of these couples in Westminster Abbey and award grants for every child they produced. The price would be well worth it: "If a twentieth part of the cost and pains were spent in measures for the improvement of the human race that is spent on the improvement of the breed of horses and cattle, what a galaxy of genius might we not create!"[4] (It is of more than passing interest, in view of his continual espousal of large families, that Galton, like Houston Stewart Chamberlain, and Madison Grant, never fathered any children of his own.)

Negative eugenics could be as innocuous as adopting a taboo against unfit couplings, similar to the one whereby marriages between cousins are frowned upon. If social sanctions proved ineffective, the state might have to step in and segregate those persons deemed unworthy of propagation (e.g., lunatics, criminals, and paupers) in monasteries and convents "for the purpose of restricting their opportunities for . . . producing low class offspring." Galton also hinted in his darker moments that the state might institute the "social arrangements" practiced by the ancient Spartans—but he quickly admitted that there was no point in pursuing that idea for now, as modern society would find infanticide "alien and repulsive."[5]

One might well ask—since Darwinian evolution posits the survival of the fittest (and the concomitant elimination of the unfittest)—why the normal operations of nature would need to be supplemented by *any* eugenic measures, either positive or negative. The answer, according to Galton, is that conditions in nineteenth-century Europe "diminish the rigour of the application of the law of natural selection." Modern society, by providing such benefits as medical care to the sick and charity to the poor, allows those weaklings who would otherwise perish to survive. These unfit people then irresponsibly propagate a new generation of unfit people, who in turn take advantage of society to create even more unfit people. Furthermore, the "weak" classes are also "imprudent": they marry at a younger age than the fitter classes, and therefore produce more children and more generations within a given period. The result is that the weak wind up outbreeding the strong, which is exactly the opposite of how natural selection is supposed to operate. "Modern industrial civilization," concluded Galton, "deteriorates the breed." To counteract this, eugenics would "sternly forbid all forms of sentimental charity that are harmful to the race." Ending charity "may seem monstrous," conceded Galton, "but it is still more monstrous that the races best fitted to play their part on the stage of life, should be crowded out by the incompetent, the ailing, and the desponding."[6]

Galton wrote nine books and some two hundred scientific papers in his lifetime, and earned a position of eminence in the international scientific community. At the age of only thirty-one he was elected a fellow of the Royal Geographical Society, and three years later he was elected to the Royal Society. He was befriended by such men as Herbert Spencer, Thomas Huxley, Sir Richard Burton, and Lord Avebury, and was knighted by the king in 1909. In addition, Galton's ideas on eugenics were fully sanctioned by his admiring cousin. In *The Descent of Man,* Charles Darwin extolled the "remarkable" and "ingenious" work of Galton, and affirmed the central tenet of eugenics, that mental traits are inherited along with the physical: "We now know, through the admirable labours of Mr. Galton, that genius . . . tends to be inherited."[7]

Furthermore, Darwin—like Galton—observed that "the reckless, degraded, and often vicious members of society, tend to increase at a quicker rate than the provident and generally virtuous members." This differential rate of reproduction worried Darwin, and he came close to endorsing negative eugenics. After all, he observed, "hardly any one is so ignorant as to allow his worst animals to breed." In fact, "in the breeding of domestic animals, the elimination of those individuals . . . which are in any marked manner inferior, is by no means an unimportant element towards success." At the very least, opined Darwin, society must take cognizance of the dysgenic effects of charity, which prevents weak individuals from dying off and therefore leads to the survival of the unfittest: "We civilised men . . . build asylums for the imbecile, the maimed, and the sick; we institute poor-laws; and our medical men exert their utmost skill to save the life of every one to the last moment. . . . Thus the weak members of civilised societies propagate their kind. No one who has attended to the breeding of domestic animals will doubt that this must be highly injurious to the race of man."[8]

To the Continent

Another scientist who endorsed Galton's ideas on eugenics, and who influenced Madison Grant, was Ernst Haeckel (1834–1919), Germany's most famous biologist in the second half of the nineteenth century.[9]

Tall, blond, blue-eyed, broad-shouldered Ernst Heinrich Philipp August Haeckel could have been the prototype of Gobineau's Aryan superman. Like Madison Grant and so many eugenicists, he had been a boy naturalist whose love of the outdoors continued through adulthood. And though (again like Grant) he suffered from arthritis, he could often be found swimming, climbing, and hiking in the German countryside, and he conducted numerous journeys to the tropics as well. Haeckel had always wanted to be a zoologist but, like Galton, he trained as a physician to please his father. In 1860, however, Haeckel abandoned his practice after he read Darwin's "epoch-making" *Origin of Species* and realized that the answers to almost all questions were to be found "in the one magic word—evolution." After pursuing studies in zoology, Haeckel was appointed professor of zoology and director of the Zoological Museum at Jena University in 1862. He held this appointment until his retirement forty-seven years later, attaining fame throughout Europe as an accomplished zoologist and tireless proponent of his theory that "ontogeny recapitulates phylogeny."[10]

In 1866, Haeckel made a pilgrimage to Down House to meet Darwin, and soon was lecturing widely to scientific and lay audiences on natural selection. His popular books on evolution sold hundreds of thousands of copies, leading Daniel Gasman to conclude that "when the Germans refer to Darwin, more often than not they in fact mean, not Darwin, but Haeckel."[11] Haeckel's most successful book, and the one that most influenced Madison Grant, was *Die Welträtsel* (published in English as *The Riddle of the Universe*), published in 1899. (Those who scoff at the power of millennialism might want to consider that Ripley's *Races of Europe,* Chamberlain's *Foundations of the Nineteenth Century,* Lapouge's *L'Aryen,* and Haeckel's *Riddle of the Universe* were all published in 1899.)

The Riddle of the Universe sold over one hundred thousand copies in the first year, was translated into twenty-five languages, and sold another half million copies when revived by the Third Reich. In the book, Haeckel expounded his philosophy of monism, in which the entire universe—man, spirit, and nature (both organic and inorganic)—was an integrated whole, unified under the rubric of evolutionary laws. In Haeckel's hands, Darwinism became an all-embracing weltanschauung that explained not just evolution but human history and politics as well. (Monism can be seen as a greatly magnified version of the social Darwinism then in vogue in England and the United States.) Haeckel claimed that in the wake of *The Origin of Species,* evolution had replaced religious dogma to become "the sure foundation of our whole world-system." "*Alles ist Natur, Natur ist Alles.*"[12]

According to *The Riddle of the Universe,* one of the laws of evolution is the inheritance of mental as well as physical characteristics. Genetic personality differences apply not just to individuals but to races as well, which differ greatly in their capacities. Haeckel believed that at the summit of humanity stood the tall, blond, blue-eyed Aryans, mystically connected to their native German landscape. In the book he cites Gobineau to show that all other races are intellectually inferior to the Aryans. Among the lower races are the Jews, who are incapable of understanding Germanic culture. Taking aim at those sentimentalists who tout the Jewish contributions to Western civilization, Haeckel counters that "recent historical investigation" has proved that both Paul and Jesus were only half Jewish (Christ's father was actually a Roman officer who seduced Mary). Furthermore, for Haeckel (who was yet unaware of Houston Stewart Chamberlain's writings on the subject), the non-Jewish ancestry of the Messiah is confirmed "when we make a careful anthropological study of the personality of Christ," for the "characteristics which distinguish his high and noble personality . . . are certainly not Semitical; they are rather features of the higher Aryan race."[13]

As a zoologist, Haeckel was concerned that the dilution of German blood by inferior types was causing the degeneration of the Aryan race. He insisted that the state step in to enforce the laws of evolution and maintain the biological purity of the German people. Authoritarianism would have to replace bourgeois liberalism, so that civil rights and constitutional principles could be dispensed with and the good of the state finally take precedence over the rights of the individual. As a first step, he called for halting the immigration of the "filthy" eastern Jews, and hinted at more extreme measures: since the inferior races are, he said, "nearer to the mammals (apes and dogs) than to civilized Europeans, we must, therefore, assign a totally different value to their lives."[14]

In fact, like Houston Stewart Chamberlain, Georges Vacher de Lapouge, and Sir Francis Galton, Haeckel drew the attention of his countrymen to the example set by ancient Sparta, where "all newly-born children were subject to a careful examination and selection. All those that were weak, sickly, or affected with any bodily infirmity, were killed. Only the perfectly healthy and strong children were allowed to live, and they alone propagated the race."[15]

Although the lack of empirical support for many of his ideas eventually tarnished Haeckel's reputation in the scientific community, he maintained a huge personal following among the German public, who considered him a messiah of national and racial regeneration. He also had a number of followers abroad: one of the more prominent converts to monism was dancer Isadora Duncan, and another was G. Vacher de Lapouge, who translated Haeckel's books into French. Madison Grant, of course, greatly admired Haeckel (and it is interesting that Grant's rival, Franz Boas, was mentored in physical anthropology by Haeckel's rival, Rudolf Virchow).

Sir Francis Galton and Ernst Haeckel agreed on the desirability of enacting biological measures to counteract the dysgenic effects of modern civilization. But there was still a major obstacle to the implementation of any program of eugenics: the belief by most people in the inheritance of acquired characteristics. Even Haeckel accepted to a degree the neo-Lamarckian notion that traits acquired during the life of the parent could be passed on to the offspring. And if this was so, then there was no need for eugenics. To improve its genetic stock, a nation could simply institute a comprehensive program of environmental reforms (e.g., better nutrition, medicine, education, housing) that would raise the physical and moral nature of its citizens, and the resultant improvements would be passed on to succeeding generations.

In the 1880s, however, University of Freiburg zoologist August Weismann challenged the neo-Lamarckian doctrine. Crediting Sir Francis Galton with having originally propounded these ideas, Weismann differentiated the "soma plasm" (the cells of the body) from the "germ plasm" (which we know today as the genes inside the egg and the sperm cells). He demonstrated (by cutting off the tails of generation after generation of mice and observing that their progeny always grew normal tails) that changes to the body have no effect on the germ plasm. Traits acquired during the lifetime of a parent are not passed on to the children.

Scientists hotly debated the import of Weismann's work. Proponents of nurture resisted Weismann's findings, which implied that no amount of improvement in education or environment could be passed on to future generations. Advocates of nature, on the other hand, welcomed Weismann's rejection of the inheritance of acquired characteristics, though they could not explain how the germ plasm transmitted traits from one generation to the next. In 1891, Henry Fairfield Osborn (himself a neo-Lamarckian) expressed the rather desperate hope that some miracle would occur to cut this biological Gordian knot. "Is it not possible that . . . we shall discover some new factor of evolution which will work as great a surprise and revolution in our ideas as did the theory of natural selection in 1858?"[16]

And, of course, the hoped-for revolution did take place, in the year 1900, when Gregor Mendel's obscure, thirty-five-year-old paper in the *Proceedings of the Natural History Society of Brünn* was finally "discovered." The paper showed exactly how Weismann's germ plasm operates, at least for peas: each trait (e.g., height) is determined by the joining of two hereditary "elements" (genes). One gene comes from the mother and one from the father. But genes do not blend together; rather, some are dominant (and are expressed) while others are recessive (and are phenotypically invisible). And genes cannot be modified: if a yellow pea is painted green, it will still transmit a "yellow" gene to its progeny; if a tall plant is stunted by lack of water, it will still transmit a "tall" gene to its children.

Researchers in the new field of "genetics" quickly established that Mendel's laws of heredity applied not just to the traits of the pea plant but throughout the plant and animal kingdoms. And it seemed reasonable to assume that Mendel's laws might explain human heredity as well. Eugenicists were quick to draw the logical conclusion from the research of Weismann and Mendel: intelligence and other human traits are determined by genes, which are impervious to environmental influences. It was now clear to any scientifically literate person that social reforms, however well intended, could not improve the inherited nature of man. Society could spend great amounts of time and money educating inferior persons, but their germ plasm would remain unaffected, and hence the improvements would not be passed on to their offspring. Therefore, reformers who were serious about improving the human race, and not just about acquiring reputations as do-gooders, would do better to devote their efforts to eugenic programs that strove to eliminate defective germ plasm from the population. Nature had finally vanquished nurture.

The first decade of the twentieth century therefore witnessed a remarkable efflorescence of eugenic organizations in over thirty countries worldwide. In Britain, for example, the Eugenics Education Society was formed in 1907, with a rejuvenated Francis Galton as its first president and then Major Leonard Darwin (son of Charles) as his successor. In Germany, Alfred Ploetz and Ernst

Rüdin (both future members of the Nazi Party) founded the Gesellschaft für Rassenhygiene in 1905, with Ernst Haeckel and August Weismann as honorary chairmen. And the following year, Haeckel himself founded the German Monist League to push for a society restructured along eugenic lines. With thousands of members in Germany and Austria, the Monist League called for the isolation of the feebleminded in asylums where they could be prevented from breeding, and the application of negative eugenic measures to the deformed. "We are not bound under all circumstance to maintain and prolong life," stated Haeckel, and he reassured his fellow Monists that administering poison to unfit specimens of humanity was comparable to destroying "weeds" in one's well-cultivated garden.[17]

To the United States

At the same time that Ernst Haeckel was organizing the German Monist League, the eugenics movement in the United States was given form by Galton's leading American disciple, Charles Benedict Davenport. Davenport (1866–1944) was born one year after Madison Grant into a wealthy family descended from New England Puritans. Davenport was fiercely proud of his roots (his father had traced the family's ancestry back to the Norman conquest) and often lamented that "the best of that grand New England stock is dying out through failure to reproduce."[18]

An avid outdoorsman and inveterate walker, Davenport was one of the many zoologists who started out with a childhood interest in natural history. Against the wishes of his authoritarian father, he earned a doctorate in zoology from Harvard and began climbing the academic ranks (as his wife perused the death notices in *Science* every week to see what better positions might open up for her husband). In 1904 Davenport became director of the Station for Experimental Evolution at Cold Spring Harbor, Long Island (just down the road from Theodore Roosevelt's home at Oyster Bay). Davenport and his staff discovered—through a series of extensive experiments on everything from houseflies to sheep—a number of traits that were inherited in animals according to Mendelian ratios.

With Davenport's solid WASP credentials, expertise in natural history, devotion to Mendelian genetics, and proclivity for organizing (he was on the board of ten scientific associations and belonged to fifty-four others), it was inevitable that Davenport and Madison Grant would become lifelong friends and comrades. Davenport joined the New York Zoological Society in 1904, served on the Aquarium Committee, and was a frequent guest at the voyages of the Half-Moon Club.

Around the time that Madison Grant was moving from mammals to man, Davenport, spurred by an encounter with Francis Galton in England, decided to expand his research to embrace human heredity and eugenics. He saw nothing incongruous about a zoologist studying human heredity. To the contrary, Davenport wholeheartedly believed that the methods used by agronomists to improve domesticated plants and animals could and should play a role—the key role—in solving society's problems. "Man," explained Davenport, "is an organism—an animal; and the laws of improvement of corn and of race horses hold true for him also. Unless people accept this simple truth and let it influence marriage selection, human progress will cease."[19]

As a true progressive, Davenport believed that the mission of the eugenicist was "to improve the race by inducing young people . . . to fall in love intelligently." No sooner should a woman accept a husband "without knowing his biologico-genealogical history" than a stockbreeder should accept "as a sire for his colts an animal without a pedigree." After all, declared Davenport, "Marriage is an experiment in breeding," and "the success of a marriage from the standpoint of eugenics is measured by the number of disease-resistant, cultivable offspring that come from it." Sounding like the Gifford Pinchot of biology, Davenport often referred to human babies as the "human harvest . . . the human product . . . the world's most valuable crop."[20]

Obviously, there was no way that one could conduct direct breeding experiments on human beings. But Davenport reasoned that if he collected detailed family pedigrees of thousands of people, he should be able to discern patterns that would reveal, once and for all, which physical and mental traits are heritable. He hoped that eugenic marriages could then be planned scientifically and dysgenic pairings prevented, and the "national protoplasm" thereby saved from degradation.

Accordingly, Davenport decided to create a eugenics organization as an adjunct to the Station for Experimental Evolution at Cold Spring Harbor. He persuaded young Mary Harriman, who had spent a summer at Cold Spring Harbor while a student at Barnard College, of the enormous potential of eugenics to permanently reform society. Mary (whose brother Averell would be FDR's ambassador to the Soviet Union) was a friend of Eleanor Roosevelt and Francis Perkins (with whom she later shared a house in Georgetown) and active with them in the settlement movement. She agreed that eugenics was a promising reform movement and arranged a meeting between Davenport and her mother, Mary W. Harriman, recent widow of railroad magnate E. H. Harriman, whose estate was estimated at between $70 million and $100 million. Mrs. Harriman admitted that her aristocratic familiarity with well-bred racehorses gave her an appreciation of the importance of proper matings, and she agreed in 1910 to underwrite Davenport's project. An elated Davenport remarked in his diary that it was "A Red Letter Day for humanity!"[21]

Over the next eight years, Mrs. Harriman would pour hundreds of thousands of dollars into Davenport's coffers. With this extraordinary largesse, plus another $22,000 from John D. Rockefeller, Jr., and then massive support from the

Half of the Big Four of scientific racism: Charles Benedict Davenport and his indefatigable lieutenant, Harry H. Laughlin. Davenport often dressed in white, as did many other eugenicists (e.g., John Harvey Kellogg), who were obsessed with purity. Courtesy of the Harry H. Laughlin Papers, Special Collections Department, Pickler Memorial Library, Truman State University.

Carnegie Institution of Washington, Davenport created the Eugenics Record Office (ERO) at Cold Spring Harbor. (The Carnegie, Harriman, and Rockefeller families had all contributed to the conservation activities of Madison Grant, and now they were aiding the eugenic ambitions of Charles Benedict Davenport. The conservation and eugenics movements did not want for friends in high places in the early twentieth century.) Davenport wrote to the aged Galton in 1910: "So you see the seed sown by you is still sprouting in distant countries." And he assured the founder of eugenics that "as the years go by, humanity will more and more appreciate its debt to you. In this country we have run 'charity' mad. Now, a revulsion of feeling is coming about, and people are turning to your teaching."[22]

For the position of superintendent of the Eugenics Record Office, Davenport selected Harry H. Laughlin (1880–1943), a graduate of Princeton and—in yet another connection between husbandry and eugenics—an expert on the breeding of thoroughbred horses and an instructor in agriculture at North Missouri State Normal School. Like the parents of so many American eugenicists, Laughlin's father was a reverend and his mother wore the white ribbon of the Women's Christian Temperance Union. As superintendent of the ERO, Laughlin would assume an ever larger role in the eugenics movement and would soon join Madison Grant, Charles Benedict Davenport, and Henry Fairfield Osborn as the Big Four of scientific racism in the United States.[23]

Every summer, Davenport and the childless Laughlin trained a cadre of (mostly female) students from the East's most prestigious colleges to be field workers for the ERO. The field workers were then sent out to hospitals, asylums, prisons, charity organizations, schools for the deaf and blind, and institutions for the feebleminded to obtain anthropometric measurements and collect family histories. Davenport eventually amassed in his fireproof vault at Cold Spring Harbor a multigenerational database of the U.S. population recorded on 750,000 cards, which provided most of the material for his articles and books on eugenics. Davenport's most famous book, and the one that Madison Grant cited most often, was *Heredity in Relation to Eugenics,* published in 1911.

As the first major scientific work in the United States on the subject of eugenics, *Heredity in Relation to Eugenics* introduced the public to the thesis that the rules of Mendelian genetics apply to human beings. Using the data collected by his field workers, Davenport reported that many physical traits (e.g., eye color, hair color, and stature) and a number of diseases (e.g., epilepsy, Huntington's chorea, and certain types of insanity) were inherited in humans as "unit characters" (that is, each characteristic, like the height of the pea plant, was determined by a specific and independently operating gene). He also showed that many personality traits were similarly attributable to the action of a single gene. Musical ability, for instance, was apparently caused by a single recessive gene. Literary ability was likewise recessive, as were mathematical ability, memory, handwriting, shiftlessness, criminality, addictiveness, feeblemindedness, and pauperism. Indeed, Davenport claimed that thousands of human traits—physical, mental, and moral—were inherited as unit characters.

It would soon become apparent to professional geneticists, of course, that most characteristics do not have such a simple Mendelian explanation. The vast majority of human traits are determined by the complex interaction of many genes, first with each other and then with a host of environmental factors. In his work with peas, Mendel had fortuitously looked at traits that appeared to be determined by single genes, and thus his followers—in their enthusiasm over the monk's having discovered the key to life—prematurely embraced a simplistic version of genetics. Whenever Davenport's field workers, for instance, came across a family that had three well-behaved children and one incorrigible child, Davenport pounced on this as evidence of a "Mendelian ratio" and induced that criminality was a recessive unit character. (In other words, three of the children had inherited one recessive "criminalistic" gene and one dominant "law-abiding" gene and were therefore law-abiding, but the fourth child had inherited two recessive criminalistic genes and was thus a criminal.)

Given Davenport's belief that traits like criminality, pauperism, and feeblemindedness were genetic, *Heredity in Relation to Eugenics* proposed a number of eugenic solutions. First, the government should collect a detailed family pedigree of each citizen. The government "has not only the right, but the duty, to make a thorough study of all the families in the state and to know their good and bad traits." (And there was certainly a hope on Davenport's part that his Eugenics Record Office would be the government-financed bureau that would process all this data.)[24]

Second, since the state has a "duty" to prevent the mating of people with "antisocial" protoplasm, those with a dangerous genetic endowment should either be segregated in unisex institutions during their reproductive lives or castrated. "Concerning the power of the state to operate on selected persons there can be little doubt," wrote Davenport. "Society," railed the descendant of Puritan divines, "must annihilate the hideous serpent of hopelessly vicious protoplasm." (And as a Puritan, Davenport preferred castration to vasectomy, since the latter procedure, by rendering the victim sterile yet still potent, would do nothing "to safeguard female honor.")[25]

Third, society should consider adopting a policy of executing those with defective germ plasm. Capital punishment may be "crude," but "it is infinitely superior to [setting] the feeble-minded and criminalistic . . . loose upon society and permitting them to perpetuate in their offspring their animal traits." Echoing the protofascism that Haeckel's Monist League was concurrently developing in Germany, Davenport claimed: "The commonwealth is greater than any individual in it. Hence the rights of society over the life, the reproduction, the behavior and the traits of the individuals that compose it are . . . limitless, and society may take life, may sterilize, may segregate so as to prevent marriage, may restrict liberty in a hundred ways."[26]

And if the state has a duty to regulate the private life of its own citizens, it certainly has the right to bar the entry of biologically dangerous foreigners. Davenport was not against immigration, per se. The early immigrants to North America, for instance, "were men of courage, independence, and love of liberty" who contributed much to the germ plasm of the nation. (Davenport named Madison Grant's ancestor Robert Treat Paine, signer of the Declaration of Independence, as an example of distinguished protoplasmic character.) But regarding the New Immigration, "no patriotic American can contemplate this vast annual addition to our kinds of germ plasm without [questioning] the probable eugenic effect on our nation of this constant influx of new blood." According to Davenport, each immigrant carried in his loins the particular genetic traits of his country of origin. Thus, the Germans brought genes for thrift, honesty, and a love of songbirds. Scandinavians brought independence, self-control, and a facility for agriculture. The Italians, on the other hand, were a mixed blessing: they had a welcome willingness to work as day laborers but also a "tendency to crimes of personal violence." And as for the Hebrews: "There is no question that . . . the hordes of Jews that are now coming to us . . . represent the opposite extreme" from the original settlers, who had brought the ideals of "advancement by the sweat of the brow, and the uprearing of families in the fear of God and the love of country." The Jews admittedly had a high rate of literacy, but Davenport was suspicious of their "highly developed sense of personal rights," and his puritanical streak was offended by their supposed propensity to engage in prostitution, that "lowest of crimes."[27]

Looking to the future, Davenport predicted that "the population of the United States will, on account of the great influx of blood from Southeastern Europe, rapidly become darker in pigmentation, smaller in stature, more mercurial, more attached to music and art, more given to crimes of larceny, kidnapping, assault, murder, rape and sex-immorality." Also, "it seems probable that the ratio of insanity in the population will rapidly increase."[28]

Heredity in Relation to Eugenics became a popular college text for years to come, and Charles Benedict Davenport was acknowledged as a major figure in American science. In 1912, the year following the publication of *Heredity in Relation to Eugenics,* Davenport was elected to the National Academy of Sciences. He went on to publish some four hundred papers and books in his lifetime, and served on the editorial board of a number of scientific journals, including *Genetics, Growth and Human Biology,* the *Journal of Experimental Zoology,* the *Journal of Physical Anthropology,* and *Psyche.* Charles Rosenberg reminds us that "there were few American geneticists of note in the first three decades of the century who did not spend at least a summer at Cold Spring Harbor," and it was through Davenport that Madison Grant became acquainted, both personally and professionally, with such scientists as William E. Castle of Harvard and E. G. Conklin of Princeton. Grant learned a great deal about genetics from these researchers, and it would not be long before Charles Benedict Davenport would view Grant as his scientific equal, and begin accepting advice and criticism from him on eugenic matters.[29]

Just one year after the appearance of *Heredity in Relation to Eugenics,* Davenport's theories were strongly reinforced when his friend Henry H. Goddard published *The Kallikak Family,* an enormously influential work that seemed to prove that feeblemindedness was indeed hereditary. After studying psychology with G. Stanley Hall at Clark University, Goddard taught briefly at the University of Southern California (where, among other things, he was the first coach of the Trojan football team). He then joined the staff of the Training School for Backward and Feeble-minded Children in Vineland, New Jersey. While there, he investigated six generations of the pseudonymous Kallikak family, which had the curious distinction of numbering among its members both respected professionals and lowly criminals. Goddard (or rather his field workers, led by Elizabeth Kite who, like Goddard, had been raised a Quaker) traced the root of the problem back to Martin Kallikak, a young man of good stock in colonial New Jersey, who had joined the militia at the outbreak of the American Revolution. When Martin Kallikak's company stopped at a tavern one day, he met a feebleminded girl and, in "an unguarded moment," fathered a feebleminded son. That bastard son had 480 descendants, of whom 143 were judged by Goddard's field workers to be paupers, horse thieves, prostitutes, drunkards, delinquents, and epileptics. One hundred forty-three feebleminded persons in a sample of 480 was as good a Mendelian ratio as one was likely to find in the real world. After the war, Martin Kallikak "straightened up" and married "a respectable girl" with whom he produced a highly moral family whose 496 descendants were all appraised by Henry H. Goddard's field workers to be completely "normal." (Hence the alias Kallikak, from the Greek words *kalós* [good] and *kakós* [bad].)[30]

The Kallikak Family seemed to be an admirable work of science, complete with a testable hypothesis (that Mendelian genetics applies to human beings), an objective set of data (the number of normal and feebleminded descendants of Martin Kallikak), a control group (the good side of the Kallikak family, which was raised in the same New Jersey environment as the bad side), and a conclusion (feeblemindedness is a recessive trait inherited along Mendelian lines, and no amount of education or good environment can alter its transmission to future generations). As one review of *The Kallikak Family* put it: "It would be hardly possible to devise in the laboratory experimental conditions better adapted to produce a clear and decisive influence of heredity; nor could there be a more impressive lesson of the far-reaching and never-ending injury done to society by a single sin."[31]

Goddard's book went through twelve editions, and "Kallikaks" immediately entered the nation's vocabulary as a byword for the menace of the feebleminded. The book was equally influential abroad; it was translated into German in 1914 and enjoyed even greater popularity when the Nazis brought out a second edition in 1933. Madison Grant was impressed with Goddard's assertion that segregation or sterilization of the nameless tavern girl would have saved the state of New Jersey untold misery, vice, and tax dollars over the centuries. He probably also noticed that Goddard, like so many eugenicists, viewed people with defective genes as "filthy," "vulgar," and "repulsive" creatures, akin to "vermin" and "sores" on the body politic. (Goddard was funded at Vineland by philanthropist Samuel S. Fels, president of the company that manufactured Naphtha soap.) And certainly there was no mistaking in Goddard—the childless descendant of a governor of Plymouth colony—the streak of puritanism so common to eugenicists. (Goddard was raised by his mother, a Quaker minister,

A German pamphlet from 1924 demonstrates that when Martin Kallikak dallied with the feebleminded tavern girl (on the right) the result was hundreds of miscreants. Martin then married a worthy Quakeress (on the left) and the result was hundreds of the highest types of human beings. United States Holocaust Memorial Museum.

The views or opinions expressed in this book, and the context in which the images are used, do not necessarily reflect the views or policy of, nor imply approval or endorsement by, the United States Holocaust Memorial Museum.

his father having died when he was very young.) There is hardly a paragraph in *The Kallikak Family* that does not decry alcoholism, vice, prostitution, syphilis, "sexual immorality," "promiscuous behavior," "immoral acts," "base instincts," "animal appetites," "loose manners," or "depraved conduct."

Sir Francis Galton, Ernst Haeckel, Charles Benedict Davenport, Henry H. Goddard—these were the eugenicists who most influenced Madison Grant. He admired and appreciated them for a number of reasons. They were all highly esteemed scientists and leaders in their fields, with lengthy lists of publications to their credit. Also, with backgrounds in medicine or zoology, they were committed to viewing man from a biological perspective, as a mammal like any other, and they consistently drew their examples from animal husbandry and agronomy. In addition, they were enthusiastic Darwinists, to the point of making a secular religion of natural selection. They were progressives, in the sense of wanting to use the state to enact scientific reforms that would utilize the nation's germ plasm more efficiently. They were also the literal or spiritual descendants of Puritans, who were certain that man's inner nature was more important than his nurture and who wanted to harness the power of the state to eliminate sin from society (even at the cost of suppressing individual rights). And finally, they were outdoorsmen and aristocrats who feared the breeding power of the urbanized *Üntermenschen* and were deeply troubled that modern civilization permitted the survival of the unfittest.

In reviewing the influence of these scientists and the growth of the eugenics movement toward the end of his life, Sir Francis Galton stated with no small satisfaction: "I kindled the feeble flame." It is an interesting choice of words. A few years later, Charles Benedict Davenport, three decades before the Nazis constructed their crematoria, presciently provided his own variation on the theme (in a letter to his patron, Mrs. Harriman): "What a fire you have kindled! It is going to be a purifying *conflagration* some day!"[32]

The Appeal of Eugenics

Eugenics, I believe, appealed to Madison Grant on four levels: personal, spiritual, professional, and political. On a personal level, eugenics spoke to the condition of a patrician sensing that the position of his class was declining. In the evenings, Grant could retreat into the plush and comforting confines of his exclusive men's clubs, but in the harsh light of day there was no escaping the realities of the modern world: immigration, urbanization, industrialization, commercialization, socialism, feminism, and race suicide. Grant feared that his time had passed. And that is why at the end of the nineteenth century a penumbra of fatalism began to darken his natural aristocratic confidence. He embodied that fascinating contradiction, characteristic of post-1789 aristocrats, in which the inherent optimism of "the Yale man" was combined with pessimistic forebodings regarding the future of society. But Grant never succumbed to the enervating pessimism of a Henry Adams or Count de Gobineau, and the reason is that eugenics—by reassuring Grant that his class comprised superior protoplasm, and by offering a program to ensure the continuance of that protoplasm—provided hope. The aristocracy, by definition, believed that talent was hereditary and that the lower breeds were inferior, and eugenics supplied a scientific rationale and emotional reinforcement for those beliefs.

Secondly, eugenics provided solace to Grant upon the deaths of his saintly sister and revered father. Though he was nominally an Episcopalian, Grant's instincts were secular and scientific, and when his family members died, it was not the pastor at St. Bartholomew's that answered his existential questions but the new science of eugenics. Eugenics offered the promise of eternal life through the survival of one's indestructible genes (what Weismann called the "immortality of the germ plasm"). The *Eugenical News* explained that the "gonads . . . of man are perhaps his most precious possession, since it is on them he depends for the realization of his longings for immortality." And Leon

F. Whitney of the American Eugenics Society intoned that "though we ourselves shall not live on, the germ-plasm that created us will go on creating our children and our children's children. The Immortal Germ-Plasm! . . . we *do* have immortality."[33]

Eugenics offered the postmillennial hope that, through good breeding, the victory of the righteous would be assured and the perfect kingdom could be established on earth. We sense this in the American Eugenics Society's proclamation that "the discovery that man is able to guide his own evolution by means of eugenics" is the "most momentous" of human achievements, ranking ahead of the discoveries of fire, speech, tools, and writing. "Today," announced eugenicist Ellsworth Huntington in the midst of the Great Depression, "we are beginning to thrill with the feeling that we stand on the brink of an evolutionary epoch whose limits no man can possibly foretell." Thanks to the new dispensation, the dysgenic ravages of modern civilization were going to be reversed, and significant steps taken toward achieving human perfection—not through the action of some unseen god but through the proper selection of (equally unseen) genes. (One of the reasons for the current resurgence of hereditarianism is that we can now actually see the genes whose existence Madison Grant had to take on faith.) Eugenicists had an unshakable belief in the redeeming power of the invisible germ plasm, and Grant expressed the millennial hope that, via eugenics, "mankind will continue to ascend until . . . he will control his own destiny and attain moral heights as yet unimagined."[34]

The herald of this secular faith was Sir Francis Galton. In fact, Galton expressed the hope on many occasions that eugenics, by catering to the spiritual needs of men immersed in a scientific age, would become the religion of the twentieth century. (Indeed, just as Saint Ignatius founded the Society of Jesus to propagate the teachings of the Messiah, so Grant would go on to found the Galton Society to disseminate the precepts of *his* savior.) The scriptural texts of the new dispensation were the writings of Galton, along with those of Darwin, Haeckel, Davenport, and Goddard. The priests were the biologists who understood the mysteries of Mendelian genetics and instructed the populace in the proper application of these precepts to matters of marriage and procreation. "Eugenical truth," declared Grant's disciple Albert E. Wiggam, "is the highest truth men will ever know." He explicitly referred to the genetics laboratory as "the new Mount Sinai" and announced that the findings of eugenics were the Ten Commandments of science.[35]

The Elect were the genetically fit—the Nordics—who were hereditarily predestined to rule the earth. And the damned were the feebleminded, condemned by their defective protoplasm to lives of pauperism, criminality, and insanity. Surely, the Judeo-Christian belief that the iniquities of the fathers are visited upon the sons had its equivalent in Martin Kallikak's ill-advised dalliance with the feebleminded barmaid. To the eugenicists, the unfit were tainted, and no amount of environmental reform could overcome the original sin of a dysgenic mating. Atonement could come only through the cleansing sacrifice of castration. These are just some of the ways that eugenics served the function of a religion for Madison Grant and his colleagues. If religion was the opium of the people, eugenics was the religion of the aristocrats.

I believe that eugenics beguiled Grant for a third reason: as a conservationist he found that eugenics harmonized with his concurrent development of wildlife management. There was no duality in Madison Grant's life, no basic conflict between his espousal of conservation and his preaching on behalf of eugenics and immigration restriction. In the 1910s, Grant saw that the protected animals on his wildlife refuges were dangerously increasing in numbers, and he exhorted preservationists to accept the techniques of wildlife management to control them. At the same time, he saw that the inferior races in America were dangerously increasing in numbers, and he exhorted the public to accept the techniques of eugenics to control *them.* Grant simply took the concepts he was developing in wildlife management and applied them to the human population. Once he made the philosophical and moral decision that it was acceptable to eliminate "surplus" members of the wildlife population, it was not difficult for him to decide that such measures could and should be practiced on the expendable members of the human race.

Thus, whereas wildlife managers divided their animals into the *breeding* stock that must be preserved and the *superfluous* stock that was expendable, so Grant separated the human population into the "producing classes" that must be conserved and the "worthless types" that could be dispensed with. Whereas wildlife managers felt that the survival of the species as a whole was more important than the lives of a few individuals, so Grant preached that the fate of the race outweighed that of a few particular humans who were "of no value to the community." Whereas wildlife managers emphasized the need to reject misguided sentimentalism so that the old, the sick, and the deformed members of the herd could be culled, so Grant urged the public to move beyond its "sentimental belief in the sanctity of human life" so that negative measures could be applied to the "weak" and "unfit" members of society.[36]

Wildlife managers understood that when predators were eliminated from the Kaibab ecosystem, the deer population exploded because the weak and infirm individuals were no longer culled from the herd. Similarly, Grant could see that when modern civilization eliminated pestilence and starvation, unfit groups such as the Jews began to engage in "indiscriminate reproduction." If the solution to the growing deer population was castration and licensed hunting, the solution to the growing numbers of Polish Jews would seem fairly clear.

The fourth reason why Grant embraced eugenics was that the movement appealed to his progressive instincts. Eugenics offered a modern, rational, technocratic method of reforming society and improving the human species. Poverty, insanity, alcoholism, criminality, and genetic disease could all be eliminated if the tenets of eugenics were systematically applied. After Darwin, Mendel, and Weismann, we could no longer afford to leave the propagation of the human race to sentimental amateurs. In place of the wasteful and damaging system of ad hoc marriages and unplanned births, a cadre of professional experts—biologists, psychologists, anthropologists, sociologists, physicians, social workers—would be called on to determine procreative policy. As Harry H. Laughlin put it: eugenics was "the warp in the fabric of national efficiency." From now on, marriage and breeding would be planned, orderly, and productive. There was a great deal of optimism among progressives and eugenicists that the practical application of science would go a long way toward solving most of society's problems. Charles R. Van Hise, president of the University of Wisconsin (and a leading conservationist) expressed this in 1914: "We know enough about agriculture so that the agricultural production of the country could be doubled if the knowledge were applied; we know enough about disease so that if the knowledge were utilized, infectious and contagious diseases would be substantially destroyed in the United States within a score of years; we know enough about eugenics so that if the knowledge were applied, the defective classes would disappear within a generation."[37]

Eugenicists like Van Hise fully regarded themselves as progressive reformers. Madison Grant, we know, was involved in such typical progressive endeavors as the conservation movement and the reform of municipal government. His fellow eugenicists could be found in the leadership of such kindred movements as Prohibition, birth control, public health, child welfare, prison reform, factory safety, amelioration of poverty, decent housing, proper care of the insane, and world peace. And every one of them believed that eugenics was the preeminent reform. As Grant's disciple Lewis Terman, president of the American Psychological Association, put it: "The ordinary social and political issues which engross mankind are of trivial importance in comparison with the issues which relate to eugenics." And Charles Benedict Davenport spoke for them all when he declared: "Today, as never before, we realize that at the bottom of real social progress lies the germ plasm."[38]

If one truly had concern for the epileptic, the alcoholic, the pauper, and the criminal, surely the answer lay not in building more and more institutions to warehouse them but in making sure that such persons did not contaminate future generations with their defective germ plasm. Eugenics was the supreme preventive medicine; it was the preeminent philanthropy. As Davenport insisted, "Vastly more effective than ten million dollars to 'charity' would be ten million dollars to eugenics. He who, by such a gift, should redeem mankind from vice, imbecility and suffering would be the world's wisest philanthropist."[39]

Like all progressive reform movements, eugenics called for widespread education and state action (and this is where the eugenicists parted company with the social Darwinists). The eugenicists needed an activist state—so providentially strengthened under TR and the progressives—to intervene in society if their programs were to succeed. Positive eugenics could perhaps rely on moral suasion (though it would be helpful if the state provided financial incentives to large families, and if the educational system disseminated eugenic propaganda). But negative eugenics definitely required the coercive power of the state to prevent dysgenic marriages, segregate the unfit, restrict immigration, and implement involuntary sterilization.

Thus, eugenics meshed well with the scientific and reformist ethos of American progressivism. It was not an accident that soon after Grant and his colleagues in the interlocking directorate formulated the original principles of wildlife management, Grant and another group of colleagues created the organized eugenics movement in the United States. For if wildlife management was the penultimate progressive idea, eugenics was the *ultimate* progressive idea.

The Coming of the Passing

Until Madison Grant, eugenicists were concerned with unfit *individuals,* not inferior races. (The Kallikaks, after all, were of pure Anglo-Saxon stock.) To be sure, Galton, Darwin, Haeckel, Davenport, and Goddard all felt that northern Europeans were preferable to other races. But in devising their eugenic programs they strove to raise the hereditary endowment of *all* mankind, and hoped that eventually everyone could reach the intellectual level of the upper classes. (Davenport even acknowledged that there might be "good Greeks or Serbians and undesirable Norwegians and English.") Madison Grant's major contribution to eugenics, therefore, was to advance it from a skirmish against individuals who were socially unfit into a war against groups who were *racially* unfit.[40]

As for the anthropologists and the Aryanists discussed in chapter 5, they—by definition—focused on races, and posited the superiority of the dolichocephalics. But they had no program to ensure the continued dominance of the Aryans. All they could do was continue to measure skulls, and despair. Madison Grant, by injecting eugenics into anthropology, provided the means to ensure the proper outcome of the racial struggle.

In sum, what Grant did was to combine seven disciplines—wildlife management, anthropology, paleoanthropology, the study of race suicide, Aryanism, eugenics, and genetics—into an amalgam that we call *scientific racism.*

At its simplest level, scientific racism involved three axioms:

1. The human species is divided into biologically distinct and hierarchical races, with the Nordic race at the apex.
2. The intellectual, moral, temperamental, and cultural traits of each race are correlated to, and inherited with, its physical traits; and the genes for those traits are unaffected by the environment.

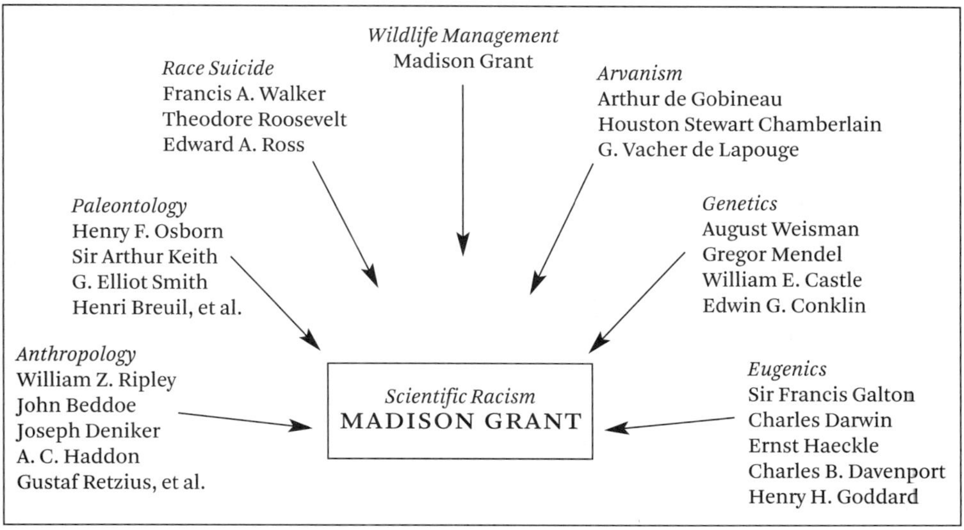

Scientific Racism.

3. If a member of an inferior race mates with a member of a superior race, the result is a reversion to a primitive type; hence positive and negative eugenic measures must be taken to prevent the degeneration of the superior race.

Scientific racism is to be distinguished from *popular* racism, which is simply the irrational hatred of the ethnographic "other." The popular racist fears and despises other races because they are phenotypically different; but he or she has no scientific theory to explain this revulsion, other than baseless ethnocentrism. The scientific racist, on the other hand, can employ the findings of physical anthropology, Darwinian evolution, and Mendelian genetics to explain *why* other races are biologically inferior. Whatever Madison Grant may have thought privately about "smelly dagos," "greasy spics," or "dumb micks," his contempt was always couched in the language of biological determinism. (The strictly anthropological nature of Grant's anti-Semitism is born out by the fact that there was no place in his philosophy for international Jewish conspiracies. Grant had no concern about the Jewish stranglehold on the world's financial markets, or plots by Israelites to abduct Christian babies and use their blood to make matzo.)

An anti-Semite who hates Jews and fervently tries to convert them to Christianity is a *popular*—but not a scientific—racist. The very fact that a bigot thinks that Jews can be converted proves that she does not believe there is too much biologically amiss with Jews. But a racist like Grant, familiar with the laws of Mendel and Weismann, understood that racial characteristics were inherited and immutable. The strange and repellent customs and morals of the Polish

Jews of the Lower East Side were not a matter of culture but of genetics. Converting a Jew was futile, because you could never alter his protoplasm. The holy water of baptism had no effect on his chromosomes. The children of a Jew who had converted to Christianity were still Jews. To Grant, the races of man were like the Society of Colonial Wars: if your great, great, great grandfather served in King Philip's War, then you were eligible to join; but the actual events of your own life were irrelevant. There was no action you could take that would modify the facts of your heredity. A popular racist might admit (and hope) that a Jew could become a Christian, but a scientific racist would point out that a Jew could never become a Nordic.

In the early 1910s, while running the Bronx Zoo and the New York Aquarium, overseeing the Boone and Crockett Club and the Society of Colonial Wars, putting a stop to the market hunters and the plume traders, serving as a trustee of the American Bison Society and the American Museum of Natural History, creating the Bronx River Parkway and Glacier National Park, and lobbying for gun control and wildlife refuges, Grant set to work researching and writing the bible of scientific racism, *The Passing of the Great Race.* It would be a diabolic masterpiece alerting the elite public that unless it embraced the tenets of eugenics, the Nordic race (the "Great Race" of the title) would soon pass from the scene as a result of intermixture with inferior immigrant stocks.

At the same time, Grant's friend Henry Fairfield Osborn commenced work on *his* great anthropological work, *Men of the Old Stone Age.* Osborn finished his tome one year ahead of Grant, and in 1915 *Men of the Old Stone Age* was published to much acclaim. (One of the reasons that Osborn was able to beat Grant to the publishing punch was that—as usual—little of the substantive research or writing had been performed by Osborn himself.) *Men of the Old Stone Age* was a massive, impressively illustrated, 512-page compendium of the most current knowledge about Paleolithic man. It was Osborn's most popular book, and went through thirteen printings in his lifetime. The book's protagonists were the dolichocephalic Cro-Magnons, a vigorous, inventive race of hunters lucky enough to live in what was, apparently, a most congenial environment. The glaciers were retreating, the forests were full of game, the rivers were teeming with fish, the hunters were strong, tall, and honest . . . to Osborn, life during the Old Stone Age was one vast Boone and Crockett Club.

Men of the Old Stone Age ended with the unwelcome demise of the Cro-Magnons and the invasion of the European continent by the Teutons, Alpines, and Mediterraneans at the end of the Paleolithic period. Osborn's book can thus be seen as the opening of a two-part survey of the history of the white races, of which Madison Grant's *Passing of the Great Race,* published one year later, would prove to be the second—and even more influential—installment.

II

Conserving the Nordics

All is race, there is no other truth.

Benjamin Disraeli

7

The Passing of the Great Race

There is a peculiar kind of vehemence, a main ingredient of our literature, which can be achieved only by Americans disillusioned with America.

Perry Miller

Madison Grant completed the manuscript for *The Passing of the Great Race* in the spring of 1916 and showed it to three of his closest friends, Charles Stewart Davison, Moses Taylor Pyne, and Henry Fairfield Osborn, all of whom suggested numerous corrections.

Charles Stewart Davison, yet another member of Grant's circle who was descended from the Puritans, received an excellent education (he earned a B.A. and a master's from Cambridge University and an LL.B. from Columbia University) and then embarked on a successful Manhattan legal career, aided in no small measure by his extremely distinguished countenance, which featured a perfectly trimmed beard and majestic mustache. Davison had first become acquainted with Grant and his brother DeForest in the municipal reform movement of 1894 and was a member of all the same social clubs as the Grants. An avid outdoorsman, Davison worked with the Grant brothers in many conservation organizations (including the Boone and Crockett Club, the American Bison Society, the American Society of Mammalogists, and the National Council on Parks and Forests). The rabidly anti-Semitic Davison (who, like Grant, never married) had utter contempt for the masses, leading Grant to declare proudly that Davison "is saturated with my point of view."[1] Davison served with Grant on the boards of a number of right-wing groups, including the American Defense Society and the Immigration Restriction League, and later collaborated with Grant in editing two anti-immigration books, *The Founders of the Republic on Immigration* (1928) and *The Alien in Our Midst* (1930). Grant turned to Davison for advice on his manuscript of *The Passing of the Great Race* because Davison was a talented writer and quite familiar with animal husbandry (he owned a farm in Massachusetts where he bred Dutch belted cattle and Hampshire swine).

Grant also showed a draft of his book to his friend Moses Taylor Pyne. Pyne was the son of Percy Rivington Pyne, who had been one of the leading financiers in New York City and a founder of the American Museum of Natural History. Moses Taylor Pyne graduated from Princeton in 1877 (where he and Henry Fairfield Osborn were classmates and close friends), received his LL.B. from Columbia University in 1879, and went on to become a prominent lawyer and financier with numerous railroad, banking, insurance, and industrial interests in New York and New Jersey. Pyne was involved in many benevolent activities, including years of service to Princeton University. He was an influential trustee of the university for thirty-six years and once declined an offer of the presidency. (He is commemorated on the campus today by Pyne Hall.) The impossibly handsome and refined Pyne, with his classic Nordic visage, piercing blue eyes, blond hair, and clipped mustache, was a mainstay of Grant's social world. He also had an amateur interest in scientific matters (he was a charter member of the Half-Moon Club) and was sufficiently knowledgeable that Grant turned to him for assistance with his manuscript.

Grant's third adviser, Henry Fairfield Osborn, not only submitted suggestions but also supplied a three-page preface that lauded *The Passing of the Great Race* for launching "a new and fascinating field of study," to wit, the interpretation of history in terms of race. "There is no gainsaying that this is the correct scientific method of approaching the problem of the past," wrote America's foremost evolutionist. But the true importance of Grant's book, according to Osborn, was not so much its elucidation of the past as its relevance to "*our* day and generation"—and to the future. Osborn was confident that the application of Grant's eugenic teachings would ensure the "conservation of that race [the Nordic race] which has given us the true spirit of Americanism." Osborn, the former neo-Lamarckian, congratulated Grant for extending the work of Galton and Weismann, and for compelling us to recognize that heredity is "more enduring and potent than environment."[2]

After incorporating the suggestions of Davison, Pyne, and Osborn, Grant submitted his manuscript to Charles Scribner's Sons. Though Grant clumsily attempted to convince "old CS" (Charles Scribner II) that he was simultaneously negotiating with other publishing houses, there was never any doubt as to who his publisher would be. For one thing, as the most traditional of all U.S. publishing houses, the genteel Charles Scribner's Sons specialized in chroniclers of the decaying patriciate (e.g., Henry Adams, Henry Cabot Lodge, Edith Wharton, and John Galsworthy). For another, the firm already handled a number of Boone and Crockett types, including Theodore Roosevelt, Francis Parkman, Rudyard Kipling, Charles Sheldon, Roy Chapman Andrews, and William T. Hornaday. And, most importantly, Scribner's had just published

Men of the Old Stone Age; hence it was logical that it would now publish Grant's effort, which was essentially a sequel to Osborn's book. (And, subsequent to the success of *The Passing of the Great Race,* the firm would take on many of Grant's eugenicist colleagues, including E. G. Conklin, Edward M. East, Charles W. Gould, Ellsworth Huntington, William McDougall, John C. Merriam, Stewart Paton, Lothrop Stoddard, and Albert E. Wiggam.)

Grant's editor at Scribner's was the soon to be legendary Maxwell Perkins, who would go on to discover and nourish some of the most prominent American writers of the century, including F. Scott Fitzgerald, Ernest Hemingway, and Thomas Wolfe. (Fitzgerald and Wolfe shared Grant's anti-Semitic attitudes. So did Hemingway—but then there were not a lot of people that Hemingway *did* like.) After seeing Grant's manuscript, Perkins confided to Theodore Roosevelt that the book "seems to me one of unusual importance."[3] Scribner's received the manuscript in June 1916, and by October *The Passing of the Great Race* was already in the science section of bookstores, with a dedication page that read: "To MY FATHER."

The Manifesto of Scientific Racism

The Passing of the Great Race is an extraordinary overview of Western history as seen through the eyes of a scientific racist. The book consists of twenty-one chapters, distributed into two parts. Part I is a general discussion, with numerous tangents, of the main principles of scientific racism; part II is a historical survey of the evolution of the races of Europe from the Eolithic period through the present. (The deductive method of scientific racism is thus inherent in the very structure of the book: first we learn the theory, and then we examine the evidence.) Grant's opening sentence explains that the book is "an attempt to elucidate the meaning of history in terms of race." And in the succeeding 228 pages he proceeds to do just that. With confidence and dynamism, he propounds the great lesson of "modern anthropology," which is that "race lies today at the base of all the phenomena of modern society, just as it has done throughout the unrecorded eons of the past."[4]

Whereas other historians have looked at the past and seen everything from nations clashing to genders attaining consciousness, Grant's gaze penetrates beneath those surface irruptions to perceive that the history of mankind is actually a tale of the evolution, migration, and confrontation of races. Thus, for example, he explains that the empire of Alexander crumbled when the pure Macedonian blood mixed with Asiatic blood; he shows that the division of Roman society into patricians and plebeians was actually a manifestation of the racial conflict between Nordics and Mediterraneans; he demonstrates that the long decline of the empire of Spain was caused by the progressive dilution of the germ plasm of the Gothic race; and so forth. Indeed, the more Grant contemplates the *longue durée,* the clearer he sees that "the lesson is always the same, namely, that race is everything."[5]

Many of Grant's passages sparkle with energy and wit, and his rhetoric, while occasionally banal, is often exciting and even majestic. On the other hand, the book is not without its problems. The text is often scattered and has a curiously disjointed feel to it, revealing its origins as a pastiche of ideas plucked from the various sources discussed in chapters 5 and 6 of this book. The disorganization, I think, was only exacerbated by the editing process, in which Davison, Pyne, and Osborn each had their own ideas about revising, relocating, or removing particular passages, and Grant possessed neither the temperament nor the scientific expertise to veto their (often conflicting) suggestions. And certainly the absence of editor par excellence Maxwell Perkins for much of the summer of 1916 (he was doing reserve duty with the U.S. Cavalry on the Mexican border) could not have helped matters.

While the range of topics in *The Passing of the Great Race*—from Aachen to the Zendavesta—is astonishing, it is doubtful whether the majority of Grant's audience paid strict attention to some of his more arcane and complicated meanderings through the thickets of comparative philology and prehistoric ethnology. As was the case with Osborn's best-selling *Men of the Old Stone Age,* vast swaths of Grant's book (especially the chapters in part II devoted to European paleontology) were assuredly read with something less than strict attention by the American public. However the very denseness of the later chapters, along with the complex charts and sophisticated, four-color, foldout maps (with titles like "Expansion of the Teutonic Nordics and Slavic Alpines, 100 B.C.–1100 A.D."), only added to the apparent erudition of the work and, for the average reader, probably legitimized its political and racial arguments. Passages like the following, to pick one at random, may be pedantic nonsense, but they certainly give the impression that the author is in command of his facts:

> The Goidels were of bronze culture. When they reached Britain they must have found there a population preponderantly of Mediterranean type with numerous remains of still earlier races of Paleolithic times, and also some round skull Alpines of the Round Barrows, who have since faded from the living population. When the next invasion, the Cymric, occurred, the Goidels had been very largely absorbed by these underlying Mediterranean aborigines who had accepted the Goidelic form of Celtic speech, just as on the continent the Gauls had mixed with Alpine and Mediterranean natives though imposing upon the conquered their own tongue. In fact, in Britain, Gaul, and Spain the Goidels and Gauls were chiefly a ruling, military class, while the great bulk of the population remained unchanged, although Aryanized in speech. The Brythonic or Cymric tribes, or "P Celts," followed about five hundred years later, driving the Goidels westward through Germany, Gaul, and

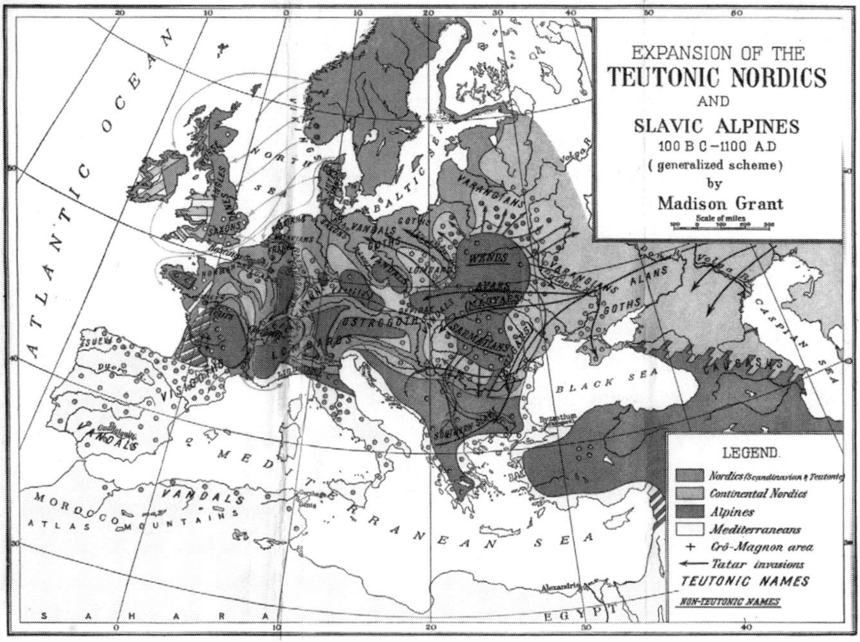

Grant spent much time and effort compiling the maps for his book. All the reviewers were impressed except Franz Boas, who (correctly) contended that Grant's maps were "entirely fanciful in their details."

Britain, as is proved by the distribution of place names, and this movement of population was still going on when Caesar crossed the Channel. The Brythonic group gave rise to the modern Cornish, extinct within a century, the Cymric of Wales, and the Armorican of Brittany.[6]

Despite these excursions down the alleyways of European pseudoarcheology—and although the subtitle of the book is *The Racial Basis of European History*—the text pertains to the United States as much as it does to the Old World. This is not only because Grant views the United States, demographically speaking, as the western outpost of Europe (and hence spends more than a few pages discussing race in America) but also because the reader is meant to extract even from those chapters that are nominally about European history lessons directly applicable to the contemporary United States. Whether Grant is explaining the rise of the Ligurians or the fall of the Livonians, we understand that it is actually the fate of the Americans that is being discussed.

Anthropologically, Grant tells us, the Americans are "Nordics." And the Nordic race, he explains, is the classic European type, "the *Homo europœus,* the white man par excellence."[7] Unlike the Alpines and Mediterraneans, who migrated into Europe at the close of the Paleolithic period from Asia and Africa, re-

Table 1. *The three races of Europe*

	Nordics	Alpines	Mediterraneans
Came from:	Northern Europe	Southwestern Asia	North Africa
Current habitat:	Shores of the North and the Baltic Seas	Mountainous terrain of eastern and central Europe	Shores of the Mediterranean and coast of western Europe
Cephalic index:	Dolichocephalic	Brachycephalic	Dolichocephalic (but smaller capacity than the Nordics)
Stature:	Tall	Medium	Stunted
Hair color:	Light, often blond	Dark	Dark
Eye color:	Light, often blue	Dark (sometimes grayish)	Dark
Skin color:	Fair	Fair to dark	Swarthy
Nose:	Narrow, aquiline	Broad, coarse	Broad
Build:	Muscular	Stocky	Feeble

spectively, the autochthonous Nordics developed in the forests and steppes of northern Europe. Grant's extended discussion of the physical differences between the native Nordic race and the two interlopers is summarized in table 1. With the exception, perhaps, of skin color, these physical traits are all immutable unit characters and hence are permanent attributes of the European races. In addition to their physical traits, Grant makes the crucial claim that the races are endowed with specific *mental* traits as well. "Moral, intellectual, and spiritual attributes are as persistent as physical characters, and are transmitted unchanged from generation to generation."[8]

Unlike the sluggish Mediterraneans and the servile Alpines (who have the mentality of serfs), the admirable Nordics are an impressive, energetic race comprising hunters, explorers, adventurers, sailors, and soldiers. The evolutionary explanation for their splendor is that the proto-Nordics evolved in the north of Europe, where climatic conditions were such "as to impose a rigid elimination of defectives through the agency of hard winters and the necessity of industry and foresight in providing the year's food, clothing, and shelter during the short summer. Such demands on energy . . . produce[d] a strong, virile, and self-contained race which would inevitably overwhelm in battle nations whose weaker elements had not been purged by the conditions of an equally severe environment." Grant invests his masterful Nordics with overwhelmingly masculine attributes: "When the isolation and exacting climate of the north had done their work and produced the vigorous Nordic type, these men burst upon the southern races, conquering east, south, and west. They brought with them from the north the hardihood and vigor acquired under the rigorous selection of a long winter season, and vanquished in battle the inhabitants of older and feebler civilizations." The virile Nordics are constantly "expanding . . . invading . . . occupying . . . conquering . . . exterminating . . . swarming . . . driving . . . ravaging . . . sweeping across . . . pushing in . . . bursting out . . . and pouring through." There is no doubt, in short, that the Nordic race is "the Master Race."[9]

The fiercely warlike males of the Nordic race traditionally have provided the officer corps of the armies of Europe. This was true in the days of Ajax ("Both the Trojans and the Greeks were commanded by huge blond princes, the heroes of Homer"), and it is true today ("From a race point of view, [World War I] is essentially a civil war," as all the belligerent states "have sent to the front their fighting Nordic element, and the loss of life . . . will fall much more heavily on the blond giant than on the little brunet"). These supposed facts lead Grant to restate a thesis propounded by Darwin, elaborated by Lapouge, and perfected by his friend Vernon Kellogg, that war is a dysgenic endeavor, as in battle "the tall Nordic strain" is always killed off in greater proportions than the other races. From an evolutionary point of view, the only true victor in wartime is "the little dark man."[10]

In addition to virility, other traits that are peculiarly Nordic are loyalty, chivalry, and veracity, as well as a love of efficiency, a predilection for organization, and proficiency in marksmanship (the degree to which Grant projects onto the Nordics his own self-image is wondrous to behold). The Nordics are inherently individualistic, self-reliant, and jealous of their personal freedom, which explains why they are overwhelmingly found in the pews of Protestant congregations. In comparison with the other races, Nordics excel in literature and in scientific research. "In fact," declares Grant in Gobinesque fashion, "in the Europe of today the amount of Nordic blood in each nation is a very fair measure of its strength in war and standing in civilization."[11]

In Grant's framework, race and class are inseparably intertwined, and his defense of the Nordics is always a defense of the patricians. Thus we are informed that throughout European history the Nordics have comprised the aristocracy, which is why to this day from Finland to Sicily the ruling class is invariably taller and blonder than the peasantry. Even nonscientists have an unconscious understanding of this identity between the Nordics and the nobility: "Most ancient tapestries," for example, "show a blond earl on horseback and a dark haired churl holding the bridle." Similarly, "In church pictures today all angels are blonds, while the denizens of the lower regions revel in deep brunetness." And certainly no competent novelist would describe his hero as anything other than tall, blond, and honest—which, of course, are all Nordic traits.[12]

Like modern-day Afrocentrists who perceive negroid features in Minoan frescoes and Hittite amulets, Grant sees the Nordics everywhere. Representative Nordics from the past include Alexander the Great (recognizable by his "Nordic features, aquiline nose, [and] gently curling yellow hair"), Dante Alighieri, and all the chief men of the Renaissance, including Raphael, Titian, Michelangelo, and Leonardo da Vinci ("a fact easily recognized by a close inspection of busts or portraits"). Practically all of the forty-niners who sought adventure in California were Nordics. King David was probably Nordic (as indicated by biblical references to his fairness), and as to Jesus, Grant assures us that he possessed all the "physical and moral attributes" of a Nordic. This tradition is "so deeply rooted in everyday consciousness that . . . in depicting the crucifixion no artist hesitates to make the two thieves brunet in contrast to the blond Saviour."[13]

According to *The Passing of the Great Race,* the Founding Fathers of the United States were purely Nordic. (It is true that Great Britain, the mother country, was composed of both Nordic and Mediterranean strains, but the Nordic proclivity for sailing and adventure ensured that it was members of the Master Race who crossed the Atlantic to found the thirteen colonies.) Those early Americans were "one of the most gifted and vigorous stocks on earth, a stock free from the diseases, physical and moral," that beset the less favored subspecies of mankind. And for two and a half centuries, the Nordic blood was kept pure in the New World because the settlers had a strongly developed sense of "race consciousness." Like the sensible southerners of Grant's day, the proud Americans would not intermarry with inferior strains, and as a result they had "the greatest opportunity in recorded history to produce in the isolation of a continent a powerful and racially homogeneous people."[14]

And then, in a fit of humanitarian madness, the old-stock Americans threw it all away. The Civil War "put a severe, perhaps fatal, check to the development and expansion of this splendid type." The reasons are threefold. First, the rise of sentimentalism during the antislavery agitation proved inimical to Nordic racial consciousness and weakened taboos against miscegenation. Second, the war itself, like all wars, was dysgenic; it destroyed "great numbers of the best breeding stock on both sides." And third, the prosperity that followed the war attracted to the United States "hordes of immigrants of inferior racial value," who no longer faced a difficult ocean crossing. "The transportation lines advertised America as a land flowing with milk and honey, and the European governments took the opportunity to unload upon careless, wealthy, and hospitable America the sweepings of their jails and asylums. The result was that the new immigration . . . contained a large and increasing number of the weak, the broken, and the mentally crippled of all races drawn from the lowest stratum of the Mediterranean basin and the Balkans, together with hordes of the wretched, submerged populations of the Polish Ghettos."[15]

The consequence, Grant remarks bitterly, is that today "Our jails, insane asylums and almshouses are filled with this human flotsam and the whole tone of American life, social, moral and political has been lowered and vulgarized by them." The man of the old stock is being "elbowed out of his own home" and "literally driven off the streets of New York City by the swarms of Polish Jews." These

Hebrew immigrants cunningly "adopt the language of the native American; they wear his clothes; they steal his name"; and now (and here Grant's sexual anxieties come to the fore) "*they are beginning to take his women.*" Since the days of Rome, Alexandria, and Byzantium, large cities have always attracted the ethnic dregs of the world, but modern Manhattan, writes Grant at his floral best, is becoming "a *cloaca gentium* which will produce many amazing racial hybrids and some ethnic horrors that will be beyond the powers of future anthropologists to unravel."[16]

In a passage that is one part jeremiad and one part restrictionist propaganda, Grant cries "Shame!" on those Americans who actually encourage this immigration "in order to purchase a few generations of ease and luxury." He warns that the continued refusal of the native-born American to work with his hands "when he can hire or import serfs to do manual labor for him is the prelude to his extinction, and the immigrant laborers are now breeding out their masters and killing by filth and by crowding as effectively as by the sword." Grant understands that factory owners have a vested interested in encouraging the New Immigration, but he is dumbfounded by the naive sentimentalists who actually welcome the influx of these "social discards" and provide them with all manner of charitable assistance to prolong their pathetic lives. As a consequence, society is being burdened "with an ever increasing number of moral perverts, mental defectives, and hereditary cripples" (wrote the man who himself would be crippled five years hence).[17]

Well-meaning humanitarians, having fallen under the spell of Christianity with its unfortunate and wholly unscientific bias in favor of "the weaker elements," have preserved large numbers of individuals of inferior mentality, "whereas in the savage state of society the backward members [were] allowed to perish and the race [was] carried on by the vigorous and not by the weaklings." We can now see that social workers and their ilk have done "more injury to the race than black death or smallpox."[18]

The chief failing of the sentimentalists is their obstinate and irrational refusal to face "inevitable facts, if such facts appear cruel." But we as a society must accept that "efforts to indiscriminately preserve babies among the lower classes often result in serious injury to the race" and that permanent harm is done to the community "by the perpetuation of worthless types." Scientists have long understood that "Nature cares not for the individual. . . . She is concerned only with the perpetuation of the species or type." In clear, sober language that is indistinguishable from the official dogma of National Socialism, the charming Park Avenue conservationist instructs us that "the laws of nature require the obliteration of the unfit, and human life is valuable only when it is of use to the community or race."[19]

According to Grant, the sob sisters who work in settlement houses, hoping to "Americanize" the immigrants, simply do not comprehend the laws of genetics, which uphold the primacy of nature over nurture. "There exists today a widespread and fatuous belief in the power of environment, as well as of education and opportunity to alter heredity"; but the public needs to accept, once and for all, that the unit characters of each race are immutable and cannot be altered by environmental reforms.[20]

Grant was fully aware that a study by anthropologist Franz Boas seemed to support the environmentalists and their melting-pot fetish. In 1911, after measuring almost eighteen thousand immigrants and their children in New York City, Boas published a report titled "Changes in Bodily Form of Descendants of Immigrants," showing that the American children of dolichocephalic immigrants were often brachycephalic, while the children of brachycephalic immigrants were often dolichocephalic. Thus even the most "immutable" of hereditary traits, the cephalic index, was susceptible to environmental influence. If living conditions in the United States were affecting supposedly permanent racial characteristics in this manner, Boas could only conclude that nurture was just as effective as nature in molding the bodies, and maybe the minds, of humans.[21]

But Boas stood alone in those days, and Grant questioned his motives by explaining that immigrant Jews like Boas have a vested interest in overstating the power of the environment to counteract the effects of their inferior heredity. Grant points to the history of Africans in the New World as proof of the immutability of genes: "It has taken us fifty years to learn that speaking English, wearing good clothes, and going to school and to church, does not transform a negro into a white man." And, Grant predicts with scientific assurance, "We shall have a similar experience with the Polish Jew, whose dwarf stature, peculiar mentality, and ruthless concentration on self-interest are being engrafted upon the stock of the nation."[22]

Grant is appalled when social scientists like Boas irresponsibly declaim that the continued intermingling of the races in the United States could have a benign genetic outcome. To the contrary, as far as the Nordics are concerned, miscegenation is the first step toward extinction. For the principle of reversion states that when two races interbreed, the characters of the higher race are lost. Thus, even though the Nordics truly are the Master Race, their reversion to a lower type is always a possibility, since they evolved more recently than the other races and hence their characters are still relatively unstable. (For Grant, Nordicism is like a recessive gene.) The specialized traits of Nordic man—his great height and fair skin, his blond hair and blue eyes, and his "splendid fighting and moral qualities"—cannot survive genetically if mixed with the darker and more primitive elements of the Alpines and the Mediterraneans. In the most famous passage in all of racist literature, Grant summarizes the principle of reversion thus:

> Whether we like to admit it or not, the result of the mixture of two races, in the long run, gives us a race reverting to the more ancient, generalized and lower type. The cross between a white man and an Indian is an Indian; the

> cross between a white man and a negro is a negro; the cross between a white man and a Hindu is a Hindu; and the cross between any of the three European races and a Jew is a Jew.[23]

Indeed, wherever one looks in the world, the Nordics appear to be an endangered species. In Europe, for example, the Nordics formerly occupied a much larger swath of the continent and comprised a much higher proportion of the population than they do now; but miscegenation is allowing the short brunets, after "centuries of obscurity," to reassert themselves. Like a subterranean gas that in the dead of night seeps to the surface, spreads over the landscape, and silently poisons all who breathe it in, the primitive blood of the Alpines and Mediterraneans is insidiously sapping the vigor of the Master Race. The demographic decline of the European Nordics is hastened by the fact that they are currently killing each other off in the fratricidal Great War, which is nothing less than "class suicide on a gigantic scale." No race can stand the loss of so much good blood, and the casualty figures from the front are "the evidence, if such be needed, of the actual passing of the Great Race."[24]

The situation is equally bleak in Africa, Asia, and South America, where the imperialistic Nordics are being outbred by their darker colonial subjects. Having evolved amidst the cold fogs and long winter nights of the north, the Nordics simply cannot endure the tropical sun's direct actinic rays, which disrupt their "delicate nervous organization." Grant's sexual anxieties and his Puritan aversion to sensuality (the word "purity" appears some thirty times in *The Passing of the Great Race*) contribute to his fear of warm, enervating climes. Nordics, he tells us, "revel" in the "blasts of the northern winter," but their vigor is diluted when transferred to "the *softer* conditions" of the tropics, where they "grow listless and cease to breed." (Grant's theory, we might note, did not prevent him from heading to Boca Raton every winter, where he endured the 'blasts of the northern winter' poolside at the Hotel Coquina.)[25]

Even in North America, the habitat to which they are so well acclimated, the Nordics are passing from the scene. "Survival of the fittest," after all, means the survival of the type best adapted to prevailing environmental conditions. In colonial times, the environment that confronted the settlers was an untamed continent, and survival entailed clearing the forests and fighting the Indians—tasks for which the Nordics were eminently suited. But the United States has changed from an agricultural to a manufacturing society, and "the type of man that flourishes in the fields is not the type of man that thrives in the factory." The truth is that the dark, little immigrants can operate a machine and navigate a sweatshop far better than "the big, clumsy, and somewhat heavy Nordic blond, who needs exercise, meat, and air, and cannot live under Ghetto conditions." It is with great pain that Grant is forced to admit that, "from the point of view of race," the environment of his homeland is leading to the "survival of the unfit."[26]

The Passing of the Great Race

The Consensus

1940s

"The bible of the Nordics, the book had an enormous influence . . . . [It] contains the frankest and the most clear-cut statement of the racist ideology ever published in this country."

Carey McWilliams

1950s

"This, at last, was racism."

John Higham

1960s

"Easily the most significant single piece of racist literature produced in this period by an American."

I. A. Newby

1970s

"It is nearly impossible to overstate the profound influence this book of Madison Grant's was to have—and continues to have—on American history and public policies."

Allan Chase

1980s

"A monumental work in the history of American racism."

Ronald M. Pavalko

1990s

"Racialism's American bible."

Frederic Cople Jaher

2000s

"A bestseller of great and lasting cultural importance."

Matthew Guterl

Little wonder that American patricians are refusing to bring children into a society where they must compete with the Italians, the Slovaks, and the Jews. And, as with the Red Deer of Moritzburg Castle, when the fittest males do not breed, the result is racial degeneration. The old-stock American is "withdrawing from the scene, abandoning to these aliens the land which he conquered and developed."[27]

In sum, all over the globe the Great Race is heading toward destruction as a result of miscegenation, civil war, imperialism, industrialization, urbanization, and race suicide. And so, Grant concedes, "Those who read these pages will feel that there is little hope for humanity."[28]

But, of course, there is hope, and it is provided by the new faith of eugenics. Unfortunately, so long as the United States is a democracy, it will be extremely difficult to enact a eugenic program. Ever since "the loose thinkers of the French Revolution and their American mimics" inflicted on us "the dogma of the brotherhood of man," Americans have had a perverse fondness for democracy. The consequences of republican government were not overly detrimental as long as the electorate was predominantly Nordic. But in the late nineteenth century the country permitted the beaten men of beaten races to enter its portals, and then carelessly granted political rights to these incoming "plebeians." The effect of universal suffrage has been to secure "the transfer of power . . . from a Nordic aristocracy to lower classes predominantly of Alpine and Mediterranean extraction." And it is difficult to see how the enfranchised "helots," indoctrinated by "the assumption that environment and not heredity is the controlling factor in human development," will ever allow the government they now control to enact eugenic measures.[29]

Grant yearns for the days when the nation was led by "a native American aristocracy." As a patrician progressive, he posits that in "a *true* republic" the people are led by "the wisest and best"—the aristocrats, who are the "experts" best qualified by "antecedents, character and education" to head the nation. Human society is like a long serpent dragging on the ground: the head of the serpent is the patriciate, while the body and tail are the lower classes that obediently follow wherever the head leads. It is only natural that "the head is always thrust a little in advance and a little elevated above the earth." Grant's choice of image would seem to reveal his uneasiness over the declining potency of his class. He employs another—equally phallic—simile to explain that in an aristocratic, as opposed to a democratic, society "the intellectual and talented classes form the point of the lance, while the massive shaft represents the body of the population and adds by its bulk and weight to the penetrative impact of the tip."[30]

Unfortunately, the day of the aristocrat has seemingly come to an end. "In America," says Grant, describing a situation that most of his countrymen approved but that he laments, "we have nearly succeeded in destroying the privilege of birth." The disciples of equality are even now plotting to establish "cacocracy and the rule of the worst and put an end to progress."[31]

And yet Grant, as always, is optimistic. The democrats will not always be in the saddle—nature will not permit it. In an antiegalitarian, patriarchal screed that had not been heard from a mainstream American since the heyday of John C. Calhoun, Grant decrees the "basic truth" that "inequality and not equality is the law of nature." Anticipating the rise of fascism, he predicts that the spread of scientific literacy will enable us to see that "the basis of the government of man is now and always has been, and always will be, force and not sentiment." Once democracy is disposed of and power returned to the patricians—once the "minute minority" is once more "called upon to supply brains for the unthinking mass of the community"—then a program of eugenics can be instituted.[32]

Such a program will attack the crisis of America's racial decline on three fronts. First, birth control will be legalized in order to reduce the number of offspring created by "the undesirable classes."[33]

Second, antimiscegenation laws will be passed, and steps taken to permanently segregate the races. History teaches us that when two races are located side by side, one of two things can happen: either one race exterminates the other, as the Nordics did the American Indians, or else they interbreed "and form a population of race bastards in which the lower type ultimately preponderates." And when that happens, a civilization is doomed. Fortunately, once eugenicists spread the word that the children of mixed marriages belong to the lower type, "to bring halfbreeds into the world will be regarded as a social and racial crime of the first magnitude."[34]

Third, sterilization will be instituted on a massive scale, so that inferior types who are "of no value to the community" will be "deprived of the capacity to procreate their defective strain." Grant envisions an ever-expanding program to remove "those who are weak or unfit." As a first step, he proposes the immediate sterilization of what he elsewhere referred to as "the submerged tenth." After all, "it would not be a matter of great difficulty to secure a general consensus of public opinion as to the least desirable . . . ten per cent of the community." Then, once the public has seen the benefits that accrue from eliminating this eugenic antipode of Du Bois's Talented Tenth, the program will be widened. "When this unemployed and unemployable human residuum has been eliminated, together with the great mass of crime, poverty, alcoholism, and feeblemindedness associated therewith, it would be easy to consider the advisability of further restricting the perpetuation of the then remaining least valuable types." Grant has faith that his program will be implemented, as soon as native-born Americans turn a deaf ear to the apostles of social uplift with their "sentimental belief in the sanctity of human life."[35]

And so, to those despairing over the passing of the Great Race, Grant offers the rational, efficient, and scientific remedy of eugenics. It should be underscored that Grant is not advocating the liquidation of undesirables, just their sterilization. And yet modern readers, knowing that the theorists of the Third

Reich read and admired *The Passing of the Great Race,* cannot help but feel a chill when they read Grant's assessment of eugenics: "This is a practical, merciful, and *inevitable solution* of the whole problem, and can be applied to an ever widening circle of social discards, beginning always with the criminal, the diseased, and the insane, and extending gradually to types which may be called weaklings rather than defectives, *and perhaps ultimately to worthless race types.*"[36]

Grant concludes his incredible book with an oft-quoted paragraph in which he solemnly warns his countrymen of the dangers posed by continued immigration and egalitarianism:

> We Americans must realize that the altruistic ideals which have controlled our social development during the past century, and the maudlin sentimentalism that has made America "an asylum for the oppressed," are sweeping the nation toward a racial abyss. If the Melting Pot is allowed to boil without control, and we continue to follow our national motto and deliberately blind ourselves to all "distinctions of race, creed, or color," the type of native American of Colonial descent will become as extinct as the Athenian of the age of Pericles, and the Viking of the days of Rollo.[37]

The Critics Applaud

The Passing of the Great Race was a tour de force. As Perry Miller said of another Puritan (the seventeenth-century theologian Gershom Bulkeley—incidentally a good friend of one of Grant's ancestors, Governor Robert Treat): "Energy and incisive language can, after all, flow from a reactionary pen."[38]

It should be clear, however (*pace* Osborn's claim in the preface that Grant's work is "wholly original") that almost nothing in *The Passing of the Great Race* is original. (But then, who would know less about originality than Henry Fairfield Osborn?) *The Passing of the Great Race* is a compendium of the work of other scholars, and almost every paragraph can be directly traced to one of the authors discussed in chapters 5 and 6 of this book. The primacy of race as the guiding force of history, the three subspecies of Europeans, the immutability of unit characters, the heritability of mental as well as physical traits, the dysgenic effect of war, Christ's Nordic heritage, the survival of the unfittest, the folly of charity, the fear of race suicide, the dominance of heredity over environment, the Nordic nature of the aristocracy, the principle of reversion, miscegenation as the cause of cultural degeneration, the dangers of the tropical climate, an obsession with purity, the horrors of urbanization, the dangers of industrialization, the evils of democracy, the need for aristocrats to rule society, the subservience of the individual to the community, protofascism, anti-Semitism, negative eugenics—we have seen each of these before. But no one had brought them all together in one place and presented the whole with such esprit, audacity, and clarity. As a result, what had been the province of a few obscure academics was now made accessible to the general reader. After *The Passing of the Great Race,* the biological threat posed by inferior races was no longer a speculative theory held by a few, but a palpable danger feared by all. And as we shall see, like all good sermons, the book galvanized the congregation into action. Entertaining, passionate, erudite—*The Passing of the Great Race* did for scientific racism what *The Communist Manifesto* did for scientific socialism. Fortunately for Marx and Grant, they both died before they could see the horrors that resulted when a regime embraced their philosophy and tried to remake society in its name.[39]

When *The Passing of the Great Race* was published (in blue cloth with a gold seal of Rollo the Norseman on the cover), Grant's friends were awestruck. From big-game hunter Carl Akeley to novelist John Galsworthy, they were astonished by Grant's scholarship and moved by his stirring call to save the endangered Nordics. Letters from readers poured in to Grant from all over the world, and continued to arrive for years afterward. One of Grant's favorite fan letters came from a stranger in Oregon who wrote to say how much he was indebted—literally indebted—to Grant. Ever since reading *The Passing of the Great Race,* the devotee had adopted the strategy of investing in the stocks of only 100-percent-Nordic nations, and the results had been highly remunerative. "Jugo-Slavia, Russia, Turkey, Italy, France, Czecho-Slovakia, Brazil, Mexico, Japan, China, et al, have been in my index expurgatorious ever since. The record of the investment market is the best vindication of your book within my observation. . . . You have saved me a lot of money and I take pleasure in writing this letter of appreciation."[40]

Grant's conservationist colleagues were equally enthusiastic. They knew that Grant had dedicated his life to saving endangered fauna, endangered flora, and endangered natural resources, and it did not seem at all strange that he was now trying to save his own endangered race. Theodore Roosevelt, for instance, who had been alerted by Henry Fairfield Osborn that Grant was working on something special, eagerly anticipated the book's publication. When it arrived at Sagamore Hill, he at once began "not just to read it, but to study it." He then sent Grant a very lengthy letter detailing his reactions. (Like so much of Grant's correspondence, that letter has disappeared from the archives, and all that remains is a blurb excerpted by Scribner's in its promotional materials.) "This book," Roosevelt told Grant, "is a capital book; in purpose, in vision, in grasp of the facts our people most need to realize. It shows an extraordinary range of reading and a wide scholarship. It shows a habit of singular serious thought on the subjects of most commanding importance. It shows a fine fearlessness in assailing the popular and mischievous sentimentalities and attractive and corroding falsehoods which few men dare assail. It is the work of an American scholar and gentleman; and all Americans should be sincerely grateful to you for writing it."[41]

The reviewers were as beguiled as TR by Grant's scholarship. The daily newspapers and popular periodicals were especially fulsome in their praise (although about a third of them called the book "The Passing of *a* Great Race," which rather misses the point). The *New York Herald* labeled the book "a profound study of world history from the ethnological standpoint" and predicted that it was "likely to excite wide-spread interest." The *Nation* waxed enthusiastic about Grant's "distinct qualities of originality, conviction, and courage." And the *New York Sun* was blunt: "It is an important book. . . . Get the book and read it."[42]

The more scholarly journals (e.g., *Science,* the *Journal of Heredity,* the *Geographical Review, Man,* the *International Journal of Ethics,* the *Annals of the American Academy of Political and Social Science*) also gave positive, albeit somewhat tempered, reviews. Few of them disagreed with Grant's premises or his conclusions, but they did feel that he was too partisan in tone, and they were all disappointed by the lack of footnotes. Those caveats aside, the vast majority of popular and scholarly opinion was impressed and convinced by Grant's book.

Wellesley economist Emily Greene Balch was one of the few non-Jews to criticize Grant's work. And, incredibly, her piece in the *Survey* was the only one to object to Grant's statement that negative measures should ultimately be applied to "worthless race types." Balch was, by any definition, one of those "do-gooders" of whom Grant despaired. A colleague of Jane Addams, Balch had been a member of the first graduating class at Bryn Mawr College in 1889, had founded a settlement house in Boston, and was a leader of the women's peace movement (she was dismissed from Wellesley in 1918 for opposing the United States' entry into World War I but was rewarded with the Nobel Peace Prize in 1946). And despite Madison Grant's opposition to war, his doubts about imperialism, and his support of birth control, Balch rightly felt that she had little in common with Grant. "Rash is the man," she said of Grant, "who passes lightly from skull measurements to vast unprovable sociological and historical generalizations. The pseudo-science of Gobineau and Houston Chamberlain and the ethical theory which makes it the function of the strong man to purge the world of the weak, might, one hoped, by this time have gone out of date."[43]

The remaining critiques were, as eugenicist Frederick Adams Woods pointedly noted, "signed by persons of non-Nordic race." For example, Horace Kallen, the immigrant son of an orthodox rabbi, attacked "blond, blue-eyed" Grant in the pages of the *Dial.* Kallen was soon to move to New York City to help found the New School for Social Research, but at the time he was an instructor of philosophy and psychology at the University of Wisconsin. Kallen bitingly noted: "The publisher's announcement heralds this stuff as an entirely new and original recasting of history on a purely scientific basis. This may be so, but if it is, the science is so pure that it is altogether imperceptible."[44] But Kallen's mocking review did not directly dispute any of Grant's statements; in the face of Grant's claims of Nordic superiority, Kallen offered no concrete critiques, only ridicule. This was, in fact, the modus operandi of all of Grant's critics. We now realize, of course, that they were absolutely correct to denigrate his Nordicism, but at the time none of them had any hard data to support their dissent. They were indignant not because Grant was a charlatan but because he opposed their humanistic ideals. Their motivation for resisting Grant's book, in short, was just as rooted in ideology as Grant's reason for writing the book in the first place. And Grant dismissed Kallen's attack as "that of a Jew and just what I expected from the followers of Boas."[45]

Which brings us to the most serious assault on Grant, which came from Franz Boas himself in the *New Republic.* It is a testament to the relative standing of the two men at the time (a position that in due course would be totally reversed) that Boas was compelled to avow, right at the beginning of his review, that he was "reluctant" to criticize Grant's book, as "the people of New York are indebted to the author . . . for much valuable scientific work that has been done by the Zoological Garden and the American Museum of Natural History, to both of which institutions he has given much of his time and energy." Nonetheless, Boas felt obliged to criticize because "the opinions expressed in this book are . . . so dangerous, that the very fact that the author is well known on account of his scientific interest, and that the book is introduced by a man so eminent as President Henry Fairfield Osborn, makes it necessary to expose the fallacies on which it is built up."[46]

In general, Boas charged that the deductive Grant was more interested in trumpeting his prejudiced theory than in reaching unbiased conclusions based on observed facts. Grant's history, according to Boas, consisted of a "vast amount of fanciful reconstruction," and "the supposed scientific data on which the author's conclusions are based are dogmatic assumptions which cannot endure criticism." He indicted the book (in words that Grant would have readily endorsed) as "practically a modern edition of Gobineau, and a reflex of the opinions of Chamberlain. It is a dithyrambic praise of the blond, blue-eyed white and of his achievements; a Cassandric prophecy of all the ills that will befall us on account of the increase of dark-eyed types."[47]

Specifically, Boas attacked Grant on three issues. First, he dismissed Grant's hierarchy of races and derided his attempts to claim for the Nordics any and all cultural achievements. Second, he criticized Grant's denial of the power of environment to influence various traits. And third, Boas criticized Grant's "naive" and "dogmatic" belief that unit characters can explain human heredity.[48] And so the gauntlet was laid down. Boas's Petrine denial of the doctrine of racial inequality, the predominance of nature over nurture, and the inheritance of unit characters amounted to a complete rejection of scientific racism. As we shall see, it was a challenge from which Grant would not shy away. But in the meantime, both men knew that Grant controlled the high ground. In concluding his review, Boas stated: "Nobody has so far succeeded in proving racial superiority." But that was misplacing the burden of proof. It was already obvious to most people familiar with the findings of modern science that the races were *unequal*; if Boas thought differently, it was up to him to prove it. And neither genetics nor anthropology had advanced to the point where they could substantiate Boas's contentions. Even A. L. Kroeber, Boas's disciple, begrudgingly admitted that "real proof, to be sure, is as wanting on one side as on the other."[49]

As a beleaguered patrician, Madison Grant had a fundamental belief in racial inequality, and he proceeded to construct an elaborate pseudoscience to justify it. As a persecuted immigrant, Boas had an a priori belief in equality of opportunity, and he therefore rose to challenge scientific racism. But with the exception of his study of immigrant head form, he had little data on which to base his opposition. "Political motivations," Robert Proctor reminds us, "can be as important in justifying correct views in science as they are in justifying false views." When Boas accused Grant in 1917 of deductive reasoning, he may as well have been looking in the mirror.[50]

A Fine-Tooth Comb

Despite its generally positive reviews, *The Passing of the Great Race* was only a modest success. While *Wikipedia* claims that the book sold 1,600,000 copies, it sold only 17,000 in the United States. Realistically it could not have sold more. The book had the misfortune to appear just as the country was preparing to declare war and the anti-German propaganda machine was shifting into high gear. A public that heard every day of the raping of nuns, the bombing of cathedrals, and the pillaging of peaceful villages was not about to purchase a book claiming that the Nordics were responsible for all the advances in civilization since the Neolithic period. Secondly, the antidemocratic (not to mention anti-Christian) message of the book was not calculated to win favor among a citizenry now swelled with patriotic pride about its superior system of government. People whose sons were dying to save the world for democracy certainly did not want to hear that democracy was a pathetic sham that empowered illiterate helots. Similarly, Grant's hereditarian philosophy ran counter to the traditional American faith in education, hard work, and initiative. Members of a society with a powerful faith that anyone could pull himself up by his bootstraps were not going to embrace a book telling them that their station in life had been predetermined by their genes. (Diane Paul has pointed out the incongruity that anyone would ever think eugenics could flourish "in the land of Horatio Alger.") Also, with all transatlantic shipping allocated to the war effort, immigration declined to almost nothing after 1914, and Grant's fulminations about the foreign peril suddenly seemed overblown if not obsolete. Finally, we should remember that his book was categorized and publicized by Scribner's as a work of science; it was accordingly reviewed in the "science" section of journals and placed on the science shelves of bookstores, and so it never really had a chance at mass popularity.[51]

Given all this, the fact that the book sold as many copies as it did is actually rather impressive. "No one would have believed," remarked Charles Stewart

Davison some years later, that "a book on racial descent and racial characteristics throughout the world would have been purchased and read in the United States." Novelist Gertrude Atherton agreed, and noted in 1922 that "Mr. Grant's book has had an astonishing popularity for a scientific work." Indeed, Scribner's was quite pleased with the book's reception and reprinted it in December of 1916, just two months after the initial printing. A delighted Maxwell Perkins congratulated Grant: "It is a pleasure to us to be able to say that your book, which was undoubtedly one of the most successful books addressed to the thoughtful public published at the same period, in a commercial sense as well as one of the most widely discussed and favorably commented upon, continues up to the present to sell very steadily."[52]

In 1918, Scribner's decided that the entrance of the United States into the Great War on the side of the Allies necessitated a revised edition. Grant exorcised from the second edition almost all associations between the Teutons (who were that portion of the Nordic race that had settled in Germany) and the original white settlers of America. Thus, for example, the Dutch colonizers of New York, who in the first edition were "purely Teutonic," in the second edition are "purely Nordic." Also, the second edition minimizes the geographical extent and cultural achievements of the Teutons, and reiterates that the present inhabitants of Germany are only "to a *limited* extent descendants of the ancient Teutonic tribes, being very largely Alpines." Grant now assures his readers that "with the United States in the field the balance of *pure* Nordic blood will be heavily *against* the Central Powers." Also, now that the United States was fighting to make the world safe for democracy, Grant's assertion that "the basis of the government of man is now and always has been, and always will be, force and not sentiment" had to be removed as sounding far too Germanic.[53] (It was perhaps to bolster his anti-German credentials that Grant now became a trustee of the American Defense Society, the rabidly anti-German patriotic organization that sprang up during the Great War to foster military preparedness and suppress "disloyalty." The chairman of the ADS was Grant's collaborator Charles Stewart Davison.)

Grant also took advantage of the second edition to correct particular points of the historical record. He numbered among his correspondents scores of scientists in the United States and Europe who were enthusiastic about the book and anxious that it be as accurate as possible. Scholars like John Beddoe, James Breasted, Henri Breuil, A. C. Haddon, T. Rice Holmes, Harry H. Johnston, Sir Arthur Keith, John Dyneley Prince, Sir William Ridgeway, G. Elliot Smith, William J. Sollas, H.G.F. Spurrell, and A. S. Woodward were continually updating Grant on issues of ethnographic significance, and he included their corrections in the revised document. Theodore Roosevelt also had a number of suggestions he wanted Grant to incorporate into the second edition, and he corresponded with Grant for many months about possible alterations.

Grant wrote seventeen drafts of the new edition and then drove out to Cold Spring Harbor, where he and Charles Benedict Davenport of the Eugenics Record Office (who had hopes that the book would be used as a textbook in colleges—hopes that would be realized) went over the manuscript with a fine-tooth comb. Grant also consulted with Charles Stewart Davison, William E. Castle (chair of zoology at Harvard), and E. G. Conklin (chair of biology at Princeton). They knew that *The Passing of the Great Race* was going to be used as a weapon by the proponents of nature, and with the specter of Franz Boas always hovering in the background, Grant and his friends were determined, as Grant said, "to make this second edition fool-proof, so that it can be used and quoted without hesitation by those of us who have the courage to do so."[54]

Scientifically, the most important change they made to the text was the removal of all references to "unit characters." Even as the first edition was being published, modern geneticists like T. H. Morgan were abandoning the concept of unit characters, and Grant knew that he had to drop the term if he was going to maintain scientific credibility. This was a major concession, as the implementation of a program of eugenics had been predicated on the ease with which defective characters could be identified and eliminated from the population. If geneticists were now correct that each human trait was the product of multitudinous genes, then exorcising alcoholism, pauperism, feeblemindedness, and so on from the germ plasm was going to be far more difficult than the early eugenicists foresaw. The textual change from "unit characters" to "characters," however, did not alter Grant's conviction that human personality traits—no matter how many genes it takes to create them—are determined by heredity, and that the role of the environment in their expression is minimal.

In certain respects, therefore, the second edition of *The Passing of the Great Race* is more accurate than the first. On the other hand, Grant cleverly revised the text in all sorts of subtle ways to strengthen its value as propaganda. The word "negro," for example, is now capitalized—not as a sign of respect but as a sly way of indicating that Negroes are a separate species of mankind. And instead of claiming that Mediterraneans are "stunted" in height compared with Nordics, Grant now says that their height is "distinctly less" than that of Nordics—since the term *stunted* implies that environmental factors may have played a role in their diminished stature.[55]

Grant also used the second edition to respond, obliquely, to Franz Boas. Without directly referring to the Jews, Grant explains to his readers that it is only to be expected that certain types would object to his claim that heredity is more important than environment, as "those engaged in social uplift and in revolutionary movements are . . . usually very intolerant of the limitations imposed by heredity." Using phrases that everyone understood as code words for Jews (and that would be echoed by later anti-Semites who referred to Jews as "rootless cosmopolitans"), Grant claims that his espousal of nature over nurture is naturally offensive "to the advocates of the obliteration, under the guise of internationalism, of all existing distinctions based on nationalism, language, race, religion and class. Those individuals who have neither country, nor flag, nor language, nor class, nor even surnames of their own and who can only acquire them by gift or assumption, very naturally decry and sneer at the value of these attributes of the higher types."[56] Grant warns that these outcast breeds are only going to grow stronger in America as its democratic institutions continue to effect the transference of power "from the higher to the lower races." And he predicts that as democracy advances, the United States will witness the florescence of its two illegitimate offspring: socialism and "obsolete religious forms"—a thinly veiled reference to Judaism. Grant alerts his readers that as in ancient Rome, where the emperors placated the mob by erecting shrines to strange gods utterly unknown to the original Romans, so in America today, "strange temples, which would have been abhorrent to our Colonial ancestors, are multiplying."[57]

These indirect attacks on Boas and the Jews are typical of Grant, whose anti-Semitism is almost always expressed in the most delicate manner. Indeed, it is a curious fact that in all of *The Passing of the Great Race* only five sentences cast direct aspersions on Jews. And yet Grant is (rightly) considered to be one of the leading anti-Semites in American history (the description by Allan Chase is typical: "no cause was dearer to the heart of Madison Grant than the total annihilation of the Jews"). It is a testament to the cleverness of Grant's presentation that those five sentences were able to affect history with such force.[58]

Two years later, in 1920, yet a third edition of *The Passing of the Great Race* was produced. The changes in the new edition were very minor but Grant insisted they be made, and Scribner's consented, as it was keenly interested in keeping Grant happy. His name brought prestige to the firm, and Maxwell Perkins could only declare his and his colleagues' "great sympathy with your desire to improve nearly as possible to the point of perfection, a book which has been so valuable to us and in which we have all here felt the deepest interest since the day we undertook to publish it."[59] It was in 1920 that Edgar Lee Masters (whose *Spoon River Anthology* had made him the most popular poet in the country) reflected the public's growing appreciation of Grant:

"The Great Race Passes,"
by Edgar Lee Masters

They were the fair-haired Achaeans,
Who won the Trojan war;
They were the Vikings who sailed to Iceland
And America.
They became the bone of England,

And the fire of Normandy,
And the will of Holland and Germany,
And the builders of America.

They were chosen for might in battle;
For blue eyes and white flesh,
For clean blood, for strength, for class.
They went to the wars
And left the little breeds
To stay with the women,
Trading and plowing.
They perished in battle
All the way along the stretch of centuries,
And left the little breeds to possess the earth—
The Great Race is passing.

Their blood flowed into the veins of David,
And the veins of Jesus,
Homer and Aeschylus,
Dante and Michael Angelo,
Alexander and Caesar,
William of Orange and Washington.
They sang the songs,
They won the wars.

On State street throngs crowd and push,
Wriggle and writhe like maggots.
Their noses are flat,
Their faces are broad,
Their heads are like gourds,
Their eyes are dull,
Their mouths are open—
The Great Race is passing. . . .[60]

After being shown the poem, Charles Scribner mused that he ought to sue for infringement of copyright.[61]

Just one year later, in 1921, Scribner's brought out the famous fourth edition of *The Passing of the Great Race,* whose chief feature was the long-awaited documentary supplement. Maxwell Perkins understood that Scribner's was probably not going to make a profit on this new edition (which sold for $3.50—a hefty price in those days). In fact, he was convinced that bringing out four editions in fewer than five years had "very distinct disadvantages from the purely commercial point of view." But he nonetheless green-lighted the project, as *The Passing of the Great Race* was a book "that we have all of us here come to have a peculiarly strong interest in and regard for," and the documentary supplement would "give the book a greater value from the point of view of scholarship."[62]

The documentary supplement provides references for—and elaborations on—many of the claims made by Grant in the body of the book. It was included in response to the persistent demands of the critics for footnotes, and because Grant hoped that it would lessen the text's "appearance of dogmatism." The documentary supplement is not the equivalent of endnotes. It does not necessarily supply the sources of Grant's statements (this is obvious if only from the fact that one-third of the works cited in the supplement were published *after* Grant's book was written). It is, instead, exactly what it says it is: a *supplement* to the text that serves as an ex post facto bibliography, cum suggestions for further reading, cum commentary on the text. The supplement is 139 pages in length, causing Henry Fairfield Osborn to remark, "The tail is now bigger than the dog!"[63] If *The Passing of the Great Race* is the Bible of scientific racism, then the documentary supplement is the Mishnah, and in the fourth edition Grant expresses the hope that the supplement "will be of assistance to students of anthropology and to those who care to inquire further into the subjects under discussion."[64]

The fourth edition also included a new introduction and a greatly expanded 24-page bibliography from which Grant removed the names of Chamberlain and Gobineau, who damaged Grant's claim to scientific respectability and who had become political liabilities in the English-speaking world after the Great War. *The Passing of the Great Race* now weighed in at 476 pages, almost double its original size. It was reprinted several times (twice in 1922 and 1923, and once in 1924, 1926, 1930, 1932, and 1936), but the fourth was the final revision. As the years passed and sales began to dwindle, so did the royalties. Once, after receiving a check from Scribner's in the amount of $33, Grant assured Maxwell Perkins, with the usual twinkle in his eye: "I shall invest this large sum of money in some security that will afford me a handsome income. I will, however, abandon any attempt to live exclusively on the moneys derived from my literary activities."[65]

* * *

In the new introduction to the fourth edition of *The Passing of the Great Race,* Grant triumphantly points to the effect of his book on U.S. society in the five years since it first appeared. The work's original intention, he recalls, was to rouse his fellow Americans "to the overwhelming importance of race and to the folly of the 'Melting Pot' theory, even at the expense of bitter controversy. This purpose," Grant now states with pride, "has been accomplished thoroughly." The truth of that claim will be verified in the next three chapters, as we enumerate the role played by *The Passing of the Great Race* in the growth of scientific racism and the implementation in the 1920s of the infamous programs of immigration restriction, sterilization, and antimiscegenation.

8 Grant's Disciples

We used to think our fate was in the stars. Now we know, in large measure, our fate is in our genes.

James Watson

The Passing of the Great Race appeared just before the United States entered the Great War, and hence the book was somewhat overlooked when first published. But the resumption of large-scale immigration after the armistice, the nation's postwar disillusionment with democratic crusades, and the Red Scare's legacy of intolerance created fertile ground for the racist and elitist message of Madison Grant's book. While *The Passing of the Great Race* was never a best seller, its ideas began percolating throughout U.S. society soon after the war, and became part of the common intellectual currency of the 1920s.

During the half decade between the end of the war in 1918 and the enactment of the Immigration Restriction Act of 1924, which all but ended immigration to the United States, numerous references to Madison Grant and his book appeared in scholarly works by prominent scientists. Grant's ideas also materialized in the speeches of politicians and the sermons of ministers. His words showed up in the pages of ladies' magazines and the pamphlets of the Ku Klux Klan. His theories were incorporated into paintings and into poems. His findings were cited by birth-control advocates on the left and white supremacists on the right.

As with the Boone and Crockett Club, Grant was able to change history by convincing a small but well-connected group of influential figures of the rectitude of his ideas. His book may have been read only by thousands, but the works of his disciples were read and seen by millions. And as a result, race consciousness among America's Nordics was revived to the level of antebellum days. (Few people realized that the very term "Nordic," which was universally accepted and employed by laymen and scientists alike, was a neologism introduced by Grant in 1916.) By 1922, paleontologist William K. Gregory was marveling that *The Passing of the Great Race* "simply by its own inherent force [had] stimulated anthropological investigation, aroused widespread interest in the subject of race, and given a powerful impetus to the eugenics movement in this country and abroad; [and] it has unquestionably influenced the Congress of the United States." Charles Stewart Davison concurred that when it came to the public's acceptance of scientific racism, Grant's book had "marked the turning point." Referring to him as if he were a religious prophet, the *Eugenical News* in 1927 noted with reverence: "The new way was opened up by the great conservationist, Madison Grant, in his *Passing of the Great Race*."[1]

The Evangelists

After *The Passing of the Great Race* was published, a host of the nation's leading academics endorsed its findings. I have demonstrated elsewhere[2] the direct influence that Madison Grant had on the work of important biologists (such as Samuel J. Holmes and E. G. Conklin), geneticists (e.g., William E. Castle and Edward M. East), zoologists (e.g., Vernon Kellogg and Horatio H. Newman), sociologists (e.g., Henry Pratt Fairchild and Edward A. Ross), psychologists (e.g., Kimball Young and William McDougall), anthropologists (e.g., Clark Wissler and Albert E. Jenks), historians (e.g., Wallace Thompson and Hamilton J. Eckenrode), university presidents (e.g., George Barton Cutten and Jacob Gould Schurman), and geographers (e.g., Ellsworth Huntington). These were well-established scholars at institutions such as Harvard, Princeton, Stanford, and Yale who incorporated Grant's racist theories into their books, recommended *The Passing of the Great Race* to their students, and often explicitly endorsed Grant's positions on immigration restriction, sterilization, and miscegenation.

As American scholars converted to eugenics, they soon began teaching courses on the subject. Before *The Passing of the Great Race* was published, fewer than 9 percent of the nation's colleges and universities offered courses on eugenics; by the late 1920s, eugenics was being taught at 75 percent of these institutions, including Harvard, Columbia, Brown, Cornell, Wisconsin, Northwestern, and the University of California.[3] To service these courses, a number of eugenics textbooks were produced, of which the most popular was *Applied Eugenics*, written by two friends of Madison Grant: Paul Popenoe (editor of the *Journal of Heredity*) and Roswell H. Johnson (who had been a student of Charles Benedict Davenport's at Harvard and was a professor of eugenics at the University of Pittsburgh). Their textbook, which was translated into German and Japanese, referred students to *The Passing of the Great Race* for an analysis of the racial makeup of the United States.

In addition to professional scientists, there was another group of authors who were influenced by Grant. This was the cohort of popular writers and journalists who were largely untrained in the sciences and received most of their appreciation of eugenics from *The Passing of the Great Race.* They were usually younger than Grant, looked up to him as an older and wiser mentor, and became his enthusiastic disciples in the 1920s. Unencumbered by scholarly caution, these converts unabashedly preached the Grantian creed to the public at large and, through their best-selling works, became quite famous and influential.

Among their number was Seth K. Humphrey, an unmarried descendant of Puritans, who recapitulated *The Passing of the Great Race* in his book *The Racial Prospect* (1920). Clinton Stoddard Burr did the same in *America's Race Heritage* (1922), as did Charles W. Gould, whose *America: A Family Matter* (1920) traced the decline of Egypt, Greece, and Rome to the infusion of blood from degenerate races and warned that America was not immune to a similar fall from grace. Like so many racists, Gould likened the presence of the New Immigrants to a cancer: "Ten million malignant cancers," he wrote, "gnaw the vitals of our body politic," and his book ends with the warning: "Americans, the Philistines are upon us." Gould and Grant were both alumni of Yale and Columbia Law School, and they were very good friends. They corresponded regularly and dined together quite often, which they had time to do since both were childless.[4]

Another leading Grantian was Albert E. Wiggam, popular speaker and best-selling author of *The New Decalogue of Science* (1922). "I have always been more of an evangelist than a scientist," Wiggam admitted. He extolled Madison Grant, by contrast, as "an eminent student of anthropology" and heartily recommended *The Passing of the Great Race* to his readers. Wiggam's *Fruit of the Family Tree* (1924) and *The Next Age of Man* (1927) were also very popular, and his articles appeared regularly in *Good Housekeeping, Ladies' Home Journal, Cosmopolitan, McCall's, Century,* and *Reader's Digest.* As such, he was one of the most well-known advocates of eugenics in America—and he too passed on without creating any children of his own.[5]

Another popularizer of Madison Grant was William Sadler, the author of dozens of popular "self-help" books including that quintessential book of the 1920s: *The Elements of Pep.* After reading *The Passing of the Great Race,* Sadler was converted to eugenics and went on to publish books with titles such as *Race Decadence, The Truth about Heredity,* and *Long Heads and Round Heads.* In the latter work, Sadler wrestled with the question of how Germany, one of the most cultured nations in the world, could have acted so barbarously in the Great War. Following Madison Grant's lead, Sadler posited that the explanation lay in "the anthropology of the Germanic peoples," to wit, the destruction of the Nordic element by the roundheaded Alpines during the Thirty Years' War. Sadler's book was quite an homage to *The Passing of the Great Race.* A comparison of passages from the two books gives a sense of Sadler's rather liberal method of intellectual borrowing (a technique occasionally emulated by modern-day undergraduates):

Madison Grant	*William Sadler*
What the Melting Pot actually does in practice, can be seen in Mexico, where the absorption of the blood of the original Spanish conquerors by the native Indian population has produced the racial mixture which we call Mexican, and which is now engaged in demonstrating its incapacity for self-government. The world has seen many such mixtures of races, and the character of a mongrel race is only just beginning to be understood at its true value. It must be borne in mind that the specializations which characterize the higher races are of relatively recent development, are highly unstable and when mixed with generalized or primitive characters, tend to disappear. Whether we like to admit it or not, the result of the mixture of two races, in the long run, gives us a race reverting to the more ancient, generalized and lower type. The cross between a white man and an Indian is an Indian; the cross between a white man and a negro is a negro; the cross between a white man and a Hindu is a Hindu; and the cross between any of the three European races and a Jew is a Jew.	What the "melting pot" actually does, and what it threatens to do in this country, can be better seen by an ethnic study of Mexico and its people. Here the blood of the original Spanish conquerors (who, like all the world's explorers, were Nordics) has been absorbed by the inferior native Indian population, resulting in a race admixture which we now observe in the present-day inferior Mexican people; and from the days of Rome down to the present, these mongrel types have always represented retrograde movements in the civilization of the day. It must be borne in mind that some of the more desirable specifications in the civilized races are of relative recent origin, and that when two greatly dissimilar races mix, the usual result is a quick gravitation downward to the more ancient, primitive, and lower type of man. The cross between a white man and a negro is not a white man, but a negro. The cross between a white man and a Hindu is a Hindu; and a cross between any of the three more modern European races and a Jew is always a Jew.[6]

In scores of books, hundreds of articles, and thousands of speeches delivered to women's clubs, businessmen's luncheons, fraternal organizations, and reform groups, Grant's disciples spread the good word of scientific racism throughout the land in the early 1920s. They preached that inequality was a biological fact, and that the Nordics were the superior race. They upheld the primacy of nature over nurture, and claimed that social ills such as poverty and crime were the result of inferior heredity. They taught that modern civilization artificially keeps alive the physically and mentally weak, that the defective types were consequently outreproducing the fitter types, and that applying the lessons of Mendelian genetics to human beings was both sensible and necessary. And they urged that, until a full-blown program of eugenics could be instituted, steps must be taken—especially immigration restriction—to lessen the danger of miscegenation and the consequent possibility of reversion.

The Grantians were quite successful in preaching the eugenic gospel, and their message reached both ends of the political spectrum. Hiram Wesley Evans, for example, the imperial wizard and emperor of the Knights of the Ku Klux Klan, announced that he was exceedingly pleased that "modern research is finding scientific backing" for the platform of the Klan. Like Houston Stewart Chamberlain, he noted that the Klan did not really require hard evidence to justify its prejudices, for anti-Semitism and anti-Catholicism were inherent in the soul of the native-born American. Still, Evans specifically lauded Madison Grant—an admirable man "of great education and mind"—for his uncanny ability to express the grievances of the typical Klansman. In an address at the Texas State Fair in 1923, Evans called for a complete ban on the immigration of non-Nordic types. He cited the Grantians to warn that the Jews, like the Catholics and the Negroes, were an "absolutely unblendable element" who were "alien and unassimilable." No amount of environmental reform could alter the genetic makeup of the Jew: "Were the melting pot to burn hundreds and hundreds of years, Jew and Gentile would each emerge as he is today." Thus, in formulating the nation's immigration policy, the imperial wizard pleaded with Congress to take account of "all the established truth with respect to anthropology" in order to prevent the "tainted blood" of the immigrant hordes from turning America's cities into "modern Sodoms and Gomorrahs."[7]

On the opposite end of the spectrum, we find liberal defense attorney Clarence Darrow arguing in the Leopold and Loeb case of 1924 that it was his clients' genes, and not their environment, that made them criminals. "This terrible crime was inherent in [Loeb's] organism," intoned Darrow before the jury. "I do not know what remote ancestor may have sent down the seed that corrupted . . . Dicky Loeb. All I know is, it is true, and there is not a biologist in the world who will not say I am right." Darrow, of course, did not believe a word of what he was saying. But he knew that American jurors of the 1920s, having been exposed for half a decade to the scientific racism of Madison Grant and his cohorts, would accept this hereditarian argument as scientific gospel.[8]

Grant's most famous protégé, and, as such, the second most influential racist in the country, was Lothrop Stoddard (1883–1950). Stoddard came from an old New England family and was a descendant of Solomon Stoddard, the seventeenth-century pastor of Northampton (and grandfather of Jonathan Edwards). In keeping with family tradition, Lothrop Stoddard's father had studied to become a minister (although he withdrew after losing his faith). As a boy, Lothrop was sent to school in Dresden, the city of Richard Wagner, Houston Stewart Chamberlain, Theodore Roosevelt, and Madison Grant. Stoddard re-

Lothrop Stoddard, the second most influential racist in America.

turned to the United States to earn a law degree but, like Grant, he practiced only briefly. At the age of twenty-six, he quit the legal profession to study history at Harvard where, as his biographer put it, he "embraced Grant's racial cosmogony in its entirety." Stoddard's dissertation was subsequently published as *The French Revolution in San Domingo,* in which he recounted the "annihilation" of the white colonists of Haiti by their slaves, who were inflamed by the egalitarian ideals of the French Revolution. Stoddard saw in that eighteenth-century clash between white supremacy and black equality a parable for the modern world. Ten years earlier, W.E.B. Du Bois had declared that "the problem of the Twentieth Century is the problem of the color line," and in the opening sentence of his book Stoddard globalized that sentiment by announcing that "the world-wide struggle between the primary races of mankind . . . bids fair to be the fundamental problem of the twentieth century."[9]

Stoddard earned his Ph.D. in 1914, and spent the rest of his life as a journalist, author, and devotee of Madison Grant. Stoddard was in constant contact with Grant, and he borrowed almost all his ideas (and many of his sentences) from the master. His relationship with Grant was probably the closest one of his life, with the notable exception of that with his mother—with whom he continued to live until he finally married at the age of forty-three. (When Stoddard violated what was almost a rule among eugenicists and actually managed to sire two children, a delighted Robert DeCourcy Ward congratulated him on creating "more young Americans of the right stock!" Stoddard's beloved mother, on the other hand, was greatly perturbed by his marriage and promptly joined a religious cult.) Soon after their marriage, Stoddard's wife became an ever more devout Christian Scientist, and Stoddard briefly emulated her. He bitterly renounced the religion, however, when his wife died gruesomely from cancer after refusing medical assistance.[10]

Stoddard honored Madison Grant as "one of our most eminent patriots," and he revered *The Passing of the Great Race,* which, he later remembered, "made a great sensation and played an important part in America's awakening to national and racial realities." Entire passages from *The Passing of the Great Race* appeared almost verbatim in Stoddard's writings, while those ideas of Grant's that were perhaps overly complex were rephrased by Stoddard in his more journalistic and accessible style.[11]

The names of Grant and Stoddard were naturally linked together by the reading public, which hailed Grant as the prophet and Stoddard the apostle of scientific racism. They were a formidable combination, and they complemented each other perfectly. Like Grant, Stoddard captured the imagination of his loyal readers (and his enthusiastic reviewers) with his striking style, breadth of vision, and pose of expertise. Stoddard also cultivated the Calvinist pessimism of Grant's biological determinism as well as the patrician optimism of his faith in the redeeming qualities of the Nordic race. But while Grant concentrated on the internal threats to the Nordics in America, Stoddard focused on the international situation and the dangers posed by the multiplication of the black and yellow races abroad. Over the years, Stoddard published twenty-two books on world affairs, but his most popular works were the sensationalist tetralogy produced in a rush of creativity in the first half of the 1920s: *The Rising Tide of Color* (1920) (for which Grant wrote the lengthy introduction), *The New World of Islam* (1921), *The Revolt against Civilization* (1922), and *Racial Realities in Europe* (1924). The central theme of these Grant-inspired best sellers was that the Nordic race, which is "always and everywhere . . . a race of warriors, sailors, pioneers, and explorers," was on the brink of being inundated by the inferior colored races, and for the usual Grantian reasons: low birthrate, indiscriminate philanthropy, immigration, mongrelization, reversion, and the dysgenic effects of the Great War (which Stoddard called the "White Civil War"). Sounding much like his ancestors in the Northampton pulpit, Stoddard pleaded with his brethren in postwar America not to become so distracted with "domestic dissensions" that they failed to notice the impending "cataclysm" being wrought by the forces of darkness abroad, and he urged his readers to adopt the Grantian program of immigration restriction, sterilization, and antimiscegenation laws.[12]

Most of the nation's leading newspapers and magazines accepted the writings of Grant and Stoddard as holy writ. The two men received the official endorsement, for example, of the exceedingly influential George Horace Lorimer, who had attended Yale at the same time as Grant and was now the editor of the most widely read magazine in the United States, the *Saturday Evening Post.* In a series of editorials in the spring of 1921, Lorimer informed the nation that it was time to relinquish "the rose-colored myth" of a "magical melting pot" that was going "to make Americans out of any racial scrap humanity cast into it." He endorsed immigration restriction, and recommended "two books in particular that every American should read if he wishes to understand the full gravity of our present immigration problem: Mr. Madison Grant's *The Passing of the Great Race* and Dr. Lothrop Stoddard's *The Rising Tide of Color.*" Lorimer, an ardent hunter and conservationist who was also well known for his crusades to regulate big business and end the evils of child labor, upheld Grant and Stoddard as "scientific" writers who "base their theses upon rather recent advances in the study of heredity and other life sciences." He pointed to the example of Luther Burbank who, "unswerved by sentiment," employed only selected strains when creating new breeds of plants and "ruthlessly suppressed" any "degenerate types." And Lorimer concluded that "we cannot say we have not been warned when authorities of Mr. Madison Grant's standing" provide "overwhelming evidence" to show that the old-stock American "will become as extinct as the Athenian of the age of Pericles and the Viking of the days of Rollo."[13]

As an enthusiastic convert to scientific racism, Lorimer assigned future Pulitzer Prize recipient Kenneth Roberts to write a series of anti-immigration articles for the *Saturday Evening Post* in the early 1920s. Roberts proclaimed: "Every American who has at heart the future of America . . . owes it to himself and to his children to get and read carefully *The Passing of the Great Race,* by Madison Grant [and] *The Rising Tide of Color,* by Lothrop Stoddard." He then borrowed extensively from *The Passing of the Great Race* to show that while the United States was "founded and developed by the Nordic race," if immigration were allowed to continue and "a few more million members of the Alpine, Mediterranean and Semitic races" were "poured among us, the result must inevitably be a hybrid race of people as worthless and futile as the good-for-nothing mongrels of Central America and Southeastern Europe."[14]

The eagerly anti-Semitic Roberts was an effective propagandist, and his widely read articles—which would be quoted often in Congress during the debates on immigration restriction—gave a particularly unflattering view of the "ruthless" and "underhanded" Jews of Europe, who in Roberts's view were nothing more than "human parasites." Roberts was especially revolted by the "howling, shrieking, pushing, squirming mass" of "wretched" Jews who crowded into the filthy Warsaw ghetto, with their long beards, hooked noses, and greasy gabardines, and he imparted his revulsion to the readers of the *Saturday Evening*

Post on a regular (some might say obsessive) basis. Roberts found the "stench" of Polish Jews to be "sickening," and he professed to be frustrated that mere words could not convey the absolute "undesirability of these peculiar people."[15]

In a chilling passage in "The Existence of an Emergency," Roberts explained in 1921 that European Jews who obtained visas for the United States were prevented from boarding their steamships until they had been certifiably deloused at the port of embarkation under the watchful eye of a medical officer of the U.S. Public Health Service. The most efficient delousing operation had been set up at the Troyl Emigrant Camp at the port of Danzig, to which Polish Jews were conveyed on special sealed trains from Warsaw. The camp, composed of rows of one-story barracks surrounded by barbed wire, was run and operated by Germans officials. When the Jews arrived at the camp, they were separated by sex, shaved of their hair, stripped of their clothing, and made to line up outside the showers. Roberts interviewed the German guards, who were proud of the camp's efficiency and boasted of their ability to delouse hundreds of Jews in a single day. While the Jews were in the showers, their clothing was placed in airtight rooms and poisonous gas was piped in "in order to destroy the lice and the germs that are usually present." Roberts mocked the hysterical reactions of the Jews when they were ordered to undress in front of the shower rooms, and he snidely remarked that from all their "moans and tears" you would think it was the Jews themselves, rather than their clothing, that were being gassed by the Germans.[16]

Roberts's articles were packaged in book form in 1922 as the best-selling *Why Europe Leaves Home* (the frontispiece of which was a photograph of a horde of America-bound Jews lined up at the visa office in Warsaw). One of the few negative reviews came from the *New Republic*, which regretted that Roberts had chosen Madison Grant as "his prophet" and lamented that Roberts's articles in the *Saturday Evening Post* had "a greater American circulation in a week than that which, say Bertrand Russell's *Roads to Freedom* can hope to attain in ten years." (To which Grant's colleague E. G. Conklin pointed out that the *New Republic* were better titled the "Jew Republic.")[17]

Kenneth Roberts died, childless, in 1957 (thus living long enough to see the practices of the Troyl Emigrant Camp systematized during the Holocaust). Though known today as one of the twentieth century's most popular writers of historical fiction (*Arundel, Northwest Passage, Rabble in Arms,* et al.), it was his anti-immigration articles for the *Saturday Evening Post* that first made him a household name in the United States.

In addition to Kenneth Roberts, the team of Grant and Stoddard influenced many other writers of fiction. Edgar Rice Burroughs, according to Richard Slotkin, "was directly influenced" by their work, as was popular novelist Gertrude Atherton, who declared that "human nature is largely a matter of the cephalic index" and staunchly acknowledged her debt to Madison Grant's "remarkable work, with its warning of tremendous import to civilization." In an article in the *Bookman,* Atherton perspicaciously attributed the unfortunate vogue for naturalism in American fiction to the preponderance of brachycephalic writers, and explained that "the early literature of this country . . . was written by Nordics. Fancy a round-head writing *The Scarlet Letter*!" Atherton's chief regret was that the Nordics had so far been "too unenlightened to sterilize such groups and exterminate them."[18]

Peter B. Kyne was another popular writer who incorporated scientific racism into his novels. Kyne's racial views are best expressed by the character in *Never the Twain Shall Meet* who sagely warns the naïve protagonist, momentarily lovestruck by the come-hither beauty of a half-breed from the South Pacific: "You cannot dodge the Mendelian law, my boy. Like begets like, but in a union of opposites we get throwbacks. . . . You're not going to run the risk of mongrelizing the species, are you?"[19]

Madison Grant and Lothrop Stoddard were sufficiently established in the public's consciousness in the early 1920s that their Scribner's stablemate, F. Scott Fitzgerald, could conflate their names in his novel *The Great Gatsby* and gently mock them through his character Tom Buchanan:

> "Civilization is going to pieces," broke out Tom violently. "I've gotten to be a terrible pessimist about things. Have you read *The Rise of the Coloured Empires* by this man Goddard?"
>
> "Why, no," I answered, rather surprised by his tone.
>
> "Well, it's a fine book, and everybody ought to read it. The idea is if we don't look out the white race will be—will be utterly submerged. It's all scientific stuff; it's been proved."
>
> "Tom's getting very profound," said Daisy, with an expression of unthoughtful sadness. "He reads deep books with long words in them. What was that word we—"
>
> "Well, these books are all scientific," insisted Tom, glancing at her impatiently. "This fellow has worked out the whole thing. It's up to us, who are the dominant race, to watch out or these other races will have control of things. . . .
>
> "This idea is that we're Nordics. I am, and you are, and you are, and—" After an infinitesimal hesitation he included Daisy with a slight nod, and she winked at me again. "—And we've produced all the things that go to make civilization—oh, science and art, and all that. Do you see?"[20]

One of the reasons that we sometimes sense an ideological affinity between the works of such otherwise divergent writers as Kenneth Roberts and Edgar Rice Burroughs, is that their books sported similar covers and illustrations. And that was because they were illustrated by the same man, artist N. C. Wyeth. While better known to a later generation as the father of painter Andrew Wyeth, resolute New Englander N. C. Wyeth was one of the country's foremost artists in the 1920s and 1930s. Over the years, his art appeared on the covers of all the leading magazines, including *Good Housekeeping, Harper's, Ladies' Home Journal,* and the *Saturday Evening Post.* Wyeth also produced a lot of work for Scribner's, where he shared the same editor (Maxwell Perkins) as Madison Grant and F. Scott Fitzgerald. In fact, Wyeth often socialized with Scott and Zelda Fitzgerald, and it was through them that the artist fell under the spell of *The Passing of the Great Race.* In 1919, Wyeth told his mother that the "monumental" *Passing of the Great Race* had "absorbed my attention tremendously," as it was "a startlingly new but convincing argument based upon profound scientific knowledge." Wyeth was especially impressed that Grant's book had "aroused the deepest attention of the Great Roosevelt (the loss of whom I'm feeling more and more is a calamity to the whole world!)."[21]

Wyeth's post-*Passing* work included illustrations for *The Return of Tarzan* by Edgar Rice Burroughs and for *Arundel, The Lively Lady, Rabble in Arms,* and *Trending into Maine* by Kenneth Roberts. Wyeth similarly envisioned the colonial period for James Fenimore Cooper's *Last of the Mohicans* and *The Deerslayer,* Henry Wadsworth Longfellow's *Song of Hiawatha* and *The Courtship of Miles Standish,* and Washington Irving's *Rip Van Winkle.* Wyeth showed Americans how to perceive the West in his illustrations for Buffalo Bill Cody's *The Great West That Was,* Helen Hunt Jackson's *Ramona,* and Charles Kingsley's *Westward Ho!* Boone and Crocketteers Francis Parkman (*The Oregon Trail*) and George Bird Grinnell (*Blackfeet Indian Stories*) also relied on Wyeth to bring the Old West to life. In addition, Wyeth's paintings appeared in numerous children's books and in many basic textbooks of American history, including James Truslow Adams's *History of the United States,* Charles Beard's *Basic History of the United States,* and Grace A. Turkington's *My Country.* Among several generations of Americans, when anyone tried to picture a scene from U.S. history, it was more than likely a painting by the Grantian N. C. Wyeth that came to mind.

With Nordicism becoming so popular in American society, it did not take long for leading politicians to associate themselves with the movement. In 1921, the Grantian position was endorsed both by the vice president of the United States and then by the president of the United States. In an article in *Good Housekeeping* titled "Whose Country Is This?" Calvin Coolidge (who attended the same Northampton church where Lothrop Stoddard's Puritan ancestors had preached) declared that the United States must cease to be regarded as a "dumping ground" for "the advancing hordes of aliens" that were not just an economic but also a biological threat to the well-being of the populace. Reaching, as always, for a business metaphor, the vice president explained: "It would be suicidal for us to let down the bars for the inflowing of cheap manhood, just as commercially, it would be unsound for this country to allow her markets to be overflooded with cheap goods." He then invoked a pithy slogan: "There is no room either for the cheap man or the cheap goods."[22]

Coolidge's boss, President Warren G. Harding, let it be known that he was mulling over a plan to appoint a blue-ribbon committee to study the race question. Senator James K. Vardaman of Mississippi endorsed the idea but cautioned that the president must be very careful about whom he appointed to the committee. Vardaman recommended a panel headed by Madison Grant and Lothrop Stoddard. He also urged the president to read *The Passing of the Great Race* and *The Rising Tide of Color.* Although the blue-ribbon committee was never created, Harding and Stoddard struck up an acquaintance, and the president invited Stoddard to spend an evening in his private study at the White House. ("I can promise you some pretty fair cigars," wrote the always genial Harding, "and I somehow don't think you take the Eighteenth Amendment too seriously!" Stoddard assured the president that he was, indeed, "no friend of Mr. Volstead.")[23]

In October of 1921, President Harding publicly endorsed Stoddard's *Rising Tide of Color* in a major speech in Birmingham, Alabama. The president initially gave his southern audience a start by declaring that the American Negro was entitled to full economic and political rights—but he hastened to add that this did not imply "*social* equality." Indeed, he warned the "colored people" in attendance that social equality was a dream that would never be realized. "Racial amalgamation there cannot be," declared the president (whose ancestors, it had been rumored during the 1920 campaign, had not been adverse to a little amalgamating themselves). "Whoever will take the time to read and ponder Mr. Lothrop Stoddard's book on *The Rising Tide of Color* . . . must realize that" there was "a fundamental, eternal, inescapable difference" between the races. The black man needed to understand that he belonged to "a distinct race, with a heredity . . . all its own." Consequently, there must be "absolute divergence in things social and racial. . . . A black man cannot be a white man."[24]

In the ensuing weeks, Stoddard's book began to wend its way through the administration. Secretary of the Interior Albert Fall read it and then forwarded it to Secretary of State Charles Evans Hughes. Secretary of Labor James J. Davis soon adopted the Grantian line, and wrote that since America was a Nordic country, "we should bar from our shores all [other] races which are . . . physically, mentally, morally and spiritually undesirable, and who constitute a menace to our civilization." Testifying before Congress in 1924, Davis explained that "the historians and scientists tell me that all the great civilizations of the past have fallen, not through hostile invasion, but through the peaceful penetration of alien peoples, usually entering their gates as workers or slaves." And he ominously warned that "more foreigners have passed through Ellis Island within a few months than there were in the hosts of the Huns and Vandals who utterly destroyed the boasted civilization of the Roman Empire."[25]

In sum, within a few short years of its publication, the thesis of *The Passing of the Great Race* was being repeated in the speeches and writings of scholars, editors, popular writers, and politicians. But as far as Madison Grant was concerned, the point of anthropology was not to interpret the world but to change it. And for Grant, good progressive that he was, changing the world meant forming an organization.

The American Eugenics Society

The largest and best-funded eugenic organization in America was the Eugenics Record Office, founded at Cold Spring Harbor in 1910 by Charles Benedict Davenport, with Harry H. Laughlin as superintendent. As we saw in chapter 6, the ERO sponsored research in eugenics, operated a summer school for eugenic field workers, and served as a repository for the hereditary data collected by those field workers. The ERO also published the *Eugenical News,* a small but important monthly edited by Laughlin that described eugenic research being conducted at Cold Spring Harbor and elsewhere.

In 1913, Davenport created a new organization to supplement the work of the Eugenics Record Office. Named the Eugenics Research Association, its main function was to hold an annual summer conference in Cold Spring Harbor at which scholars interested in eugenics could exchange information and coordinate plans for future research. The ERA was affiliated with the American Association for the Advancement of Science (fittingly through Section F: "The Zoological Sciences") and had two seats on the Council of the AAAS.

In 1918, Madison Grant was elected president of the Eugenics Research Association. It was a measure of the reputation of *The Passing of the Great Race* that Grant, with no scholarly credentials, was named to this position (the previous presidents of the ERA had all been eminent scientists). Under Grant's presidency, the membership of the ERA quadrupled to some four hundred persons, of whom slightly more than half were "active" members (meaning they were scientists engaged in eugenic research), while the rest were "supporting" members (i.e., amateurs like Grant who provided financial support for such research).

The ERA eventually assumed responsibility for publishing the *Eugenical News,* which evolved dramatically under Grant's influence. In the early years, the journal had trumpeted eugenics as a means for society to preserve superior—and eliminate inferior—persons. The focus was solely on *individuals.* But after Grant injected racism into eugenics, the *Eugenical News* was dominated by stories about the inherent incapacities of various racial groups and the dangers of miscegenation.

Grant's term as president of the ERA expired in 1919, but he remained an important member of the executive committee for many years, and the officers often held their meetings at his house. But since the ERA and the ERO were so-

Table 2. *Major eugenics organizations in the United States*

Founded	Name	Purpose	Founder(s)
1910	Eugenics Record Office (ERO)	Conduct research	Davenport
1913	Eugenics Research Association (ERA)	Facilitate exchange of scholarly information	Davenport
1922	Eugenics Committee of the U.S.A. (ECUSA)	Organize the AES	Grant, Davenport, Fisher, Little, Olson, Osborn
1926	American Eugenics Society (AES)	Propaganda and lobbying	Grant, Crampton, Fisher, Laughlin, Osborn

cieties primarily of and for scholars, Grant felt there was a need for another organization whose tasks would be to disseminate eugenic propaganda to the layman and to lobby for the enactment of eugenic legislation. The organization Grant had in mind would come to be known, eventually, as the American Eugenics Society (see table 2). We are going to take a close look at the formation of the AES, as it reveals a great deal about the methods and the mind-set of the eugenics movement in the 1920s.

Step One: The ECUSA

In February 1922, six leading eugenicists met at the American Museum of Natural History and formed the Eugenics Committee of the United States of America (ECUSA), an ad interim committee charged with creating a permanent eugenics organization in the United States. The six were Madison Grant, Henry Fairfield Osborn, Charles Benedict Davenport, C. C. Little (assistant director of the Department of Genetics of the Carnegie Institution of Washington, and future president of the University of Maine and the University of Michigan), Irving Fisher (famous economist, ardent conservationist, active Prohibitionist, and friend of Madison Grant since their days at Yale), and Harry Olson (chief justice of the Chicago Municipal Court, who was determined to stop "promiscuous breeding" among criminals by sterilizing "defectives" at as early an age as possible and thereby "stem the poison that is now being poured constantly into the American blood-stream").[26]

The ECUSA met bimonthly, usually in Madison Grant's home, where Grant and his colleagues methodically set about creating a central eugenics organization for the United States. They proceeded carefully and patiently, and with an optimistic sense that by bringing forth a new organization they could disseminate the eugenics ideal throughout society and thereby alter the course of history.

Step Two: The Advisory Council

The second step was to create an advisory council, as Grant decided that the ECUSA (like the Bronx Zoo, the New York Aquarium, and the American Bison Society) should be assisted by a committee of scientific experts. Grant devoted most of the spring of 1922 to forming the Advisory Council of the ECUSA. He and his associates sent letters to prominent scholars and public figures asking them to serve on the council and thereby protect America "against indiscriminate immigration, criminal degenerates, and . . . race suicide."[27] Ninety-nine figures responded to the call, including the usual stable of Grantians (e.g., Charles W. Gould, Seth K. Humphrey, Ellsworth Huntington, Roswell Johnson, David Starr Jordan, John Harvey Kellogg, Harry H. Laughlin, Paul Popenoe, Lothrop Stoddard, Robert DeC. Ward, A. E. Wiggam, and Frederick Adams Woods), and also many of the leading scientists in the nation. (For a selected list, see appendix C.)

These were authors of major textbooks, editors of important journals, and quite often presidents of their professional organizations. Many of them served important roles in the National Academy of Sciences and the National Research Council, and six were past or future presidents of the American Association for the Advancement of Science. Furthermore, the Advisory Council included the presidents of Antioch College (Arthur E. Morgan), Boston University (Lemuel H. Murlin), the University of California (William W. Campbell), Cornell University (Livingston Farrand), Stanford University (Ray Lyman Wilbur), and the president emeritus of Harvard (Charles W. Eliot). Grant was also anxious to have representatives from women's colleges, and he enticed onto the Advisory Council the dean of Barnard (Virginia Gildersleeve) and the presidents of Smith (William A. Neilson) and Wellesley (Ellen Fitz Pendleton).[28]

In addition, Grant shrewdly convinced Frederick S. Bigelow (the editorial writer for the *Saturday Evening Post*) to serve on the Advisory Council, as well as philanthropists such as Frank L. Babbott, Raymond B. Fosdick, Mrs. E. H. Harriman, Helen Hartley Jenkins, and Mary Harriman Rumsey. The Advisory Council also featured two of the most important clergymen in America, Harry Emerson Fosdick (of the Riverside Church) and William Lawrence (the Episcopal bishop of Massachusetts), as well as such politicians as Senator Royal S. Copeland (R-New York), Senator Robert L. Owen (D-Oklahoma), Rep. Albert Johnson (R-Washington), Gov. John Clayton Phillips (R-Arkansas), and Gov. Gifford Pinchot (R-Pennsylvania).

Almost every one of the ninety-nine members of the Advisory Council of the ECUSA was a highly educated, wealthy, influential American of old Puritan stock. They were, in general, progressive Republicans who were involved in numerous charitable causes and reform movements. Their desire to serve on the Advisory Council is an indication of the stature of eugenics in the early 1920s, and of the social and scholarly connections of Grant, Osborn, Davenport, and the other members of the executive committee.

Step Three: Membership

Unlike most of Grant's organizations, the Advisory Council was not just a list of names added to the letterhead for prestige value. Many of its members actively helped to define the priorities of the ECUSA, drafted its reports, and shaped its legislative program. The Advisory Council, in short, actually did *advise*. Still, the cachet of those ninety-nine names cannot be denied, and once they were placed on the letterhead, the executive committee felt confident enough to take step three, which was to send out a letter soliciting charter memberships. Prospective members were urged to join the "strong public movement to stem the tide of threatened racial degeneracy" and ward off "the destruction of the white race." The executive committee explained that eugenics not only represented "the highest form of patriotism and humanitarianism" but also offered to reduce "our burden of taxes . . . by decreasing the number of degenerates, delinquents, and defectives supported in public institutions."[29]

Response was favorable, and by 1924 twelve hundred members from forty-five states (and many foreign countries) had sent their dues to the ECUSA, including social workers, physicians, teachers, lawyers, ministers, conservationists, scientists (including Luther Burbank and future Nobel Prize winner Hermann J. Muller), librarians (including Melvil Dewey, who not only shared Madison Grant's mania for classification but also his anti-Semitism, having founded a resort on Lake Placid that excluded "victims of contagious diseases, cripples, and Jews"), and at least one football coach (Yale's legendary Walter Camp, who was also the brother-in-law of William Graham Sumner). Altogether they comprised what one observer called "a veritable blue book of prominent and wealthy men and women."[30]

Step Four: Fund-raising

All during this period, Grant's colleagues were publishing their best-selling books on eugenics, and each month brought more converts to the movement. The Finance Committee (of which Madison Grant was a member) was collecting more money than it could spend, as reform-minded philanthropists, impressed by the argument that "eugenics is a charity to lessen charity," were beginning to open their wallets. Early in 1925, George Eastman (who was also a benefactor of William T. Hornaday's Permanent Wild Life Protection Fund and Madison Grant's American Defense Society, and who distributed hundreds of copies of A. E. Wiggam's *New Decalogue of Science* to his Kodak employees) contributed $10,000 to the ECUSA, and he continued to give an equivalent sum until his death in 1932.[31] One week later, the Finance Committee asked John D. Rockefeller, Jr., if he would like to make a contribution, as "it is the greatest service any person living today can render." Rockefeller agreed that the matter of eugenics was "a profoundly important one to the future of this country" and donated $5,000 every year for the rest of the decade. Several other philanthropists made contributions of over $1,000, and as a result the organization's budget grew from $4,000 in 1923 to $26,000 in 1926.[32]

Step Five: Incorporating the AES

At this point, with a distinguished Advisory Council in place, a growing membership list, and a swelling treasury, the ECUSA finally determined that it had laid the proper foundations for a viable organization that would launch "a far-reaching eugenic campaign" to rid society of "disease, disability, defectiveness, degeneracy, delinquency, vice, and crime." Accordingly, on January 30, 1926, the ECUSA was dissolved, and the American Eugenics Society was officially incorporated in the parlor of Madison Grant's Manhattan town house. With expectations that the eugenics movement was "destined to become increasingly significant in molding the history—or more specifically the germ plasm—of future generations of mankind," the incorporators (Madison Grant, Henry Fairfield Osborn, Harry H. Laughlin, Irving Fisher, and Henry E. Crampton) adopted a constitution for the AES (prepared largely by Madison Grant). It had been more than four years since they had begun planning the organization. The Grantians had taken their time because they did not view eugenics as a passing fad or an ameliorative reform, but a doctrine that held out the promise of improving the human race forever.[33]

While the officers continued to meet regularly at Grant's home, the AES established offices in New York, Chicago, Ann Arbor, and New Haven and hired a full-time office staff of eight employees, augmented by several volunteer workers. The most important employee was executive secretary Leon F. Whitney. An alumnus of the field worker course at Cold Spring Harbor, Whitney was an expert on dog breeding and president of the Fruit Growers Supply Company in Massachusetts. He was also the author of *Pigeon City,* a eugenic allegory for children in which three young boys, assisted by an expert with the surname Grant, decide to raise homing pigeons. The boys allow only the best American pigeons into their flock, and they wisely refuse to accept birds from Europe until the bloodlines of the immigrants can be verified. When the boys have to move out of town for a year, they entrust their birds to two girls who almost destroy the flock because, owing to their naïve sentimentalism, they allow low-grade mongrels to mate with the purebreds. And so a valuable lesson is learned: when it comes to raising homing pigeons, training is not as important as good heredity, and undesirable individuals must be eliminated if the flock as a whole is to prosper. Leon Whitney's *Pigeon City,* with a foreword by A. E. Wiggam, was accepted by the Junior Literary Guild as its book of the month for boys in March 1931.[34]

There were now three major eugenics groups in the United States: the Eugenics Record Office (whose function was to conduct scientific research), the Eugenics Research Association (whose task was to facilitate the exchange of information among eugenicists), and the American Eugenics Society (whose purpose was to disseminate propaganda aimed at the layman and the legislator). While each group had its specific task to perform, they featured similar boards and overlapping membership lists, and not only did the AES and the ERA decide to jointly publish the *Eugenical News,* but they also held their annual meetings concurrently in Cold Spring Harbor.

The members of the American Eugenics Society were extremely active in the 1920s. They were missionaries, and their goal was a society permeated with the eugenic ideal. They held meetings, wrote reports, made surveys, published pamphlets, provided articles to newspapers, sponsored lectures, and served enthusiastically on the various committees of the AES. The most important committees (and their chairpersons) are listed below, grouped according to function.[35]

Committees of the American Eugenics Society

Organization
Finance (Leon F. Whitney)
History of the Eugenics Movement (S. J. Holmes)
Organization (Arthur E. Morgan)

Research
Anthropometry (Clark Wissler)
Biologic Genealogy (Ellsworth Huntington)
Eugenic Birth Control (Robert L. Dickinson)
Psychometry (Edward L. Thorndike)
Research (Charles Benedict Davenport)

Propaganda
Cooperation with Clergymen (Henry S. Huntington)
Cooperation with Physicians (Harold Bowditch)
Cooperation with Social Workers (Henry P. Fairchild)
Editorial Committee (Charles Benedict Davenport)
Formal or Scholastic Education (C. C. Little)
Popular Education (Mrs. Mary T. Watts)

Legislation
Crime Prevention (Chief Justice Harry Olson)
Legislation (Roswell H. Johnson)
Selective Immigration (Madison Grant)

A sampling of the activities of some of these committees provides an idea of the scope of the American Eugenics Society. The Committee on Legislation, for example, lobbied the states to pass eugenic marriage regulations. By the mid-1930s, four states had laws prohibiting the marriage of alcoholics, seventeen states prohibited the marriage of epileptics, and forty-one states prohibited the marriage of the feebleminded and the insane. The committee also endorsed legislation forcing newlyweds to post bond promising not to reproduce if they had relatives who suffered from blindness, deafness, epilepsy, feeblemindedness, or insanity.

The Committee on Popular Education sponsored traveling exhibits on eugenics that were set up at Kiwanis meetings, PTA conventions, museums, movie theaters, and state fairs across the country. They included a poster showing the pedigree of the Kallikak family, a marionette show (called the "Mendel Theater") that demonstrated the principles of Mendelian genetics, and a very effective display consisting of a series of flashing lights mounted on a large board, titled "Some People Are Born to Be a Burden on the Rest." One light flashed every forty-eight seconds (indicating the birth of a "defective" person), while another light flashed only every seven and a half minutes (indicating the birth of a "high grade person"). Another light informed the viewer that "every 50 seconds a person is committed to jail," with a reminder that "very few normal people ever go to jail." And in case moral outrage was not sufficient to spur the viewer to embrace eugenics, the light at the top of the board flashed regularly to indicate that "every 15 seconds $100 of your money goes for the care of persons with bad heredity such as the insane, feeble-minded, criminals, and other defectives."[36]

The AES received front-page coverage in local newspapers for its annual Fitter Families Competitions held at state and county fairs. Mary T. Watts, chairman of the society's Committee on Popular Education, had been running Better Babies contests for years. But after Charles Benedict Davenport admonished her that "a prize winner at two might be an epileptic at ten," she realized that phenotype was not nearly as important in evaluating the value of a baby as genotype. It was then that she decided to organize the Fitter Families Competitions, in which an entire family entered as a unit and was rated for eugenic fitness by a team composed of a historian, a pediatrician, a psychiatrist, a psychologist, a dentist, a clinical pathologist, and an ear, nose, and throat specialist. The first competition was held at the Kansas Free Fair in 1921, where Mrs. Watts explained: "while the stock judges are testing the Holsteins, Jerseys, and whitefaces in the stock pavilion, we are judging the Joneses, Smiths, and the Johnsons." Indeed, it was explicitly hoped that the competition would teach the public that humans were as subject to the laws of inheritance as domestic animals, and the director of the fair concurred that "human stock should be given attention quite as much as live stock." At the climax of the fair, when the prize livestock were paraded in front of the crowd, the winning family was driven in an automobile under a great banner reading "Kansas' Best Crop."[37]

From this beginning, the Fitter Families Competitions spread in just a few

Dolichocephalics all: The winner of the Fitter Family Contest at the Kansas State Fair in 1925. The pleasure of the father at siring six Nordic offspring seems somewhat greater than that of the mother. American Philosophical Society.

years to fairs throughout the country. The American Eugenics Society supplied the instructions, rules, scorecards, and equipment, and convinced important politicians to present the winners with a large bronze medal (designed by Madison Grant) depicting the logo of the AES: a man and wife passing on the stream of life to their child, under the biblical quotation, "Yea, I have a goodly heritage." (In Sinclair Lewis's novel *Arrowsmith,* the Eugenic Family at an Iowa fair, none of whom "smoked, drank, spit upon pavements, used foul language, or ate meat," is exposed as the criminal Holton gang.)[38]

The AES's Committee on Cooperation with Clergymen, guided by the premise that "men do not gather figs from thistles," organized the annual Eugenic Sermon Contest, which awarded prizes of $200, $300, and $500 to the pastors that most effectively encouraged the fitter members of their congregations to outbreed the "undesirable elements in the community." William Lyon Phelps, Madison Grant's classmate from Yale who by now was a nationally known literary critic, agreed to be one of the judges. In 1926, three hundred sermons were inspired by the contest, of which sixty made it to the final round. One of them was by Rabbi Harry H. Mayer, who was commended for praying in his Kansas City Temple during a special Mother's Day service: "May we do nothing to permit our blood to be adulterated by infusion of blood of inferior grade." The first prize, however, was awarded to Reverend Phillip E. Osgood of Minneapolis, who reminded his congregation that while it was important to restrict immigration from Europe, it was equally important to regulate "immigration from Heaven" (i.e., the birth of babies). After all, "worthy citizens do not spring from the loins of the unfit."[39]

The Committee on Cooperation with Clergymen also distributed thousands of copies of *A Eugenics Catechism* to ministers and college professors, featuring Socratically helpful information such as:

Q. Does eugenics contradict the Bible?
A. *The Bible is one of the finest eugenics books. The connection of most of the great men mentioned therein with great families is carefully recorded.*
Q. What makes slums?
A. *People. Slums have been cleaned up and the people move on and make other slums.*
Q. What is the most precious thing in the world?
A. *The human germ plasm.*[40]

One of the largest committees of the AES was the Committee on Cooperation with Physicians. Co-opting the medical community was important to the eugenicists (as it would be to the Nazis), not just because doctors would be the ones called on to implement any program of population control, but also because physicians played an important symbolic role for the Grantians. Much like Richard Wagner and Houston Stewart Chamberlain, the eugenicists viewed the dysgenic elements of the population as akin to cancer cells that had to be cut out of society. Lothrop Stoddard, for example, warned in *The Revolt against Civilization* that the feebleminded in America were "spreading like cancerous growths . . . and infecting the blood of whole communities." Ellsworth Huntington likewise explained that "hereditary weaklings" were threatening society in the same way that "new cancer cells remorselessly penetrate into sound tissues," and he wrote that "the defectives . . . may be compared to an insidious disease affecting the body politic."[41]

This fear of the biomedical dangers posed by the feebleminded extended to the New Immigrants, who were invading U.S. society as malignantly as any virus ever attacked a weakened cell. Madison Grant wrote that America's liberal immigration policy was "introducing the seeds of fatal disease into the body politic." Novelist Gertrude Atherton likened the immigrants to a "flu which attacks high and low alike," and Congressman Albert Johnson wondered how the republic could endure "if there be steady deterioration of standards by ever-recurring new foci of infection arriving in the land."[42] (It is interesting that the derogatory term "wog," which is used in Australia to designate a person of southern European ancestry, is also routinely used as a synonym for "germ.")

When they were not compared to cancer cells and flu germs, the immigrants were described as "human weeds." "Weeds! . . . Millions of them!" shouted Ellsworth Huntington and Leon F. Whitney in *The Builders of America* (which was revised by Madison Grant before going to press). "Genuine human weeds, whole shiploads of them, from almost every nation in Europe." Henry Fairfield Osborn bewailed the fact that "there is no port in this country through which you can bring in a diseased animal or a noxious animal or weed," and yet "noxious human beings" were daily entering New York from abroad. And the surgeon general of the U.S. Navy, E. R. Stitt, explained that just as our quarantine laws forbade the importation of injurious plants and animals, so we needed to curtail the importation of dangerous immigrants.[43]

The revulsion of the eugenicists regarding the immigrants was related to their Puritan anxieties about contamination. Prescott F. Hall of the Immigration Restriction League cautioned that the New Immigrants would "pollute the Yankee blood," and Owen Wister similarly warned that the United States was being "polluted" by alien minorities every hour of every day. Kenneth Roberts wrote in the *Saturday Evening Post* that immigration was like a "poison working in the veins of America," and George Creel told the readers of *Collier's* that, as a result of unrestricted immigration, "the wells of our national life have been poisoned and will continue to be poisoned unless sentiment is cast aside in favor of common sense." The American Eugenics Society promised that eugenics would "do for the race what personal hygiene does for the individual," and it is not an accident that the German word for eugenics translates literally as "race hygiene."[44]

The dirtiest immigrants of all were the Jews, who—in contrast to the clean and nature-loving Nordics—seemed to have a bizarre predisposition to swarm in the crowded fetidness of large cities. Charles E. Woodruff, a physician whose books were read and admired by Madison Grant, warned that the Jews were "harmful parasites" and "bacilli." Their continued immigration was like an "infectious disease" invading the homeland, and to persecute them was to engage in a needed "process of disinfection." Prescott F. Hall similarly told Grant that the Jews were akin to "germs of infectious disease" and that they should be dealt with in the same manner as "noxious weeds" or "insect pests." Grant agreed that Jewish immigrants threatened to "poison the blood" of the old-stock Americans, and Kenneth Roberts wrote that the Jews were "the true human parasites."[45]

Like many anti-Semites, the Grantians had an exorbitant concern with matters of digestion. Kenneth Roberts, for example, announced that the Jews "present one of our most indigestible problems." He compared the New Immigrants to spoiled food that was giving indigestion to the American public, and pleaded with Congress to pass immigration restriction legislation and thereby give the country "a chance to digest the millions of unassimilated, unwelcome and unwanted aliens that rest so heavy in her." The national commander of the American Legion warned that the nation was suffering from "indigestion of immigration," and historian John Burgess doubted the ability of Uncle Sam to "digest and assimilate such a morsel." After observing the immigration situation, *World's Work* not so delicately explained: "We have prattled about the 'melting pot' and have wakened to find the stomach of the body politic filled to bursting with peoples swallowed whole whom our digestive juices do not digest. Wise doctors have compounded a prescription called 'Americanism' which we are assiduously pouring down our throat in the hope that it will disintegrate these knots that give us such pain and allow us to absorb the meal we have gorged ourselves with."[46]

Psychiatrist William Sadler exemplified as well as any of Grant's disciples the connection between the fear of inferior races and an obsession with maintaining a clean intestinal tract. Indeed, to help promote personal as well as racial health, Sadler thoughtfully included in his book *Race Decadence* an appendix listing three eugenic diets. The first was an "Anti-Constipation Diet"; the second was a diet for those seeking to prevent flatulency, "intestinal toxemia," and "catarrh of the bowel"; and the third was a diet for those wanting "to bring about a change in the intestinal flora" by "driving out the more vicious and harmful bacteria so commonly inhabiting the bowel tract." This last diet necessitated drinking a full glass of whole milk—plus the juice of one-half lemon and one whole orange—every forty-five minutes from 7:30 A.M. until 7:30 P.M. (Sadler—probably futilely—urged his readers not to be deterred by the nausea that often accompanied such a regime.)[47]

Flirting with Sanger

With regard to birth control, the position of the American Eugenics Society was that the Comstock Law should be repealed so that cheap, simple, and safe methods of contraception would be readily available, particularly to immigrants and the lower classes. To be sure, the AES was somewhat conflicted on the issue. Some of Grant's colleagues, notably Henry Fairfield Osborn and Charles Benedict Davenport, frowned on endorsing contraception. Their main concern was that family planning would be dysgenic, since the inferior breeds were too stupid, lazy, or poor to use birth-control devices, leaving only the middle and upper classes to practice contraception. As Samuel J. Holmes put it: "The trouble with birth control is that it is practiced least where it should be practiced most." In addition, as good Puritans, Grant's friends were afraid that birth control would lead to sexual promiscuity. And as respectable Republicans, many Grantians were repulsed by the radicals and libertines who attached themselves to the birth control movement, including such disreputable free lovers as Havelock Ellis, Olive Schreiner, and Victoria Woodhull (who, it so happens, had been an early supporter of eugenics in the United States). As Paul Popenoe (who strongly supported birth control but was wary of the birth control *movement*) once explained to Grant: "If it is desirable for us to make a campaign in favor of contraception, we are abundantly able to do so on our own account, without enrolling a lot of sob sisters, grandstand players, and anarchists to help us."[48]

Despite these fears, many Grantians were actively and prominently involved in the birth-control movement. Their ranks included such mainstays of the AES as Edward M. East, Irving Fisher, C. M. Goethe, Samuel J. Holmes, Ellsworth Huntington, Roswell H. Johnson, David Starr Jordan, Harry H. Laughlin, C. C. Little, William McDougall, Frederick Osborn, Edward A. Ross, Lothrop Stoddard, Robert DeC. Ward, A. E. Wiggam, Clark Wissler, and Robert M. Yerkes. Franklin H. Giddings explained his support of birth control by stating: "The question is really quite simple. . . . Do we want to fill the earth with imbeciles, paranoiacs, snake worshippers, fundamentalists, . . . holy rollers and Tennessee statesmen, or with normal men and women, sound in mind and body?[49]

Soon after the Great War, the eugenics movement and the birth-control movement began drawing ever closer together, until by the mid–1920s they overlapped considerably in both goals and personnel. The prime agent pushing the two groups together was Margaret Sanger, a New York nurse who is now venerated as the founder of the American birth-control movement. Sanger's father, an Irish immigrant with a fondness for phrenology and the proletariat, had raised Sanger to revere Eugene Debs, and in 1911 she joined the Socialist Party and began writing articles on hygiene for the *Call,* New York's socialist daily. However, frustrated with the party's tactical timidity and antifeminist culture, Sanger drifted into the camp of the IWW, where she associated with the leading radicals of the day, including Big Bill Haywood, John Reed, and Emma Goldman. As a nurse ministering to the poor immigrant women of the Lower East Side, Sanger realized that birth control could be the key factor improving the health and living conditions of the working class.

But Sanger quickly became disillusioned with the workers, who proved to be stubbornly uninterested in limiting the size of their families and, in any event, were a meager base on which to build her birth-control movement, given their lack of time and resources. In an effort to secure support from persons of means, Sanger began espousing the eugenic effects that family planning would have on society. She started pointing to the menace of "reckless procreation" and "indiscriminate breeding" among the lower classes, and *Birth Control Review,* the magazine founded by Sanger, ran an editorial in 1919 declaring that "more children from the fit, less from the unfit—that is the chief issue of birth control."[50]

Sanger, who eventually became a registered Republican, was concerned that "civilized nations are penalizing talent and genius, the bearers of the torch of civilization, to coddle and perpetuate the choking human undergrowth, which, as all authorities tell us, is escaping control and threatens to overrun the whole garden of humanity." She repeatedly warned that these "human weeds" must be "extirpated" if society was to survive, and she explained that to criminalize birth control was "to abandon the garden to the weeds."[51]

Margaret Sanger in the 1930s. She was a great American heroine, but two decades later the founder of Planned Parenthood was still advocating sterilization to weed out "the unfit." Courtesy Margaret Sanger Lampe.

Sanger tirelessly pushed for an alliance between birth control and eugenics, as "both are seeking a single end." She wooed the Grantians by explaining that her desire was "to breed a race of human thoroughbreds," and that the legalization of contraception would serve to compensate for the fact that "the slums of Europe dumped their submerged inhabitants into America." As Grant had done in *The Passing of the Great Race,* Sanger pointed out that since the upper classes already knew about, and had access to, birth control, it behooved the eugenicists to cooperate with her in getting birth-control devices into the hands of the masses and, in so doing, to begin the process of "weeding out the unfit."[52]

As Sanger barnstormed the country, references to Karl Marx and Emma Goldman disappeared from her speeches, and instead she began to quote Sir Francis Galton, Henry H. Goddard, and Charles Benedict Davenport. She claimed that birth control was the most rational and efficient solution to race suicide, and her advocacy of science to regulate the most personal and important of human activities was in perfect harmony with Madison Grant's ideal of a society organized along technocratic lines. Indeed, Sanger had first learned of eugenics from her mentor (and lover) Havelock Ellis, who, along with Grant, preached that society could be transformed by a scientifically educated elite. Like many Fabians, Ellis was an ardent admirer of Sir Francis Galton, and Ellis nicely summed up the eugenic position when he wrote: "The superficially sympathetic man flings a coin to the beggar; the more deeply sympathetic man builds an almshouse for him so that he need no longer beg; but perhaps the most radically sympathetic of all is the man who arranges that the beggar shall not be born."[53] (Ellis did his part to make sure the beggar was not born, since he himself was childless.)

Sanger agreed with Ellis and Grant that sentimentalism was dysgenic, and denounced society's misguided policy of "indiscriminate charity" toward "the very types which in all kindness should be obliterated from the human stock." In a speech that could just as easily have been delivered by Madison Grant, Sanger warned an audience at Vassar College in 1926 that the "defectives and insane" were "being multiplied with breakneck rapidity and increasing far out of proportion to the normal and intelligent classes." She complained that the government was spending $9 billion a year to care for and perpetuate these "undesirables." As a result, the American public was "taxed, heavily taxed, to maintain an increasing race of morons," who were threatening "the very foundations of our civilization." And the political costs were just as devastating as the financial: "When we . . . realize that a moron's vote is as good as [that of] an intelligent, educated, thinking citizen, we may well pause and ask ourselves: 'Is America really safe for democracy?'"[54]

As Sanger courted the eugenicists, *Birth Control Review* began to run articles bearing such titles as "The Menace of the Feebleminded," "Sterilization of the Unfit," "The Sins of the Fathers," and "Birth Control or Racial Degeneration—Which?" And the magazine routinely printed favorable reviews of the books by all the major Grantians. Havelock Ellis, for example, wrote an enthusiastic critique of Lothrop Stoddard's *Rising Tide of Color,* in which he commended Stoddard for relying on the expertise of Madison Grant. (No other book—not even those by Sanger—received such a lengthy review in the magazine.)[55]

Given such a congenial environment, important eugenicists began contributing articles to *Birth Control Review.* In its third issue, for example, just three pages away from an encomium to "Comrade Sanger" from Eugene Debs, Paul Popenoe pledged "the active support" of the eugenicists to Sanger's cause, since "a spread of birth control to the less capable part of the population will be an important advance for eugenics in cutting down the racial contribution of inferior stocks." In exchange, the *Review* devoted a special issue to "Eugenics," another one to "Sterilization," and eventually lauded immigration restriction as both "reasonable and eugenic." By 1926, Harry H. Laughlin was happy to report in the pages of *Birth Control Review* that "the efforts of eugenics and of Birth Control are tending, more and more, to work for the common end."[56]

As she raised the eugenic banner, Margaret Sanger was increasingly willing to endorse coercion to forestall the "peril" of the feebleminded. "Surely," she wrote, "it is an amazing and discouraging phenomenon that [the state dares] not attempt to restrain, either by force or persuasion, the moron and the imbecile from producing his large family of feeble-minded offspring." She openly and forcefully called for the segregation of degenerates and the sterilization of defectives, and when *Birth Control Review* published for the edification of its readers a "Selected Reading List on Sterilization," it consisted of twenty-two eugenic works that included, of course, *The Passing of the Great Race* by Madison Grant.[57]

"When we think of eugenics," writes Diane Paul, "it is usually not Margaret Sanger or Havelock Ellis who comes to mind but Madison Grant or Adolf Hitler." True enough. But eugenic arguments dominated birth-control propaganda in the 1920s in both the United States and Europe (and as late as 1937 Ellis was still defending the Nazi Sterilization Law). Marie Stopes, leader of the British birth-control movement, was extremely enthusiastic about eugenics; she urged the sterilization of "the careless, the feeble-minded, the very lowest and worst members of the community" so they could not continue to produce "innumerable tens of thousands of warped and inferior infants."[58] Unlike Stopes, who was blatantly anti-Semitic (and sent Hitler a volume of her poems in 1939 as a token of her admiration), Margaret Sanger was not particularly racist. Although she was reticent to admit it, Sanger's first husband was the son of Jewish immigrants, and she always understood that human beings were the product of the interaction between nature and nurture. While it is futile to deny that Sanger adopted eugenics partly out of conviction, it is also true that her conversion was mainly a pragmatic move on her part to garner the scientific respectability and the resources that an alliance with the eugenicists brought to her movement.

Regardless of her motives, David M. Kennedy points out that as a result of Sanger's embrace of eugenics, the birth-control movement was transformed "from a radical program of social disruption to a conservative program of social control." This shift was personified by Sanger's pragmatic marriage in 1922 to second husband J. Noah Slee, wealthy founder of the 3-in-One Oil Company. (One wonders how many women in the 1920s realized that their vaginal diaphragms had been smuggled into the United States in 3-in-One Oil containers.) As Sanger intended, Slee became by far the largest benefactor of the American Birth Control League, and 3-in-One Oil was also the most important sponsor of *Birth Control Review* (where its ads, with no intended irony, informed readers of the magazine devoted to contraception that "Little Drops Do Big Things"). In addition, the two main financial supporters of the American Eugenics Society—George Eastman and John D. Rockefeller, Jr.—annually contributed to Margaret Sanger the same amounts they gave to the AES. Rockefeller generously overlooked the fact that there had been a time, in her Greenwich Village days, when Sanger had called for his assassination, calling Rockefeller a "blackhearted plutocrat whose soft, flabby hands carry no standard but that of greed."[59]

By the early 1920s, the national council of the American Birth Control League (which eventually became Planned Parenthood) included not only such expected names as Herbert Croly, Theodore Dreiser, and Sinclair Lewis, but also leading members of the American Eugenics Society such as Edward M. East, Samuel J. Holmes, Roswell H. Johnson, William McDougall, C. C. Little, and Lothrop Stoddard.

In 1923, Margaret Sanger told Madison Grant that it would give her "great pleasure" if he would deliver a paper at the Birth Control Conference held in Chicago. Sanger invited many other eugenicists, for as she explained to Harry H. Laughlin: "I believe that this conference is going to do much to unite the Eugenic Movement and the Birth Control movement, for after all they should be and are the right and left hand of one body." Because of illness, Grant could not travel to Chicago, but it is doubtful that he would have attended the conference even if healthy. Like Henry Fairfield Osborn and Charles Benedict Davenport, Grant was one of the old-guard eugenicists who never felt comfortable with Sanger's radical origins, and he definitely feared that an alliance with her organization would discredit his own movement. Davenport pointed out that while agitators like Margaret Sanger actually gloried in being arrested, "No police have closed the doors of a eugenics meeting yet." And Grant himself explained to Leon F. Whitney that although he had "nothing but cordial sympathy for the birth control movement, especially if it can be applied to undesirable strains and races," he thought it wise to "keep clear" of the ABCL, as it was "a feminist movement and would bring us a lot of unnecessary enemies. I am not afraid of enemies, but I like to make them myself. Besides, the movement is hardly respectable and can only injure the scientific standing of the American Eugenics Society."[60]

Grant's instincts were correct. In the 1920s, after all, the former Wobbly Margaret Sanger needed respected conservationist Madison Grant much more than he needed her, and Grant had little to gain by going to Chicago and sitting on a dais with the likes of Sanger, Jane Addams, and Harold Ickes. Nonetheless, Sanger never stopped pressing for closer ties with the American Eugenics Society; she repeatedly proposed, for example, that the American Birth Control League and the AES merge their journals. But Madison Grant was "definitely opposed" to any connection between the AES and Sanger, and the union never occurred.[61]

While Madison Grant was wary of Margaret Sanger as an individual, his enthusiasm for birth control as a eugenic tool never slackened. Still, he was a realist, and programs that annually distributed a few hundred condoms in the

Lower East Side were not going to be very effective if every year one million inferior New Immigrants invaded the United States. And that is why, as far as Grant and the rest of the leaders of the American Eugenics Society were concerned, the organization's most important committee (and the one that received the most publicity) was the Committee on Selective Immigration, of which Grant was the chair.

The task of the Committee on Selective Immigration was to convince Congress to solve the nation's "immigration problem" through legislation. The New York Zoological Society had put a boundary around an obscure section of the Bronx and transformed it into a preserve for the threatened mammals of North America; the American Bison Society had erected a fence around a portion of the Montana prairie and turned it into the Montana National Bison Range; and now Grant would raise a barrier, in the form of laws to restrict immigration, around the United States and thereby create a haven for the Nordics where they could breed in safety.

In 1920, a somewhat distraught Charles Benedict Davenport asked Grant: "Can we build a wall high enough around this country, so as to keep out these cheaper races?" If not, Davenport despaired that the Nordics would have to "abandon the country to the blacks, browns and yellows, and seek an asylum in New Zealand." Grant saw no need to flee south of the equator. He was a patrician and he was a progressive, which is to say that he had complete confidence that through his connections he could bring about the necessary legislation that would enable him to respond to Davenport's question with an affirmative: Yes, we *can* build a wall high enough around this country. And it is to Grant's central role in constructing that wall that we now turn.[62]

9

Creating the Refuge

In 1883, Emma Lazarus penned the famous poem embronzed at the Statue of Liberty inviting Europe to send its unwanted masses to America.

"The New Colossus" (Emma Lazarus, 1883)

Give me your tired, your poor,
Your huddled masses yearning to breathe free,
The wretched refuse of your teeming shore.
Send these, the homeless, tempest-tost to me.

Madison Grant was dismayed by the attitude of Lazarus. The very idea that we would actually invite "wretched refuse" to enter and defile our homeland sickened him. More to Grant's liking was the 1892 poem "Unguarded Gates" by Thomas Bailey Aldrich, editor of the *Atlantic Monthly.* Aldrich's poem was also about the Statue of Liberty, but it expressed a quite different philosophy.

"Unguarded Gates" (Thomas Bailey Aldrich, 1892)

Wide open and unguarded stand our gates,
And through them presses a wild motley throng—
O Liberty, white Goddess! is it well
To leave the gates unguarded?

The difference between the two poems reflects the difference not only between the Jewish Lazarus and the Nordic Aldrich but also between the 1880s and the 1890s. For it was in the latter decade that the New Immigrants started to arrive in droves. And they kept coming, in ever larger numbers, well into the twentieth century.

It was then that Madison Grant began quoting Gouverneur Morris, who said in 1787: "Every society from a great nation down to a club has the right of declaring the conditions on which new members should be admitted." From the Society of Colonial Wars to the American Eugenics Society, Grant had sternly controlled admission to all his clubs, and now he was intent on doing the same with his nation. His unflinching and determined effort to preserve the Nordic character of the United States involved three legislative steps that progressively and severely restricted immigration from southern and eastern Europe: the literacy test of 1917, the Emergency Quota Act of 1921, and the Immigration Restriction Act of 1924. It was an undertaking that one scholar has referred to as "America's most ambitious program of biological engineering."[1]

Step One: The Literacy Test

The main proponent of the literacy test was the Immigration Restriction League, which was founded by a group of Boston Brahmins from the Harvard class of 1889 (the same class that produced Charles Benedict Davenport). The three original leaders of the IRL were Prescott F. Hall, an anti-Semitic lawyer with a love for the music of Richard Wagner and the writings of Houston Stewart Chamberlain; Robert DeCourcy Ward, yet another eugenicist who spent his childhood in Dresden and who became America's leading professor of climatology at Harvard; and Charles Warren, who served as assistant attorney general in the Wilson administration (during which he drafted the Espionage Act of 1917) and went on to become one of the century's leading constitutional historians (Warren's *Supreme Court in United States History* won the Pulitzer Prize in 1923). Hall, Ward, and Warren were approximately the same age as Madison Grant, and they all shared his intense unease that the country that was their birthright was being overrun by ungrateful and undeserving foreigners. In the spring of 1894 they gathered in Charles Warren's State Street office and formed the Immigration Restriction League, with chief financial backing from Joseph Lee, heir to a New England banking fortune and an overseer of Harvard University. John Fiske, foremost exponent of the Anglo-Saxon tradition among American historians, agreed to serve as the league's first honorary president, and the vice presidents included publisher Henry Holt, novelist Owen Wister, and Bowdoin College president William DeWitt Hyde. Within two years, the league had a membership of almost seven hundred blue-blooded New Englanders, featuring various Endicotts, Houghtons, Lowells, Lymans, and Saltonstalls.[2]

The goal of the Immigration Restriction League, of course, was to convince Congress to pass legislation restricting undesirable immigration. But political considerations demanded that the effort not be aimed too obviously at any particular ethnic group. In those pre–*Passing of the Great Race* days, the league's leaders could not openly state that they wanted to bar Greeks, Italians, Russians, and Jews—although that was precisely their desire. Therefore, they hit upon the idea of requiring immigrants to pass a literacy test. Such a provision would not only lower the total number of immigrants entering the country every year; it would also—without having to mention any nationality by name—discriminate against the New Immigration from southern and eastern Europe, where literacy was much lower than in northwestern Europe.

In 1895, the Immigration Restriction League convinced Representative Samuel McCall of Massachusetts to sponsor its literacy bill in the House, while Henry Cabot Lodge did the same in the Senate. The bill was opposed by a coalition of immigrant organizations, business interests (e.g., steamship and railroad lines) that depended on immigrant customers, industrial groups (e.g., the National Association of Manufacturers) that favored immigration as a source of cheap labor, and by what Madison Grant disdainfully called "the refuge-of-the-oppressed idealists."[3] On the other hand, the literacy bill was supported by nativist and patriotic groups (who feared the social character of the New Immigrants), progressives (who saw immigrants as a source of crime and political corruption), and organized labor (e.g., the American Federation of Labor, led by Samuel Gompers, himself a Jewish immigrant, who was concerned about low wages and unemployment).

In deference to party leaders who did not want to antagonize foreign-born members of the electorate, Congress waited until after the election of 1896 and then, in December, passed the literacy bill by an overwhelming majority. But President Grover Cleveland, who perceived the discriminatory intentions behind the bill, vetoed it in his last months in office, leading sociologist Edward A. Ross to call Cleveland's veto "one of the most disastrous actions ever taken by an American president." Madison Grant agreed, and later groused to Elihu Root that "we would have been spared the curse of our present flood of East Side Jews if the Hon. Grover Cleveland had not vetoed the literacy test."[4]

The Immigration Restriction League reintroduced the literacy bill year after year in Congress, but the return of prosperity in the late 1890s, the pressure of the increasingly powerful immigrant groups, and the opposition of Speaker Joe Cannon (who, in Grant's estimation, was "hopelessly behind the times") prevented its passage. In the meantime, the league sought to educate the public by publishing pamphlets and articles, sending letters to editors and ministers, and providing pro-restriction speakers to civic and business groups. But, for all that, the IRL never became a well-known or broad-based group, and it funneled most of its resources into lobbying Congress. To that end, it hired a full-time Washington lobbyist, James H. Patten, an extraordinarily anti-Semitic figure who was convinced that "practically all the gangsters, gamblers, and gunmen" in the country were Jews, as were most of the "white slavers" and "a great preponderance" of the inmates of insane asylums. Patten intensely lobbied selected congressmen, especially members of the House and Senate Immigration Committees, and spent a good deal of the league's treasure underwriting campaigns to defeat politicians who opposed the literacy bill.[5]

In 1909, Madison Grant was asked to serve as one of the nine vice presidents of the IRL (and one of the few who had not attended Harvard), and he immediately accepted. To Grant, politics was always the art of the possible, and he agreed with the IRL that the literacy test was the most viable method of restricting immigration. At the very least, Grant told his old friend the congressman John F. Lacey, the literacy test "would serve to keep out a great mass of worthless Jews and Syrians who are flooding our cities." With Grant on board, the leaders of the league embraced eugenics, and were delighted to discover that while sentimentalism was on the side of the liberals, science was on the side of the restrictionists. The league learned that Alpines and Mediterraneans were biologically inferior to Nordics, that the American environment would have no ameliorating effect on the inferior germ plasm of the southeastern Europeans, and that miscegenation between native-born Americans and New Immigrants would cause reversion. Prescott F. Hall (who maintained that "there is considerable evidence that one or both of the parents of Jesus were not Jews at all") was thrilled that his Brahmin anti-Semitism was scientifically sanctioned, and even proposed changing the organization's name to the Eugenic Immigration League (an idea rejected by the traditionalists on the board).[6]

Under Grant's tutelage, the IRL began to employ explicitly racial arguments in favor of the literacy test, and pro-restriction articles by Robert DeC. Ward and Prescott F. Hall began appearing in scholarly venues like *Science, Scientific Monthly,* and the *Journal of Heredity.* Grant also brought many of his eugenicist friends into the league, including Charles Stewart Davison, Henry Pratt Fairchild, Jeremiah Jenks, C. C. Little, John Dyneley Prince, Lothrop Stoddard, Frederick Adams Woods, and his by now inseparable accomplice Henry Fairfield Osborn. Grant also placed some of his conservationist colleagues (e.g., George Shiras III and John C. Phillips) on the National Committee of the IRL.

With the public increasingly alerted to the dysgenic effects of immigration, and with the 1911 report of the congressionally appointed Dillingham Commission, which recommended a reading test as "the most feasible method of restricting undesirable immigration," the IRL reintroduced its literacy bill in 1912.[7] Madison Grant, who was by far the most active of the non-Boston members of the league, worked hard for the literacy test. He organized a petition drive urging New York's congressmen to support the bill, and he met with his old friend the senator Elihu Root to plot legislative strategy. Grant was not reticent about revealing the anti-Semitic intent behind the measure: he told Root that he was determined to see the literacy bill passed so as to "stop the draining off into this country of the great swamp of human misery and degradation which has centered around Warsaw, and which has ruined Poland and terribly impaired the development of Russia."[8] Grant also took his campaign to the White House. Speaking as "one who has given some considerable time to the study of zoology and more particularly anthropology," Grant warned President William Howard Taft that "vast floods of utterly alien races and types are pouring in, and the great cities are being swamped by the Polish Jews from Eastern Europe. Anyone who scientifically faces the facts can understand the extremely inferior and immoral structure of these latter, and it is universally admitted and deplored in private conversation." Grant authoritatively explained to the president that "the old theological views in regard to the unity of the human race and its relatively recent origin (some six thousand years ago), is giving away to the knowledge that man as such dates back two or three hundred thousand years, and that consequently the line of cleavage between the so called races of mankind is fundamental and cannot be modified by any change of environment in the life time of a nation." Therefore, speaking as a "scientist," Grant asked Taft to stand up to the steamship companies, the industrial interests, and the immigrant organizations and "preserve the native American stock" by taking a "brave" stand in favor of immigration restriction.[9]

After the election of 1912, the literacy bill sailed through Congress, but President Taft—like Cleveland before him—shocked the IRL by vetoing it in his last months in office. Grant was bitterly disappointed, and was convinced that the president had fallen prey to "Jewish influence." But George W. Wickersham, Taft's attorney general and a close friend of Grant, had a simpler explanation for the president's position: "My dear Grant, if the manual laborer is shut out, we will soon be in a condition where we will have nobody to dig our ditches!"[10]

Still, the Immigration Restriction League had high hopes for Taft's successor as president. In 1908, in his *History of the American People,* Woodrow Wilson had written: "Throughout the [nineteenth] century men of sturdy stocks of the north of Europe had made up the main strain of foreign blood which was every year added to the vital working force of the country . . . but now there came multitudes of men of the lowest class from the south of Italy and men of the meaner sort out of Hungary and Poland, men out of the ranks where there was neither skill nor energy nor any initiative of quick intelligence; and they came in numbers which increased from year to year, as if the countries of the south of Europe were disburdening themselves of the more sordid and hapless elements of their population."[11] In the election of 1912, Wilson's Republican opponents gleefully distributed this passage in immigrant neighborhoods, and Wilson was forced to repudiate his words at many campaign stops. Despite the candidate's backtracking, Prescott F. Hall of the IRL felt that the Wilson administration represented "the greatest chance that native Americans have had" since Cleveland's veto to "keep the Jews out." Before reintroducing the literacy bill in Congress, Hall met with Madison Grant many times in New York to plot lobbying strategy. Among other things, they decided to send a letter—addressed to congressmen only from the South—warning that the New Immigrants "have not the same objections to interbreeding with the negroes that the northern races have. . . . The one serious result of immigration of Mediterranean stock to the southern states would be an increase of negro half breeds."[12]

To ensure President Wilson's support for the literacy bill, numerous league executives wrote him lengthy letters, and four different delegations of IRL officials visited the president in the White House. Grant—who correctly understood that "the way to get at President Wilson is through Colonel House"—had many of his well-connected friends intervene with the president's alter ego. Still, the league began to sense that Wilson—under pressure from Jewish and other immigrant groups—was wavering on the literacy issue, and that he might feel constrained by the various pledges he had made in immigrant neighborhoods while on the campaign trail. The league's executive committee asked Grant himself to write to the president and explain the urgency of supporting the bill. Grant did so, and reminded the president that if Grover Cleveland had supported the literacy test in 1897, "it would have kept out a vast number of undesirable immigrants who have choked up the slums of our large cities, and are becoming a menace to the institutions of the country." Grant explained to Wilson that there were numerous economic and social reasons to bar the gates, "but I desire to base my recommendation solely on the corruption of blood which will result from a prolonged immigration." He went on to claim that if the "melting pot" were allowed to continue to boil, "a racial chaos and ruin will be the result, and the race types which are now appearing in the East Side of New York are horribly suggestive of what the future has in store for us unless immediate action is taken."[13]

The president, however, was more concerned with the campaign promises he had made to the "race types" of the East Side than with the threat they posed to the nation's germ plasm, and after Congress passed the literacy bill in 1915, he vetoed it.

Despite this third straight presidential veto, Grant—ever the patient strategist—counseled his colleagues in the IRL to maintain the faith. Science was on the side of the restrictionists, and it was only a question of time before the literacy bill would be enacted. "Aside from the Jews," he told Theodore Roosevelt, "our chief opponents are the wishy-washy sentimentalists," and fortunately they were "as usual wrong in this, and in almost everything else."[14]

The election of 1916 presented something of a dilemma to the IRL. They now knew that Wilson would veto any literacy bill, and so they looked to the Republicans to nominate a restrictionist. Theodore Roosevelt would have been the ideal candidate and, during a lengthy meeting at Oyster Bay, TR reassured Grant (as only TR could) that in his opinion "the national gizzard cannot masticate more." But, alas, the Republicans shunned the Bull Mooser and nominated Charles Evans Hughes, who immediately dismayed the Immigration Restriction League by declaring that liberty should be available to all regardless of race or creed, and by vowing to "*welcome* those who seek the opportunities of American freedom." The league's officials were certain that the latter was a reference to the literacy test, and they exchanged letters filled with consternation.

Still, most members of the IRL held their noses and voted for Hughes, the Republican, in November.[15]

Woodrow Wilson was reelected anyway, but all was not lost. Quite the contrary—with most congressmen outraged by reports of German atrocities in the Great War, Madison Grant felt (correctly) that there was finally enough antiforeign feeling in Congress to override a veto by Wilson. Grant urged the league to "take advantage of the crisis and the feeling against letting in Germans to close all gates. . . . It is the only chance of our life time to shut out the Jews. It is now or never." Accordingly, in 1917 the literacy bill yet again was proposed in Congress. Some representatives argued that by this time the measure was superfluous, as the European war had already reduced immigration to practically nothing. But the league countered that after the conflict "the flotsam and jetsam of Europe" would be "washed into this country on a living wave," and it was only prudent to have the law already in place. Senator Borah (R-Idaho) agreed that "we ought to have our fences up before the hordes" started arriving.[16]

With aid from the American Federation of Labor, which had made immigration restriction its chief legislative demand, the league's bill was quickly approved by Congress, whereupon it was once more vetoed by President Wilson. But this time, with war raging in Europe, Congress—just as Grant had predicted—mustered the strength to override the veto. In the same month that Congress created Denali National Park to protect the wildlife of Alaska, it finally passed the literacy bill to protect the Nordics of the United States, and as of February 5, 1917, all adult immigrants were required to pass a reading test before entering the country.

Ironically, the literacy test, once put into practice, proved to be remarkably irrelevant. The literacy rate in southern and eastern Europe had risen dramatically since the test was first proposed by the IRL in the 1890s, and only about 1,500 immigrants were debarred annually for illiteracy between 1918 and 1921. Furthermore, the most radical immigrants often turned out to be the most literate ones, and hence the test did nothing to save the country from the dreaded Bolshies.

Accordingly, Madison Grant began formulating plans to implement step two of his campaign to restrict immigration. He knew that he would have to play the leading role in the coming struggle, as Prescott F. Hall of the IRL was unavailable, having suffered a nervous breakdown in 1918 that left him increasingly incapacitated until his death three years later. Nevertheless, Grant was, as always, fully confident that he would succeed, and he vowed to Hall that before he was done "this shall be a white man's country."[17]

Step Two: The Emergency Quota Act of 1921

Madison Grant understood that the key to getting a tougher immigration law was convincing the House Committee on Immigration and Naturalization to favorably report such a measure to the full House. He therefore launched a concerted effort to woo the chairman of that committee, Albert Johnson (R-Washington), to the side of the restrictionists. It was not a difficult task. Johnson was a stocky, heavy-drinking, small-town newspaper editor from Hoquiam, Washington, who was first elected to Congress in 1912 by railing against radicals, city slickers, and the slant-eyed and suspiciously hardworking immigrants from the Land of the Rising Sun. When *The Passing of the Great Race* was published, someone sent Johnson a copy and he was suitably impressed. He contacted Madison Grant, and they struck up a casual correspondence. Grant sent Johnson abstracts of his writings and excerpts from the works of G. Vacher de Lapouge ("the most distinguished anthropologist in France"). In addition to tutoring Johnson on the basics of racial anthropology, Grant also introduced the congressman to the tenets of wildlife conservation, and Johnson became a faithful supporter of that cause as well.[18]

In March 1919, the Republicans assumed control of the House, and Albert Johnson—thanks to some shrewd lobbying by the Immigration Restriction League—found himself appointed chairman of the House Committee on Immigration and Naturalization. It was now that Grant brought to bear against the overmatched congressman the full weight of his charm and his connections. He began actively corresponding with Johnson and made a point of dropping in on him whenever he was in Washington, D.C. Johnson, in turn, began taking the train up to Manhattan whenever possible, where Grant entertained Johnson in his home and showed him around his exclusive clubs. After viewing lower Manhattan through Grant's eyes, Johnson agreed that the Nordics had great "reason for apprehension" over "the aliens creeping up on New York City like locusts a block or two at a time."[19]

Grant saw to it that the congressman was made a member of the Eugenics Research Association, the ECUSA, and the Galton Society (an anthropological organization recently founded by Grant), and Johnson was flattered to find out that his West Coast brand of popular racism was actually cutting-edge science. Grant introduced Johnson to his disciples in the eugenics movement, including his friend John B. Trevor, who would have a great influence on the congressman. Trevor was a wealthy New York lawyer who had been a Columbia Law School classmate of FDR (Trevor's wife, Caroline Wilmerding Trevor, was one of Eleanor Roosevelt's oldest friends). Like Grant, Trevor was a member of the Society of Colonial Wars, and his law office was right next to Grant's in the financial district. Trevor was also a member of the Eugenics Research Association, the Immigration Restriction League, and the American Defense Society, and he was a trustee of the American Museum of Natural History. During the war, he had been chief of the Military Intelligence Division of the U.S. Army in New York City, and as such was responsible for keeping an eye on the disloyal aliens of the Lower East Side. Trevor was now the chief lobbyist for a number of patriotic societies, and he admitted to Albert Johnson that he experienced "convulsive shivers" just thinking about all the Polish Jews who were emigrating to the United States.[20]

Congressman Johnson derived enormous pleasure from associating with the eugenicists, and the Grantians seemed to take a genuine liking to the congressman. While never accused of being a deep thinker, Johnson was an undeniably amiable fellow (in the backslapping, ruddy-faced manner of small-town politicos), and Grant made it clear to all that the representative from Washington's Third District could do their cause a lot of good. Still, flattery was raised almost to the level of satire when Grant arranged for Johnson to be elected president of the Eugenics Research Association in 1923—a group that, for all of its biases, was still a scientific organization committed to genuine research. (It was as if Joseph McCarthy had been elected president of the American Political Science Association in 1953. To be sure, Madison Grant himself had been president of the ERA in 1918, but he at least had a background in zoology and a quasi-scholarly interest in anthropological matters. Putting a politician in charge of the ERA showed how completely the propagandists had taken over the eugenics movement.)

At the instigation of Grant, Representative Johnson now made the acquaintance of Harry H. Laughlin, supervisor of the Eugenics Record Office. Johnson was impressed by Laughlin's scholarly manner and asked him to appear before the House Committee on Immigration to present the eugenic argument for immigration restriction. In April 1920, Laughlin testified for two days before the committee and informed the highly attentive congressmen that since "the character of a nation is determined primarily by its racial qualities," it was incumbent on them to take eugenics into account when formulating immigration policy. Speaking as a scientist, Laughlin taught the committee that heredity was more important than environment, that charity was a "biologically unfortunate" custom, and that continued immigration would inevitably lead to miscegenation since "wherever two races come in contact, it is found that the women of the lower race are not, as a rule, adverse to intercourse with men of the higher."[21]

Albert Johnson was exceedingly pleased by Laughlin's presentation and immediately appointed him the "Expert Eugenics Agent" of the House Immigration Committee (a position he would hold for the next eleven years) with instructions to conduct scientific studies of the immigration problem and report regularly to Congress. Laughlin was delighted by the appointment and blushingly called Johnson "the great American watchdog whose job it is to protect the blood of the American people from contamination and degeneracy."[22]

Thanks to the machinations of Madison Grant, Albert Johnson was now surrounded by a kitchen cabinet of eugenicists consisting of Laughlin, John B. Trevor, James H. Patten (the IRL's lobbyist), Robert DeC. Ward, Henry Pratt Fairchild, Charles W. Gould, Kenneth Roberts, and Lothrop Stoddard. The head chef in the kitchen cabinet, of course, was Grant himself, who by now was a de facto member of the House Immigration Committee (a status that Johnson subconsciously acknowledged by addressing his letters to "The Honorable Madison Grant"). The kitchen cabinet met often and regularly with Johnson to plot legislative strategy and supply him with statistics, position papers, and generous quantities of imported wines.

In the latter half of 1920, Grant determined that the time was propitious to introduce a "drastic" immigration bill that would move far beyond the literacy test and slow down "the hordes of inferior stock that have poured in during the last generation." With demobilization completed, the steamship lines were fully operative, and immigration was edging toward its pre-war levels. Up to seventy-five thousand new immigrants a month were being deposited at Ellis Island, from whence they ventured out into an economy suffering from the postwar recession and a labor market already glutted by the returning doughboys. In November, the commissioner of immigration at Ellis Island warned in a well-publicized address that "whole races of Europe are preparing to remove to the United States. Never since the early days of barbarian Europe has there been such wholesale migrations of population." Such predictions did not sit well with an isolationist public that was not particularly keen to welcome any more foreigners into the country. Thanks to the Red Scare and the propaganda of the antiradical American Defense Society (of which Madison Grant was a trustee), indignation toward hyphenated Americans and fear of alien socialists was running high, and people were clamoring for 100 percent Americanism. Even industrialists, who normally encouraged the influx of cheap workers, had been convinced by the Red Scare that most immigrants were dangerous radicals, and they now saw restriction as a way of forestalling labor unrest. As far as most Americans were concerned, the melting pot had coagulated, and the case for restriction was only strengthened when epidemics of typhus and cholera broke out in central Europe and the surgeon general (a future member of the American Eugenics Society) warned that immigrants were a definite menace to public health.[23]

"Severe restriction is now in the air," gloated Madison Grant, and in the fall of 1920 he headed to the nation's capital to huddle with Albert Johnson. They decided to push a bill that would establish a quota system for the next year. After being amended several times during its legislative journey, the proposed quota limited the number of immigrants from any European country to 3 percent of the immigrants from that country who currently resided in the United States (based on the most recent census, which was that of 1910). Such a formula would not only cut total allowable admissions to 355,000 (in the last year before the war, immigration had hit 1.2 million), but, equally important, it would completely reverse a thirty-year trend by slashing immigration from southern and eastern Europe from 78 percent of the European total to only 43 percent. The system, Grant explained to Prescott F. Hall, would finally favor the beloved Scandinavians over the "Jewish tailors and Greek banana vendors."[24]

Grant and Johnson agreed to sell their Emergency Quota Bill as a one-year stopgap measure to meet the "emergency" of postwar immigration, especially the impending influx of Polish Jews, which Grant characterized as "by far the most serious immigration matter that now confronts us." In December 1920, the House Immigration Committee held hearings on the bill, and the congressmen were jolted when Chairman Johnson read into the record scores of cables from U.S. consular officials in Europe warning that up to one million Polish Jews "of the usual ghetto type" were preparing to emigrate to the United States in the upcoming year. The Jews, warned the foreign service officials, were "filthy and ignorant and the majority are verminous." They were "abnormally twisted, . . . un-American, and often dangerous in their habits." In sum, the Jews comprised a "thoroughly undesirable" class of immigrant and were unquestionably "unassimilable."[25]

The Immigration Committee had heard enough, and favorably reported the quota bill to the full House in early December. The majority report warned that the nation was being deluged with left-wing radicals and cheap laborers, of whom the largest percentage "by far" were "peoples of Jewish extraction." The two Jews on the committee, Isaac Siegel (R–New York) and Adolph Sabath (D-Illinois), issued a minority report insinuating that the majority had been motivated by anti-Semitism and lamenting that the Emergency Quota Bill was "so drastic a change in the historic policy of the United States as to be startling. . . . It would be a sorry day in American history if our country that has heretofore been an asylum for the persecuted, were to slam its doors in the faces of those who have been and continue to be the victims of oppression, persecution, and discrimination in the lands in which they live."[26]

The minority report was ignored by the full House, but during the floor debate some Republican congressmen did express their concern that the quota measure might make it harder to obtain household servants. But the bill's backers continually assured everyone that it was simply a temporary measure designed to hold off the invasion from Europe for one year, until America could recover from the postwar recession and Europe could come to grips with the typhus and cholera epidemics.[27]

When opponents of the bill pointed out that every American had originally been, or had descended from, an immigrant from somewhere else, Congressman William N. Vaile (R-Colorado, and a member of the Immigration Committee who had been well tutored by the kitchen cabinet) retorted that such comments ignored "the great fundamental truth" that, unlike the current tide of immigrants, the brave people who built this country were members of "the Nordic race." The bill's managers provided to their colleagues a chart that explicitly showed that the quotas would cut in half the number of New Immigrants. In addition, Albert Johnson gave a lengthy speech in Congress (using figures provided him by Grant and Laughlin) on the racial history of the U.S.

population, and deduced that the American Nordic was headed for serious trouble unless immigration was severely curtailed. "Put up the bars!" was his concluding exhortation.[28]

Burton J. Hendrick publicly admitted in 1923 that the Emergency Quota Act of 1921 was "chiefly intended—it is just as well to be frank about the matter—to restrict the entrance of Jews from eastern Europe." Overall, however, race was not the predominant issue in the debate. Instead, the economy and the Red Scare were on most people's minds. Supporters of the Johnson bill argued that restriction was needed to save the American workingman from cheap European labor, and to protect the American way of life from "Anarchists, Bolsheviks, Communists, and radicals"—the majority of whom, stated one congressman to great applause, could not even be bothered "to sing or appreciate 'America,' or 'The Star-Spangled Banner.'"[29]

Interestingly, a number of speakers, led by Albert Johnson, defended the quota bill by claiming that the immigration restriction movement was really a branch of the conservation movement. Johnson had seen a draft of an article Grant had written for the *Journal of the National Institute of Social Sciences* (titled "Restriction of Immigration: Racial Aspects") in which Grant warned that the number of "wretched outcasts" coming here from Europe threatened the conservation of our natural resources. By all means, wrote Grant, open wide the gates, "if all the valleys of the Sierras [i.e., Hetch Hetchy] are to be drowned to irrigate deserts, . . . if all our rivers and streams must be stripped of their fish and turned into sewers to carry off waste materials for factories, if the land must be gridironed with railroads and highways." Johnson introduced Grant's article, in its entirety, into the *Congressional Record.* He also quoted on the floor of the House an editorial in the *Saturday Evening Post* that parroted the Grant line and described the destruction that "these new hordes" would do to our natural resources.[30]

Other congressmen now took up the conservation banner to argue for immigration restriction, including J. Will Taylor (R–Tennessee, and a member of the Immigration Committee), who bluntly summed up the situation by stating that "the issue, stripped of its frills and furbelows and without any varnish or veneer, is simply this: Shall we preserve this country, handed down to us by a noble and illustrious ancestry, for Americans, and transmit it to our posterity as our forefathers intended; or shall we permit it to be overrun and submerged by a heterogeneous, hodgepodge, polyglot aggregation of aliens, most of whom are the scum, the offal, and the excrescence of the earth?"[31]

With such oratory ringing in their ears, the members of the lower house enthusiastically approved the Emergency Quota Bill with bipartisan support from all geographical regions. The measure was then sent to the Senate, where it faced an uncertain future, mainly because the chair of the Senate Immigration Committee, LeBaron B. Colt (R-Rhode Island), was known to be cool toward immigration restriction. (Rhode Island was not only home to a number of large manufacturing firms but also had, proportionately, the next-to-largest foreign-born population in the country.)

Throughout the early weeks of 1921, the kitchen cabinet mercilessly hit Colt, the Senate Immigration Committee, and the American public with a barrage of propaganda. The *Saturday Evening Post* fired the first shot with a timely editorial on January 8 endorsing the Johnson bill as a desperately needed form of "quarantine," which would post "No Admittance signs on every frontier." The *Post* then published a stream of articles by Kenneth Roberts, who kept insisting that immigration restriction had become "a matter of life and death for the American people." Alongside photographs of "howling, fighting, frantic crowds" of verminous Jews besieging the U.S. Consulate in Warsaw for visas, Roberts pleaded with the Senate to save America's Nordics from "the filth peril of Eastern Europe." Biologist E. G. Conklin of Princeton then informed the readers of *Scribner's Magazine* that the principles of genetics mandated a stricter immigration policy to prevent the "amalgamation of superior hereditary types" with the "inferior races." In addition, the *New York Times* declared that "the need of restriction" was "manifest" and was due to "the swarms of aliens" menacing the economy, the institutions, and the health of the nation. The paper urged the Senate to ignore "the interests" and instead follow the example of the House, which had acted "with exemplary wisdom and dispatch" in approving the Johnson bill.[32]

Albert Johnson himself then strode across the Capitol to testify before the Senate Immigration Committee on behalf of his bill, and he brought with him new State Department cables warning that those Jews preparing to sail to America were "evasive, dishonest, and . . . do not have the moral qualifications for American citizenship," and that, in any event, "most of them are infected with lice and other vermin." Madison Grant then saw to it that the committee received a statement from Charles Benedict Davenport, who cautioned: "Every day there pass through the portals of Ellis Island persons who . . . bear in their germ plasm family defects which will recur again and again in their descendants."[33]

Grant then brought in the American Defense Society to push for restriction. As an executive of both the Immigration Restriction League and the ADS, Grant understood that the two groups had much in common: one fought to keep the foreign enemy outside the country, and the other fought to suppress the enemy already within. When the Red Scare began to show diminishing returns in late 1920, Grant gave the ADS new life by convincing it to take up the fight against immigration. Accordingly, the society formed a Committee on Immigration composed of Grant, Charles Stewart Davison, Charles W. Gould, Jeremiah Jenks, and Francis H. Kinnicutt. Kinnicutt, yet another bachelor attorney within Grant's circle of friends, was a lobbyist for the Allied Patriotic Societies (a coalition of fifty groups, including the American Defense Society, the National Security League, the Daughters of the American Revolution, and the Boy

Scouts). In early 1921, with Prescott F. Hall of the IRL seriously ailing, the Immigration Committee of the ADS took up much of the lobbying slack, and Grant sent its members to Capitol Hill to warn the Senate Immigration Committee that the New Immigrants supplied "a large proportion of the radicals, the terrorists, and a number of the criminal classes."[34]

And finally, alerted that immigrant anthropologist Franz Boas had been in contact with Senator Colt, Madison Grant himself hastened to Washington, D.C. After two days of private meetings, he succeeded in convincing Chairman Colt and the members of the Senate Immigration Committee to support the Emergency Quota Bill. The measure then received the quick approval of the full Senate and was sent to the White House. The entire legislative process, from the time Albert Johnson first introduced the bill until it arrived on President Wilson's desk in February 1921, had taken just eight weeks.

The president, however, killed the measure with a pocket veto in his last days in office. And yet, Wilson's action barely perturbed the restrictionists. Grant had already traveled to Marion, Ohio, to commune with president-elect Warren G. Harding, and had determined that Harding was "thoroughly sound on the immigration question." The kitchen cabinet calmly waited for Wilson's term to expire in March 1921, and then, the day following Harding's inauguration, Johnson reintroduced the Emergency Quota Bill in the House, which debated it for a perfunctory four hours. Johnson read aloud to the House fresh consular cables provided by the State Department, warning that the majority of European Jews embarking for the United States were "subnormal," "twisted," "deteriorated," and full of "perverted ideas. . . . These are not those who hewed the forests, . . . conquered the wastes, and built America. These are beaten folk" who, "besides being as a class economic parasites, . . . are impregnated with Bolshevism." In short, according to the U.S. State Department, "this type of immigrant is not desirable from any point of view at this time."[35]

During the debate, Representative Isaac Siegel (one of the two Jews on the House Immigration Committee) demanded that the new census of 1920, which was just being tabulated, be used to compute the 3 percent quotas, rather than the now outdated 1910 census. The restrictionists scrambled to defeat Siegel's motion, and in doing so they were quite open in admitting that they preferred the 1910 census because it would admit fewer immigrants from southern and eastern Europe. With the Siegel amendment disposed of, the Emergency Quota Bill was quickly approved by the House (276–33) and then the Senate (78–1), and sent to President Harding, who signed it on May 19, 1921. The quota system was finally the law of the land.

Unlike the literacy test, the Emergency Quota Act of 1921 had a drastic effect on immigration. In the year before the law went into effect, 805,228 immigrants had been admitted to the United States; in the year after, only 309,556 immigrants made it in. Immigration from Italy was cut from 222,260 in 1921 to 40,319

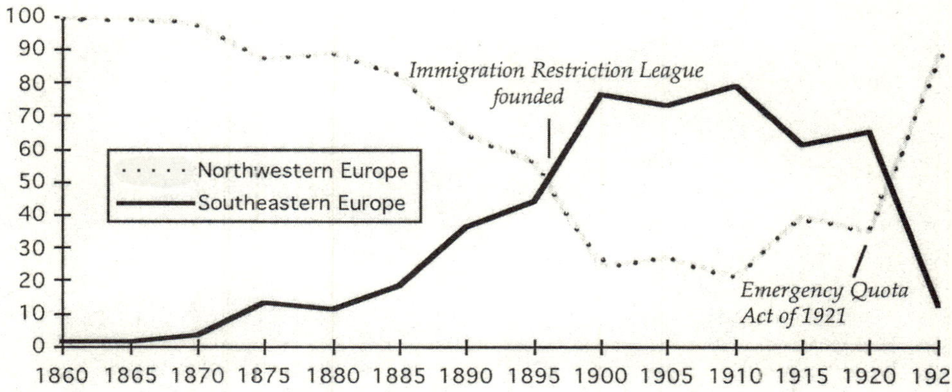

Source of European immigration.

in 1922; immigration from Greece was reduced from 28,503 to only 3,457; and immigration from Turkey dropped from 11,735 to 1,998. (Because of the number of immigrants who returned to their homeland, net immigration for the year following the law's enactment amounted to only 50,090.) Sociologist Henry Pratt Fairchild called the Quota Act "an epoch-making piece of legislation."[36] The fact that it permitted entry to far more Germans—with whom the United States was legally still at war—than to immigrants from nations that had been allied with the United States in the late conflict was indicative of how quickly Americans had replaced hatred of the Hun with fear of the New Immigrant.

During the congressional debate over the Emergency Quota Bill, some of its opponents charged (quite correctly) that "while purporting to be a temporary measure . . . this bill is really intended to pave the way to permanent exclusion." Apparently, however, many congressmen sincerely believed that the Emergency Quota Act of 1921 was only a temporary measure that would expire in one year when, hopefully, the European typhus epidemic would be under control, U.S. employment on the rise, and the radicals on the run. But Madison Grant had known from the beginning that what was supposedly a "temporary" measure would eventually become accepted as permanent policy. To that end, he embarked on an extended and well-coordinated campaign that would not only make the quota system enduring but also render it as discriminatory as possible against the New Immigrants.[37]

Science in the Service of Politics

The Second International Eugenics Congress

Madison Grant, Henry Fairfield Osborn, and Charles Benedict Davenport decided that an international conference on eugenics, held in New York City, would garner tremendous publicity for the cause of eugenics in general and immigration restriction in particular. Their model was the First International Eugenics Congress held at the University of London in 1912, where Arthur Balfour delivered the inaugural address to an audience that included First Lord of the Admiralty Winston Churchill and scientists from England, Switzerland, Germany, Italy, France, and the United States. (For those who doubt that eugenics is a quasi religion, I will simply note that the exhibit hall at the conference displayed "relics" of Darwin, Galton, and Mendel.) The meeting was a great success, and plans were made to hold a second gathering in 1915. With the outbreak of the Great War, however, the congress had to be canceled.

Once the war ended, Grant, Osborn, and Davenport began planning for the Second International Eugenics Congress, to be held in the fall of 1921 at the American Museum of Natural History. To increase the scientific prestige of the congress, they secured the official sanction of the National Research Council, which agreed that Madison Grant should be the treasurer and Henry Fairfield Osborn the president of the Eugenics Congress. (A. E. Wiggam later stated that "had Jesus been among us, he would have been president" of the Eugenics Congress. The NRC rightly predicted that since the Messiah was not available, Henry Fairfield Osborn would be more than happy to fill His sandals.)[38]

Grant and Osborn met continuously throughout 1921 to organize the congress. If they felt that the key to success was organization, they knew that the key to organization was an impressive letterhead. To that end, they named Alexander Graham Bell honorary president of the congress and tabbed an international array of scientists to serve as vice presidents, including Leonard Darwin, Edward M. East, Herbert S. Jennings, Thomas Hunt Morgan, and Raymond Pearl.

As treasurer, Grant collected thousands of dollars from his wealthy friends to help underwrite the congress. Among those who contributed were Mrs. E. H. Harriman, Charles W. Gould, Archer M. Huntington, John Harvey Kellogg, and Herbert Hoover. Grant thanked them all for contributing to "one of the most important causes for furthering true patriotism" and assured them that the International Eugenics Congress would serve as a catalyst to stem the "floods of aliens" poisoning "the blood and the morale" of the United States.[39]

With the organization, the stationery, and the funding in place, Grant, Osborn, and Charles Benedict Davenport worked to ensure that all the important eugenicists attend the congress. They sent out an announcement explaining that as a result of the Great War the "finest racial stocks" had been depleted to the point of extinction, and therefore it was urgent that the world's scientists convene to share ideas on how to preserve the Great Race. Osborn visited Europe with the especial purpose of securing the cooperation of the leading foreign scientists. (As Osborn was departing, Davenport admonished him to keep "crackpots" out of the congress—only "scientific men" such as Madison Grant should be invited to speak.)[40]

The International Eugenics Congress formally opened on September 22,

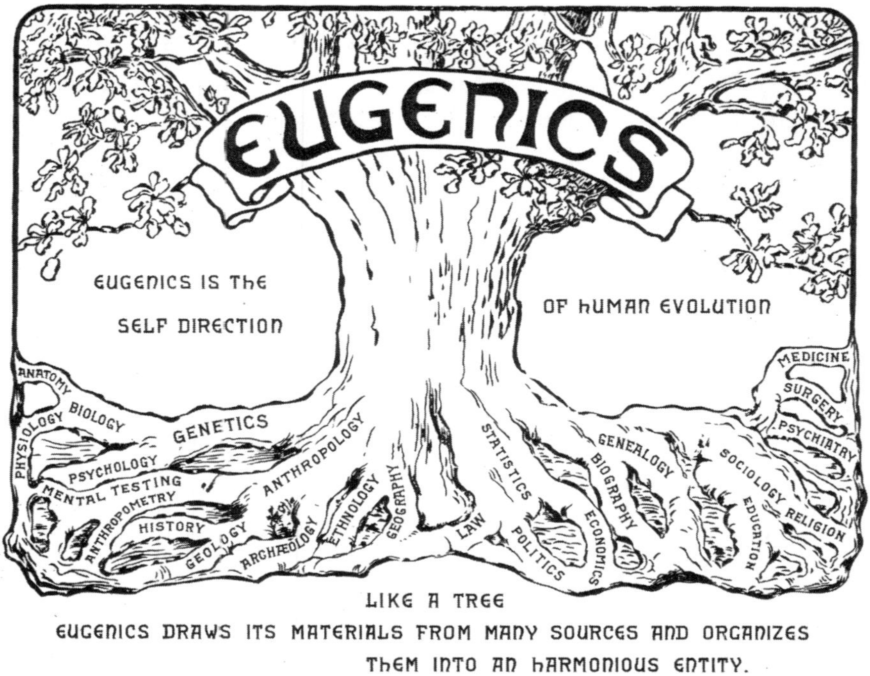

Logo of the Second International Eugenics Congress, showing the all-encompassing nature of eugenics.

1921 (just six weeks before Margaret Sanger convened the First National Birth Control Conference), with over three hundred delegates in attendance. Eugenics was still a mainstream science in the early 1920s, and this was not a gathering of cranks but a convocation of many of the world's foremost researchers. Representatives from most of the leading American universities were there, as were delegates from numerous scholarly associations, including the American Anthropological Association, the American Association for the Advancement of Science, the American Genetic Association, the American Neurological Association, the American Pediatric Society, the American Philosophical Society, the American Public Health Association, the American Social Hygiene Association, and the American Sociological Society. European scientists came from England, Scotland, France, Belgium, Italy, the Netherlands, Norway, Sweden, Denmark, and Czechoslovakia; the Americas were represented by Argentina, Brazil, Canada, Colombia, Cuba, Chile, El Salvador, Guatemala, Mexico, Nicaragua, Venezuela, and Uruguay; and researchers also arrived from Japan, India, Siam, Australia, New Zealand, and Tunisia.

Henry Fairfield Osborn delivered the address of welcome, in which he hoped the congress would drum up support for permanent immigration restriction as well as teach the country's leaders about the need to prevent "the multiplication of worthless members of society." After all, argued Osborn (borrowing a notion employed in wildlife management), the preservation of the "type" must take precedence over the life of any individual.[41]

Following the opening speeches, delegates toured the 131 exhibits assembled by Harry H. Laughlin in the American Museum of Natural History. Field workers from the ERO were on hand to explain the displays, which included anthropometric instruments, plaster casts showing that the brains of Negro fetuses were smaller than those of white fetuses, and charts revealing the reckless fecundity of immigrants. Laughlin had been anxious to show the European delegates examples of "the most thorough-going ethnological and anthropological studies being carried on in the United States," and hence he prevailed on Madison Grant to sponsor Exhibit #51, where the just-published fourth edition of *The Passing of the Great Race* was displayed, along with enlarged copies of the book's maps showing the distribution of European races.[42] (It is a bizarre fluke of history that *The Passing of the Great Race* would also be Exhibit #51 at the Nuremberg Military Tribunal.) Charles Scribner's Sons had its own booth displaying a number of its eugenic offerings, including E. G. Conklin's *Direction of Human Evolution,* William McDougall's *Is America Safe for Democracy?* Lothrop Stoddard's *Rising Tide of Color,* and the ubiquitous *Passing of the Great Race.*

The work of the congress was divided into four sections that met over the next seven days (with the exception of Sunday, September 25, when delegates toured the Bronx Zoo). Section I (presided over by geneticist Herbert S. Jennings) was devoted to the latest research in genetics; section II (presided over by eugenicist Roswell H. Johnson) heard papers dealing with birth control (including a paper from Margaret Sanger, who declared that the ideals of birth control were indistinguishable from those of eugenics); section III (presided over by anthropologists Ales Hrdlicka and Clark Wissler) was concerned with "human racial differences"; and section IV (presided over by eugenicist Irving Fisher) dealt with the relation of eugenics to the state. All told, more than one hundred scientific papers were read at the congress, including presentations by such respected researchers as L. C. Dunn, Ronald A. Fisher, Sewall Wright, and future Nobel laureate H. J. Muller. Indeed, a fair amount of real science took place at the congress. But the daily reports in the press—thanks to the efforts of Lothrop Stoddard (who was head of the publicity committee)—focused their attention on section III (Human Racial Differences), where presenters had been advised by the conference's organizers that their papers should be modeled on the work of Madison Grant.

The keynote address in section III was delivered by the French eugenicist G. Vacher de Lapouge, who had agreed to come to the United States at the urging of his friend Madison Grant (and would have stayed in Grant's house were it not being replastered). His speech recapitulated their ideas about the dangers that miscegenation posed to the Nordics, and argued that it was up to the United

States to provide a refuge for the Master Race: "America, I solemnly declare that it depends on you to save civilization and to produce a race of demigods." Lapouge feared that if the eugenic movement in the United States did not succeed in saving "the whites, the wealthy, and the intellectually superior elements," then mankind would "return to the barbarism of the days of the mammoths."[43]

The message of Lapouge was disseminated by the press, as were the calls for negative eugenics emanating from all the other Grantians at the conference. On September 23, for example, the *New York Times* informed its readers that the delegates to the congress were extremely concerned about "the future of the human race because of the threat of race degeneration" and that most of the researchers had endorsed "strict laws rejecting the unfit immigrant as a necessity for the healthy racial progress of the American people." Two days later, the headline in the *Times* gave Madison Grant great satisfaction:

> EUGENISTS DREAD TAINTED ALIENS
> Believe Immigration Restriction Essential to
> Prevent Deterioration of Race Here
> MELTING POT FALSE THEORY
> Racial Mixture Liable to Lower the Quality of the Stock
> —Prof. Osborn's Views

The article explained that science had concluded that "severe restriction of immigration is essential to prevent the deterioration of American civilization," and that the congress "had vigorously combated . . . the theory held by some eminent anthropologists that all races have an equal capacity for development."[44]

At the conclusion of the congress, an ecstatic Charles Benedict Davenport told Grant he was so impressed by his efforts at organizing the event that Grant should run for president of the United States. Grant modestly declined to act on the suggestion, but he admitted that he was thrilled with the "really brilliant success of the Congress."[45] The proceedings—and their relevance to immigration restriction—had been extensively covered in the news and editorial columns of the national press; important politicians such as Herbert Hoover and the members of the House and Senate Committees on Immigration had conspicuously attended the sessions; and thousands of persons had visited the exhibit hall. The message of the congress reached many levels of American society. Actress Lillian Russell, for example, now announced to her legions of adoring fans that "alien infiltration" had ruined Rome and that it was time to face the fact that the American melting pot had failed as well. She called for stringent restriction of immigration, for "if we don't put up the bars and make them higher and stronger there no longer will be an America for Americans." Little wonder that a satisfied Henry Fairfield Osborn called the Eugenics Congress "perhaps the most important scientific meeting ever held in the Museum."[46]

When the congress ended, the more relevant exhibits, including Grant's maps from *The Passing of the Great Race,* were shipped to Washington, D.C., and displayed for several months in congressional meeting rooms so that the nation's solons might draw the necessary conclusions. And in May of 1922, just before the Emergency Quota Act was set to expire, Congress voted to extend the law for two more years until 1924, with the active encouragement of Grant, Samuel Gompers, Kenneth Roberts (who testified that "masses" of "unspeakably filthy" Polish Jews would invade America if the restrictions expired), the *Saturday Evening Post* (which explained that extending the Quota Act would prevent "our old original stock" from becoming "hopelessly bogged down in the mire of mongrelization"), and George Creel (whose timely articles in *Collier's*—"Melting Pot or Dumping Ground?" and "Close the Gates!"—warned that "several millions of immigrants" were "gathered thick on the shores of the Old World, swarming like flies [and] feverishly awaiting opportunity to come to America"). These New Immigrants promised to bring "disease, crime, insanity, and parasitism" to our shores and thereby threaten "the Nordic stock.")[47]

An Analysis of America's Modern Melting Pot

With the immigration quotas extended for two more years, Madison Grant and the kitchen cabinet went to work to convince Congress and the American public that an even more stringent law would have to be passed when the Emergency Quota Act expired in 1924. After all, as Henry Fairfield Osborn remarked to Grant, the 3 percent law was still admitting far too many "Jews and other undesirables." During this period, *The Passing of the Great Race* was being reprinted every six months, the ERA was providing position papers to selected politicians, the ECUSA was disseminating its propaganda, the National Research Council (as we shall see in chapter 12) was sponsoring scientific studies proving that immigration was a grave danger, and Grant's disciples were publishing widely read articles and lecturing constantly to ever larger crowds. In the October 1922 issue of *World's Work,* for instance, Stanford psychologist Lewis Terman, who felt that "the greatest problem of conservation relates not to forests or mines, but to the . . . proper utilization of human talent," declared authoritatively that the nation faced a "biological cataclysm" because "the immigrants who have recently come to us in such large numbers from Southern and Southeastern Europe are distinctly inferior mentally to the Nordic and Alpine strains."[48]

The following month, on November 21, 1922, Harry H. Laughlin, Congress's "Expert Eugenics Agent," returned to the House Immigration Committee, covered the walls with enlarged maps from *The Passing of the Great Race,* and announced that science had now proven that the new racial stocks invading America were genetically inferior to the old. Using the congressional franking privilege, Laughlin had surveyed 445 of the 667 state and federal institutions for the mentally and physically handicapped, and discovered that immigrants from southern and eastern Europe made up a significantly disproportionate percentage of the inmates of the country's asylums. Laughlin assured the committee that "the researches were not meant to support any preconceived idea; they are simply measures of the facts." But he felt compelled to warn the committee that the data revealed that America was rapidly becoming "a custodial asylum for degenerates," and he concluded that the congressmen should permanently restrict immigration posthaste to protect the blood of the native-born Americans.[49]

Along with his testimony, Laughlin submitted a massive report titled *An Analysis of America's Modern Melting Pot,* filled with impressively complex formulas, graphs, and tables that again proved that the New Immigrants were far more degenerate than the Nordics. Chairman Albert Johnson assured his colleagues that he had examined Laughlin's figures and vouched that "they are both biologically and statistically thorough." (As president-elect of the ERA, Johnson was highly qualified to assess such matters.) Johnson explained to the committee that "The value of first-hand field data skillfully collected and scientifically analyzed is very great. Facts of this nature are the basis upon which the American people must develop their permanent immigration policy." Of course, he had the causal sequence reversed: the "facts" were not the *basis* of immigration policy, but rather the justification for a policy already decided upon years ago in Madison Grant's parlor.[50]

Nevertheless, *An Analysis of America's Modern Melting Pot* was extremely effective, as it enabled congressmen for years to point to Laughlin's findings as scientific justification for immigration restriction. Albert Johnson called Laughlin's report "probably the most important document that has been issued by this Committee for years," and he ordered Laughlin to conduct further research on the "racial damage" wrought by immigration. In the meantime, excerpts and summaries of *An Analysis of America's Modern Melting Pot* appeared in newspapers and periodicals all over the country. Kenneth Roberts, Laughlin's kitchen cabinet colleague, summarized Laughlin's report for a lead story in the *Saturday Evening Post,* and concluded: "If the farmer doesn't keep out the weeds by his own toil, his crops will be choked and stunted. If America doesn't keep out the queer, alien, mongrelized people of Southeastern Europe, her crop of citizens will eventually be dwarfed and mongrelized in turn."[51]

A Study of American Intelligence

Of equal impact to Laughlin's report was the publication the following year of Carl C. Brigham's *Study of American Intelligence.* Brigham was an assistant professor of psychology at Princeton University and scion of a prominent New England family (whose progenitor was the fourth signer of the Mayflower Compact). His book was an analysis of the army intelligence tests administered during World War I. Those tests, writes George Stocking, Jr., "provided the most important single scientific buttress for the racism of the 1920s."[52]

Intelligence tests had only recently been adapted for use in the Untied States. When Congress declared war, Robert M. Yerkes of Harvard, who had studied zoology with eugenicists Charles Benedict Davenport and William E. Castle and was now president of the American Psychological Association, saw an opportunity to validate the practical value of psychological testing. Yerkes, along with two other psychologists active in the eugenics movement, Henry H. Goddard (of *Kallikak* fame) and Lewis M. Terman (whose intellectual hero was Sir Francis Galton and who had modified the IQ test into the Stanford-Binet test), met at the Vineland Training School to devise the intelligence tests that they ultimately administered to 1,726,966 inductees to help the army make personnel and placement decisions.

As Congress was debating whether to enact stricter immigration laws, it occurred to Madison Grant's disciple Charles W. Gould that the data from the army mental tests, if manipulated carefully, could be used to quantify intelligence differences among the races and justify discrimination against Alpines and Mediterraneans. Gould immediately invited Grant, Yerkes, and Carl C. Brigham to dine with him in order "to talk over certain features" of the army intelligence tests (adding with a wink: "I trust that notwithstanding the Volstead Act, the dinner will not be quite as dry as the above statement would seem to indicate"). The men agreed that Brigham would conduct a study, to be underwritten by the wealthy Gould, analyzing the intelligence tests along racial lines. Heretofore, the cephalic index had been employed by racial researchers because the head was measurable and intelligence was not. But now that it was possible to quantify intelligence itself, mental testing replaced calipers as the instrument of choice among scientific racists. After a great deal of work, Brigham presented his findings in 1923 in *A Study of American Intelligence,* with monumental consequences.[53]

Brigham's analysis of the data showed, first of all, that Negroes were vastly inferior to whites, the former possessing an average mental age of only ten. Interestingly, northern Negroes performed better on the intelligence tests than southern Negroes, which might have indicated the beneficial impact of environment on IQ. But Brigham explained that the higher scores of the northerners resulted from the more intelligent Negroes migrating to the North, as well as the higher percentage of white blood among northern blacks.

Brigham then determined that among white Americans, native-born whites were intellectually superior to foreign-born whites. Given that the army tests included questions such as

The Wyandotte is a kind of _horse _fowl _cattle _granite
The Knight engine is used in the _Packard _Stearns _Lozier _Pierce Arrow
Isaac Pitman was most famous in _physics _shorthand _railroading _electricity

"There's a reason" is an ad for a _drink _revolver _flour _cleanser
Five hundred is played with _rackets _pins _cards _dice
Ensilage is a term used in _fishing _athletics _farming _hunting
Bud Fisher is famous as an _actor _author _baseball player _comic artist

it should not have been particularly surprising that the natives scored higher than the immigrants. And Brigham did concede that the scores of the foreign-born increased according to the number of years they had lived in the United States. In fact, immigrants who had been in the country twenty years or more earned scores identical to those of the native-born draftees. Today, we would immediately conclude that these increasing scores reflect the immigrants' growing acculturation to American language and society. But Brigham was under the deductive spell of *The Passing of the Great Race,* and in a bit of logical legerdemain that is marvelous to behold, he decided that the real reason IQ scores *seemed* to go up with increasing years of residence in the United States was that the intelligence of previous waves of immigrants had been higher than that of recent immigrants. In other words, new arrivals did not test as well as those who had been here for a long time because, as Madison Grant had been claiming for years, the New Immigration was intellectually inferior to the old. Such a conclusion found favor with Grant, Yerkes, and Gould—whose money was making this project possible.

Furthermore, Brigham postulated that the reason the intelligence of recent immigrants was so low was because the percentage of Nordics among the immigrants had declined. Unfortunately, with the exception of Negroes, the army had not classified its inductees by race. But it had fortuitously noted their country of origin. Therefore, using figures supplied to him by Grant, Brigham estimated the proportion of Nordics, Alpines, and Mediterraneans in every European nation, and claimed he could thereby break down the immigrant test scores by "race" instead of nationality. The results proved, yet again, that Madison Grant had been right: the immigrants from countries that had high percentages of Alpines and Mediterraneans scored significantly lower on the mental tests than immigrants from Nordic countries. As if astonished to pull this rabbit from his statistical hat, Brigham stated in *A Study of American Intelligence* that "in a very definite way the results . . . support Mr. Madison Grant's thesis of the superiority of the Nordic type."[54]

Brigham buttressed his argument with lengthy quotations from *The Passing of the Great Race* (although he advised his readers that "the entire book should be read to appreciate the soundness of Mr. Grant's position and the compelling force of his arguments"). Speaking as an authority on psychology and mental testing, Brigham predicted that no one ever again would be able to "deny the fact that [racial] differences exist," and he triumphantly reiterated, in clear, unmistakable language, and with the imprimatur of Princeton University Press, that "the intellectual superiority of our Nordic group over the Alpine, Mediterranean, and negro groups has been demonstrated."[55]

Before penning the conclusion of his book, Brigham had met with the kitchen cabinet (Grant, Gould, Osborn, Kinnicutt, and Patten), and their influence is obvious in Brigham's call on the final page for Congress to enact a much more restrictive immigration law to reduce the incoming stream of Alpines and Mediterraneans.

Carl C. Brigham's *Study of American Intelligence* was a major achievement in the history of scientific racism. "Few works in the history of American psychology," writes Leon J. Kamin, "have had so significant an impact." Henry H. Goddard declared that the analysis of the Army mental tests was "probably the most valuable piece of information which mankind has ever acquired about itself." In much the same way that Goddard's *Kallikak Family* had provided scientific confirmation in 1912 of Davenport's thesis that feeblemindedness was hereditary, so Brigham's book, with its impressive charts, graphs, and tables, confirmed the validity of Madison Grant's Nordicism.[56]

Robert M. Yerkes wrote an enthusiastic foreword for *A Study of American Intelligence* in which he affirmed that Brigham "presents not theories or opinions but facts." Yerkes was well aware of the political implications of Brigham's work, and he declared that "Mr. Brigham has rendered a notable service to psychology, to sociology, and above all *to our law-makers,*" who assuredly could no longer "afford to ignore the menace of race deterioration." Even before the book was published, Yerkes had alerted the chairmen of both the House and Senate Immigration Committees that Brigham's analysis of the army tests indicated the mental inferiority of eastern and southern Europeans. And an impatient Yerkes kept urging Princeton University Press to expedite Brigham's book so it would appear before Congress voted on whether to extend the immigration quotas. When Yerkes reviewed the book for *Atlantic Monthly,* he declared that science had shown that "men of foreign birth" were "markedly inferior in mental alertness to the native-born American," and he remarked that "the race differences" were "so pronounced" that they would have "obvious practical significance" on immigration policy.[57]

Most of Yerkes's colleagues in the psychology establishment were equally enthusiastic about Brigham's study. The *Journal of Educational Psychology,* for example, asserted that Brigham's thesis that "the Nordic stock is more intelligent than the Alpine or Mediterranean . . . is carefully worked up to by a logical and careful analysis," and declared that "we shall certainly be in hearty agreement with him when he demands a more selective policy for future immigration." Lewis Terman of Stanford, speaking as president of the American Psychological Association, was pleased to report that psychology had become "the beacon light of the eugenics movement" and that the army tests were being "appealed to by congressmen in the reshaping of national policy on immigration." And

Henry Fairfield Osborn, after perusing what Brigham, Yerkes, Gould, and Grant had done with the army data from World War I, made the incredible statement that "those tests were worth what the war cost, even in human life," since they clearly showed that the New Immigrants were "far inferior" to the race that built this nation. Brigham's work secured his reputation in the psychology profession. He was soon elected secretary of the American Psychological Association and went on to develop the SAT exam for the College Entrance Examination Board.[58]

As intended, Brigham's interpretation of the army intelligence tests carried great weight with Congress and helped to crystallize the sentiment in favor of more extreme restrictions. At a hearing before the Senate Immigration Committee, Francis Kinnicutt referred extensively to the just-published *Study of American Intelligence,* which he called "the most important book that has ever been written on this subject." The committee's chair, Senator Colt, announced: "I think every member of the committee ought to read that book," whereupon Kinnicutt obligingly presented a personal copy to each of the senators.[59]

By 1923, thanks to the Second International Eugenics Congress, Harry H. Laughlin's *Analysis of America's Modern Melting Pot,* and Carl C. Brigham's *Study of American Intelligence,* all congressmen were aware that science had proved that southern and eastern Europeans were biologically and intellectually inferior to the Nordics, and that the genetic health of the nation would be jeopardized if any more New Immigrants were permitted to enter the country. Madison Grant noted in the *North American Review* that there had been "a great change of public opinion during the last decade. Ten years ago . . . the old American idea, inherited from the nineteenth century, of the efficacy of the Melting Pot, still prevailed." But he was happy to see that his countrymen had awakened to the disconcerting fact that the U.S. population was now "a jumbled-up mass of undigested race material" and that immigration restriction had become legislative priority number one.[60]

Indeed, scientific racism, by providing a scholarly rationale for restriction that went beyond mere nativism or economic self-interest, would play the dominant role in the passage of the epochal—and highly discriminatory—Immigration Restriction Act of 1924.

Step Three: The Immigration Restriction Act of 1924

In one of his last public appearances, President Warren G. Harding told an audience in Oregon, "I choose quality rather than quantity in future immigration." The president, in this, reflected well the mood of the public, the press, and Congress, amongst all of whom sentiment in favor of tighter immigration restriction was waxing in the fall of 1923. Referring to the Emergency Quota Act of 1921, *Current Opinion* stated: "Although successful in shutting off a large part of the turgid stream of undesirable and unassimilable human 'offscourings' from southern and eastern Europe, this measure did not go far enough. The conviction is growing that if the tall, big-boned, blue-eyed, old-fashioned 'white' American is not to be bred out entirely by little dark peoples, Uncle Sam must not simply continue the temporary quota law in operation, but must make its revisions more stringent."[61]

With the sixty-eighth Congress due to open in December 1923, Madison Grant informed Albert Johnson that the moment had arrived to replace the Emergency Quota Act with a more permanent, and stricter, law. "I believe you have the country behind you and a most popular cause. It is growing in strength every day and it is only a question of time when even greater restrictive measures will be put through." The two men then sat down and mapped out their strategy. The Emergency Quota Act had imposed a quota on each nation of 3 percent of the immigrants from that country who resided in the United States as of the census of 1910. Grant and Johnson decided to lower the percentage from 3 percent to 2 percent, and to move back the base year from 1910 to 1890 (when there had been were far fewer immigrants from eastern and southern Europe). These two seemingly minor changes would severely cut the total annual allowance from 355,000 to 165,000, and would reduce immigration from southern and eastern Europe to a negligible 12 percent of that total. The immigration to the United States of Greeks, Hungarians, Italians, and most importantly Jews, would effectively be ended. Grant's colleagues were enthused about the plan. George Horace Lorimer of the *Saturday Evening Post* fully endorsed the base year switch to 1890, "that being a year," he explained to Kenneth Roberts, "when Nordic immigration was strong and the low-grade stuff hadn't begun to come to us in volume." An editorial in the *Post* titled "Back to 1890" declared that moving the base year "would be a fine and well-deserved tribute to those immigrants of a past generation to whom the country owes so much."[62]

Grant and Johnson knew that at least twenty proposals dealing with immigration were going to be introduced in the upcoming session of Congress, but they were determined that their own plan rule the roost. On December 5, 1923, the opening day of Congress, Johnson introduced their bill. On the following day President Coolidge, presenting his first annual message to Congress, declared that "America must be kept American. For this purpose, it is necessary to continue a policy of restricted immigration." Bolstered by the president's statement, the House Committee on Immigration proceeded to hold hearings on the Johnson bill. It was the beginning of a debate that would occupy the attention of Congress for the next six months.[63]

By now, the House Committee on Immigration was entirely conversant with the major tenets of Grantian eugenics, and an overwhelming majority of its members were in favor of a bill that discriminated on the basis of race. Kenneth Roberts was a constant presence in the committee room throughout the hearings, available to answer any questions the congressmen might have, and Grant made sure that, behind the scenes, Chairman Johnson and his colleagues had frequent visits from John B. Trevor, Francis Kinnicutt, Harry H. Laughlin, and others. With the outcome of the committee's vote a foregone conclusion, the hearings were designed mainly to give a pretense of fair play. A few lobbyists for big industry, big agriculture, and the steamships did show up to oppose the bill, but they were much less active than they had been during the debate over the Emergency Quota Act of 1921. A number of immigrant groups testified against restriction, but they did so in a halfhearted, almost sullen, manner. Aware that "science" was on the side of the restrictionists, they were resigned to the inevitability of some permanent measure based on race making it through Congress. The most they could hope for was to convince the committee to retain 1910 as the base year and not embrace the blatantly discriminatory move to 1890. In fact, halfway through the hearings, Albert Johnson noted that "few witnesses" were bothering to advocate "anything like our former liberal immigration policy. Restriction," he observed with satisfaction, "is here to stay."[64]

Still, Madison Grant was attacked many times during the hearings by those he referred to as "the noisy racial interests." Even though Grant himself had never appeared publicly before the House Immigration Committee, everyone knew that he was the éminence grise of restriction. Thus, for example, Rabbi Stephen S. Wise, head of the American Jewish Congress, told the committee that he was distressed by the American public's recent obsession with Grantian anthropology. " I remind you, gentlemen, that anthropology is an upstart science; it is one of the newest of sciences; and [yet] we sit here listening to dissertations upon anthropology." He recalled for the congressmen that there had never even been a "Nordic race" until the "insolent" Grant came along and invented it, and he remarked that Grant's position as head of the New York Zoological Society "hardly constitutes adequate qualification for a decision as to the worthiness of one group of aliens and the unworthiness of another group of aliens."[65]

Similarly, Gedalia Bublick, editor of the *Jewish Daily News* (a leading New York paper), went before the committee and blamed Madison Grant for being "the discoverer of this race business. He is the Moses of race hatred in the United States." "Read what this man says," pleaded Bublick. "He wants only men with long skulls and blond hair, and he says that if a man has a different kind of skull and a different kind of hair . . . we do not want him." Bublick paused, and movingly added: "No man has the right to say to me that . . . I am of a low race, because I am not a Nordic." Bublick predicted that one day the books of Grant and his disciples would be seen as "a shame to America in her history."[66]

Louis Marshall, an extremely influential figure in the Jewish community, one of the most eloquent lawyers of his generation, and a man to whom even rubes like Albert Johnson could not help but show respect, also assailed Grant in his testimony. "We have heard a great deal about anthropology," Marshall told the House Immigration Committee in January 1924. "There has been more anthropology published in the press and in printed books during the last 10 years than has ever before been conceived by the mind of man." Marshall conceded that the most important of these books, a book that he had read "quite religiously," was *The Passing of the Great Race.* But he lampooned Grant for inventing this "new race for us"—the Nordic race, a race that "nobody ever heard of" before Grant. And he scoffed at Grant for claiming on the one hand that the Nordics were the superior race and on the other that they were in danger of disappearing. Apparently, argued Marshall, Grant "was not a real scientist, after all," for if he were he would know that Darwin's theory of the survival of the fittest posited that the superior race should be dominating, not passing.[67]

Republican Congressman William N. Vaile immediately interjected to refute Marshall—and in doing so demonstrated just how well the committee had absorbed Grant's teachings. Vaile pointed out that "survival of the fittest" did not mean survival of the *best,* but rather the survival of the type most adapted to the current environment. And then—because he always had his well-marked copy of *The Passing of the Great Race* handy—Vaile promptly supported his statement by opening the book and reading a passage from page 82: "The 'survival of the fittest' means the survival of the type best adapted to existing conditions of environment, today the tenement and factory, as in Colonial times they were the clearing of forests, fighting Indians, farming the fields, and sailing the Seven Seas. From the point of view of race it were better described as the 'survival of the unfit.'" A flummoxed Marshall conceded the point, and gave up trying to argue science with the restrictionists.[68]

Louis Marshall's momentary gaffe notwithstanding, the opponents of Grant who testified in 1924 were often quite moving in their pleas that the United States not abandon its historic ideal of opportunity for all. And yet, almost to a man, they used Grantian language to concede that the exclusion of *non*-whites should be continued. Minority groups such as the Jews hoped that by showing themselves to be loyal enemies of the nonwhites, they would be permitted to join the Nordics and other old-stock Americans of European descent in a coalition known as "the white race." Thus, for example, Joshua Kantrowitz of B'nai B'rith and Judge B. A. Rosenblatt of the American Jewish Congress argued eloquently against immigration restriction, but they assured the committee that they wanted immigration to be limited to "races that can be assimilated" (a code phrase for the exclusion of Asians). Similarly, William Edlin, editor of the *Yiddish Day* and chair of the United Foreign Language Newspaper Publishers and Editors, gave a poignant speech against any form of restriction, but then added: "I am speaking here on behalf of the Caucasian race, *to which we all belong.* I do not want to take in those races which do not assimilate readily. . . .

After all, . . . the Chinese, Hindus, and other races do not have those things that *we* call civilization, and I look upon those people as too far from *us* for assimilation purposes." And O. D. Koreff, editor of the *National News,* a Slovak newspaper, was adamantly opposed to restriction—except, of course, restriction of the Japanese, because "they are not assimilable. They are of the yellow race. . . . and *we* are white."[69]

We will leave it for later (chapter 13) to evaluate the effectiveness of the "we're all whites" tactic. In the meantime, to counteract the antirestrictionists, Grant sent a parade of eugenicists up Capitol Hill in 1923 and early 1924 to testify that the 2 percent/1890 formula was the only eugenically sound plan for conserving the germ plasm of the Nordics. Lothrop Stoddard, Robert DeC. Ward, John B. Trevor, James H. Patten, and Elon H. Hooker (chairman of the American Defense Society) all testified for the Johnson bill, as did Francis Kinnicutt, who reminded the congressmen that New York City was already overrun by Jews and that the nation simply could not absorb any more, and Harry H. Laughlin, who—in over two hundred pages of testimony—presented data, based on Carl C. Brigham's *Study of American Intelligence,* showing that Nordics rank at the top and Jews at the bottom of the intelligence scale. Laughlin argued that Congress must employ eugenics to formulate an immigration statute that would do "for the human stock the same thing that our Department of Agriculture does for protecting and improving our best domestic animals."[70]

Most members of the committee fawned over the eugenicists and constantly praised the scholarly nature of their testimony. In fact, twelve more Grantians were waiting in the wings to testify in favor of the bill, but they were not called, as Johnson saw no point in beating a dead horse. All this led an overwhelmed Samuel Dickstein (D–New York), a freshman committee member (and Jewish immigrant from Russia), to complain that his colleagues were "infected with the germ of the Nordic superior race theory."[71]

Incredibly, the author of that theory—Madison Grant—could not personally participate in this, his moment of triumph. Grant had spent a lifetime affecting public policy by working behind the scenes. He was well aware of—and quite comfortable with—the fact that he possessed a certain amount of power and influence, and he did not need a spotlight shining on him to confirm this. Still, there were times—the opening of the Bronx River Parkway, the dedication of a redwood park (chapter 11)—when he felt that he should make a public appearance. And certainly the debate on the Immigration Restriction Act of 1924 was one of those times. Albert Johnson asked Grant to testify for the bill, but Grant, now fifty-eight years old, had been struck down by arthritis and was confined to his bed. But in a testament to the power of malevolent ideas, Grant's physical presence was not really required on Capitol Hill in 1924. He had already done his part, from the publication eight years earlier of *The Passing of the Great Race* through the lobbying of the Immigration Restriction League, the propaganda of the American Defense Society, the work of the ERA and the ECUSA, the seduction of Albert Johnson, the sessions of the International Eugenics Congress, the manipulation of data in *A Study of American Intelligence*—it had all ensured that scientific racism was part of the atmosphere that Americans were breathing in 1924. Whereas economic and political considerations had been quite important in the passage of the 1921 Emergency Quota Act, the debate in 1924 was, as Franz Boas admitted, primarily about "the idea of the racial superiority of the 'Nordic.'" And Madison Grant owned the patent to that idea.[72]

In lieu of a personal appearance, Grant sent a lengthy statement to the House Immigration Committee defending the adoption of the 2 percent/1890 formula as "scientific and thoroughly just" and the best means of preventing the immigration of foreign races who threatened to "displace native Americans and replace them by lower types." Johnson read the letter aloud to the committee and entered it into the record.[73]

As soon as the hearings on the Johnson bill concluded, the House committee, to no one's surprise, reported the measure favorably on March 24, 1924, by a vote of 15–2, with the two Jewish members, Samuel Dickstein and Adolph J. Sabath, dissenting. They charged in their minority report that the bill was "a palpable injustice" based on a supposed anthropological theory that was, in fact, "pure fiction."[74] The attacks on Grant carried over into the floor debate, where Dickstein and Sabath were joined by a small group of young, urban Jewish representatives (Emanuel Celler, Meyer Jacobstein, Nathan Perlman), Catholics (William Patrick Connery, Jr., James Gallivan, Charles Anthony Mooney, John Joseph O'Connor, Patrick O'Sullivan, Peter Francis Tague), and both (Fiorello La Guardia). And just as Grant was the chief ideologist of the restrictionist forces, so his lifelong foe, Franz Boas, now took an active role in advising the antirestrictionist bloc.

For example, Grant's chief inquisitor during the floor debate was freshman representative Emanuel Celler (D–New York), who, after consulting with Franz Boas, despaired that "we have grown accustomed to hear a great deal of loose thinking, senseless jargon, and pompous jumble concerning Nordic superiority." Celler contemptuously placed the blame on Madison Grant: "The fallacy of 'Nordic supremacy' was made popular by one, Madison Grant, who wrote a book called *The Passing of the Great Race.* This book had a great vogue, and correspondingly it has created a great mischief. The opinions expressed in his book are most dangerous. The opinions are rendered more dangerous because they come from a man who has contributed a great deal that was good to the subject of zoology." Celler charged that *The Passing of the Great Race* "is about as fine an example of dogmatic piffle as has ever been written," and he denounced the "Nordic myth" as being "outrageously absurd." He lamented that Albert Johnson and the House Immigration Committee had fallen under Grant's conjurations, and suggested that the committee should have heard from a serious scientist like Franz Boas, but "No; the committee only wanted those who believed in 'Nordic' superiority; men who deal in buncombe, like Grant and Stoddard." Celler then attempted to offset Grant's influence by reading to the House excerpts from Boas's *Mind of Primitive Man.*[75]

Like Celler, Charles Anthony Mooney (D-Ohio) also invoked Boas to counteract the "Nordicologists."

> Suddenly a new word made its way into the English language—"Nordic," "Nordic," "Nordic"—everywhere you turned. There is not a fifth-rate extension lecturer but does not speak of it with scientific exactness. Newspaper editorials, magazine articles, know exactly what the word means, what it implies. . . .
>
> [But] it is a thing of common knowledge among ethnologists and anthropologists that this talk of racial superiority is largely verbiage. Prof. Franz Boas, America's leading authority on anthropology, in his book *The Mind of Primitive Man,* shows that notion to be a most ridiculous one."[76]

Adolph Sabath, the Bohemian immigrant who (despite the best efforts of the Immigration Restriction League) represented Chicago's polyglot Fifth District in the House from 1907 to 1952, also took a leading role in the deliberations. One of the founders of the Anti-Defamation League, Sabath rarely participated in floor debates, owing to his heavy accent and his tendency to break into Czech when excited. But in 1924 he indicted the restrictionists for being duped by the "unfounded anthropological theory" that native-born Americans "are the progeny of fictitious and hitherto unsuspected Nordic ancestors," and as a member of the Immigration Committee he assured the House that "no scientific evidence worthy of consideration was introduced to substantiate this pseudoscientific proposition. It is pure invention and the creation of a journalistic imagination."[77]

Faced with these intemperate attacks, the bedridden Grant, who was in contact with Albert Johnson by phone or letter at least once—and sometimes twice and even thrice—daily, struck back in the first few months of 1924 with a multipronged offensive whose goal was to convince Congress that, no matter what Franz Boas and his Jewish and Catholic adherents were saying, restriction had the backing of science.

First, Grant and his minions "bombarded" (as Johnson put it) the members of Congress with telegrams and letters. The Grantians also took to the hustings to speak in favor of the immigration bill, and they invariably cited the army intelligence tests and Harry H. Laughlin's congressional reports as proof that the nation could not afford to continue "the wholesale importation of low-grade people" from Europe.[78]

And then the restrictionists spent the month of March permeating the country's most influential journals. Owen Wister entered the lists by defending immigration restriction in the *American Magazine.* Lothrop Stoddard then wrote "Racial Realities in Europe" (which was simply a five-page précis—often verbatim—of *The Passing of the Great Race*) for the *Saturday Evening Post.* Master propagandist Kenneth Roberts was sent into action and contributed three well-publicized articles to the *Saturday Evening Post* aimed at bolstering the proponents of restriction. The articles by Roberts, illustrated with the requisite photos of Jews lining up in Warsaw to get visas to the United States, insisted that it was imperative that Congress adopt the 2 percent/1890 formula to save America from the "parasite races" of eastern Europe who—as confirmed by the army intelligence tests—were genetically inferior. Why, just walking into a room "with a crowd of emigrants from Poland was like coming up on the leeward side of a glue factory on a warm day." Congress, declared the future Pulitzer Prize winner, must embrace restriction to guard "against the slum dwellers of Europe; against the human scrubs and runts and culls that will otherwise have a part in future generations of Americans; against the incompetent, unreliable, unintelligent masses."[79]

And then Madison Grant himself placed an article ("The Racial Transformation of America") in *North American Review* in which he lambasted the liberals for welcoming "undesirable races and classes" into the Nordic sanctuary, where they were allowed to "breed recklessly" and threaten the old-stock American. Like his colleagues, Grant bolstered his argument for Nordic superiority by referring to the army intelligence tests and the testimony of Harry H. Laughlin. And thus the circle was complete: Grant cited Laughlin, who had based his analysis on Brigham's statistics, which were in turn based on Grant's calculations of the racial composition of the European population. What seemed, in other words, to unaware observers to be a plethora of independent studies by reputable scientists was actually a series of self-referential claims that, like the worm Ouroboros, constantly fed upon itself. The American Defense Society reprinted "The Racial Transformation of America" as a pamphlet and distributed it where it would do the most good. Within days, Grant's words were being echoed in both houses of Congress, leading an exasperated congressman, James Gallivan, to exclaim: "And so the Nordic hysteria continues!"[80]

In addition to carrying articles by the Grantian elite, the *Saturday Evening Post* also churned out pro-restriction editorials on a regular basis during the immigration debate. The editorials, by George Horace Lorimer and Frederick S. Bigelow (who was also a member of the Advisory Council of the ECUSA) praised Albert Johnson and his colleagues on the House Immigration Committee as a "group of devoted and patriotic men in Congress . . . who are working night and day . . . to forestall the day when we shall have a completely mongrelized America."[81]

Along with the *Saturday Evening Post,* the *New York Times* became practically a newsletter for the Grantians during this period. On January 17, the *Times* printed "One Hundred Years of Immigration," a full-page article in which Secretary of Labor James J. Davis (himself an immigrant from Wales) cited the army intelligence tests and Laughlin's reports to Congress and declared that Americans must protect their "present and future population from the evil mental, moral and physical influence" of the New Immigrants. Davis predicted that if the Johnson bill failed, immigrants "would eventually submerge and absorb the American people, as the old Roman civilization was completely submerged by the hordes which once migrated to that fair land."[82]

The *Times* also published a series of letters from the usual Grantians (e.g., Roswell H. Johnson, Francis Kinnicutt, Henry Pratt Fairchild, Clinton Stoddard Burr, and Charles Stewart Davison), all providing eugenic arguments in favor of the immigration act.[83] Probably the most talked-about letter was the one on April 8 from Henry Fairfield Osborn, who was fed up with congressmen of Italian heritage who pandered to their hyphenated constituents by holding up Christopher Columbus as proof that Nordics did not have a monopoly on greatness. After consulting with Madison Grant, Osborn composed his letter, which, after recapitulating *The Passing of the Great Race* on the characteristics of the Nordic race, gleefully proclaimed that "Columbus, from his portraits and from his busts, authentic or not, was clearly of Nordic ancestry." In fact, Osborn announced that (among others) Leonardo da Vinci, Galileo, Raphael, Donatello, Botticelli, Petrarch, and Dante were all Nordics as well. As for those Polish-Americans who pitifully boasted about the achievements of Kosciusko and Pulaski during the American Revolution, it was Osborn's scientific duty to disclose that, since those two worthies were members of the Polish nobility, they too were probably Nordic.[84]

The most influential piece of propaganda issued by the Grant camp was the detailed report of the Committee on Selective Immigration of the ECUSA. The committee, chaired by Madison Grant, had six other prominent members: Charles W. Gould, Lucien Howe, Francis Kinnicutt, Harry H. Laughlin, Robert DeC. Ward, and Congressman Albert Johnson. Their report, citing (what else?) Brigham's analysis of the army intelligence tests and Laughlin's "very thorough investigation" of the nation's mental institutions, called on Congress to adopt the 2 percent/1890 formula, for such a move would heavily favor the Nordics, with their "higher grade of intelligence," over the "peddlers, sweatshop workers, fruit-stand keepers, [and] boot-blacks" of southeastern Europe who were, after all, "non-essential members of the community."[85]

The report of the Committee on Selective Immigration of the ECUSA was sent to members of President Coolidge's cabinet, to numerous public libraries, and to 2,885 journals and newspapers, including the *New York Times,* which quoted extensively from the document. In addition, the report was distributed to key members of Congress. (We thus witness the somewhat superfluous act of Albert Johnson sending himself a report, signed by Johnson, which urges Johnson to vote for the Johnson bill).

Most congressmen may not have known what exactly the ECUSA did or what it stood for, but it certainly sounded scientific, and the members of its Committee on Selective Immigration were all well-known and highly respected figures. Of this group, Grant, Howe, Johnson, and Ward were also leading members of the Immigration Restriction League, and Grant, Gould, and Kinnicutt were also on the American Defense Society's Committee on Immigration. It should be apparent by now that in the early 1920s there was, for all intents and purposes, an interlocking directorate of scientific racism, just as there was one for wildlife conservation. This interlocking directorate consisted of half a dozen organizations, and some three dozen persons, all cooperating on an informal basis to bring about immigration restriction legislation and thus save America from the scourge of the Jews (see appendix D).

Every day, congressmen received more and more material from the ADS, the IRL, the ERA, the ERO, and the ECUSA, all of it clamoring for restriction, and it therefore seemed as if the immigration restriction movement had the support of a number of important scientific organizations. What may not have been clear to the congressmen, however, unless they looked carefully at the letterheads, was that the same elite coterie was running each organization. What was also not obvious to the congressmen, unless they took the time to make a textual comparison of the statements of the ADS and the ECUSA, the articles of Stoddard and Roberts, the editorials in the *Saturday Evening Post* and the *New York Times,* the speeches of Brigham and Trevor, and the letters of Kinnicutt and Davison, was that they were all drawing on a common well of influence. For at the center of all these people and organizations, like a spider perched in the middle of its increasingly intricate web, was one figure: Madison Grant. His joints painfully inflamed by arthritis, the crippled Grant lay in his bed in Manhattan and used the telephone, the telegraph, and the U.S. mail to masterfully coordinate the activities of the interlocking directorate of scientific racism.

The antirestrictionists in Congress were totally frustrated by the propaganda blitz of the interlocking directorate. Adolph Sabath complained that "the advocates of restriction . . . send out from day to day statements, interviews, and misleading articles to prejudice the minds of the American people against the so-called newer immigration." Emanuel Celler grumbled to Franz Boas that Madison Grant had "immeasurably influenced the members of the Immigration Committee," and Celler charged on the floor that ever since the bill was referred out of committee, its proponents "have indulged in a veritable paean of hate against our alien population, and there has been let loose the dogs of racial and religious hatred and animosity." Fiorello La Guardia had a pretty good sense of the forces arrayed against him, and hinted at "the dirty work" being done "by the secret influences back of this bill" who were motivated by animus against Italians and Jews. The legislation, cried La Guardia, was "inspired, prompted, and urged by influences who dare not come out in the open, . . . but who believe that it is proper to take vengeance upon these humble, harmless, helpless immigrants."[86]

After listening for months to the complaints of immigrant groups that shifting the base year from 1910 to 1890 would be an act of "deliberate discrimination," Grant's supporters issued their proud plea: guilty as charged. The *Saturday Evening Post* unabashedly announced that "if the Johnson bill is discriminatory, so much the better. If there is one thing we need more than another it is a little discrimination in our immigration policy." The restrictionists charged that for years European countries had been deliberately issuing passports to Jews to "purify and homogenize their own racial composition," and Francis Kinnicutt declared that thanks to "a long and well-established policy of organized international Jewry," America had become nothing less than the "Jewish land of refuge." Accordingly, Representative Samuel D. McReynolds (D-Tennessee) stood up in the House and proclaimed, to great applause, that "if the eastern and southern European countries are discriminated against they have brought it upon themselves, and we are justified in discriminating against them." And Congressman J. Will Taylor brought his colleagues to their feet with his statement that "if it takes arbitrary, discriminatory, or even *despotic* legislation to protect America and American institutions, in the name of God, let us have them!"[87]

In fact, the restrictionists were so fired up that they managed to turn on its head the accusation of "discrimination." After all, they pointed out, Nordics comprised 87 percent of the total white population of the United States, yet they had made up only 24 percent of the immigration in 1910. Therefore, as Francis Kinnicutt explained in a letter to the *New York Times,* continuing to use the census of 1910 "amounts to a clear discrimination against the very peoples who principally settled the American Colonies and founded our civilization." (Kinnicutt's letter was, perhaps, the first time an angry white male levied the charge of "reverse discrimination.") Clinton Stoddard Burr and Charles Stewart Davison followed up with letters of their own (Burr, in fact, claimed that the only fair census for establishing quotas would be that of 1790), and within days the outcry from Nordic congressman—suddenly conscious that they were the victims of racial prejudice—was deafening. An outraged Senator David Reed (R-Pennsylvania) charged that the use of quotas based on 1910 was "a great discrimination against us, the American born." It was now clear to Representative Vaile (who later that year sponsored legislation legalizing birth control) that using 1910 as the base year was "a very gross discrimination" against the Nordics, and he lambasted the "special and unequal privileges" hitherto enjoyed by Italians and Jews in the United States. And Congressman Riley Wilson (D-Louisiana) similarly declared: "It is high time we stopped the long-continued discrimination against the native-born American of the old stock."[88]

The *New York Times,* while supporting restriction, had until now opposed the switch from 1910 to 1890. But Madison Grant and the ADS Committee on Immigration put a lot of effort into convincing Nicholas Roosevelt, who was responsible for the paper's immigration editorials, to endorse the new party line. In March, the *Times* (whose Jewish publisher, Adolph Ochs, constantly endeavored not to antagonize old-stock Americans) conceded that while 1890 *might* be discriminatory against the New Immigrants, the 1910 census *definitely* discriminated against the *Nordics.*[89]

The debate in the House was reaching its climax, and so was the rhetoric of the politicians. Many congressmen echoed the claim of the eugenicists that the New Immigrants were akin to germs that were invading the national body. Immigrants were routinely described as "infectious," "poisonous," and "parasites." The congressmen also shared with the eugenicists an obsession with digestion: Senator William Bruce (D-Maryland) was one of scores of speakers who claimed that immigrants were "indigestible lumps" in the "national stomach" (as well as "insoluble blood clots in the national circulation"). And as in 1921, the restrictionists again characterized themselves as being part of the conservation movement. Representative Robsion of Kentucky, for example, stated that Americans must bar foreigners in order to "protect our institutions and *conserve our resources*. . . . Let us save something for our posterity." Albert Johnson wrapped up the debate over his bill by proclaiming that "a restrictive immigration act is as truly a *conservation* measure as any dealing with natural resources."[90]

And with that, the immigration bill, with the 2 percent/1890 formula, sailed through the House on April 12, 1924 (by a vote of 323–71). Six days later, under the sponsorship of Senator Reed, the Senate passed a modified version of the bill (on a 62–6 vote). However, rather than base the quotas on the number of immigrants resident in the country in a particular census year, the Reed bill used the "national origins" principle, which calculated quotas according to the original homelands of the entire current population. Thus, if 13 percent of the U.S. population was of French ancestry, then France would be allotted 13 percent of the annual immigration total. Although there were conflicting motives for introducing this new principle, the "national origins" concept was soon embraced by the antirestrictionists. While the new formula would result in roughly the same quotas as using the census of 1890, the antirestrictionists knew that some form of restriction was going to pass anyway in 1924, and they were counting on the fact that it would take years for the government to determine the national origins of 120 million Americans—especially those whose ancestors had been here for hundreds of years and had intermarried with people of other nationalities untold times (not to mention the difficulties caused by the fact that numerous nations had disappeared or come into existence in the recent past).[91]

Madison Grant realized that the "national origins" formula was a stalling tactic. He explained to Albert Johnson that it was a "subtly dangerous suggestion" that would delay immigration restriction for many years.[92] Suitably warned, the Grantians outfoxed the antirestrictionists, and in the conference committee

Table 3. *The immigration restriction laws*

Law	Formula	Total annual quota
Emergency Quota Act of 1921	3% of foreign-born in 1910	355,000
Immigration Restriction Act of 1924	2% of foreign-born in 1890	165,000
National Origins Act of 1929	proportion of entire population in 1920	150,000

they simply combined the Johnson bill and the Reed bill. In brief, the Johnson-Reed bill limited European immigration to a yearly total of 150,000, apportioned according to the national origins of the population in 1920. *However,* until the national origins of the population could be tabulated (a process that took until 1929; see table 3), immigration from each nation would be limited—as Grant had originally wanted—to 2 percent of the foreign-born of each nationality according to the census of 1890.

Having co-opted the enemy's plan, the House approved the Johnson-Reed bill (308–69) on May 15, and later that day the Senate did likewise (69–9). Support cut across party lines and geographical divisions, with the only opposition coming from the urban centers of the Northeast. Immigrant leaders tried to meet with President Coolidge to plead for a veto, but he declined to see them, and on May 26 the Johnson-Reed Act, a.k.a. the Immigration Restriction Act of 1924, was signed into law by the president. Forty-one years after Emma Lazarus invited Europe to send its huddled masses to America, America withdrew the invitation.

On the West Coast, the *Los Angeles Times* hailed the "Nordic Victory." On the East Coast, the *Boston Globe* exulted: "All signs point to the junk-heap for the melting-pot." And in the Midwest, the *Chicago Tribune* called the act "the most momentous domestic event since the Civil War, . . . not less significant and epoch-making for America and the world than the Declaration of 1776."[93]

The inflow of New Immigrants, which had approached one million per year before World War I and had been reduced to 158,000 by the Emergency Quota Act of 1921, was now slashed to only 20,477 a year (12 percent of the European total). The annual Polish quota declined to a negligible 5,982; the Italian to 3,845; the Russian to 2,248; and the Greek to 100. These were numbers with which even the most diehard racist could live.[94]

Furthermore, the Immigration Restriction Act banned completely all immigration from Asia. The government of Japan protested that it had not been granted a quota like those of the European nations; anti-American demonstrations immediately broke out in Japan and lasted through the summer. One Japanese newspaper labeled the Immigration Act "the greatest insult in our history," and the American ambassador to Japan resigned in protest over the law. The Tokyo press declared July 1 a national day of mourning, and one protester disemboweled himself near the American embassy. Admiral Yamamoto (who would lead the attack on Pearl Harbor seventeen years later) declared that it would take many years for Japan to forgive the bitter insult to her honor, and Shinkishi Uyesugi predicted that as a result of the Immigration Act, military conflict between Japan and the United States was inevitable.[95]

Nevertheless, in the estimation of Madison Grant, the Immigration Restriction Act of 1924 was "one of the greatest steps forward in the history of this country." He called it "an amazing triumph" and breathed a sigh of relief that "we have closed the doors just in time to prevent our Nordic population being overrun by the lower races." His disciples were ecstatic. Robert DeC. Ward declared that the passage of the act was "a turning point in American civilization." Lothrop Stoddard labeled the event "epoch-making" and exulted that "America is saved!" Henry Cabot Lodge pronounced it "a very great measure, one of the most important if not the most important, that Congress has ever passed." And the imperial wizard of the Ku Klux Klan was glad that "the chief of Mr. Grant's demands, that the un-American alien be barred out, has [been] accomplished."[96]

Great praise was heaped on those involved in the effort. Henry Fairfield Osborn warmly congratulated Albert Johnson "on the practically unanimous approval of the Immigration Bill. I regard this as one of the most important steps taken in the whole history of our country." Frank L. Babbott agreed, saying that "few congressmen have ever rendered more important service." And *World's Work* claimed that "Albert Johnson's immigration bill has saved the nation."[97]

Secretary of Labor James J. Davis told President Coolidge that "history will record it as one of the greatest acts of your administration," Senator Shortridge of California agreed that "the President has rendered a great service to our country and to civilization," and Robert DeC. Ward told Coolidge that "future generations of Americans owe you a very deep debt of gratitude." But, for all the acclamation for Johnson and Coolidge, Harry H. Laughlin knew who had been operating behind the scenes and deserved most of the credit: "Madison Grant," he wrote in his private papers, "was *the* instrumental force in the framing of the Johnson Restriction Bill of 1924."[98]

Thanks to Grant and scientific racism, the nation was now a refuge for the Nordics, where they could breed in peace, unmolested by alien strains. Grant expressed to Robert DeC. Ward of the IRL his joy that they had finally put a stop to the foreign incursion, and he referred to the thirty years they had devoted to the cause as "the long period of Egyptian night"—an interesting metaphor, since the main purpose of the Immigration Restriction Act was to keep the Jewish people wandering in their own twentieth-century Egypt, where they would soon find themselves helpless to escape the wrath of a new, Austrian-born pharaoh.[99]

10

Culling the Herd

Thou shalt not let thy cattle breed with another kind; thou shalt not sow thy field with mixed seed.

Leviticus 19:19

And so it had been done. Just sixteen years after Israel Zangwill proclaimed America the great melting pot, America abdicated the title. The next quarter century saw fewer foreigners immigrate into the United States than had entered in the single year of 1907. More importantly, the immigration of non-Nordics declined to an imperceptible trickle. By 1925, the commissioner of immigration at a suddenly quiet Ellis Island could happily report that the few immigrants landing there now looked "just like Americans."[1]

"The eugenicists," writes Stephen Jay Gould, had won "one of the greatest victories of scientific racism in American history." And as far as the American public was concerned, the immigration issue had been laid to rest. A contented citizenry now transferred its attention from quotas, percentages, and base census years to dance marathons, near beer, and the ever-rising Dow Jones Industrial Average. While the *Saturday Evening Post,* in the issue immediately following the enactment of the Immigration Restriction Act of 1924, did carry an article by eugenicist Kenneth Roberts, the subject this time was not the lice-ridden Jews of Poland but the can't-miss real estate bargains just waiting to be snatched up in Florida.[2]

In Madison Grant's mind, however, the Immigration Restriction Act of 1924 was not the end of the story. To the contrary, it was only the opening battle of the campaign to save the Nordics. For having repelled the invasion of foreigners, the nation now needed to concentrate on purifying the population within its borders. Grant understood that Congress and the American people were exhausted from the battle over the Johnson-Reed Act, but he told Robert DeC. Ward: "Personally, I always believe that the best way to hold ground once gained is to renew the attack and try to take more ground." There were, after all, still millions of racially inferior people in the country, including a sizable group of Negroes who had been ominously migrating to the North ever since the war. And even within the Nordic community, there were a large number of degenerate individuals whose germ plasm had to be removed from the breeding stock. As Harvard anthropologist E. A. Hooton put it: the country still needed to do some "biological housecleaning." It was time, therefore, to implement the full-fledged eugenics program outlined in *The Passing of the Great Race,* which called for banning miscegenation and sterilizing the defectives. The refuge had been secured; it was time to cull the herd.[3]

Sterilization

Margaret Sanger agreed. In an address at Vassar College in 1926, she hailed the country's efforts to improve the quality of the population through immigration restriction but explained that an organized program of sterilization would now have to be implemented to "cut down the rapid multiplication of the unfit and undesirable at home." Sanger proposed that the government set a "sensible example" to the world by offering a bonus to "unfit parents" who allowed themselves to be sterilized. Through such a policy, "a heavy burden would be lifted from the shoulders of the fit." A few years later she again recommended the sterilization or segregation of "the whole dysgenic population," and as late as 1950, in a speech that the post-Nazi world did not receive well, Sanger argued that the government should grant sterilization bonuses to couples with "defective heredity" in order to weed out the "feebleminded and unfit."[4]

Some of the more puritanical eugenicists feared that sterilization, like birth control, would lead to immoral behavior by removing the consequences of sexual activity. But most eugenicists agreed with Madison Grant that sterilization would have to be an integral part of any eugenics program. Indeed, wrote Grant, it was the eugenicists' "fundamental" duty to deprive "the unfit of the opportunity of leaving behind posterity of their own debased type." Ellsworth Huntington was even more direct: "In the old system, famine, disease and cruelty killed off the morons and their offspring. That was for the good of the race. In our own day, sterilization makes it possible . . . to do what the old system did in the way of preventing the weaker elements from passing on their weakness to future generations."[5]

The American Eugenics Society distributed a number of publications in the 1920s emphasizing that sterilization was the most economical and efficient method of reforming society. Paul Popenoe of the AES estimated that ten million Americans ought to be sterilized, a figure that matched rather nicely Madison Grant's call in *The Passing of the Great Race* for the sterilization of the Submerged Tenth. Harry H. Laughlin agreed that 10 percent of the population were worthless "culls" who belonged to "degenerate human stocks" and should be sterilized for the overall good. "Cutting off the lower levels of the human breeding stock," wrote Laughlin, was really a matter of "conservation."[6] And once the lower tenth was eliminated, Laughlin hoped that the lowest tenth that *then* remained would be eliminated, and then *another* 10 percent, and so on, until a race of supermen remained. "Continuous decimal elimination," he stated, "should become a part of the eugenics creed of civilized people."[7]

The eugenicists never questioned the right of the government to forcibly sterilize the defective classes. Laughlin was typical in his insistence that "society must look upon germ-plasm as belonging to society and not solely to the individual who carries it." This philosophy, which sounds so similar to the assumption of wildlife managers that the individual animal is not as important as the herd, was also espoused by Laughlin's boss at Cold Spring Harbor, Charles Benedict Davenport, who stated: "The life of the commonwealth takes precedence over the right of reproduction of the individual." Similarly, biologist E. G. Conklin (who had been Laughlin's biology professor at Princeton) felt that "the freedom of the individual man is to that of society as the freedom of the single cell is to that of the human being. . . . In all organisms and in all social organizations, *the freedom of the minor units must be limited* in order that the larger unit may achieve a new and greater freedom." "It is the acme of stupidity," wrote William J. Robinson, a physician active in the eugenics and birth-control movements, "to talk in such cases of individual liberty, of the rights of the individual. *Such individuals have no rights.* They have no right in the first instance to be born, but having been born, they have no right to propagate their kind." Or, as Laughlin, the lifelong temperance man, put it: "in the long run individual effectiveness and happiness is assured only by individual subordination and occasional personal sacrifice."[8]

Laughlin, as secretary of the ERO's loquaciously named Committee to Study and to Report on the Best Practical Means of Cutting Off the Defective Germ-Plasm in the American Population, drafted a model eugenical sterilization law that called for the sterilization of the "socially inadequate classes," a phrase that encompassed ten dysgenic categories of persons: (1) the feebleminded, (2) the insane, (3) the criminalistic, (4) the inebriate (including drug addicts), (5) the chronically diseased (including those with tuberculosis, syphilis, and leprosy), (6) the blind (as well as those with seriously impaired vision), (7) the deaf (as well as those with seriously impaired hearing), (8) the deformed (including the crippled), (9) the dependent (including orphans, ne'er-do-wells, tramps, paupers, and the homeless), and (10) the epileptic.[9] (The latter category was perhaps an unfortunate choice, as Laughlin would soon begin experiencing the grand mal seizures signaling the onset of his own epilepsy.) The Committee on Legislation of the AES worked closely with state legislators to enact laws based on Laughlin's model sterilization law. Many states began passing such measures that, in general, obliged the authorities to sterilize the inmates of prisons, hospitals, and mental institutions who belonged to one of Laughlin's ten dysgenic categories. Few operations were performed at first, however, as there was some question as to the constitutionality of compulsory sterilization. But when the Virginia Sterilization Law of 1924 was challenged in the courts, the constitutionality of eugenical sterilization was upheld by the U.S. Supreme Court in the infamous case of *Buck v. Bell.*

The case involved a Virginia girl named Carrie Buck, who had been raised in a foster home since she was three years old. In 1924, at the age of seventeen, Carrie gave birth to an illegitimate baby. Her baby was immediately taken away from her and placed in another foster home, and Carrie herself was committed to the State Colony for Epileptics and Feeble-minded in Lynchburg, Virginia (a few miles from where Ota Benga had committed suicide). It is not clear whether Carrie's foster family committed her out of anger for her moral lapse or as an effort to cover up her allegation that she had been raped by their nephew. In any event, the State Colony also happened to house Carrie's biological mother, Emma. The authorities had administered the Stanford-Binet IQ test to Emma Buck and determined that she had a mental age of less than eight. When they gave the same test to Carrie Buck, they calculated her mental age as nine, whereupon the board of directors of the State Colony decided to sterilize her to protect the welfare of society.

When the sterilization procedure was challenged in the Circuit Court of Amherst County, Arthur H. Estabrook (of the ERO and the ECUSA) was called in as an expert witness for the State Colony. In testimony that resembled a symposium on Mendel's laws, Estabrook (whose degree was in zoology) explained, with many references to the Kallikaks, that feeblemindedness was a recessive unit character and that all three generations of Buck women (Emma, Carrie, and the illegitimate baby) were genetically feebleminded. A social worker for the Red Cross took the stand and concurred with Estabrook, for she had examined Carrie's seven-month-old baby and concluded that it had "a look about it that is not quite normal." Speaking "as a social worker," she further testified that the fact that Carrie had become pregnant out of wedlock indicated that she was feebleminded, and she recommended immediate sterilization, as "it would at least prevent the propagation of her kind." Eugenicist Joseph S. DeJarnette, superintendent of Western State Hospital in Staunton, testified that Buck's sterilization would "raise the standard of intelligence in the state." And Harry H. Laughlin, without ever meeting Carrie Buck (indeed, without ever leaving Cold Spring Harbor), supplied a deposition to the court declaring that Carrie was immoral, untruthful, and a low-grade moron, and that her ancestors belonged to "the shiftless, ignorant, and worthless class of anti-social whites of the South." Laughlin explained to the court that Emma, Carrie, and the baby were living proof that feeblemindedness was inherited along Mendelian lines, and with that the circuit court pronounced that Carrie was "afflicted with a hereditary form of feeble-mindedness" and ordered her to be sterilized.[10]

The sterilization order was upheld by the Virginia Supreme Court of Appeals in 1925, and two years later the case was sent to the U.S. Supreme Court as *Buck v. Bell.* In his argument, Carrie Buck's lawyer foresaw the uses to which sterilization could be put by a malevolent state, and (speaking just six years before the Nazis came to power) he predicted grave consequences if the government were permitted "to rid itself of those citizens deemed undesirable according to its standards." He warned the justices that if the sterilization statute were upheld, "A reign of doctors will be inaugurated and in the name of science new classes will be added, even *races* may be brought within the scope of such regulation, and *the worst forms of tyranny* practiced." But the Supreme Court decided (by an eight to one majority) that since Carrie Buck was congenitally feebleminded, she should be sterilized. On May 2, 1927, Justice Oliver Wendell Holmes (who seven years earlier, in *Missouri v. Holland,* had upheld Madison Grant's endeavor to have the federal government regulate the hunting of migratory birds) read the majority opinion in which he agreed with the eugenicists that "heredity plays an important part in the transmission of insanity, imbecility, etc." Therefore, while an advocate of judicial restraint, Holmes (who had been appointed by Theodore Roosevelt) felt there was a need for government to intervene in the reproductive activities of Carrie Buck "in order to prevent our being swamped with incompetence." Writing one year before the Court would rule in *Hunt v. United States* that the government had the authority to cull the deer population on the Kaibab Plateau, Holmes—sounding like an expert in wildlife management—wrote: "It is better for all the world if, instead of . . . let[ting] them starve for their imbecility, society can prevent those who are manifestly unfit from continuing their kind." After all, stated Holmes with triumphant finality: "Three generations of imbeciles are enough."[11]

A summary of Holmes's decision would later be entered into evidence by SS officer Otto Hofmann at the Nuremberg Military Tribunal. In the meantime, from the liberal Louis D. Brandeis to the conservative William Howard Taft, all the other Supreme Court justices concurred with Holmes—except for the reactionary Pierce Butler, the lone Roman Catholic on the court (and the father of eight children). The Catholic Church strongly opposed the U.S. sterilization laws. In 1929, the National Council of Catholic Women condemned eugenics as a violation of individual rights, and Pope Pius XI, in the Vatican's 1930 Encyclical on Marriage (*Casti Connubii*), directly attacked the eugenicists and repudiated sterilization (along with birth control, premarital sex, pornography, and divorce) as contrary to church teachings on the sanctity of the family. The pope explained that the root cause of degeneracy was sin, not defective germ plasm.[12]

Catholic opposition to eugenics was not just a theological proposition. The Italians and the Irish ranked pretty low on the eugenicists' evolutionary scale;

by attacking eugenics, the church was protecting its constituency in America. The difference between the two sides was laid out clearly when the pope claimed that "the family is more sacred than the state," to which Leon F. Whitney of the AES replied (with the typical eugenicist's combination of zoology and fascism): "The family is *not* paramount. . . . Just as the agriculturalist [encourages the reproduction of the most productive cow], so the state . . . must cultivate its families by seeing that the better type of individuals are preserved." Both the farmer and the state, explained Whitney, must have the courage to say: "Here is an outstandingly obnoxious weed. It must be destroyed."[13]

The opposition of the Catholic Church to eugenics only confirmed the belief of the Grantians that they were on the side of rationalism and progressivism. Justice Holmes, in fact, explicitly viewed his decision as a blow against religious fundamentalism in the United States, and he proudly wrote to Harold Laski that "the religious are astir" over *Buck v. Bell.* Replying a few days later, Laski encouraged Holmes to stay the course: "Sterilise all the unfit, among whom I include all fundamentalists."[14]

In delivering the opinion in *Buck v. Bell,* Justice Holmes had defended the state's power to compel sterilization on the grounds that "the principle that sustains compulsory vaccination is broad enough to cover cutting the fallopian tubes." His reference to vaccination reminds us of the eugenicists' notion that the dysgenic elements of the population are akin to viruses against which society needs to be inoculated. Indeed, Margaret Sanger's colleague C. O. McCormick explained to the readers of *Birth Control Review* that sterilization was a public health measure precisely analogous to vaccination for smallpox, because in both procedures society was protected from contagion. Harry H. Laughlin, writing in the same journal, similarly compared sterilizing the unfit to placing the diseased in quarantine, for in both cases "the liberty of the individual" was taken away in the interest of the "public welfare." It is somewhat revelatory that the eugenicists chose to use the term *sterilize*—"to cleanse of germs"—to describe this procedure. They could have employed a word such as "unfertilize," or "desexualize," but "sterilize" expressed better just what they hoped to accomplish by taking away the procreative power of the undesirables.[15]

In October of 1927, Carrie Buck was sterilized in the infirmary of the State Colony for Epileptics and Feeble-minded. One year later, her sixteen-year-old sister Doris was also brought to the State Colony to be sterilized, although she was told that she was being taken in for an appendectomy and did not discover until fifty years later why she and her husband had never been able to conceive a child. (Philip Reilly comments: "It is extraordinary how many 'appendectomies' were performed at some state homes for the retarded in the 1920s and 1930s.")[16]

The case of *Buck v. Bell* was of major importance. The eighty-six-year-old Holmes told a friend that by confirming the constitutionality of the sterilization law, he felt that he was "getting near to the first principle of real reform." Harry H. Laughlin was ecstatic that the highest court in the land had upheld the authority of the state to apply sterilization in a compulsory manner, "regardless of the consent or objection of the individual." He praised the decision as "by far the most important legal step thus far taken by the American people in the rise of modern eugenics," and predicted that there would now be "biological race progress instead of race deterioration." Judge Harry Olson of the AES similarly exulted: "The road is now open for [the] much wider application" of sterilization to prevent "racial degeneracy."[17]

In the wake of *Buck v. Bell,* states that had heretofore been reluctant to enforce their sterilization laws now began consistently administering such measures. They were encouraged by the taxpayers of the United States, upon whom the financial burden of those committed to asylums weighed heavily in the 1930s. Sterilization, notes Pat Shipman, was "a twentieth-century version of transportation to Australia." The Grantians were delighted that, as Laughlin put it, sterilization was finally "cutting off" the supply of defectives in America. As early as 1914, Laughlin had drooled over the thought that sterilization would make it possible to "at one fell stroke *cut off*"—again, that phrase—"practically all of the cacogenic varieties of the race." "*Cut off* the useless classes," chimed in Leon F. Whitney. Indeed, it is difficult not to notice that the leaders of the American Eugenics Society, like high priests demanding ever bloodier sacrifices for their cult, were, as Mark Haller says, "possessed by a compelling urge to castrate the unfit." It is wondrous to witness the vehemence with which such childless figures as Madison Grant, Harry H. Laughlin, Charles W. Gould, Wickliffe P. Draper, Henry H. Goddard, C. M. Goethe, A. E. Wiggam, Frederick Adams Woods, Kenneth Roberts, Seth K. Humphrey, Francis H. Kinnicutt, John Harvey Kellogg, Houston Stewart Chamberlain, and Charles Stewart Davison set about attacking the genitals of the lower breeds. (Daniel J. Kevles reports that British scientists used to refer to their colleagues in the eugenics movement as those "off-with-their-cocks boys.") And surely it is not insignificant that the most important promulgators of the faith—from the founder (Galton), to the prophet (Grant), to the ayatollah (Hitler)—were childless.[18]

At the urging of the Grantians, legislatures that had previously shied away from the issue now began passing sterilization laws of their own, and by 1931 thirty of the forty-eight states (plus Puerto Rico, as well as the provinces of Alberta, Ontario, and British Columbia) had enacted sterilization laws for the inmates of their institutions. By the time Madison Grant died in 1937, over thirty thousand Americans had been forcibly sterilized (and the number would reach sixty-five thousand by 1970).

Harry H. Laughlin was proud that "of the many thousands of operations which have been performed I have never heard a single complaint to the effect that the State had destroyed valuable breeding stock." Foreign scientists were also impressed by the U.S. sterilization program, and a number of European countries, especially those where social democratic parties had incorporated eugenics into their programs of social reform, passed sterilization laws based on Laughlin's model law. In 1928, one year after *Buck v. Bell,* the first European sterilization law was passed in the Swiss Canton de Vaud. This was followed by Denmark (1929), Germany (1933), Sweden (1934), Norway (1934), Finland (1935), Danzig (1935), Estonia (1936), and Iceland (1938). By the mid–1930s, Leon F. Whitney of the AES observed with great satisfaction that "sterilization and race betterment are indeed becoming compelling ideas among all enlightened nations today."[19]

The Negro Problem

It is often assumed that the scientific racists paid little attention to African Americans. The writings and actions of Madison Grant prove this is not so. As we have seen, a mere five sentences of *The Passing of the Great Race* are devoted to Jews, whereas Negroes are discussed in some forty paragraphs. And in *The Conquest of a Continent* (1933) Grant devotes a great deal of space to the "problem" of the Negroes: "Among the various outland elements now in the United States which threaten in different degrees our national unity," he asserts, "the most important is the Negro."[20]

Grant always maintained that the greatest mistake this nation ever made was to allow that first slave ship to dock in Virginia in 1619. Slavery, according to Grant, had been of immense benefit to the blacks themselves, as it had rescued them "from sheer savagery" and brought them into contact with white civilization. Indeed, the Negroes had "made more advance in America in two centuries than in as many thousand years in Africa." But the peculiar institution had proved to be grossly injurious to the Nordics, as it had led to miscegenation and the fratricidal Civil War, in which "hundreds of thousands of men of Nordic stock were cut off in the full vigor of manhood." And now the country was stuck with "the numbing presence" of an indigestible mass of twelve million Negroes.[21]

Of course, there was no doubt about the intellectual inequality of African Americans. Grant wrote that the "Negroes have demonstrated throughout recorded time that they are a stationary species, and that they do not possess the potentiality of progress or initiative from within." It irked him no end that naïve liberals viewed the American Negro as somewhat like "an unfortunate cousin of the white man, deeply tanned by the tropic sun," who had been thrust into his lowly station because of prejudice or an impoverished environment. The fact is that the Negro is congenitally inferior, Grant argued, and no amount of education or environmental reform is ever going to improve his lot. Some might think this an overly harsh assessment, especially coming from the son of the doctor who saved the life of General O. O. Howard, the great benefactor of Negro education. But according to Grant it was an anthropological fact that as soon as the Negro was removed from the beneficent influence of the white man, he regressed "to his ancestral grade of culture." A good illustration of what the Negro accomplished when left to himself was Haiti, where independence had caused the black inhabitants to "revert almost to barbarism."[22]

What the unscientific sentimentalists failed to realize was that the Negroes were not just a separate race—they actually belonged to a different *species* than the white man. The readers of *The Passing of the Great Race* were informed that "in the modern and scientific study of race we have long since discarded the Adamic theory that man is descended from a single pair, created a few thousand years ago in a mythical Garden of Eden." According to Grant, whites and blacks evolved independently of each other, and only "old-fashioned" thinkers still maintained that all human beings belonged to the species *Homo sapiens.* Two decades later Grant went one step further and authoritatively wrote in *The Conquest of a Continent* that "the physical differences between the Nordics and the Negroes . . . if found among the lower mammals, would be much more than sufficient to constitute not only separate species, but even *subgenera.*" He explained to Charles Benedict Davenport that anthropology had shown that the Negroes were an evolutionary "side line" and were "widely separated from the line of our own ancestors," which explained why the offspring of blacks and whites were not always fully fertile. Henry Fairfield Osborn, head of the American Museum of Natural History, concurred that Negroes belonged to a separate genus, and in 1926 he wrote in *Natural History* that "if an unbiased zoologist were to descend upon the earth from Mars and study the races of man with the same impartiality as the races of fishes, birds, and mammals, he would undoubtedly divide the existing races of man into several genera," each of which was distinguished by its own unique "spiritual, intellectual, moral, and physical characters."[23]

Grant held that when blacks and whites mated, their genes did not blend together. Rather, since the blacks were the lower, more primitive type, their traits were dominant over those of the more refined whites. As he explained to William Howard Taft: "A cross between two races partakes always of the character of the more ancient, more generalized or lower type." In other words, the mulatto was not some sort of intermediate form, halfway between black and white, but rather a full-fledged Negro. As Grant put it in his famous phrase from *The Passing of the Great Race* that was reproduced in hundreds of books and pamphlets throughout the South: "The cross between a white man and a negro is a negro."[24]

Furthermore, Grant preached (in an argument that would be taken up by many eugenicists) that miscegenation was dangerous because it led to "disharmony in the offspring." It was a "biological fact" that the mingling of widely divergent stocks caused both physical and mental aberrations. Thus, a child that inherited the large brain of its Nordic mother but the small skull of its Negro father would be condemned to a life of excruciating pain if not early death. Similarly, a mulatto who inherited the ambition of his Nordic father but the laziness of his Negro mother would suffer from years of emotional turmoil and conflict. Indeed, it was just such a disharmonious combination that accounted for the marked dissatisfaction and political radicalism of such neurotic mulattoes as Frederick Douglass and W.E.B. Du Bois. Hence, aside from any unfortunate social or cultural ramifications, for biological reasons alone miscegenation was "a frightful disgrace" that societies permitted at their evolutionary peril.[25]

Grant was concerned that one of the obstacles to banning miscegenation was mankind's "perverse predisposition to mismate." It is a well-known fact, he wrote in *The Conquest of a Continent,* that the colored races proudly "regard the possession of a blonde woman as an assertion and proof of race equality." And it was sadly but equally true that white women seemed to become uncontrollably lustful whenever a black man was in the vicinity. Grant told a friend that whenever he visited Europe he was "horrified" at the "openly expressed craze" of French women for black men, especially the "stray negroes" who traveled there from the United States. He was shocked that in Paris one could "see daily in the best restaurants white girls lunching or dining with negroes," and he suspected that the emboldened attitudes of Negroes in the United States in the 1920s was a result of their being coveted during the war by the white women of Europe. From Poitiers, anthropologist G. Vacher de Lapouge confirmed that "men of color have been particularly sought after by the women," and he was horrified that during the war "the extreme looseness of the morals of French women resulted in the production of a considerable mass of half-breeds." He lamented that France was well on its way to becoming "a melting pot," and pointed out: "It is really the very existence of white civilization and of the white races which is at stake."[26]

Madison Grant's views on the Negroes were criticized—as were his views on almost everything else—by Franz Boas. Though willing to accept that blacks *might* be inferior to whites in mental capacity, Boas was also open to the possibility that the races were *equal* in ability. What he knew for certain was that slavery and racism had prevented African Americans from attaining their full intellectual development, and that once the environments of the two races were equalized, the gap between black and white would diminish. In contrast to Grant, therefore, Boas actively opposed segregation and actually *endorsed* miscegenation, which was not "in any way dangerous" and would serve to reduce prejudice by lessening the physical differences between the races. Grant dismissed Boas's arguments as those of a self-serving Jew, who thought that by coming to the defense of Negroes he could somehow mitigate the discrimination faced by his own people. It was hardly an accident, Grant said, that the boards of the NAACP and the National Urban League were practically indistinguishable from those of the American Jewish Congress and B'nai B'rith.[27]

Grant was not the only eugenicist interested in the Negro problem. Paul Popenoe and Roswell Johnson, for example, stated in the most widely used eugenics textbook, *Applied Eugenics,* that the contribution of blacks to world civilization "must be placed very near zero on the scale." They pointed out that the intelligence of mulattoes depended on their proportion of white blood, and they urged the adoption in the United States of a rigid system of apartheid to protect the Nordic race. Almost every other important eugenicist believed that Negroes were intellectually inferior to whites and that miscegenation was a biological abomination. Harvard geneticist Edward M. East summed up their position when he stated (in *Mankind at the Crossroads*) that "the negro race as a whole is possessed of undesirable transmissible qualities both physical and mental, which seem to justify not only a line but a wide gulf to be fixed permanently between it and the white race."[28]

The definitive argument against miscegenation was *Race Crossing in Jamaica,* written by Charles Benedict Davenport and zoologist Morris Steggerda, and published under the aegis of the Carnegie Institution of Washington in 1929. Davenport and Steggerda spent a year in Jamaica subjecting three racial groups—whites, blacks, and browns (mulattos)—to seventy-seven anthropometric measurements and a battery of psychological tests, including the army intelligence tests. Their findings, presented in over five hundred pages of charts, graphs, tables, and statistics, showed that while blacks had a more highly developed musical capacity and sense of rhythm, they fell "far below" the whites in intelligence. As for the browns, the authors found serious evidence that they suffered from disharmony. Mentally, for instance, the mulattoes appeared to be "muddled and wuzzle-headed." And physically, the authors pointed to measurements showing that blacks had longer limbs than whites, and they worried what would happen if a mulatto inherited the long legs of its black father but the short arms of its white mother. Ever the humanitarian, Davenport was concerned that mulattoes would constantly have to "stoop more to pick up a thing on the ground." The preordained conclusion of *Race Crossing in Jamaica* was that blacks were inferior to whites, and that miscegenation caused disharmonious combinations that resulted in "an excessive proportion of highly ineffective persons."[29]

There was little disagreement in 1920s America that Negroes were inferior. The only question was what to do about it. Madison Grant admired "the firm resolve of the handful of white men in South Africa . . . to control and regulate the Negro population there," and he saw no reason why the Nordics of America could not be equally firm with their own Negroes. In the 1920s, therefore, at the same time that he was working to pass sterilization statutes, he also began pushing for strict antimiscegenation laws.[30]

To carry out his antimiscegenation campaign, Grant joined up with Major Earnest Sevier Cox (1880–1966), a kind and unfailingly well-mannered man who was, in the estimation of historian I. A. Newby, "perhaps the most important race theorist residing in the South in the period between World Wars I and II." Born into a family of devout Methodists near Knoxville (where his boyhood pet was a black kitten named "Nig"), Major Cox spent his adult years bouncing from place to place and profession to profession, until he finally received the calling to become a preacher. He attended Moody Bible Institute in Chicago and then Vanderbilt University's Theological School. He left Vanderbilt before getting a degree, and for the next three years roamed throughout Tennessee and Kentucky preaching at revival meetings. But chronic throat inflammation finally forced Cox to abandon the life of an evangelist. Instead, he enrolled at the University of Chicago to study sociology with Frederick Starr. Once more Cox quit school before earning a degree. Supported by timely gifts of cash from his beloved sister, he spent the years from 1910 to 1915 wandering throughout Africa, Southeast Asia, and South America, and suffering recurrent bouts of malaria. He later claimed that his peripatetic travels had been undertaken in order to research the Negro in his native habitats, but that was largely a post hoc justification. Returning to the United States in 1915, Cox billed himself as an explorer-ethnologist and an "authority on the Negro race," and gave lectures with titles such as "1800 Miles on Foot through Darkest Africa."[31]

In 1916, Cox moved to Washington, D.C., where he secured a part-time job in the Senate mailroom from Mississippi Senator James K. Vardaman (archsegregationist and avid fan of Madison Grant and Lothrop Stoddard). At the suggestion of Vardaman, Cox picked up *The Passing of the Great Race* and was struck by the passage in chapter 7 in which Grant showed that if the Negroes were not separated from the Nordics, the two races would amalgamate and the biological principle of reversion would cause an evolutionary disaster: "Where two distinct species are located side by side history and biology teach that but one of two things can happen; either one race drives the other out, as the Americans exterminated the Indians; or else they amalgamate and form a population of race bastards in which the lower type ultimately preponderates. . . . If the purity of the two races is to be maintained, they cannot continue to live side by side, and this is a problem from which there can be no escape."[32]

Cox immediately exchanged Methodism for eugenics, and for the next fifty years of his life was guided by the mantra "separation or amalgamation." Throughout 1917, he devoted every minute of his spare time (of which he had plenty) to working on the manuscript of *White America,* the book that would make him immortal among southern racists. The Gobinesque thesis of *White America,* presented in the introduction and then repeated continually over the next 389 pages, was threefold:

1. The white race has founded all civilizations.
2. The white race remaining white has not lost civilization.
3. The white race become hybrid has not retained civilization.[33]

Cox concluded that the mulatto was "a cancer that will eat deeper and deeper into the heart of the white race," and the lesson for the Nordics in the United States was plain: they must protect their purity by gathering the Negroes in concentration camps along the coast, whence they could be deported to Africa (although Cox would permit Negroes who were too old to breed to remain in the United States until they eventually died off).[34]

Work on *White America* was temporarily interrupted by World War I. Though thirty-seven years old, Cox volunteered for the army and was sent to France as a captain in the field artillery (despite scoring at the bottom of the list). Sadly, his commanding officer found him "wholly unqualified" for front-line service, and his biographer admits that "Cox did not distinguish himself while in service to his country."[35] The main problem apparently was that the mild-mannered and weak-voiced Cox could never persuade any of his men to follow his orders. Nonetheless, he was honorably discharged after the war (and continued to serve in the army reserves, where he was eventually promoted to the rank of major).

"Major" or not, Cox was now closing in on forty years of age, with no wife, no degree, no visible means of support, and no prospects other than an unfinished manuscript of dubious quality. Having already tried revivalism and racism, he now took up the third R: real estate. He moved to Virginia to work for the Lamburnum Realty Corporation of Richmond, where he remained until his retirement thirty-six years later.

But when the country was rocked by the race riots of 1919, Cox was heartened, as he hoped an anti-Negro backlash would create support for his deportation scheme. Accordingly, he resumed work on *White America,* and as soon as it was finished in 1920 he sent a copy of the manuscript to Madison Grant for appraisal. This was the beginning of a collaboration that would last for many years, thus disproving the accepted wisdom that the "scientific" racists had little interest in or involvement with the "popular" racists.[36]

Grant told the southerner that he was "heartily in sympathy with the purpose of your writings." But he was disappointed by Cox's unsophisticated handling of the material. First of all, Grant strongly advised Cox to tone down his emotionalism, as "a calm, scientific and dispassionate statement of facts" was always more convincing than "a diatribe." Secondly, he informed Cox that much of his ethnographic data had to be revised. Almost all of Cox's research had been drawn from obsolete secondary sources he had found in the Library of Congress before the war, and Grant had to explain to him, for instance, that A. H. Keane and Daniel Garrison Brinton were a full generation out of date. (Grant was also horrified that Cox had occasionally cited Franz Boas, and assured him that "Boas is now discredited.") He sent Cox a lengthy memo giving his "frank" criticisms of the manuscript and providing page by page corrections. Grant then took it upon himself to tutor Cox in the basics of contemporary anthropology. He told Cox which books he should read (e.g., Clark Wissler's *American Indian,* Alfred P. Schultz's *Race or Mongrel,* and Wallace Thompson's *People of Mexico*), and he put his devotee in touch with some of the more important leaders of scientific racism, including Clarence G. Campbell, E. G. Conklin, Charles W. Gould, G. Vacher de Lapouge, and Francis H. Kinnicutt (who agreed with Cox that it should be a crime for anyone with black blood to immigrate to the United States).[37]

Cox spent many months revising his manuscript to make it anthropologically sound. *White America* now reproduced, albeit in a more pedestrian manner, all the main ideas of *The Passing of the Great Race,* from the dysgenic effect of war to the fact that Jesus was a Nordic. In his preface, Cox acknowledged his indebtedness to Madison Grant for sharing his ethnological expertise, and he referred those readers seeking further edification to the "important" works of Grant and Lothrop Stoddard.[38]

Cox then began an extensive search for a publisher. But over the next three years he was rejected by every major firm in the United States, including Grant's own publisher, Charles Scribner's Sons. While the content of *White America* was no less egregious than that of *The Passing of the Great Race,* Madison Grant always couched his racism in a scholarly style that convinced publishers and readers alike that they were encountering a reputable work of science. Cox, on the other hand, was not clever enough to appear as anything other than what he was: a half-educated bigot."[39]

Cox was terribly distressed that he could not find a publisher for his book despite, he told Grant, having "spent years of my time and many thousands in money in a sincere effort to assist in the solution of the negro problem." Having run out of options, in 1923 Cox paid for a private printing of *White America* under the auspices of "The White America Society" (an organization that existed only in the imagination of Earnest Sevier Cox). The cover of the book reproduced the quote from Madison Grant ("If the purity of the two races is to be maintained, they cannot continue to live side by side") that had inspired the work in the first place. Grant was supposed to write the introduction for the book, but throughout the summer of 1923 he had been all but paralyzed by arthritis and was unable to pick up his pen.[40]

By the time the book was published, however, Grant was sufficiently healthy to write a long and highly favorable review for the *Richmond News Leader* in which he warned that unless Americans adopted Cox's "drastic measures" the nation was "doomed to mongrelism." Other Grantians followed the master in endorsing Cox's book. Historian H. J. Eckenrode wrote in the *Richmond Times-Dispatch* that Cox's book was "one of the most brilliant works ever written on the race problem. It is safe to say that *White America* will be a classic on the race question, quoted for generations to come." Lothrop Stoddard quoted from *White America* in his book *Re-forging America,* and Charles Benedict Davenport, in an unsigned review in the *Eugenical News,* described *White America* as a "stirring volume" by a southerner who was anxious "to save his native country from going the way" of Haiti. "If Mr. Earnest Sevier Cox can bring about [the deportation of the Negroes] he will be a greater savior of his country than George Washington. We wish him, his book and his 'White America Society' godspeed."[41]

With the exception of the eugenicists, however, *White America* garnered mostly negative reviews. The *South Atlantic Quarterly* found the book "unconsciously humorous," the *Salt Lake Tribune* felt it contained "a great deal of nonsense," and the *Norfolk Journal and Guide,* a black newspaper, cleverly noted that "we may readily infer . . . that the present civilization is, indeed, in danger of deterioration, if a work like this from a so-called trained intellect can gain the public ear." The most negative evaluation came from anthropologist Melville Herskovits, who accused Cox in the *Journal of Social Forces* of "gross errors and contradictions" and concluded that *White America* was "fallacious in its assumptions, incompetent in its handling, and loose in its logic." Herskovits was absolutely correct, and yet Cox's book would be read by thousands of influential southerners from the 1920s through the 1960s (the last edition was published in 1966), and one historian states that *White America*—which had been inspired, midwifed, and endorsed by Madison Grant—was "the most authoritative statement of intellectual-historical racism written by a Southerner."[42]

Crippled

The first symptoms of Grant's disease appeared in the summer of 1921, when he began to complain of a lingering sore throat. He would never be healthy again. A panoply of befuddled physicians dithered for six months and then used X-rays to burn out Grant's tonsils, but the ache then jumped from his throat to his joints, which in the following months became more and more swollen and almost unbearably painful. His knees in particular were severely inflamed, and it became increasingly difficult for him to walk. By 1922 Grant was an invalid, confined to his bed for weeks at a time. His doctors desperately prescribed a series of ineffective vaccinations and dubious serums that, Grant told Albert Johnson, "unfortunately have the result of crippling me worse than ever." The simple act of grasping a pen became too painful for Grant, and his secretary had to sign his dictated letters for him. It was a "sad hour," wrote Major Frederick Russell Burnham, "when, in the midst of a great and useful work, fate dealt Grant a heavy blow. For to him came the certain knowledge that he was doomed to a living death."[43]

Grant was being progressively crippled by what I am sure was infectious arthritis. This disease is usually the result of a previous infection elsewhere in the body—often in the throat. The germ then spreads via the bloodstream to the joints, where it produces great pain and inflammation. Since animals can carry the disease-producing germs, people who work closely with animals are particularly susceptible to this form of arthritis. Infectious arthritis is usually not a long-term illness these days. In most cases, it can be cured if treated promptly with antibiotics—which, of course, did not exist in the early 1920s.

For the rest of his life, Grant would be wracked with pain, and often he could not stand up at all. When he did venture out, it was usually in a wheelchair, though he occasionally managed on crutches (with heavily bandaged legs). For a man who lived to chase wild game and to climb distant mountains, the disease was a devastating blow. Yet he faced his situation with patrician equanimity and never complained or revealed to any but his closest friends the extent of his suffering. Henry Fairfield Osborn told him, "I have the greatest admiration for the heroic fortitude with which you have borne . . . this almost intolerable illness," and added: "Certainly you require all your Nordic courage to sustain your cheerfulness." Grant's colleagues were astounded by his bravery; whenever somebody pressed Grant on the progress of his disease, he usually warded off the inquiries with a casual remark such as: "While not yet prepared for football, I manage to hobble around." Conservationist W. Redmond Cross wrote that "of Madison Grant's many admirable characteristics, none was more outstanding than his courage, and this flame never burned more clearly than in his closing years when, wasted by disease, he continued steadfastly on his way, determined that his spirit would not be conquered by the physical degeneration which he could no longer control."[44]

For years Grant engaged in a fruitless search for relief from the chronic pain. He underwent long courses of treatment in fashionable spas and exclusive resorts from California to Cuba. Occasionally a particular hot spring or exercise regimen would provide him with temporary relief, and for a few weeks he would lead a normal life. Often, in fact, he would be fooled into thinking that he was on the road to recovery and excitedly plan a hunting trip to the Andes, an excursion to Morocco, or an expedition to Tibet, only to have to cancel at the last minute when the agonizing inflammation returned.

Henry Fairfield Osborn was always bragging to Grant about the number of books he was publishing, the conferences he was attending, the lectures he was giving; and while this was due partly to Osborn's gargantuan ego, I suspect that it was also an attempt by Osborn to goad Grant into forgetting about his knees and getting on with his own work. And, indeed, though he was never free from pain, Grant continued to work: to write, to lobby for immigration restriction, to press for wildlife conservation, to preside over the Galton Society, to play an active role at the American Museum of Natural History and the American Defense Society, to advise the Save-the-Redwoods League and the American Bison Society, to direct the American Eugenics Society and the Eugenics Research Association, to run the Zoological Society and the Boone and Crockett Club, and to head the Taconic Park Commission with Franklin Roosevelt.

Grant and Roosevelt had been appointed to the commission by Governor Al Smith to design the Taconic State Parkway, the 105-mile parkway that winds through some of the most scenic areas of New York State. The Taconic Parkway was modeled on the Bronx River Parkway, the world's first modern parkway, which had been conceived and built by Grant in order to rehabilitate the Bronx River Valley, one of the most polluted areas in the state. In 1925, after two decades of Herculean effort and the expenditure of $16.5 million, the Bronx River Parkway—"a wonder of the world" in the estimation of Robert Caro—was finally completed, and universally hailed as the most beautiful road in America.[45] The motoring public immediately went parkway mad, and similar projects sprang up all over the country, of which one of the most attractive was the Taconic State Parkway. As they worked together through the 1920s, a touching friendship developed between Grant and FDR, a relationship that transcended—at least for a few years—their ideological differences. Roosevelt admired Grant's "mighty mind," and his letters to "My dear Madison" were reciprocated with letters to "My dear Frank." The two aristocrats—who had both been struck down in 1921—often shared tips on possible cures for their respective handicaps. Roosevelt continually tried to coax Grant down to Warm Springs, where the "wonderful water relaxes muscles." Grant congratulated FDR on discovering "your Georgian Lourdes," but he was optimistic about a sanatorium he had heard about in Battle Creek, Michigan, run by the internationally acclaimed health reformer Dr. John Harvey Kellogg.[46]

Raised as a Seventh-day Adventist, John Harvey Kellogg devoted his life to spreading the message that a vegetarian diet, good posture, regular outdoor exercise, and above all a clean colon were the keys to a long and healthy life. Kellogg was convinced that the body poisons itself by absorbing toxic wastes from decomposing food in the lower intestine. In order to remove decaying matter from the digestive tract he insisted that his patients practice "colon hygiene" (the title of one of his best sellers) by drinking one glass of water each hour and performing one bowel movement upon arising in the morning, one before retiring in the evening, and one following each meal in between. It was Kellogg's quest for colonic health that led him to create a morning repast of flaked corn (though he left it to his younger brother, W. K. Kellogg, to run the family's cereal company).

Dr. Kellogg's obsession with the state of the lower intestine—"a place where more bad weeds grow than any other I know of"—raised the daily examination of the feces into the realm of religious ritual. It was absolutely vital that the noxious toxins polluting the bowels be destroyed. Kellogg, who always dressed completely in white from his hat to his shoes, also demanded total abstinence from the poisons of tea, coffee, chocolate, tobacco, and alcohol. His wife worked closely with Frances Willard in the Women's Christian Temperance Union, serving as that organization's national superintendent of hygiene and head of its Social Purity Department. One would be correct in surmising that Dr. Kellogg and his wife practiced a celibate marriage and had no children.[47]

After formally breaking with the Seventh-day Adventist Church, it was not long before Dr. Kellogg embraced the eugenics movement, where his obsession with purity and his jeremiads against "pernicious bacteria" found a congenial home. Among the friends and disciples of Madison Grant who put themselves under the ministrations of Kellogg at the Battle Creek Sanitarium were Charles Benedict Davenport, Will Durant, David Fairchild, Henry Ford, David Starr Jordan, William Sadler, George W. Wickersham, Luther Burbank (who swore that he owed his life to Kellogg), John D. Rockefeller, Jr. (who declared himself Kellogg's "ardent apostle"), and Irving Fisher (the president of the American Eugenics Society, who checked in at the sanitarium no fewer than nineteen times and became one of Kellogg's closest friends). Kellogg also defended the putrefied duodenums of a number of Grant's conservationist friends, including Roy Chapman Andrews, John Burroughs, and Gifford and Amos Pinchot, as well as big-game hunter Carl Akeley and Arctic explorer Roald Amundsen. And Battle Creek was practically the official sanatorium of the New York Zoological Society, as both William T. Hornaday of the zoo and Charles H. Townsend of the aquarium were devoted patients (and Hornaday wedded his wife in the Seventh-day Adventist Tabernacle in Battle Creek).[48]

In 1906, the same year that Ernst Haeckel formed the Monist League, Kellogg formed the Race Betterment Foundation to spread eugenic propaganda. Kellogg observed that farmers were developing "wonderful new races of horses, cows, and pigs," and promised that eugenics would similarly create "a new species of man . . . in not more than six generations." The Race Betterment Foundation sponsored three national conferences on eugenics (in 1914, 1915, and 1928), where leading scientists, social workers, and educators heard presentations on immigration restriction, sterilization, personal hygiene, venereal disease, and the evils of such "race poisons" as alcohol and tobacco. After observing the proceedings at the third Race Betterment Conference, Leon F. Whitney commented admiringly that "Battle Creek surely has 'got eugenics,' which to my way of thinking is a better thing to get than religion."[49]

In the summer of 1923, at the insistence of Charles Benedict Davenport, Madison Grant agreed to go to Battle Creek and put his arthritic limbs in the hands of Dr. Kellogg. Davenport alerted Kellogg to expect "one of the most useful citizens of the United States . . . . We want to see him continue his activities and that is why we are all hoping great things from his sojourn at Battle Creek Sanitarium."[50]

Grant spent two full months at Battle Creek in 1923, and he and Kellogg got along famously. Kellogg was, to be sure, an inveterate name-dropper and a relentless booster, and certainly Grant cringed whenever the doctor began shouting at him that he must "Clear the colon with repeated enemas!"[51] But Kellogg was also a fascinating man who was always open to new ideas, and he and Grant spent hours discussing medicine, politics, and their memories of Dresden (where Kellogg had studied hygiene). In October, Charles Benedict Davenport joined Grant at Battle Creek, and when those three divines—Davenport, Grant, and Kellogg—convened in the acidophilus milk bar to discuss the need to maintain sexual and racial purity, the atmosphere must have been similar to when the magistrates of Boston met in 1660 to decide the fate of Mary Dyer.

During Grant's stay at Battle Creek, Dr. Kellogg wrote to William T. Hornaday to say that "we esteem it a great pleasure as well as an honor to have an opportunity to be of service to so eminent and useful a man as Mr. Madison Grant." While Kellogg admitted that "it is too much to expect that he will be restored to his original physical soundness," he held out hope that if Grant adhered "very close to the straight and narrow way . . . his disease may be arrested and . . . we shall be able to patch him up very considerably." Kellogg forced Grant to participate in a daily exercise regimen that did help (temporarily) to loosen his joints and relieve his discomfort. Grant even managed, albeit with great effort, to get out of his wheelchair for a bit. But absent the invention of antibiotics, there was really very little that Kellogg or anybody else could have done about Grant's condition.[52]

At the suggestion of Kellogg, Grant began wintering in Florida, where he could soak up the sun's rays at the Boca Raton Club; and rather than go to Newport and Bar Harbor in the summers, as he had done for years, he usually journeyed to his "exile" in Battle Creek. Despite his fondness for Dr. Kellogg, the truth is that Grant detested the place. Even worse than the colonic cleansing was the fact that the Midwest bored him out of his mind. But he was in tremendous physical pain, and in desperation he kept returning to Battle Creek "to be overhauled."[53]

It was partly to take his mind off his swollen limbs that in 1924 Grant returned to the antimiscegenation effort—with tragic consequences for American society in the twentieth century.

The Racial Integrity Act

In the early 1920s, Earnest Sevier Cox, author of *White America,* told Madison Grant: "I stand ready to offer any assistance in my power if you could use me here in Richmond." Grant had been expecting such an offer, and he knew exactly how he wanted to use Cox.[54]

In the eighteenth century, the Commonwealth of Virginia had decreed that persons who were less than three-fourths white were to be considered Negroes. The state later raised the requirement so that persons who were less than seven-eighths white were Negroes, and in 1910 the ratio was raised still further, to fifteen-sixteenths. In other words, if "only" fourteen of a person's sixteen great-great-grandparents were white, the person was a Negro. He or she was not allowed to marry other whites, could not attend white schools, and was subject to all the other restrictions of Jim Crow. But Madison Grant thought it was "absurd" that someone who was fifteen-sixteenths white was considered to be white. He was determined that the state legislature revise the law so that only those with 100 percent pure white blood could claim to be white, and be allowed to mate with other whites.[55]

As head of the Bronx Zoo and several other conservation organizations in the North, Grant could not publicly press Virginia to alter its racial integrity laws, so he convinced Earnest Sevier Cox to spearhead the effort for him. Cox then turned for assistance to a number of Richmond racists, the two most important being W. A. Plecker and John Powell. Dr. Walter Ashby Plecker had been the Virginia State registrar of vital statistics (the official in charge of recording births, deaths, and marriages) since 1912. He was a very thorough public servant, and was especially vigilant about catching light-colored Negroes who tried to register their babies as white, even though they were less than fifteen-sixteenths white. When Plecker read *The Passing of the Great Race,* he was greatly pleased to discover that his popular racism could be rationalized by "indisputable scientific fact." He immediately converted to eugenics, and took it upon himself to spread the gospel that American society could save itself from ruin only by "turning a deaf ear to those who would interpret Christian brotherhood to mean racial equality." He expressed the hope "that the people of the whole country are becoming ready to accept the guidance of Madison Grant, Lothrop Stoddard, [and] Earnest Sevier Cox," and he began "strongly" recommending Grant's *Passing of the Great Race,* Stoddard's *Rising Tide of Color,* and Cox's *White America* in all his official correspondence. Plecker gave speeches and wrote articles that were extended paraphrases of the writings of Grant, and declared in an address to the annual meeting of the American Public Health Association in Detroit that the antimiscegenation effort was "a struggle more titanic, and of far greater importance" than even the Great War. It was, in fact, "the struggle which means the life or death of our civilization."[56]

After consulting with Madison Grant and Earnest Sevier Cox, Plecker drafted "A Bill to Preserve Racial Integrity," which historian Peggy Pascoe calls "the most draconian miscegenation law in American history." The bill repealed the fifteen-sixteenths rule and legally classified as "white" only those persons with "no trace whatsoever of any blood other than Caucasian." The bill outlawed marriages between whites and nonwhites in Virginia, and made it a felony, punishable by one year in the penitentiary, for anyone to mislead the authorities about his or her racial heritage.[57]

Plecker and Cox now asked John Powell to organize public support for the "one-drop" measure. Powell, an aristocratic native of Richmond, was a fanatical white supremacist (and according to some sources a former member of the Ku Klux Klan) who was particularly obsessed with racial purity. He was also one of America's leading classical pianists. After studying in Vienna under Theodore Leschetizky (probably the most influential piano teacher in the world), Powell had performed in all the leading cities of Europe and America, and he was the poster boy for Steinway pianos. He also acquired a solid reputation as a composer and wrote numerous hymns, choral settings, and orchestral works, including the aptly titled *Sonata teutonica.*

Powell eagerly absorbed Grant's eugenic teachings and, like W. A. Plecker, he began recommending the writings of the prophet to all his correspondents. Powell also read Cox's *White America* and labeled it "the most important contribution ever made to the study of the Negro Problem." Like Grant, Powell was horrified that Virginia permitted persons who were only fifteen-sixteenths white to call themselves white. He reminded his fellow Virginians that "history, ethnology and biology, all bear out the Anglo-Saxon instinctive conviction that 'one drop of Negro blood makes the Negro,'" and he warned that if the whites let down their guard, "our civilization and our race will be swallowed up in the quagmire of mongrelization." At the end of 1922, Cox helped Powell create the all-male Anglo-Saxon Clubs of America to lobby for the Racial Integrity Bill. Powell was a forceful lecturer and efficient organizer, and within months thirty-one chapters of the Anglo-Saxon Clubs had been established throughout Virginia, with a particularly strong presence at the University of Virginia (which, according to J. Douglas Smith, had become "a hotbed of eugenical studies").[58]

In the meantime, Grant introduced Cox, Plecker, and Powell to his disciples Lothrop Stoddard and Harry H. Laughlin, and they all became quite friendly. Just as Margaret Sanger had cozied up to the eugenicists to achieve respectability, so the southern segregationists used the eugenicists to gain scientific credibility. For years to come, the three northern scientific racists (Grant, Stoddard, and Laughlin) provided advice and acted as mentors to the three southern popular racists (Cox, Plecker, and Powell).

It is difficult not to notice that five of these six men—so obsessed with sterilizing the lower breeds and maintaining the purity of their own kind—were childless. (The lone exception was Lothrop Stoddard, who—as we have seen—finally married at the age of forty-three and actually managed to sire two children.) Madison Grant never married and was childless. E. S. Cox, who (like Grant) was extremely close with his sister, never married and was childless. While John Powell was also very devoted to his sister, he finally did marry at the age of forty-six, but he too had no children. Harry H. Laughlin and W. A. Plecker were both married but produced no offspring. And yet the childless Plecker was typical of the group in that he continually instructed the young persons of Virginia that it was their duty to marry and produce four children each in order to preserve the white race. Plecker (who was an avid conservationist) was imbued with the puritanical attitudes so common among racists, and his antimiscegenation crusade was fraught with sexual anxiety. He often warned of the "intense desire" of the mulattoes to "thrust themselves" into the white race, but he vowed to resist their "intrusions." He also lived in constant dread that women were engaging in masturbation. As a physician with over twenty-five years of experience, he publicly admonished the women of Virginia that "the method of Onan" was "not only a violation of Divine and human law" but was harmful to their health, as it led directly to "neurasthenia."[59]

Throughout 1923 and 1924, the three northerners worked behind the scenes mapping strategy, while Cox, Plecker, and Powell operated through the Anglo-Saxon Clubs to lobby the Virginia General Assembly to pass the Racial Integrity Bill. The measure was introduced into the legislature in February 1924 and had the backing of the state's leading newspaper. But things immediately hit a snag when state senator James S. Barron, whose family had always proudly claimed to be descended from the union of John Rolfe and Pocahontas, realized that the bill would eliminate him from the white race. He insisted that the bill be amended so that "descendants of Pocahontas" could be classified as white. Plecker and Cox hastily consulted with Grant, who thought the whole incident "screamingly funny" and—like Hitler granting honorary Aryanship to the Japanese—made the official ruling that Virginians with one-sixteenth or less of Indian blood could be considered white. With the support of a grateful (and now officially white) Senator Barron, the Virginia Act to Preserve Racial Integrity was passed just four weeks after being introduced, on March 8, 1924—the same day that the Virginia Sterilization Law was enacted. In just eight more weeks, the U.S. Congress would pass the Immigration Restriction Act of 1924. When Congressman Emanuel Celler denounced the Immigration Restriction Act as being "not inconsistent with the tenets of the Ku Klux Klan," he had no idea just how close the ties really were between northern restrictionists and southern racists. Immigration restriction, antimiscegenation, sterilization, scientific racism, popular racism—they were all part and parcel of the Grantian program to preserve the Nordic race.[60]

From Cold Spring Harbor, Charles Benedict Davenport congratulated Grant on the passage of the Racial Integrity Act and noted that while Grant had managed successfully to stay out of the spotlight, those with "an interest in race in America will see your hand in this legislation." Lothrop Stoddard was "delighted" with the legislature's action: "That is FINE," he exulted to his new friends Cox, Plecker, and Powell. "Your group has deserved well of the Republic." Cox called the Racial Integrity Act "the most important eugenical effort that has been made during the past four thousand years," while W. A. Plecker showed proper Puritan restraint by murmuring that the law would at least ensure that "complete ruin can probably be held off for several centuries longer."[61]

The Racial Integrity law required that Plecker, as the registrar of vital statistics, verify that all birth, death, and marriage certificates reflect the new definition of who was white and who was nonwhite. With a zeal for racial purity that would be matched only by the Nazis when they came to power nine years later, Plecker increased the size of his staff from four to forty, and set about determining the true racial ancestry of the citizens of Virginia by combing through nineteenth-century death certificates, eighteenth-century marriage licenses, seventeenth-century baptismal records, old tax rolls, musty census records, obscure historical archives, and even the field notes of Carter G. Woodson. Plecker assumed unprecedented powers and, until his retirement after World War II at the age of eighty-five, he waged an all-out war on miscegenation. He worked tirelessly with court clerks, local registrars, physicians, teachers, midwives, and undertakers to track down thousands of persons who were only fifteen-sixteenths white but were trying to "pass" as full-blooded Caucasians. "Few have escaped us" was Plecker's proud boast. In fact, people who honestly thought that they were white, whose families had been white for as long as they could remember, whose neighbors testified that they were white, who even had in hand yellowed court decisions certifying that their forebears were white, suddenly discovered—thanks to W. A. Plecker's genealogical detective work—that because some hitherto anonymous ancestor had been partly nonwhite, *they* were not white, hence they could not marry a white person, attend a white school, or ride on a white train.

Plecker conceded to Madison Grant that his Javertian efforts had "aroused the determined opposition of these mongrels," but he was proud that for more than a generation his office constituted the "insurmountable barrier" between miscegenation and pure whiteness. The Grantians marveled at Plecker's devotion to the cause, and Lothrop Stoddard congratulated the registrar for "smoking out the colored gentlemen in the white woodpile."[62]

As it turned out, the mixed breeds were almost as clever as W. A. Plecker. For within weeks of the legislature's passing the Racial Integrity Act, there was a sudden upsurge in the number of citizens who admitted that while they were only fifteen-sixteenths white, the remaining one-sixteenth was of Indian blood. Thus, they claimed full membership in the white race by virtue of "the Pocahontas Exception." Counties that for centuries had not had a single resident Indian now had hundreds. "Indians are springing up all over the state as if by spontaneous generation," shrieked an alarmed John Powell. A situation that Grant had once found "screamingly funny" he now deemed "grotesque," and the Grantians were forced to beat back this incursion by assuring state officials that there were practically no Indians in Virginia whose ancestors had not intermarried at some time with Negroes; hence almost anyone who claimed to be one-sixteenth Indian was by definition admitting to be part Negro as well, and therefore ineligible to join the white race. Virginia's Indians, of course, were gravely insulted by this slur. "I will tie a stone around my neck and jump in the James River rather than be classed as a Negro," cried Chief George Cook of the Pamunkey Indians. Nevertheless, the authorities concurred with the eugenicists that there were almost no "unpolluted" Indians left in Virginia, and that round was won by the Grantians.[63]

Stymied by the antimiscegenationists, many betrothed couples left Virginia to be married in states where the racial laws were less strict. As a consequence, the legislature passed a law making it a criminal offense to leave Virginia for the purpose of holding a wedding that would not be allowed in Virginia, and when such persons returned to the state warrants were issued for their arrest. W. A. Plecker admitted that his zealousness was causing much pain to individuals, but his fascistic response was: "We cannot consider the individual but the State."[64]

At this point, respectable citizens were reduced to testifying in public that they were illegitimate. Confronted with Plecker's evidence that they had a great-great-grandfather who was black, they claimed that their great-great-grandmother must have been promiscuous and had an affair with a white man, who was their true progenitor, so they were not genetically related to their black ancestor. While this meant they were descended from a bastard, it at least meant that they were still white. Plecker devised a catch–22 to forestall this tactic. Postulating that "white people are ashamed to talk or write about their mothers and grandmothers being of loose character," he reasoned that only a nonwhite person would be unabashed to claim to be descended from the product of an illicit liaison, and he thus decreed that the very attempt to prove whiteness-on-account-of-illegitimacy was in itself proof of nonwhite ancestry. And with that, the last loophole was closed, and the Nordic race in the Old Dominion was saved from the scourge of miscegenation.[65]

In 1925 Plecker proudly wrote in the *American Journal of Public Health* that "Virginia has made the first serious attempt to stay or postpone the evil day when this is no longer a white man's country." It was now up to the rest of the states to do their part, and pass their own antimiscegenation laws. (Or, just as important, those states that had already banned miscegenation needed to reject the pressure from "fanatical and noisy zealots" like Franz Boas to repeal those laws.) Plecker's office mailed literature to the legislators of all the states, appealing to them to join Virginia "in a united move to preserve America as a White nation," and Governor Trinkle of Virginia sent a copy of the Racial Integrity Act to his fellow governors for their consideration. More importantly, Madison Grant pledged to his southern friends the aid of the interlocking directorate in the national antimiscegenation campaign; under his prodding the AES supported antimiscegenation bills in a number of states, and Grant intervened personally in several midwestern legislatures. The effort soon bore fruit, and by World War II thirty U.S. states—from Georgia to California—had passed (or beaten back attempts to rescind) legislation banning marriages between whites and Negroes.[66]

Looking in the Mirror: Grant and Garvey

Leon F. Whitney of the American Eugenics Society once facetiously said to Raymond Pearl: "How would it be if you and I spoke up and persuaded Congress to ship all [the Negroes] back to Africa?" Little did they know that for a long time Madison Grant had been seriously mulling over that very proposal.[67]

On March 18, 1924, one week after Virginia enacted the Racial Integrity Act, Madison Grant sent a letter to Earnest Sevier Cox signaling that it was time to implement the permanent solution to the Negro problem. The two men were in full agreement that this meant deporting the Negroes to Africa, but they differed over means. Cox was more than willing to use force, and pointed to the example of the Trail of Tears, when those Indians who were "in the way of the advancement of civilization [but] who did not wish to move were *made* to move." Cox's analogy suggested that he envisioned using the U.S. Army to round up millions of Negroes and send them in chains across the Atlantic, a prospect that did not appeal to Madison Grant. It is true that Grant had once told Cox, "I am interested in seeing the revival of the Ku Klux Klan all over the country." But temperamentally, Grant was a peaceful man, and he recognized that the only way to remove the Negroes without bloodshed was to gain the cooperation of the Negroes themselves. Accordingly, he sent Cox a clipping from the *New York Herald* about a black leader named Marcus Garvey who had just given a fiery speech in Madison Square Garden condemning miscegenation and calling on the government to solve the country's race problem by helping the Negroes establish their own nation in Africa. "If I were you," Grant calmly told Cox, "I should get in touch with Garvey as it might be worthwhile to back his proposition."[68]

The great black nationalist Marcus Garvey had arrived in New York City from his native Jamaica in 1916, the year *The Passing of the Great Race* was published. From the platform of Liberty Hall in Harlem (built on land once owned by Grant's seventeenth-century forefathers), Garvey spoke of the "new Negro" who took pride in his race and its accomplishments. Like a Negro Theodor Herzl, Garvey taught that blacks would be respected only when they had a homeland of their own in Africa. Claiming that integration in the United States would lead to nothing less than "race suicide," he denounced the NAACP and those African American leaders who cravenly desired to assimilate into white society. He condemned interracial marriage, explaining that "white men should be white, yellow men should be yellow, and black men should be black."[69]

Garvey launched the weekly *Negro World*, which preached racial purity and refused to carry advertisements for products that helped Negroes become more "white" by lightening their skin or straightening their hair. In 1917, he established the Harlem branch of the Universal Negro Improvement Association (UNIA), dedicated to racial pride, economic self-sufficiency, and the repatriation of America's Negroes to Africa. UNIA branches soon sprang up in the principal urban black neighborhoods of the United States, with a purported total membership of two million by 1920.

Madison Grant observed with growing interest the rapid rise of Marcus Garvey. When Garvey defended segregation by pointing to "the laws of nature," according to which animals from different species cannot produce viable offspring with each other, it was clear to Madison Grant that he and Marcus Garvey had more in common than just their initials (and the fact that both men were childless). To be sure, the charismatic Garvey thrived on public ceremony and bombastic display, while the soft-spoken Grant was comfortable only when operating behind closed doors. And while Garvey created a mass movement based on the sufferings of the lower classes, Grant formed elite groups based on the anxieties of the aristocrats. But Grant was insightful enough to understand that Garvey's message of black pride was simply the flip side of his own Nordicism. Both men preached that racial ability was congenital, that integration was dangerous, and that miscegenation would prove the ruin of humanity. Both men sought to preserve their beleaguered peoples from extinction, and urged their followers to take inspiration from the glorious accomplishments of their past. The full name of UNIA, after all, was the Universal Negro Improvement and *Conservation* Association (and certainly that organization's array of flags, emblems, and titles was reminiscent of the paraphernalia of Grant's Society of Colonial Wars). Garvey, like Grant, believed that "one drop of Negro blood makes a man a Negro," and he shared Grant's contempt for the ethnological theories of Franz Boas. Garvey even denied the Jewish ancestry of Jesus Christ.[70]

Marcus Garvey's demagoguery and his doctrine of separatism earned him the enmity of established black leaders such as A. Philip Randolph, James Weldon Johnson, and W.E.B. Du Bois. He was criticized as a swindler and ridiculed as an impostor, especially after he proclaimed himself the provisional president of Africa. The mainstream leaders were afraid that Garvey's strident nationalism would alienate well-meaning whites, but Garvey shot back that it was his critics who were endangering harmonious racial relations through their threatening quest for social equality and intermarriage with whites. And Garvey's rhetoric certainly did not alienate Madison Grant, who recognized the useful role that Marcus Garvey could play in saving the Nordic race. Obviously, however, the New York conservationist was in no position to engage in direct contact with Garvey. And that is why, as soon as the campaign to pass Virginia's Racial Integrity Act was concluded, Grant prevailed upon Earnest Sevier Cox to reach out to the black leader.

A few weeks later, when Marcus Garvey went to Richmond, Virginia, to speak

Mirror images: Marcus Garvey and Madison Grant, the two apostles of racial purity in America. Garvey: Corbis; Grant: © Wildlife Conservation Society.

at a meeting, sitting in the audience was a lone white man: Major Earnest Sevier Cox. Once Garvey began speaking, Cox recognized—as Grant knew he would—that Garvey's call for "repatriation" was no different from his own call for "deportation," and Cox identified the head of UNIA as the Black Moses who would lead America's Negroes back to Africa. Cox introduced himself to Garvey, gave him a copy of *White America,* and offered to help him in his crusade for repatriation. Garvey quickly perceived that both men were working toward the same goal and could indeed form a profitable alliance that would serve both their interests. He accepted Cox's overture, and within a surprisingly short amount of time they forged a bond of reciprocal respect. It is not just that the two men managed to contain their mutual enmity; rather, a genuine and abiding friendship—what Cox called "a spiritual understanding"—developed between the unassuming white supremacist from rural Tennessee and the pompous black nationalist from Jamaica.[71]

In 1925, when the second volume of *Philosophy and Opinions of Marcus Garvey* was published, Garvey included in the book, without cost, a full-page advertisement for Cox's *White America.* Garvey recommended the crude racist tract as an important work that showed that the "Negro Problem . . . cannot be solved except by separating the races," and he pointed out that Cox's book was endorsed by "America's outstanding authorities in ethnology and allied sciences." A UNIA member in Detroit immediately asked Cox to send him 50 copies of *White America.* When those sold out, he asked for 250 more. This kept going, until eventually the Garveyites in Detroit sold 17,000 copies of Cox's book. Similarly, a UNIA member in St. Louis assured Cox that "our members are reading and studying *White America,* and many of the members have said that *White America* should be in every Negro's home along with the Bible."[72]

That same year, when Cox's racist pamphlet *Let My People Go* was published (with a back cover featuring the usual quotes from Madison Grant), it bore a dedication to Marcus Garvey recognizing his "Herculean efforts" on behalf of the Negro race. Addressing himself "to black men who wish to keep the black race black," Cox exhorted America's Negroes to join Garvey in his repatriation crusade and urged any and all whites that opposed miscegenation to join with Garvey as well. The former preacher then closed with the intonation: "Let us labor without ceasing for we labor for a holy cause."[73]

Marcus Garvey told Cox: "I appreciate highly the effort you are making. I endorse and support your views and hope for closer cooperation." Garvey's wife, Amy Jacques Garvey, shared her husband's fondness for Major Cox; she asked the white racist to "please accept my heartfelt thanks for your efforts to place my husband and the cause he espouses in the proper light before the white people of this country." And she promptly purchased one thousand copies of Cox's *Let My People Go* to distribute to Garvey's supporters.[74]

To advance the "holy cause," Cox introduced pianist John Powell to Marcus Garvey. "I realized," wrote Powell the fanatical segregationist, "that I was in the presence of a man of the highest idealism and the noblest courage and the profoundest wisdom; a man dedicated to a noble and a sacred cause—the cause of the independence and integrity of his race." Powell announced that he was in full agreement "in every essential principle [with] the ideals and ideas of Marcus Garvey," and agreed to ally his Anglo-Saxon Clubs with the Universal Negro Improvement Association and to raise money for Garvey among the white antimiscegenation crowd. In return, Garvey arranged for Cox and Powell to speak at local UNIA meetings, and UNIA members throughout the country were encouraged to read Cox's *White America,* Stoddard's *Rising Tide of Color,* and Grant's *Passing of the Great Race.* (As late as 1975, Emory Tolbert noticed that the UNIA members he interviewed during his research still recommended those three books.)[75]

Madison Grant, confined by arthritis to his sickbed, was quite pleased with these developments. He received encouragement from Germany, where Alfred Rosenberg (the chief ideologist of the Nazi Party) commended Grant for leading the attempt to make America *negerfrei.* The Nazis, of course, were interested in the Negro repatriation scheme because they had their own racial problem for which they were seeking a solution. In truth, the Grantians had also considered repatriating the Jews, but Harry H. Laughlin grumbled to Grant that the Jews had become so politically powerful in the United States that "the deportation of four million Jews would be many times more difficult that the repatriation of three times as many Negroes." White Americans, sighed Laughlin, would have to resign themselves to the fact that "the Jew is doubtless here to stay." Charles Benedict Davenport did suggest to Grant, in jest, that as far as the Jewish problem was concerned, perhaps the eugenicists could follow the example of the Puritans who had burned their witches—but he added, "it seems to be against the mores to burn any considerable part of our population."[76]

Just as the ties between Grant's disciples and the black nationalists were beginning to solidify, Marcus Garvey was convicted of mail fraud in connection with the collapse of his Black Star steamship line. A very upset Amy Jacques Garvey blamed her husband's troubles on his enemies and explained to Earnest Sevier Cox that the judge who sentenced Garvey (Julian W. Mack) was not only a member of the NAACP but was also a Jew. The same held true for the prosecutor (Maxwell S. Mattuck) as well as the shipping broker (A. R. Silverston), whom she accused of framing Garvey. (Mrs. Garvey did not mention that among Garvey's defense attorneys were Isaac Levy and Armin Kohn.)

In 1925, Marcus Garvey was sent to the federal penitentiary in Atlanta to serve his five-year sentence. Once he was behind bars, many of his followers began deserting him, and incessant power struggles broke out among his lieutenants in UNIA. With the exception of his wife, Garvey's most loyal supporters during his prison years were the Grantians Earnest Sevier Cox and John Powell. Cox was especially distressed by Garvey's incarceration, and he and his imprisoned friend engaged in a lengthy and confidential correspondence during this period. Cox felt that Garvey was "a martyr" who had been crucified by "the products of miscegenation and the advocates of it." And he publicly exhorted Garvey's followers to maintain the faith: "You Negroes who seek independence for your race and a home in the land of your ancestors, hold fast to your purpose!" John Powell's Anglo-Saxon Clubs passed resolutions expressing indignation at Garvey's imprisonment, and in June of 1925, when Powell traveled to Atlanta to urge the legislature to enact Georgia's version of the Racial Integrity Act, he made a point of going to the penitentiary to visit Garvey. In addition, Cox, Powell, and State Registrar W. A. Plecker were all very solicitous of Amy Jacques Garvey during this period, and went out of their way to comfort and support her. Her incarcerated husband gratefully pledged his "fullest support" to his white friends and endorsed their efforts to pass antimiscegenation bills in the various states.[77]

Most black leaders were disgusted by Garvey's relationship with the white racists. W.E.B. Du Bois was so sickened that he wrote in the *Crisis* that "Marcus Garvey is, without doubt, the most dangerous enemy of the Negro race in America and the world. He is either a lunatic or a traitor." Even some of Garvey's most loyal retainers were discomfited by his newfound friends. The *Negro World* wrote an editorial questioning the need to embrace so warmly the white supremacists. From his prison cell, Garvey issued a stiff reprimand to the staff of the paper, and to emphasize his support for the Grantians, he arranged for John Powell to travel to Harlem in October of 1925 to address the UNIA at Liberty Hall. As Powell sat on the stage, one of Garvey's assistants read a letter from the Atlanta penitentiary, in which the provisional president of Africa introduced to his followers the head of the Anglo-Saxon Clubs of America. "Mr. Powell," wrote Garvey, "represents a body of men and women for whom I maintain the greatest respect because of their honesty and lack of hypocrisy. They represent the clean-cut and honest section of the white race that uncompromisingly stands for the purity of their race, even as we unhesitatingly and determinedly agitate and fight for the purity of the Negro race." Garvey then urged the members of UNIA to cooperate with the members of the Anglo-Saxon Clubs.[78]

John Powell was a captivating speaker, and he began his lengthy speech to the Garveyites by declaring: "I have never in my life been more touched and more moved, than by the cordial reception you have given me and my friends tonight." Speaking "as the descendant of slave owners," he went on to declare that the time had come to right the historical wrong of the Middle Passage by allowing the Negroes to return to their God-given homeland. To great applause, he pledged to do all in his power to make that dream a reality, and to work in the meantime for the release from prison of Marcus Garvey.[79]

Indeed, the effort to free Garvey occupied most of the time of Powell, Cox, and Amy Jacques Garvey over the next few years. Cox, for example, pleaded the case of "my friend Marcus Garvey" with James J. Davis, the secretary of labor (and an acquaintance of Madison Grant's from the immigration restriction movement). "I am sensitive to the fact that a portion of his teaching may give cause for apprehension," admitted Cox, but he explained that "in Garvey we have a powerful ally, with a great influence among his people." Both Cox and W. A. Plecker also wrote directly to President Coolidge asking him to grant clemency to Garvey, and at the end of 1927 Coolidge did finally pardon Garvey. The president announced, however, that Garvey would have to leave the country (since any alien convicted of a felony must be deported), and Garvey was placed on a boat to Jamaica.[80]

With Garvey permanently in exile, a demoralized UNIA split into rival factions and slowly began to fade from the scene. Cox and the banished Marcus Garvey continued to correspond with each other for years, and Cox maintained his friendship with Mrs. Garvey. Cox also conducted an ongoing campaign with the State Department to gain Garvey's return to the United States, but permission was never forthcoming.

Madison Grant was not pleased by Garvey's exile. By the early 1930s he resigned himself to the fact that he would never succeed in sending America's Negroes back to Africa without Garvey's charismatic personality, his proven leadership skills, and, most important, his willingness to cooperate with white racists. With reluctance, Grant called a halt to his deportation efforts and moved on to other causes. It is somewhat ironic that Madison Grant's dream of deporting the Negroes was ended by the deportation of a Negro.

Earnest Sevier Cox and W. A. Plecker never gave up the deportation struggle. In 1936 the two southerners traveled to Manhattan to huddle with an ailing Grant in his new town house at 960 Fifth Avenue (which was, and is, one of the choicest real estate locations in the world). Grant was unwilling to become directly involved in their crusade, but he did agree to create a $10,000 annual endowment to underwrite a lobbying effort in Congress to deport the country's twelve million Negroes to Africa. Grant died before he had time to set up the endowment, but the funding breach was filled by a devotee of Grant: New England textile heir Wickliffe P. Draper.

An avid big-game hunter and descendant of the Puritans, Colonel Wickliffe Preston Draper had been entranced by scientific racism ever since his undergraduate days at Harvard. Keen to spend his millions for the good of the Nordic race, Draper had anonymously funded a number of eugenics projects (including the research that resulted in *Race Crossing in Jamaica,* the classic attack on miscegenation by Charles Benedict Davenport and Morris Steggerda). In addition to amalgamation, Colonel Draper—who never married—was extremely concerned about race suicide. Like W. A. Plecker, Theodore Roosevelt, and Adolf Hitler, he feared that the Master Race would not survive unless every patriotic couple produced at least four children. To encourage that endeavor and other projects aimed at Nordic-race betterment, Draper became the primary benefactor of the infamous Pioneer Fund, the eugenic organization he founded with Harry H. Laughlin and Frederick Osborn (nephew of Henry Fairfield Osborn).[81]

Madison Grant introduced Draper to Earnest Sevier Cox and W. A. Plecker in 1936, and the Massachusetts millionaire became a lifelong friend and patron to the southern segregationists. The three of them worked with Senator Theodore G. Bilbo of Mississippi (the colorful, corrupt, race-baiting, anti-Semitic Dixie demagogue who had memorized entire passages of *The Passing of the Great Race*) to sponsor their repatriation bill in Congress year after year. I have recounted elsewhere the fascinating tale of their decades-long effort to get Congress to pass that bill;[82] suffice to say that thanks to the burgeoning civil rights movement, the repatriation scheme finally died in Congress in 1959.

And yet the connection between northern scientific racism and southern popular racism continued in force, precisely because of the growing strength of the civil rights movement. Writing in 1963, for example, Mark Haller could still state that "the works of Grant and Stoddard became standard sources, *and remain so today,* for white supremacy advocates of the South" who utilized those writings in the backlash against the civil rights movement.[83]

Grant's disciple Earnest Sevier Cox also underwent a revival in the mid–1960s. His publications were advertised in right-wing newspapers and racist journals, and a new paperback edition of *White America* was published in 1966. The *Mankind Quarterly,* the main journal of postwar scientific racism (which, not coincidentally, was bankrolled by Wickliffe P. Draper's Pioneer Fund), called *White America* "a classic book by this truly great man and respected anthropologist," and praised Cox for devising a "practical solution" for keeping America white. Cox did not enjoy his new fame, however, as he had died a few months earlier of emphysema. Having never married, he left his worldly possessions to the repatriation movement and was buried in Arlington National Cemetery. Today, almost nine decades after it was first published, *White America* is enjoying yet another comeback, as its text is posted on the websites of a number of white supremacist groups (usually alongside its northern cousin, *The Passing of the Great Race*).[84]

As for Wickliffe P. Draper (whose biographer states that Madison Grant was "in many ways the model for his life"), he spent his last years as a recluse in his enormous Manhattan penthouse festooned with his many hunting trophies.[85] During the 1960s, he secretly funneled hundreds of thousands of dollars to segregationist groups down south that infiltrated civil rights organizations and lobbied against civil rights legislation. When he finally died in 1972 at the age of eighty-one, Draper bequeathed $3.3 million to the Pioneer Fund, which provided hundreds of thousands of dollars each to hereditarian researchers such as Roger Pearson, Arthur Jensen, William Shockley, Richard Lynn, and J. Philippe Rushton, who sought to prove that whites have higher IQs than blacks. It did not go unnoticed by several critics that research underwritten by the Pioneer Fund was cited prominently in *The Bell Curve* (the cause célèbre of 1994, which famously argued that it was time for Americans to face the fact that blacks are significantly less intelligent than whites).

When liberal reviewers tried to damn *The Bell Curve* by claiming it had been inspired by Madison Grant's *Passing of the Great Race,* they thought they were being cleverly metaphoric. They did not realize that there were, in fact, longstanding personal and financial connections between the charming Manhattan lawyer of the 1930s, the southern segregationists of the 1960s, and the biological determinists of the 1990s.

11

Saving the Redwoods

They were more like gods than anything I had ever seen.

John Masefield, poet laureate of England, after seeing redwoods for the first time

In August of 1991, the California Department of Parks and Recreation received a letter from a man who had just visited Humboldt Redwoods State Park in northern California. The Humboldt redwood trees are the tallest living things on earth, towering over three hundred feet into the air—higher than the Statue of Liberty. The sight of these immense trees is truly overwhelming, and the tourist had been suitably impressed. But during his visit to the forest he had been "shocked" to come across a bronze plaque honoring Madison Grant as a founder of the Save-the-Redwoods League. His letter to the Department of Parks cited passages from *The Passing of the Great Race* and pointed out that Grant's "racist writings are so abhorrent to basic American principles that they discredit anything honorable the man may have done in his lifetime." The cogently argued letter concluded that "honoring Madison Grant with a plaque on public property is as historically bizarre as erecting a monument to Adolf Hitler for his part in founding the Volkswagen Company. Please have it removed."[1]

The issue raised by the letter is rather profound. Do the racist views of someone who lived in a different era—an era with more primitive scientific knowledge and significantly different values—discredit any positive contributions the racist may have made?

The question is pertinent because Madison Grant, founder of scientific racism, did make so many positive contributions, of which the greatest (and certainly the most improbable) was saving from the voracious saws of the powerful lumber companies the celebrated redwood trees of California.

The "Historic Camping Trip"

In August 1917, Madison Grant and Henry Fairfield Osborn traveled to California to attend the summer encampment of the exclusive Bohemian Club. This annual gathering of the Golden State's movers and shakers took place in the Bohemian Grove, the club's forest enclave sixty miles north of San Francisco. The two New Yorkers soon became engaged in conversation with a fellow member of the Boone and Crockett Club, John C. Merriam, chairman of the Department of Paleontology at the University of California. Merriam was renowned for his excavation of the La Brea tar pits and the numerous fossils of saber-toothed tigers he found there. He was also a staunch eugenicist, and Grant was quite fond of him. "I regard Dr. Merriam as one of the most delightful persons that I have met," Grant confessed to a friend, and in nominating him for the Century Association Grant declared baldly that Merriam was "one of the greatest living Americans."[2]

The three men marveled at the beauty and the height of the redwood trees that had been preserved in the Bohemian Grove. Merriam told Grant and Osborn that in the 1840s the reports of the first exploring parties describing the redwood trees were so fantastic that they had not been taken seriously. In 1854, an entire tree had been shipped to New York and put on display, but it had been considered by most to be a hoax. And he reminded them that the Baptist church in Santa Rosa they had passed en route from San Francisco, with seating room for three hundred people, was built from the lumber of a single redwood tree. At this point, an eavesdropper assured them that the more extensive but rarely seen redwood groves of Humboldt County in the northern part of the state not only had taller specimens than the Bohemian Grove but possessed a "mystery and charm unique among living works of creation." Intrigued, it did not take long for the three Boone and Crocketteers to decide to embark on an excursion to see the trees up close (an excursion that came to be known in the conservation community as the "Historic Camping Trip"). Thirty years later, the director of the National Park Service would marvel that "one of the great dramas in the history of conservation"—that is, the saving of the mighty California redwoods—had originated in that conversation in the Bohemian Grove.[3]

Grant, Osborn, and Merriam departed early the next morning, driving up the coast along a newly opened highway through Sonoma and Mendocino Counties until they arrived at a place known as Bull Creek Flat, about two hundred miles north of San Francisco in the redwood region of Humboldt County. Their driver pulled the car over to the side of the road and the men clambered out, gazing with astonishment at the size and splendor of the redwoods. They advanced on foot across a small meadow and then plunged into the shadowy, surreal world of the forest. To their chauffeur it seemed as if they had been swallowed up by the wall of giant trees towering above them.

Bounding over the soft, springy carpet of brown needles, breathing the sweet, still air, the three men were overcome by that mystical sense of wonder and exhilaration familiar to all visitors to the old-growth redwood forests. Duncan McDuffie, future president of the Save-the-Redwoods League, once wrote: "To enter the grove of Redwoods on Bull Creek Flat is to step within the portals of a cathedral dim, lofty, pillared, peaceful. But this temple which the Great Architect has been building for a score of centuries is incomparably nobler, more beautiful and more serene than any erected by the hands of man."[4]

The image of grove-as-cathedral would be invoked time and again by activists in the save-the-redwoods movement (but then, as John C. Merriam once pointed out, it is not that the forest groves are like cathedrals, but rather that the earliest cathedrals were in fact modeled after groves). Merriam recalled that the three naturalists, "who all their lives had known great forests, bared their heads in this presence." They felt the incredible sense of peacefulness and timelessness that pervades the eternal twilight of the redwood groves. Grant was convinced that "nowhere on earth does there exist a forest to compare in continuous grandeur and unqualified beauty with the Redwoods" of California. It was, quite simply, "the most magnificent forest in the world."[5]

A few stray beams of golden light managed to penetrate the canopy, illuminating the rich cinnamon-colored bark of the trees. As the three hikers' eyes adjusted further to the light, they could see that the forest floor was punctuated by the emerald leaves of sun-dappled ferns and the last fragrant clusters of the season's scarlet rhododendrons. Here and there the trunks of fallen trees, weighing up to one million pounds each, stretched on and on, their tops disappearing from sight in the distant gloom. The men felt as if somehow they had stepped back in time. When they finally sat down to eat their lunch, Merriam the paleontologist looked around at the primeval trees, the prehistoric ferns, and the steam rising where shafts of sunlight struck the wet earth, and he remarked to Grant that "if a dinosaur should stick his head around one of the tree trunks amid the ferns, he would find nothing in the landscape that was unfamiliar to him in Jurassic times."[6]

Merriam was not guilty of artistic license. For in addition to being the tallest living things on earth, the redwoods are also among the most ancient. As a genus, the redwoods have indeed been around since the Jurassic period, more than 160 million years ago. For eons, the redwoods covered much of the Northern Hemisphere, but changes in climate led to their gradual retraction until the sole survivors were confined to California. There they have hung on in two groups: the giant sequoias (*Sequoiadendron gigantea*), which are found in scattered groves on the western slopes of the Sierra Nevada, and the less massive but taller coast redwoods (*Sequoia sempervirens*), which grow along the coast in a narrow strip a few miles wide.

The range of the coast redwoods is determined by the fog bank that creeps in

An enthralled visitor (lower left) contemplates the tallest living things on earth. Photo by Howard King, courtesy Save-the-Redwoods League. *[© Permission required for reprint, sale or commercial use. Contact Save-the-Redwoods League.]*

from the Pacific Ocean to bathe the trees every afternoon in the summer; they receive critical moisture from this fog, which condenses on the needles and drips to the forest floor. Grant found it amusing that the locals, "with the usual human capacity for error, state that the Redwoods attract fog, but of course it is the moisture of the fog deposited on the tops of the Redwoods that determines their inland distribution." (In a sense, though, the locals were not totally mistaken, as the trees alter the surrounding atmosphere by transpiring huge amounts of moisture into the air—as much as five hundred gallons per tree per day).[7]

In addition to their beauty, size, and ancient lineage, the redwoods possess one other superlative: the individual trees are among the oldest living things on the planet. The average mature age of a redwood tree is five hundred to one thousand years, and some redwoods are known to be over two thousand years old. The same trees that Grant saw in 1917 had been alive in A.D. 9 when his beloved Teutonic barbarians defeated the Roman legions in another forest, at Teutoberg.

Sempervirens means "ever living," and the astounding fact is that, as one naturalist puts it, "There is no biological reason known why [a redwood tree] should ever die." The redwoods have no natural enemies. They produce a tannic acid that protects their wood from insects and fungi. Their incredibly thick, asbestos-like bark is relatively fire resistant, and the low resin and pitch content reduces the flammability of the wood even further. Death comes only from natural disasters: lightning, extremely strong winds that topple the more exposed trees, or repeated hot fires that manage eventually to burn through the bark and expose the heartwood to dry rot.[8]

But even after it seemingly has been killed, a redwood tree is not truly dead, for it has the marvelous ability to resprout from its own roots. After a tree is toppled, it is only a matter of days before new saplings, with a genetic makeup identical to that of their predecessor, pop up in a circle around the base of the parent tree and begin the millennia-long process of reaching toward the sky. It was this unique capacity for regeneration that Grant thought was perhaps the most interesting characteristic of the redwood. (But it was a great nuisance to early settlers trying to convert the forests to farmland; they had to cut and burn a patch of forest over and over again until the exhausted trees finally stopped trying to regenerate themselves. Such practices, lamented Grant, proved that "human greed" was so powerful that it could "make it impossible for even the Redwood to survive.")[9]

One scholar has stated: "You will not emerge from one of the redwood groves the same person as you first entered it." And there is no question that Madison Grant was stirred by the redwoods. He cherished their "graceful beauty" and delighted in their "truly amazing vitality." He admired their "brave resistance to axe and fire" and the fact that "even when hacked and burned and butchered" the trees still refuse to die and "persist in putting out new foliage . . . if given the slightest chance." "The Redwood," he declared, "is a beautiful, cheerful and very brave tree" and "No one who has seen these groves can fail to love them."[10]

In short, the three Boone and Crocketteers were entranced by the redwoods. Unfortunately, so were the logging companies. Those same qualities of the trees that had impressed Grant—their resistance to fire, their invulnerability to insects, and their imperviousness to rot—were, as he put it, "the unfortunate virtues that have made their lumber so valuable." Strong, beautiful, durable, lightweight, nonwarping, straight grained, a good insulator, able to hold nails and finishes: lumber from the redwood tree was a carpenter's dream.[11]

Californians were literally building their state on a foundation of redwood. The flumes and sluice boxes of the gold country, the grape stakes and wine vats of the Napa Valley, the curbs and sewers of Oakland, the piers and houses of San Francisco, the railroad ties and telegraph poles that connected those cities to the rest of the Union—all were constructed out of redwood trees. No wonder that logging was the most important industry in northern California, and that every year the companies were cutting down and dragging out of the coastal valleys another 500 million board feet of redwood lumber.

Before the gold rush, the coastal redwoods had reigned over nearly two million acres of northern California from Monterey Bay to the Oregon border. The trees grew in such dense stands that it was believed the supply would never be exhausted. But by the time of Grant's 1917 visit, a large percentage of the trees had already fallen to the ax, and almost all the surviving stands were in the hands of logging companies and "marked for massacre."[12] Walt Whitman's elegy still held true:

> Riven deep by the sharp tongues of the axes, there in the
> redwood forest dense,
> I heard the mighty tree its death-chant chanting.[13]

The lumber industry estimated that the remaining redwoods would be gone within sixty years, but Grant gave them an even shorter time. He knew that before the turn of the century, when the old-timers used handsaws, it had taken up to a week for a team of loggers to fell one of the giant redwoods. But now, with modern, mechanized logging methods, a tree that had taken five centuries to mature could be cut down and hauled out in an hour. And an entire forest that had taken millennia to create could be liquidated in a few weeks. ("Any fool can destroy trees," John Muir once wrote despondently. "They cannot defend themselves or run away.")[14]

Grant and his two companions were absolutely devastated by the havoc being wreaked by the lumbermen. They were outraged by the sight—visible everywhere along the logging roads—of single trees reduced to immense piles of thousands of grape stakes. And they were mortified by the destructive fires being set all around them. (The forests were often set on fire preceding lumbering operations so the loggers could perform their work unencumbered by brush and wild animals.)

An associate remembered that Grant "was depressed and saddened by . . . seeing acres upon acres of once beautiful Redwood forest hacked and burned and rendered a desolate waste through lumbering operations." Indeed, Grant was heartbroken over the charred stumps arrayed like tombstones on the fire-blackened hillsides, and sickened by the sight of muddy, debris-choked rivers, stagnating into lifeless swamps between recently denuded banks. It was, he told a friend, like something out of "Dante's Inferno."[15]

"The land lies worthless," cried out Grant in the *Zoological Society Bulletin.* "This example of human greed and waste can scarcely be described . . . . The devastation," he concluded, "is appalling." (But, after all, as Secretary of the Interior Stewart Udall once noted: we are the people "that made Paul Bunyan a national folklore hero.") Merriam and Osborn were as disheartened as Grant. Feeling that the redwoods were "the noblest living things upon earth," Osborn announced that their destruction would be "one of the greatest calamities in the whole history of American civilization."[16]

Madison Grant was not interested in saving the redwoods because of their ability to regulate the volume of water flowing through streams, nor because he wanted to ensure an adequate timber supply for future generations. No, the trees were worth saving because their existence was a valuable thing in and of itself. As early as 1904 Grant had remarked to President Roosevelt that "it would be little short of barbarous to allow . . . the destruction of these trees, the oldest living things on earth." TR agreed that "a grove of giant redwoods or sequoias should be kept just as we keep a great and beautiful cathedral. The extermination of the passenger-pigeon meant that mankind was just so much poorer; exactly as in the case of the destruction of the cathedral at Rheims. . . . Why, the loss is like the loss of a gallery of the masterpieces of the artists of old time." But as Grant pointed out, "Cathedrals can be rebuilt, Old Masters can conceivably be excelled by great artists in the future, . . . but [when it comes to the redwoods] it is beyond the power of man to replace these monuments of a long-vanished past."[17]

There can be little doubt that Grant identified the redwood trees with the Nordic race. And not just because both had an evolved proclivity for foggy environments. Rather, it was that the Nordics, who in their day had conquered most of the Old World, were now making their last stand in North America, where they were threatened with "a speedy extinction" at the hands of invading hordes of immigrants. And so too the redwoods, which at one time had blanketed the entire hemisphere, had been forced to retreat to their tiny redoubt on the western edge of the continent—here to make *their* last stand against the invading hordes of loggers and developers. In Grant's mind, the redwoods (which a colleague once described as "the remnant of a great race") thus joined all the other embattled native species of North America as being metonymic with himself and, by extension, with the Nordic race.[18]

John Muir had referred to "These kings of the forest, the noblest of a noble race" and insisted that "we cannot escape the responsibility as their guardians." Grant agreed. The redwoods had existed long before the first humans lived, before the first mammals of any size had evolved, before even the continents had taken their present form. Thousands of species of flora and fauna had come and gone, and yet the redwoods had lived on. But now, with the coming of modern civilization in the form of chain saws and donkey engines and mechanized logging mills and greedy lumber executives, it was all going to end. (One lumber baron stuck out his jaw, looked John C. Merriam straight in the eye, pointed at a grove of redwood trees, and exclaimed: "I *hate* 'em. I'd like to see every damn redwood down!") To mercilessly destroy these "priceless heirlooms" now, after nature had preserved them for eons in their virgin state, would be "like lighting one's pipe with a Greek manuscript to save the trouble of reaching for matches."[19]

Of immediate concern in 1917 was the fact that the state of California was constructing a new highway (U.S. 1, a.k.a. the "Redwood Highway") that ran right through the redwood forests. This not only created a wide area of devastation but also provided free and easy access to stands of heretofore remote timber, which the lumber companies were planning to cut with a vengeance. Grant recalled later that the threatened destruction of these previously safe groves "weighed heavily" upon him, and he drew the same conclusion that he did from his study of evolution: "The workings of civilization may be mysterious but there is little doubt that they are hostile to nature."[20]

Organizing the League

Two decades earlier, when someone asked John Muir to protect the coastal redwoods, he replied that he was immeasurably pained by "the ruthless destruction" of the trees, but since the groves were all in private hands there was nothing he, or anybody else, could do to save them. Madison Grant thought otherwise. Although the Historic Camping Trip had put him in a state of extreme agitation, he did not succumb to despair. Rather, Grant, Merriam, and Osborn did what any good progressives would have done: they formed an organization. It was (and is to this day) called the Save-the-Redwoods League, and from that ad hoc beginning under the trees of Humboldt County the league grew and prospered until, as one scholar has noted: "It was to become one of the greatest agencies of preservation the world has ever known."[21]

It has long been a characteristic of Americans that when confronted by a problem they join together in an organized group dedicated to solving the problem. And that is precisely what Grant did, time and time again. He was, after all, a Yale man, and never doubted that he would succeed. To be sure, such an attitude was not uncommon during the progressive period, when expansive thinking was in vogue. But in Grant's case it was also symptomatic of a certain aristocratic view of himself and his place in the world—a hubristic presumption that he could actually make a difference.

Grant would have taken quite literally Woody Guthrie's claim that "this land is my land, from the redwood forests to the New York islands." As far as Grant was concerned, this land was his land—and his father's, and his father's father's—and he intended to preserve it as they had known it. Having sounded the tocsin for impending racial depletion, he would now do the same for the forest's destruction.

Grant returned to Manhattan and spent the next few months tending to the various details involved in creating a new organization. A constitution had to be written, a statement of purpose drawn up, funds secured, offices established, executive officers recruited, a letterhead designed, legislation drafted, and so forth. Grant relished the nitty-gritty of building an organization. He labored for weeks to make sure that the branches of the trees in the Save-the-Redwoods

League's logo were tilted just right, and spent an equal amount of time arguing with John C. Merriam over whether the league's directors should be listed on the letterhead alphabetically or by seniority (Grant, of course, pushed hard for seniority). He agonized over the proper paper stock to use for the League's stationery—but then, there had always been three verities in Grant's epistemology: the need to preserve the native flora and fauna of North America, the immortality of the germ plasm, and the absolute necessity for fine stationery.

The progressives' mania for forming organizations perhaps served to salve their sense of insecurity in the face of the inexorable social transformations of the time. When planning, and classifying, and regulating, and, in a word, *organizing,* the progressives felt that they were imposing order on a highly uncertain world. Madison Grant could sit back and ponder the big picture with the best of them, but after a certain point such pondering only left him depressed. He had to "get active," as TR used to say; and, as a progressive, getting active meant organizing. At some level, therefore, Grant felt that by designing just the right letterhead he could indeed change the world.

Henry Fairfield Osborn did not participate heavily in the planning stages of the Save-the-Redwoods League; in fact, during 1918 and most of 1919, the "league" was essentially a two-man show: Madison Grant in New York and John C. Merriam in California. Though they were on opposite sides of the country, the two friends corresponded almost daily and collaborated on the minutest of organizational details. Also, Merriam was Grant's house guest in Manhattan several times every year, and much work was accomplished during those visits.

In October 1918, Merriam joined Grant in Washington, D.C., where they went to see Interior Secretary Franklin K. Lane. Grant had been lukewarm toward Lane ever since Lane (who had been city attorney of San Francisco under Mayor James D. Phelan) endorsed the wrong side during the Hetch Hetchy controversy. Furthermore, Grant felt that Lane's Interior Department was too cozy with oil and mining interests and was "commercializing" the national parks. And finally, Lane's most egregious sin was his prose, which to Grant's eye demonstrated "an affectation in classical learning." (Grant's sensibilities were outraged when Lane, in an article in *National Geographic,* referred to "Achilles touching the soil," when the proper reference—as we all know—was Anteus.)[22] Still, Grant wanted a Californian of national standing to serve as president of the Save-the-Redwoods League, and Franklin K. Lane fit the bill. Grant and Merriam assured the secretary that he would not have to do any actual work; they only wanted him for the prestige his name would lend to the new organization. That clinched it, and Lane agreed to be the figurehead president of the Save-the-Redwoods League.[23]

Grant and Merriam then drew up a list of prominent Americans to serve on the league's board of directors (where they were known as "councilors"). It was an impressive group, with national leadership roles in business, science, and politics. Several of the councilors were women, as Grant foresaw that the Nineteenth Amendment (which had been proposed in Congress) would mean that women could play a key role in such endeavors. Susan R. Schrepfer, in her masterful work *The Fight to Save the Redwoods,* has analyzed the backgrounds of the seventy-nine councilors who served between 1920 and Grant's death. As a rule, they were well educated (the plurality of councilors had attended the University of California or Yale), financially comfortable, and Protestant (Stephen T. Mather, for example, was a direct descendant of *the* Mathers of Puritan New England, and John C. Phillips was a descendant of George Phillips, who founded the first Congregational church in America in 1630). Only two of the league's councilors actually lived in the redwood counties; the rest were from urban areas, and one-third were from the East Coast. (One league official admitted, "Had it not been for our Eastern friends, I am afraid we Californians would not have waked from our trance until the last great Redwood had fallen.")[24] The councilors were progressive Republicans (albeit representing the full spectrum of the type, from reformer William Kent on the left to banker William Crocker on the right). They were members of all the elite clubs (the Metropolitan, University, and Union Clubs of New York and Washington, D.C., and the Commonwealth, Olympic, Bohemian, and Pacific Union Clubs of San Francisco). And in addition to Grant, Merriam, and Osborn, a number of them belonged to the Immigration Restriction League and/or the American Eugenics Society. No wonder one scholar in 1978 mistakenly called the Save-the-Redwoods League "a California based *eugenics* and conservation group."[25]

As he did with all his organizations, Grant insisted on an aristocratic form of governance, in which all real power was placed (and remained) in the hands not of the common members but of the councilors. But as the councilors strove with Grant and Merriam to preserve what they could of the remaining groves of redwoods, the same three questions kept arising—questions that had recurred throughout Grant's career as a conservationist (and that are still asked today): (1) In saving scenic areas for the American public, how do we protect the areas *from* the American public? (2) Why should an elite group of urban intellectuals have the right to impose their views on—and override the desires of—local residents regarding the public lands in their neighborhood? (3) Why should working people's tax dollars be spent on protecting remote wilderness areas that only the idle rich will ever be able to visit?

Grove by Grove

Madison Grant and John C. Merriam ensured that articles extolling the effort to save the redwoods were placed in national publications such as the *Saturday Evening Post, Natural History, Outlook,* and *Outing.* Grant himself wrote lengthy and effective articles for *National Geographic* and the *Zoological Society Bulletin,* in which he argued that posterity demanded that the redwoods be preserved: "After the fall of the Roman Empire the priceless works of classic art were 'needed' for lime, and statues by Phidias and Praxiteles were slaked down for this purpose, but the men who did it are today rightly dubbed 'vandals and barbarians.' What then will the next generation call us if we continue to destroy these priceless trees because lumber is 'needed' for grape stakes and railroad ties?"[26] (The mention of grape stakes was meant to capitalize on the then-popular zeal for the recently enacted Eighteenth Amendment.) But Grant's conversion to aesthetic preservationism had never prevented him from appealing to the public's economic self-interest, and he again argued (correctly) that "the value of a living tree . . . far exceeds the value of its lumber. . . . When [northern Californians] awaken to a full realization of the revolution effected by the automobiles, which will flood the country with tourists as soon as the highways are completed, they will find that a Redwood grove, such as Bull Creek Flat, is an attraction that is worth to the county many times the full net value of the timber contained in it."[27]

It was still not obvious, however, just what the Save-the-Redwoods League could do to save the trees. Grant's initial thought was to have Congress create a national park in northern California. But all the forested land was already in the hands of the logging companies. This was a problem because heretofore the national parks had all been carved out of *public* lands. When Grant created Glacier National Park and Denali National Park, they were already part of the public domain. Turning them into national parks had required only that Congress pass a resolution and the president sign it. Such situations were tailor-made for a charming aristocrat like Grant: invite the appropriate politicos over to the Metropolitan Club, pass out some cigars, discuss the situation as gentlemen, and after a glass or two of Château Lafite 1878 (and ten or twenty years of legislative wrangling) the deed was done. But with no precedent for expending federal funds to purchase privately held land for park purposes, it was evident that no Redwood National Park was going to be established any time soon.

Yet Grant was unwilling to stand by and let one of the greatest glories on the planet be wantonly destroyed. Losing the redwoods would simply be "a national catastrophe." So he returned to the West Coast and drafted a bill authorizing the state of California to launch condemnation proceedings against the lumber companies to rescue the redwood groves.[28]

Condemnation was a radical proposal. But Grant had determined that preserving the redwoods was "a case where the commercial" would "have to yield to the ideal." And hence he had few qualms about invoking eminent domain and/or instituting condemnation suits against the logging companies. Further, Grant could never transcend his bitterness over the "exceptionally conspicuous mutilation" wrought by the lumbermen in the redwood region. As far as he was concerned, justice demanded that "someone should suffer for the reckless squander of public property in the past." (Outbursts like that actually led one scholar recently to categorize Madison Grant as a "leftist progressive"!) Small wonder that Grant was the lumbermen's least favorite conservationist. When one executive realized that the man he was talking to at a dinner party was Grant's physician, he asked how the good doctor could minister to such a "wicked" man. And when the president of the Pacific Lumber Company (which Grant had labeled "the great enemy of the Redwoods") discovered that the person he had just been introduced to at the Century Club was a friend of Grant's, he immediately recoiled and sputtered bitterly: "That's the man who took my redwoods away from me!"[29]

Unfortunately, the other directors of the Save-the-Redwoods League were not comfortable with the idea of the state taking property by fiat. Joseph D. Grant ("J.D.") was particularly hostile to Grant's plan. A wealthy San Francisco entrepreneur, he argued that condemnation was an untoward interference by government in the private enterprise system.

J. D. Grant's father, Adam Grant, had immigrated from Scotland to San Francisco during the gold rush and founded the largest wholesale dry-goods concern in the West. His son J. D. increased the family fortune many times over by pioneering, in turn, the hydroelectric, the petroleum, and the steel industries in California. J. D. also served on the boards of several corporations, including the Central Pacific Railroad and the Bank of California. It is safe to say that J. D. Grant did not want for cash. When his Burlingame mansion burned to the ground in 1909 at a loss of more than $200,000 he batted nary an eye, and built the even more luxurious Villa Rose on a fifty-acre estate in the heights of exclusive Hillsborough.

Having made his millions, J. D. Grant took up the pursuits of the rich: he introduced polo to California and built the state's first golf course. (His son Douglas Grant would become a national golf champion and helped design the Pebble Beach golf course.) It goes without saying that J.D. was a leading member of the Bohemian, Olympic, and Pacific-Union Clubs, all fixtures of San Francisco society.

One of J.D.'s closest friends was Madison Grant, at whose house he always stayed while en route to and from England (where he hobnobbed with the likes of Sir Ernest Shackleton, Sir Arthur Conan Doyle, Lady Randolph Churchill, and her son Winston). The two Grants were not related, though people seeing them together on one of their "good snappy walks" around Central Park usually assumed they were—a situation that always amused J.D., who jokingly referred to his alter ego as "Cousin Madison."[30]

J. D. Grant was rather conservative in his politics. He would become quite apoplectic at the mere mention of the "tyrannical" labor unions and their "Bolshie" leaders (who had "never done an honest day's work" and who wasted their time "arm waving and orating"). In 1921, when Congress "gave way to a combination of stupidity and envy of accumulated wealth" and raised the income tax rate, he swore that the time had arrived for Cousin Madison to write a new book titled "The Passing of the Capitalists."[31]

As one might suspect, given the time when he lived and the position he held in society, J. D. Grant was firmly in favor of immigration restriction. Though he himself was only one generation removed from foreign birth, he decried the day that "we opened the floodgates and let in the alien hordes unrestricted." J.D. always had high praise for the "patriotic work" of his friend Madison Grant, who had labored mightily to stop the "distressing" flow of immigrants that had "virtually submerged the older stock."[32]

On the other hand, immigrants did have a certain place in J.D.'s world. For one thing, the newcomers—especially from the Orient—made good house servants. He particularly enjoyed being "waited upon by little sandal-footed Nipponese maids in colorful costumes." He recalled asking one such servant—"as dainty as the flower petals which adorned her kimono"—what her name was, "expecting a name which freely translated might signify 'Gentle Breeze of Morning' or 'Kiss of Night.'" But no. "'My name is Mabel,' she said, pert as any maid from Zenith, U.S.A. By the jumping frog of Calaveras! Mabel! In a few years will they all degenerate into Mabels and Mamies and Maudes? God forbid!"[33]

To forestall the rule by "inferior breeds"—a category that apparently included not just immigrants but any and all members of the Democratic Party—J. D. Grant preached that "the solution of our political ills" was for "men of affluence" to enter politics. If more capitalists like his close friend Herbert Hoover would take a larger part in public affairs, "the cult of incompetence would cease to dominate the misdirection of our commonwealth." For his part, Madison Grant thought J. D. was "a man of very broad vision."[34]

Born into wealth, J. D. Grant lived to increase that wealth, and in the process he littered the California landscape with steel mills, petroleum refineries, hydroelectric dams, and other commercial and industrial enterprises. And yet he was a supporter of the arts (numbering among his acquaintances Ambrose Bierce, Fritz Kreisler, and Ignace Paderewski) and a very generous patron of both Stanford University and the University of California.

J. D. Grant was also a true lover of nature, and an able and eloquent defender of the redwood trees, which, like the Nordics, were "an ancient race of giants making their last stand" in North America.[35] J. D. was, in turn, an indefatigable vice president, president, and, for two decades, chairman of the board of the Save-the-Redwoods League. The offices of the league have for many years been located in the Adam Grant Building off Market Street in downtown San Francisco, built on the site of the dry-goods store that was the initial source of the Grant fortune.

Madison Grant, erstwhile reactionary, argued with J.D. that saving the redwood trees was "surely one of the most glaring examples of the necessity of the state interfering with the management of private property to prevent its wasteful exploitation." Talk like that was guaranteed to make J.D. hyperventilate, but Madison went ahead and urged the California legislature to pass his condemnation bill. He appealed to the legislators' sense of shame, claiming that "the preservation of the California Redwoods is the most important conservation measure in America. It would be a disgrace to California to allow the destruction of these monuments of antiquity." And he appealed to their common sense, arguing that the wisest course for the state would be to forgo short-term profits based on logging in favor of long-term prosperity based on tourism.[36] Unfortunately, a majority of the politicians refused to look beyond the short term, and the condemnation scheme failed.

Grant and the Save-the-Redwood League now faced the maddening fact that the only way to save the redwoods was to raise a lot of money in order to purchase, at full market value, whatever scattered lands could be pried from recalcitrant—if not hostile—lumber companies. The league would then donate the areas to the state of California to be preserved in state parks. It would be like putting together a very large jigsaw puzzle, grove by grove.

And so the race was on, between the fund-raising abilities of the league and the tree-felling activities of the loggers. But at a time when most Americans still did not own an automobile, let alone envision transcontinental jet travel, it was going to be very difficult to convince them that the fate of a certain few trees in a remote corner of California—trees they would never get a chance to visit—should be of concern to them. And thus it was that Madison Grant spearheaded the creation of "memorial groves," which one writer has called "the most brilliant example of linking conservation with human sentiment ever devised."[37]

The first memorial grove was dedicated to Colonel Raynal C. Bolling, who had been killed in action by the Germans in 1918, the first high-ranking U.S. officer to lose his life in the Great War. His brother-in-law, league councilor John C. Phillips, wanted to create some sort of memorial to Bolling, and Grant suggested that Phillips purchase a redwood grove and dedicate it to the fallen hero.

Phillips, a great-nephew of abolitionist Wendell Phillips, was a naturalist and hunter who had explored the Canadian Rockies, climbed Mount Sinai, shot big game in Africa, sailed up the Nile to Khartoum, and ventured to Greenland with Admiral Peary (on the expedition that, upon its return, presented six Eskimos to Franz Boas). Phillips was, in other words, a model member of the Boone and Crockett Club. And, like many of that breed, he had made the transition from big-game hunting to conservationism. He also served with Grant on the national committee of the Immigration Restriction League.

As Phillips mulled over the proposal to establish a memorial grove for his brother-in-law, Grant encouraged him by pointing out that his action could set a precedent for other such memorials, which could be a financial windfall for the Save-the-Redwoods League. After some discussion, Phillips authorized the league to spend $15,000 to purchase a thirty-five-acre grove. "There certainly could be no more suitable memorial than these trees," Grant reassured Phillips, "the very name of which, *sempervirens*, is redolent of immortality."[38]

The Bolling Memorial Grove, in what is now Humboldt Redwoods State Park, was dedicated on the morning of August 6, 1921. The *San Francisco Chronicle* had promised that "Madison Grant, author and leader of the forces to save the *sempervirens* from the lumberman's ax," would be the principal speaker, and several hundred notables were in attendance. At the ceremonies, a bronze tablet dedicated to Colonel Bolling was unveiled by officers of the American Legion. Several telegrams were read, including messages from Henry Fairfield Osborn, Luther Burbank, and Senator Hiram Johnson.[39]

Grant was suffering greatly from the onset of his arthritis, but he came up on a special Pullman car from San Francisco. His moving speech was titled "Preserve an America Worth Fighting For." He began by commending Phillips for selecting "perhaps the most beautiful and permanent memorial ever chosen for a soldier," in contrast to the "inanimate monuments of bronze, of marble, and of granite [that] strew the land." And he urged the creation of similar memorials in the redwood region.[40]

He next recounted the circumstances surrounding the death of Colonel Bolling and remarked that "the story of how he refused to surrender and fought against overwhelming odds . . . until his pistol was empty, forms one of the stirring chapters of the Great War." Grant then moved to his main theme, which was the connection between patriotism and preservationism. He declared that the American soldiers who fell in the war had fought for "the one thing that is worth fighting for—and that is their country."

> And what is their country? It is the inheritance that God gave us of forests and fields, of rivers and streams, of mountains and plains. They did not give their lives for a field of blackened stumps, nor for rivers . . . turned into sewers of factory waste. They did not give their lives for a mountain-side rent open for minerals and coal. They gave theirs for a country that had trees on the hillsides, that had fish in the streams, that had birds in the air, that had feather and fur in the forest.

The speech then featured an inspiring peroration on the goal of preservationism:

> It is a sacred duty for Americans to guard and to preserve what little is left of this heritage our fathers so cheerfully wasted . . . . These trees are part of our national monuments, our national inheritance, of far more value to ourselves and to those who come after us than any of the works of man . . . .
>
> And it is just as much your duty and my duty to protect and preserve them, as it was the duty of Colonel Bolling and his fellow soldiers to fight for their

country. Colonel Bolling and his comrades dedicated their lives to their country. Let us dedicate ourselves to the task of keeping and preserving in its natural beauty a country which is worth fighting for.[41]

For J. D. Grant, who was there, the speech was "as stirring as a trumpet call." The Bolling Grove dedication was widely publicized, and Grant's speech was printed and distributed by the Save-the-Redwoods League for promotional purposes. Newton B. Drury (the league's executive secretary) assured Grant that "we are all deeply sensible of the value of your efforts in behalf of our cause. Your trip to California has meant much in giving an impetus and new life to the movement. We shall try to capitalize to the full the interest that has been stimulated."[42]

Indeed, the proliferation of the "memorial grove" concept became a financial windfall for the Save-the-Redwoods League, which became one of the strongest conservation organizations in the nation. The secret to convincing a potential benefactor to dedicate a grove was to have that person come to California and take a tour of the redwoods in Newton B. Drury's open car; after such an experience amid the enthralling splendor of the trees, the money was usually forthcoming. On the heels of the Bolling Grove dedication, scores of wealthy patrons donated funds to the league so that it could purchase a grove in the name of a loved one. It became (and is) the league's most successful fundraising technique: nearly one thousand memorial groves have been established so far.[43]

Two of the memorial groves are dedicated to Madison Grant. In one of the most magnificent sections of Del Norte Coast Redwood State Park, approximately 360 miles north of San Francisco, one will find the following plaque:

THE GRANT GROVE
DONATED TO THE
STATE OF CALIFORNIA
AND
DEDICATED TO THE CAUSE OF THE PRESERVATION
OF OUR
HERITAGE IN THESE ANCIENT FORESTS
BY
JOSEPH D. GRANT AND MADISON GRANT
NOVEMBER 1929

The two-hundred-acre grove, the result of a hefty donation by the two "cousins," is an awe-inspiring place: located on a bluff above a remote section of coastline, the trunks of the old-growth trees are suffused with the ethereal mist that creeps in from the ocean, while the green crowns reach into the golden sunlight hundreds of feet above. The forest floor is covered with rhododendrons, and the only sound is that of the creeks winding their way down to the Pacific. "For hundreds of years," wrote W. Redmond Cross, "these magnificent specimens . . . will stand as living monuments to the man who made their continued existence possible."[44]

After Grant died, his brother DeForest (who had taken Grant's place on the council of the Save-the-Redwoods League) resolved to create an even more dazzling redwood memorial to his sibling. He chose—at a cost of almost one million dollars—a 1,605-acre forest of spectacular redwoods now known as the Madison Grant Forest and Elk Refuge. Not only was the grove, in the opinion of the superintendent of parks, "the finest standing timber of redwoods in California," but it also contained meadows that were habitat for the last herd—three hundred strong—of Roosevelt elk in the state. These magnificently antlered animals (named for Theodore Roosevelt) are the largest land mammal in California. Teddy himself considered them "the noblest game animals in America." Conservationists had been trying for decades to create a refuge for the Roosevelt elk. They once roamed in great numbers on the West Coast, but the coming of the white man had forced the diminishing elk to retreat into this little corner of Humboldt County where, like the redwood trees, they were making their last stand on the western edge of the continent. The Madison Grant Forest and Elk Refuge thus embodied Grant's work as a preservationist by combining a redwood forest and a wildlife refuge, and it also provided an allusion to his mentor in the Boone and Crockett Club, Theodore Roosevelt. The refuge is visited by thousands of tourists every year, as it sits astride the Newton B. Drury Scenic Parkway (the road leading into the main campground of the Redwood National Park). The protected elk are now abundant in the area and can often be seen grazing in little groups among the dunes of Gold Bluffs Beach, where the redwoods meet the Pacific.[45]

Back in the 1920s, Madison Grant worked hard to persuade his wealthy friends to establish their own memorial groves. Thanks to his efforts, for instance, Edward S. Harkness (who had already given $100,000 to the Zoological Society) contributed $550,000 to the Save-the-Redwoods League. And in the spring of 1923 Grant began wooing John D. Rockefeller, Jr. He sent the philanthropist reprints of his articles on the redwoods and boldly suggested that something along the lines of a million dollars would be a nice "round" donation. (No one ever accused Madison Grant of thinking small.) Just before Thanksgiving Day in 1924, Rockefeller came to Grant's house in Manhattan and handed the famous racist a check for one million dollars to purchase the redwoods of Bull Creek Flat.[46]

And so, grove by grove, the Save-the-Redwoods League continued buying exquisite stands of trees from the logging companies to preserve them in state parks. By this time, Grant and John C. Merriam were able to curtail somewhat their detailed involvement in the organization. For Grant, the development was inevitable. He had never intended to be occupied full-time with the league; his hope (soon fulfilled) was that after a few years "the movement would gain sufficient impulse in California to run on its own momentum." Furthermore, his health no longer permitted him to travel very often to the West (though league officials continued for many years to rely on his advice and suggestions, and never stopped entreating him to visit: "We need a leader," an official told Grant during one of the many crises that required his services, "and you are the man.")[47]

As for Merriam, he departed Berkeley for Washington, D.C., having been named chairman of the National Research Council in 1919 and then president of the Carnegie Institution of Washington in 1920. Nonetheless, when Franklin K. Lane died in 1921, Merriam assumed the presidency of the Save-the-Redwoods League, which he held until 1944. He kept in touch from the East Coast via telephone and letters, and returned to California each summer to deliver his famous—and increasingly mystical—annual messages to the league, which provided philosophical guidance for the movement.

But still, the Save-the-Redwoods League was on one side of the country, and Merriam and Grant were usually on the other. Hence, the two founders gradually ascended to the role of senior strategic advisers, and the responsibility for running the league on a day-to-day basis devolved on the Californians, especially chairman of the board J. D. Grant and executive secretary Newton B. Drury.

Wooing the Peepul

As the Save-the-Redwoods League presented more and more groves to the state of California to be turned into parks in the 1920s, it became concerned about Sacramento's ability to properly manage and protect the trees. The problem was that tourists were venturing up north to gape at the glorious trees the league had saved. As early as 1922, John C. Phillips complained to Grant that "conditions in the Redwoods" were "filthy almost beyond belief," and Newton B. Drury confided that he was "much distressed to find that campers have already begun to desecrate the groves that we have saved. They show an appalling lack of judgment and ordinary decency."[48]

Furthermore, to meet the needs of the tourists, the state was constructing parking lots and entrepreneurs were building hotels that, as Drury complained to Grant, "detract from the wildness that used to be one of the chief charms of the Redwood region." And then there was the crew of misguided workmen who, when sent by the state to "clean up" one redwood grove, enthusiastically scrubbed the moss from the trunks, mowed the ferns, raked the forest floor, and carted out felled trees by the hundreds (leading the *Humboldt Standard* to applaud the effort to make the redwoods "resemble a well-kept park").[49]

These were ominous developments. The trees had to be preserved, but no one wanted Bull Creek to become Niagara Falls. The situation wounded Grant, who as an aristocratic preservationist decried the fact that "with our parks . . . the prevailing motive seems to be to build roads and hotels for the benefit of the 'peepul,' rather than to preserve nature." (Concern about environmental degradation is one of the reasons why many league officers favored immigration restriction. It is a position not unknown among today's conservationists.)[50]

In a sense, the Save-the-Redwoods League had been too successful: the tourists that its propaganda had attracted to the region now threatened to harm the trees that the league was trying to preserve. The redwood roots were so shallow that even the most respectful of hikers committed much unintentional damage by compacting the soil above them. And not-so-respectful visitors were setting fires, scattering refuse, trampling on ferns, polluting streams, and—some things never change—carving their names into trees. ("When one notes that a wastrel has cut his initials . . . in a Redwood grove," stammered an exasperated J. D. Grant, "one longs for that vanished day when such a culprit by law could be drawn and quartered.")[51]

John C. Merriam spelled out the irony that "regions set aside for protection have been subject to much heavier use and wear than would be the case had the areas not been set aside." Merriam, Grant, Drury, and J. D. Grant engaged in a great deal of soul-searching over the conundrum that in order to alert the public to the necessity of preserving the wilderness, the public had to be allowed access to the wilderness, which in turn led to the destruction of the wilderness. The paradox faced by the league was somewhat similar to that of zoos, which feel they must kidnap animals from the wild in order to educate the public about the need to preserve animals in the wild.[52]

The new redwood parks were in danger of being loved to death. And thus the league was torn between those who held that amenities such as camping facilities were harmful because they attracted the injurious public, and those who argued that such facilities were beneficial because they allowed the public to witness—and then join the movement to protect—the beauties of nature. (Campsites also concentrated the masses in designated spots, thereby preventing them from disturbing more fragile areas.) Similarly, some officials held that trails were deleterious because they attracted "the milling throngs" to previously inaccessible areas, while others felt that trails were invaluable because they prevented hikers from wandering off on their own throughout the forest.[53]

It was clear to Madison Grant that only the state had the power and the funds to provide public access to the trees (by constructing roads, building clearly marked trails, and establishing defined camping sites and rest facilities), while at the same time preserving the character of the forest as a wilderness (by fencing off endangered groves, posting uniform regulations, and hiring rangers with the authority to enforce the above). Unfortunately, California had no administrative system to implement such a program; instead, a number of overlapping state agencies (including the State Forestry Board, the State Fish and

Game Commission, the State Highway Commission, and the State Board of Control) shared a weak and ill-defined jurisdiction over the redwoods. And to make matters worse, most of these bodies were in the lap of the lumber industry.

Accordingly, the league formed a committee in 1923, headed by councilor Duncan McDuffie, charged with devising a plan whereby the incipient redwood parks could be properly regulated. After studying Madison Grant's *State Park Plan for New York* (the seminal blueprint for New York's comprehensive park system that had been issued one year earlier), McDuffie traveled to New York to consult with the master. While taking a drive along the soon to be opened Bronx River Parkway, McDuffie and Grant agreed that their desire, as progressives, for rational and centralized regulation of the redwoods was being thwarted by the welter of contending agencies claiming jurisdiction over the trees. They decided that the most effective way to protect the groves would be to consolidate all the various state park lands, including those in the redwood region, under one bureau. They concluded that a "California State Park Commission" should be created, with the power to protect and administer a "State Park System." Sounding like a character in a Samuel Hays dream, McDuffie posited that "efficiency, economy and sound administration" demanded such a statewide park plan.[54]

Returning to California, McDuffie enlisted the Redwoods League and presented to Sacramento a comprehensive legislative package that not only would create a state park system run by a State Park Commission but also would put before the voters a $6 million bond issue that would finance the commission and allow it to purchase new park lands. Under the scheme, private contributions toward the purchase of parks (e.g., donations to the Save-the-Redwoods League) would be matched dollar for dollar from the bond revenues.

In unveiling its program, the league claimed that a state park system was not a matter of "sentiment;" rather, preservation of the state's scenic and recreational attractions was vital to the development of California's tourism industry. With that, even the Chamber of Commerce joined the preservationists in endorsing the proposal, and in 1925 the league's plan for a state park system was approved by the legislature. Unfortunately, the bill was pocket vetoed by the somewhat thrifty and very conservative governor, Friend Richardson, who was swayed by the opposition of the powerful Pacific Lumber Company. (The current chief executive officer of the Pacific Lumber Company, Charles Hurwitz, is fond of saying: "There's a story about the golden rule. Those who have the gold, rule.")[55]

The league was demoralized by the veto. "The attitude of the Governor," lamented the understated league councilor Henry S. Graves, "was not encouraging." His colleague William Kent was a bit more outspoken: he called the governor a "blind porcupine" who was "entirely void of vision" and, well, "plain stupid."[56]

However, prospects brightened the next year when the governor was defeated for reelection by the progressive Clement Calhoun ("C. C.") Young. The new governor had been Newton B. Drury's English teacher at Lowell High School in San Francisco, and he also happened to be the close friend, classmate at U.C. Berkeley, and longtime business partner of league councilor Duncan McDuffie (author of the State Park Commission plan).

Reenergized, the league again sent its proposal to Sacramento, along with a copy of Grant's *State Park Plan for New York* to each legislator. In 1927 the measures creating the State Park Commission and authorizing the bond issue again passed the legislature, and this time they were signed by Governor Young. "It must be a great gratification," a thrilled J. D. Grant wrote Madison Grant, "to see the fruition of your efforts as a pioneer in this park movement."[57]

The next step was to secure voter approval of the $6 million bond in the 1928 election. John C. Merriam was very apprehensive about the outcome of the vote; he doubted that the electorate would be willing to spend money on "scenery," and he urged the league to exert "every possible effort" to win passage. The league's officers promptly met with the board of every major newspaper in the state and obtained the support of almost all of them, including William Randolph Hearst's *San Francisco Chronicle* and Harry Chandler's *Los Angeles Times* (Chandler was a longtime friend of league councilor Stephen T. Mather and a strong supporter of eugenics).[58]

The bond proposal also garnered positive coverage in the *Saturday Evening Post* (thanks to league councilor George Horace Lorimer, whose magazine had 250,000 California subscribers) and in the *New York Times* (thanks to associate editor Nicholas Roosevelt, yet another councilor of the league). As an ardent conservationist, and the son of J. West Roosevelt (Theodore's cousin), Nicholas Roosevelt had known Madison Grant for years. Reminiscing four decades later, Roosevelt recalled that "Madison Grant was deeply sensitive to natural beauty, a great outdoors man with strong prejudices, some good, some bad. When I met him . . . he was old, crotchety, and twisted with arthritis, but still fighting valiantly to save Redwoods. Because of his friendship for Theodore Roosevelt and for my father—the three were fellow members of the Boone and Crockett Club—he overlooked the fact that he and I differed on most things other than the saving of Redwoods."[59] (Roosevelt's allusion to "differences" was a tad disingenuous; by 1970 he had forgotten—or at least wanted others to forget—that his editorials in the *Times* in the 1920s had parroted the restrictionist line of Grant during the debate over immigration.)

For the 1928 bond referendum, the league organized what was at the time, according to Susan Schrepfer, "the largest publicity campaign in California's history." To allay suspicion that it wanted all the bond moneys for itself, the league set up a front organization (the State Parks Council, with McDuffie as chair and Drury as secretary) to head the pro-bond campaign. With financing from the league's leaders, regional subcommittees were organized in every part of the state to drum up public support for the bond issue. The league distributed millions of leaflets, voter guides, and window displays. It ran articles in magazines, sent letters to newspapers, and supplied the media with editorial cartoons. It organized caravans through the redwoods, sponsored radio shows, and even produced a movie short starring Mary Pickford.[60]

Back in New York, Madison Grant could only watch with bemusement and some incredulity as the organization founded by three aristocrats showed that it understood quite well how to prosper in the age of popular government and mass media.

Conservation groups and fraternal organizations (e.g., the Elks, the American Legion, the Native Sons of the Golden West, and the always popular Concatenated Order of Hoo-Hoos) joined with the league to campaign for the bonds. The league also garnered endorsements from the Automobile Club and numerous women's groups (including the Garden Club of America, the California Federation of Women's Clubs, the Native Daughters of the Golden West, the Daughters of the American Revolution, and the League of Women Voters)—proving the perspicacity of Grant's 1919 prediction: "The salvation of these great trees probably will depend on two factors just entering into active political life: one the automobilists and the other the women voters."[61]

Despite opposition to the 1928 Bond Act from lumbermen and the corporation-funded California Taxpayers' Association (which decried—as it does to this day—the "already excessive tax burden" that was "retarding development" in the state), the voters passed the $6 million bond issue by an impressive three-to-one margin.[62]

But now, having played the populist game with innovation and panache, the Save-the-Redwoods League reverted to its elitist roots and employed Grantian techniques to ensure that it would control the newly created State Park Commission—and the expenditure by that commission of the $6 million bond money. As set up by Madison Grant, the State Park Commission was a nonpartisan body, consisting of five members chosen by the governor. The commissioners were also nonsalaried—a provision inserted for reasons of economy that happened to reinforce the elitist nature of the body. Since Governor C. C. Young's closest friend was league councilor Duncan McDuffie, and since McDuffie's guru was Madison Grant, it was not a coincidence that all of the governor's appointees were sympathetic—and four of the five actually had direct ties—to Madison Grant and the Save-the-Redwoods League.

The governor's first two appointees to the State Park Commission were William E. Colby and Ray Lyman Wilbur, both councilors of the Save-the-Redwoods League. (Shortly thereafter, when Wilbur was tabbed to be secretary of the interior by fellow Stanford alum Herbert Hoover, his slot on the commission was taken by yet *another* Save-the-Redwoods League councilor, Arthur E.

Connick.) Both Colby and Wilbur were residents of the Bay Area, and Grant—who understood the political situation in California fairly well—knew the governor would now have to appoint two southern Californians to the commission. His fear was that the southern appointees would hinder his plans by insisting that for each dollar of the bond issue spent on the redwoods, one dollar should be spent in the south. His goal, therefore, was to find southerners who would not, as he put it, "be jealous of appropriations to other parts of the state." After thinking it over, Grant decided that his chum, the African adventurer Major Frederick Russell Burnham, would do nicely. Although they had known each other for only a few years, Grant and Burnham had become the closest of friends, and they often vacationed together at DeForest Grant's Bar Harbor estate. The major was one of those dashing Nordic archetypes with which Madison Grant's personal life was filled. The stunningly handsome Burnham was an internationally known explorer, soldier of fortune, and big-game hunter who had fought the Apaches, prospected for gold in the Klondike, served as a hired gun in the Arizona range wars, sought King Solomon's Mines, helped Cecil Rhodes quash Matabele uprisings, served as chief of scouts for the British during the Boer War (when he received the nickname "Hawkeye"), dined with Queen Victoria, discovered Mayan ruins, and developed the Dominguez Hills oil fields in California. He was Wyatt Earp, Douglas Fairbanks, and Sir Richard Burton all rolled into one. The British awarded Burnham the Distinguished Service Order, and his colleague-in-arms in the Boer War, Sir Robert Baden-Powell, often pointed to Burnham as a model for emulation by his Boy Scouts. Grant was hardly exaggerating when he stated that Burnham was "one of the most remarkable individuals I have met in recent years."[63]

In 1928 Grant took the train to California and persuaded Governor Young to appoint his "great friend" Major Burnham to the State Park Commission. Burnham actually declined the appointment at first, as Lord Baden-Powell was headed at any moment for that great jamboree in the sky, and Burnham was the favorite to succeed him as head of the Boy Scouts. But Burnham finally yielded to Grant's blandishments and accepted the position on the State Park Commission, leading a delighted Grant to assure the Save-the-Redwoods League that "we could not possibly have a finer man who would cooperate in every way with us." And, indeed, Burnham stated that as far as he was concerned the primary goal of the commission should be "to consolidate the long years of effort . . . the League had made to save the beautiful Redwood trees."[64]

For the other southerner, the governor was prevailed upon to pick Henry O'Melveny, the best-known attorney in Los Angeles and a friend of Major Burnham. O'Melveny privately confided that saving the redwoods was "the real motive for my acceptance."[65] Having chosen two commissioners from the Bay Area and two from southern California, the governor could consider other criteria: his final appointment to the State Park Commission was Wilbur F. Chandler, a

Major Frederick Russell Burnham, one of the many Nordic demigods befriended by Madison Grant (although Burnham had the rare distinction of being married).

former state senator from the central San Joaquin Valley who was involved in park and recreation matters.

To make the coup complete, the five commissioners then went about hiring an operations officer (known as the chief of the Division of Parks). Acting at Grant's behest, Major Burnham nominated Charles B. Wing for the position, and thus *another* councilor of the Save-the-Redwoods League was installed in the new regime. Once the personnel of the State Park Commission was set, Newton B. Drury could state, with total satisfaction and accuracy, that all the members had "a real sympathy for the [redwoods] movement."[66] Madison Grant may not have believed in democracy, but—as with his effort to enact the immigration restriction legislation of the 1920s—he certainly was an expert at manipulating the political process to achieve his ends.

The first task of the State Park Commission was to devise a long-range master plan to guide the spending of the $6 million bond issue. It therefore needed to conduct a statewide survey to evaluate and rank potential park areas, to ensure that the bond revenues went toward protecting the most scenic, historic, and/or scientifically important lands. Whoever was chosen to conduct the survey would have enormous power, as he—and the $6 million at his disposal—would determine which lands in California would be protected as state parks and which would fall prey to the developers.

Again, Grant and the Save-the-Redwoods League brought their influence to bear, and the Park Commission engaged noted landscape architect Frederick Law Olmsted, Jr., to direct the survey. Olmsted was an enthusiastic supporter of the league and felt that the redwoods should receive top priority in the state park system. This was by no means a self-evident notion, given the multitude of far more accessible lands in California that could have qualified for state park status. But Olmsted was convinced that "to preserve adequate and worthy examples of virgin redwood forest is by far the most important and urgent single duty of the State Park Commission."[67]

Olmsted conducted his two-year survey of potential state park areas with incredible energy and thoroughness. He investigated hundreds of sites all over the state. At each stage of his work, he was briefed on redwood matters by league councilors Drury and McDuffie. After winnowing out those areas he judged to be too small, too isolated, too expensive, or too local in their appeal, Olmsted submitted his report (with a cover photograph of a redwood grove), identifying 125 sites—many in the redwood region—that he felt deserved state park status.

Unfortunately, not all of the 125 sites could be acquired by the Park Commission, even with $6 million. Therefore, to prioritize its plans and to supervise the actual purchase of park sites, the commission hired an acquisition officer for the state park system. The man they tabbed to fill the position was none other than Newton B. Drury, executive secretary of the Save-the-Redwoods League. For the next eleven years, Drury worked part-time for the Park Commission and part-time for the Redwoods League—though one would be hard-pressed to differentiate between the two postings, since, as one historian notes with some understatement: "the two assignments were inter-related in crucial ways."[68]

To review, the Save-the-Redwoods League had not only conceived and brought into being the state park system, but it also directly controlled the State Park Commission, the chief of the Division of Parks, the State Park Survey, and the park acquisition officer. So much for popular control of government.

Needless to say, Acquisition Officer Newton B. Drury immediately began directing a large percentage of the bond money toward the purchase of redwood groves. And with each passing decade, more groves were turned into redwood state parks—thirty-nine at last count, scattered from the Oregon border to Big Sur—where the trees are protected in perpetuity. In 1934, Secretary of the Interior Harold Ickes offered Newton B. Drury the directorship of the National Park Service, but Drury declined the offer, realizing that he could do more for conservation in California as executive secretary of the Save-the-Redwoods League and acquisition officer of the state park system than he could in Washington as director of the National Park Service. (In 1940, however, Ickes would repeat his proposal, and this time Drury would accept. Thus, three of the first four directors of the National Park Service—Stephen T. Mather, Horace Albright, and Newton B. Drury—were all graduates of U.C. Berkeley and active officers of the Save-the-Redwoods League.)

At some point, it occurred to the Redwoods League that its rather conspicuous triumphs could engender resentment among other elements of the California conservation community. League officials thus began to counsel "a moderate" approach toward efforts to influence the State Park Commission.[69] And thus it came to pass that the commission preserved not just redwood forests but scores of the state's unique scenic attractions, historic sites, and recreational areas as well. The California state park system today includes 278 parks encompassing 1.5 million acres, from the alpine meadows of the High Sierra to the palm-lined beaches of the Pacific Ocean. (In 1928, a total of thirty-five miles of California's beaches were publicly owned; today some 295 miles—almost one-third of the coastline—are under state park system management). With 1,455 historic buildings, 3,800 miles of trails, and nearly 16,000 campsites, it is the most comprehensively planned state park system in the country, and probably the most diverse system of its kind in the world. And the California state parks are today visited by eighty million people a year.

As for the redwood trees, they are now protected forever in the thirty-nine redwood state parks and Redwood National Park (created in 1968 over the opposition of Governor Ronald Reagan, who challenged the need for another park by wondering: "How many trees can you look at?").[70] UNESCO has since designated Redwood National Park a World Heritage site (joining such other wonders as the Grand Canyon, the Great Barrier Reef, the Taj Mahal, the Great Wall of China, the Great Pyramid, and Glacier National Park as a universally recognized site deemed especially worthy of saving for future generations).

It was Madison Grant's organizational ability, political shrewdness, and commitment to preservationism that made it all possible. William T. Hornaday, who praised others in his dotage about as often as he ate Wiener schnitzel, positively gushed when he considered Grant's feat: "We hand to Mr. Grant another wreath of laurel for his years of great labor, and at last his splendid success . . . in the brilliant and spectacular rescue, in defiance of a sad handicap of ill health, of the finest remainders of giant Redwood trees of California from the axe and saw of the despoilers. Whenever you drive through those wonderful Redwood forests, just stop and think that the missionary work that saved them was performed by Madison Grant."[71]

The Save-the-Redwoods League has been described as "one of the most phenomenally successful conservation organizations in history." Since its inception, the league (which now has over fifty thousand members) has spent $130 million to acquire and then hand over to the people of California more than a quarter of a million acres of forest land. Surely, concludes one observer, the league's achievements constitute "one of the outstanding triumphs in the history of American conservation."[72]

But perhaps more significant than statistics about acres and dollars and number of parks is that a prediction Madison Grant made back in 1919 is perhaps coming to pass: "Waste of natural resources sooner or later will be checked and a proper appreciation of the value of an undefiled nature will succeed." Indeed, Grant never doubted that "the conservation impulse" would spread through time and space and "acquire such momentum" as to be unstoppable.[73]

During Madison Grant's first visit to California, he predicted that one day the tallest tree on earth would be found somewhere near Bull Creek Flat in Humboldt County.[74] In 1931, Major Frederick Russell Burnham commissioned a survey that did indeed find the tallest tree in the world, a 364-foot redwood just where Grant thought it would be, in what is now Humboldt Redwoods State Park, one of the world's most beautiful and impressive forest areas.

Later that year, the California State Park Commission approved Major Burnham's suggestion that this tree should be dedicated to the three eugenicists who founded the Save-the-Redwoods League—Madison Grant, John C. Merriam, and Henry Fairfield Osborn—"as a living monument symbolizing eternal life and the duration of our gratitude."[75]

On September 13, 1931, the Founders Tree was officially dedicated. J. D. Grant recounted for the assembled crowd the events of the Historic Camping Trip made fourteen years previously by the "three wise men of the East" led by Madison Grant, "that relentless warrior in the cause of conservation." A choir sang the "Hallelujah" chorus, and Major Burnham then delivered the main address, in which he declared: "It is an ancient and racial urge that has brought us together today in the shade of this far western forest, like the druids of old," to give thanks to Grant, Merriam, and Osborn for rescuing the redwoods from the ax.[76] He then unveiled the plaque:

THE FOUNDERS TREE
364 FEET HIGH THE WORLD'S TALLEST KNOWN TREE
DEDICATED TO THE FOUNDERS OF THE
SAVE-THE-REDWOODS LEAGUE
MADISON GRANT - JOHN C. MERRIAM - HENRY FAIRFIELD OSBORN
BY THE CALIFORNIA STATE PARK COMMISSION
SEPTEMBER 13, 1931

The Founders Tree immediately became (and remains) the most visited spot in all the redwood region and the focus of many ceremonies. The plaque at its base is the one that sparked the 1991 letter of protest that opened this chapter. Interestingly, though the early 1990s were the heyday of political correctness in the state of California, the notion of removing the plaque—as demanded by the irate tourist—was rejected by the director of the Department of Parks and Recreation (Donald Murphy), who asked, in a remarkably thoughtful letter:

> Is it "historically bizarre" to commemorate Grant's undeniable efforts on behalf of conservation in light of his undeniable racism? That might be an easier question to answer if Grant had been alive and postulating his theories in recent decades.
>
> Although he died in 1937 at age 72, Grant was a creature of the nineteenth century and as with many of his life contemporaries he held beliefs that most of us, hopefully, find both absurd and abhorrent today.
>
> Grant drew attention to his misguided deductions on race by setting them down on paper, but the sad truth is he probably did not think too differently than many others who've been "honored" for some historical role unrelated to the issue of race. I'm not sure that society can or should conduct a wholesale revision of history because the people of the past did not have a late-twentieth century vision of fairness and equality.
>
> As Director of the California Department of Parks and Recreation, I don't ordinarily wear my ethnicity on my sleeve, so to speak, but in responding to your concerns I feel compelled to note that as an African American I think I have a personal perspective on the pain and suffering, the hurt and disappointment of racism. . . .
>
> I say that only to let you know that I do not take lightly your request for removal of the plaque and that I can quite understand and appreciate the reasoning behind your request. I would hope you will understand too my decision to not have the plaque removed. Harmony among peoples comes from the true principles and attitudes of the present, not from purging the past.[77]

It is undeniably discomfiting that Madison Grant, the man who saved the most magnificent trees on the planet, was also our most significant racist, but the above letter is a marvelous rejoinder to those whose inclinations are to purge the past.

III

Extinction

Farewell my brethren,

Farewell O earth and sky, farewell ye neighboring waters,

My time has ended, my term has come.

Walt Whitman, "Song of the Redwood Tree"

12 Nordic and Anti-Nordic

There is war, not peace, in the camps of the learned.

F.J.E. Woodbridge

The lifelong hostility between Madison Grant and Franz Boas was the personification if not the core of the nature-nurture debate in the United States. Grant was the prophet of scientific racism and, in Ellsworth Huntington's phrase, the perennial "cheer leader of the Nordics in America." Boas, on the other hand, devoted a lifetime to counteracting "the vicious, pseudo scientific activity of so-called scientists" who belittled nurture and promoted "this Nordic nonsense."[1]

We should not lose sight of the fact that Grant and Boas both shared a belief in the power of science and reason to benefit humankind and transform the world. (A. L. Kroeber's remark of Boas—"It is indubitable that science was his religion"—applied equally to Madison Grant.)[2] In addition, both men were associated with the American Museum of Natural History, where they occasionally interacted and worked to modernize its anthropological collection. And both also loved the Pacific Northwest and traveled there often (in fact, unbeknownst to them, their paths crossed more than once in the region as the one tracked big game and the other conducted fieldwork).

But Franz Boas (1858–1942) was the antithesis of Madison Grant. Whereas Grant was the scion of an aristocratic American family and displayed all the attitudes and prejudices implied by such a heritage, Boas was the product of an upper-middle-class German household in which, as he put it, "the ideals of the revolution of 1848 were a living force." His progressive Jewish parents raised him with a firm belief in the dignity of the individual and the equipotentiality of all humans. As such, during his four-decade reign at Columbia University as the world's most famous anthropologist, Boas preached with increasing vigor and confidence against racial prejudice, and consciously and actively worked to thwart the dangerous influence of Grant ("that charlatan") and his disciples.[3] Boas rejected Grant's division of mankind into biologically distinct and hierarchical subspecies. He challenged not only the superiority but the very existence of the Nordic race. And he denied that there was any correlation between the physical characteristics of a population and its mental or moral traits. The latter, he asserted, were created by the "culture" in which an individual was raised, not his or her germ plasm. Where Grant proclaimed that man was a mammal like any other and that anthropology ought to be a branch of zoology, Boas took the opposite tack and, in the words of Elazar Barkan, "divorced the biological from the cultural study of humankind." In sum, Boas categorically rejected every tenet of Grant's scientific racism and actively opposed every facet of Grant's eugenic program. Of course, it was clear to Grant that the root of Boas's hostility lay in the fact that he was a Jew, and Grant explained to Maxwell Perkins that Boas "naturally does not take stock in [my version of] anthropology which relegates him and his race to the inferior position that they have occupied throughout recorded history."[4]

Interestingly, the two titans rarely attacked each other directly in public, at least in the early years. Neither felt he could afford to antagonize the other, and besides, each man invariably affected a tone of charming refinement that required he behave in a courtly manner whenever possible. Instead, for decades they engaged—like the United States and the USSR during the Cold War—in a series of proxy wars on the periphery, each of which was intended to showcase their strength and prevent their opponent from increasing his sphere of influence. This chapter explores just a sampling of those incidents, to wit: Madison Grant's attempt to establish the Galton Society as an alternative to the American Anthropological Association; the struggle for control of the National Research Council; and the contest over the *Journal of Physical Anthropology*. If, as we delve into these complicated and long-forgotten controversies, the issues sometimes seem arcane if not downright petty, it will be good to bear in mind that, like the Cold War battles over Quemoy and Matsu, a great deal more was at stake then met the eye. The lives of millions of persons depended on the struggle over the validity of scientific racism.

The American Anthropological Association

Franz Boas began publicly chipping away at the eugenics edifice in a series of books and articles in the 1910s. In his most ambitious work of physical anthropology, "Changes in Bodily Form of Descendants of Immigrants," Boas reported in 1911 that the cephalic index of children born in the United States differed from that of their immigrant parents, thus showing that environment could act to shape even those "immutable" racial traits—in this case, the shape of the skull—that were supposedly determined purely by heredity.

Thesis / Antithesis: Madison Grant vs. Franz Boas. The receding hairlines allow us to verify that Grant and Boas were indeed exemplars, respectively, of the superior longheads and inferior roundheads. Franz Boas: Collection Canadian Museum of Civilization, 79-796.

Madison Grant recognized the threat posed by these findings, and promptly sent off a number of letters to influential editors and politicians assuring them that Boas's "absurd" conclusions were "absolutely at variance with scientific anthropology." In a detailed letter to President William Howard Taft, Grant explained that "the structural differences between the main races of mankind are of immense antiquity," and hence it was hardly credible that the physical characteristics of the "undesirable races" could be affected in one generation by environmental factors. Grant accounted for Boas's "silly" claims by reminding one congressman that "Dr. Boas, himself a Jew, in this matter represents a large body of Jewish immigrants, who resent the suggestion that they do not belong to the white race." And Grant contended that even if Boas's data about the head forms of second-generation immigrants were accurate, it was probably due to immigrant mothers conducting clandestine affairs with native-born Americans. As someone who had been to Poland and seen for himself "the wretched mass of degraded human beings" living amid the "filth and unsanitary conditions" of the Warsaw ghetto, Grant informed the Reverend Percy Stickney Grant that the Jews, "like rats, have formed a race able to survive gutter conditions which quickly destroy higher types." He predicted that any further immigration of these Jews to America would be "a national calamity," and he cautioned the reverend to ignore the "misleading" and "unscientific" work of Professor Boas.[5]

Later that year, Boas renewed the attack with his most famous and widely read book, *The Mind of Primitive Man,* in which he argued, contra the eugenicists, that mental aptitude was not determined by heredity, and that given the proper conditions all races could achieve the level of civilization attained by "our own people" (i.e., "the white race," in which Boas included the Jews). Explaining that there was more variation within each race than between races, Boas concluded that environment accounted for most of the racial distinctions that did exist. As such, *The Mind of Primitive Man* was a founding statement of the notion of cultural determinism, and Boas's student Leslie Spier called the book "a Magna Carta of race equality."[6]

Grant must have been pleased to learn that one of the books targeted for burning by the Nazis in 1933 was *The Mind of Primitive Man.* Back in 1912, Grant's ungenerous reaction, expressed in a letter to Henry Fairfield Osborn, was that Boas's book was "a rather feeble effort. . . . The fact that Boas does not belong to the dominant North European race shows on every page of his book." Still, Grant's bravado masked a very genuine fear that Boas's efforts were having some effect, and he decided that somebody must "publish something that is absolutely orthodox . . . to counteract the evil effects of the Boas propaganda." The famous result, four years later, was *The Passing of the Great Race,* which the *Eugenical News* hoped would finally put a stop to "certain anthropologists, like Boas," who attempted to deny mental differences between the races.[7]

But Boas was not stopped. Just one month after Grant's book appeared in 1916, Boas parried with an article in *Scientific Monthly* charging that "the battle-cry of the eugenists, 'Nature not nurture,' has been raised to the rank of a dogma." Boas admitted that since "stock and plant breeding have shown that it is feasible, by appropriate selection, to improve the breed in almost any direction that we may choose," it was tempting to assume that similar results could be obtained in man by careful mating of appropriately selected individuals. But he warned that most human personality traits, such as alcoholism and criminality, were determined not by heredity but by conditions in the home, and therefore could not be bred out of the race. Eugenics, concluded Boas, "is not a panacea that will cure human ills, it is rather a dangerous sword."[8]

Two months later, Boas followed up with his review of *The Passing of the Great Race* in the *New Republic,* in which he excoriated the book for being "dangerous," "dogmatic," and "naive." Grant shot back that such "bitter opposition" was to be expected from members of "inferior races" who stubbornly refused to accept that their nature was determined by "fixed inherited qualities . . . which cannot be obliterated or greatly modified by a change of environment."[9]

Among academics, Boas was practically alone in those days in his opposition to Madison Grant. The eminent anthropologist William J. Sollas of Oxford, for instance, after reading *The Passing of the Great Race,* passed it on to his colleagues and then excitedly wrote to Grant to say: "I found great pleasure in following your argument in detail and I envied you the pleasure you must have had

George Bird Grinnell characterized Madison Grant as "A lighthouse of fashion." Franz Boas and his partisans had less flattering things to say about the prophet of scientific racism.

in building up your rich store of facts into a compact body of doctrine. I hope your work will be widely read and that it may have some influence on our Statesmen whose knowledge of history has not been illuminated by the teachings of anthropology."[10]

For years, however, Boas had been diligently training a cadre of professional anthropologists who shared his revulsion for the theories of Grant, so that by the end of the 1910s Boas was surrounded and supported by a growing group of scholars well positioned to use their expertise to join in the assault on eugenics. Some of the more important anthropologists who received their Ph.D. from Boas were A. L. Kroeber (who earned his degree in 1901), Robert Lowie (1908), Edward Sapir (1909), Alexander Goldenweiser (1910), Paul Radin (1911), Leslie Spier (1920), Ruth Benedict (1923), Melville Herskovits (1923), Margaret Mead (1929), and Ashley Montagu (1937). With the exception of Kroeber, Benedict, and Mead, all were Jews, many were immigrants, and several were both. (It was a poorly kept secret that Ashley Montagu, the son of a Polish-born Jewish tailor, had been Moses Israel Ehrenberg before metamorphosing into Montagu Francis Ashley-Montagu.)[11] By the early 1920s, the members of the first generation of Boas's students were devising the intellectual weapons and amassing the ethnographic data they would need to combat the disciples of Grant.

On a theoretical level the debate between the Grantians and the Boasians pitted the defenders of heredity against the proponents of environment. Intellectually, the split was a disagreement between adherents of polygenesis, who were obsessed with the classification of races, and adherents of monogenesis, who were fairly certain that races were socially constructed myths. And professionally, it was a conflict between an older generation of physical anthropologists (often gentleman amateurs with no academic affiliation or perhaps an association with a museum) and the newer generation of cultural anthropologists (usually trained professionals with full-time positions in academia). But for all that, it was difficult not to notice that at heart it was a confrontation between the ethos of native Protestants and the zeitgeist of immigrant Jews.[12]

The older generation of amateurs were aristocratic WASPs with the money and leisure time to ponder fossils as an avocation, whereas the younger generation of professionals were immigrant Jews who saw higher education as a route to social respectability and jobs in academia as a means of economic survival. (Not that university trustees were clamoring for Jewish professors in those days. It is salient to note that as late as 1927 Boas was one of only two Jews on the faculty at Columbia, and the same was true of Robert Lowie at Berkeley.)[13]

The gap between the two sides was all but insurmountable. When the Grantians looked at the cultural anthropologists, they saw a group of bearded, Jewish, socialist aliens who lacked any appreciation of the importance of evolution and the laws of biology. At a time when all patriotic researchers should have been examining the morphology of the strange immigrants flooding the country, the cultural anthropologists were wasting their time studying American Indians: American Indians, for heaven's sake, who were practically an extinct people and who were no longer a threat to anybody. The nation's germ plasm was under siege, and the Boasians were off in the New Mexico desert tracing exogamous marriage patterns. Henry Fairfield Osborn expressed the common view when he confided disdainfully to a colleague: "Between ourselves, much anthropology is merely opinion, or the gossip of natives. It is many years from being a science."[14]

The culturalists were well aware that their work was viewed as trivial and unscientific. And their response—with Boas leading the way—was to professionalize their discipline. They understood that by transforming anthropology from an amateur hobby into a professional vocation, they would garner not only respect but also academic positions (and funding) that would be distributed on the basis of merit rather than through the anti-Semitic old-boy network. (One of the main reasons why Boas had emigrated to the United States was because he recognized that his chances for a professorship in Germany would be thwarted by anti-Semitism.)

The Boasians therefore worked to reconstitute the American Anthropological Association, heretofore composed to a large extent of wealthy, untrained amateurs, into an organization of professionally qualified scholars. Their efforts paid off when Boas was elected president of the AAA in 1907. His former students soon began attaining seats on its governing council, and by the 1910s the American Anthropological Association had evolved into a respected society of academic anthropologists, with the Boasians in the majority. They then moved to take control of the association's journal (*American Anthropologist*), and by 1915 Boas's ally Pliny Goddard had been installed as editor and Boasian loyalist Robert Lowie had been placed in charge of the journal's reviews. Under Goddard and Lowie, biological determinism was banished from the pages of *American Anthropologist,* and the culture idea was well on its way to becoming the predominant thesis in the profession. A bewildered Grant could only observe that these developments confirmed his "belief that you must have at the head of any anthropological work a member of the North European race, who has no bias in favor of helots or mongrels."[15]

It is easy to forget that there was no Platonic dichotomy of nature versus nurture at the beginning of the twentieth century. There was no intrinsic reason why scientists could not have agreed that human personality is the product of both heredity *and* environment. And, to be sure, most Boasians were quite willing to recognize that human nature is the result of an interaction between both biology and culture. The Grantians, however, were far less catholic (for fear of giving ammunition to the enemy), and A. L. Kroeber expressed a universally accepted assumption when he declared that between the two sides there stood an "eternal chasm." The explanation for the seeming inevitability of the hostility between the two camps must in large part lie in the different social backgrounds of Grant and Boas. For them, the nature-nurture debate was much more than an intellectual exercise; it was a matter of personal survival. As an aristocrat witnessing the demographic passing of his race, Grant clung to the life raft of nature, which confirmed that he and his kind were biologically superior, no matter what the birth and immigration statistics might indicate. And as a Jew who recognized that one of the underlying aims of eugenics was the destruction of his kind, and whose face bore the scars of duels fought with anti-Semites in his university days in Germany, Boas increasingly felt that he could not afford to compromise on the issue of nurture, which taught that the condition of racial minorities in the United States had been environmentally determined, and that there was no biological reason why they had to settle for the status of second-class citizens. The result, writes Derek Freeman, was "an unrelenting struggle" between "two fervently held half-truths"—the one "overestimating biology and the other overvaluing culture."[16]

The assault on the Grantians, the professionalization of anthropology, the fight against anti-Semitism, and the espousal of "culture" over "germ plasm" were all intertwined. It is interesting, in this context, that (as noted in chapter 4) the professionals who began entering the conservation movement in the 1910s were Madison Grant's intellectual heirs. They respected him personally, honored him publicly, and were content to continue following his lead well into the 1930s. They were also exclusively WASPs. In contrast, the professionals taking over the field of anthropology in the 1910s were Grant's intellectual *opponents.* They rejected his theories, denigrated his methods, and publicly castigated his work. They were also overwhelmingly Jews.

The Galton Society

The situation within American anthropology was no longer tenable. The Boasians were in the saddle, and something had to be done. And as far as Madison Grant was concerned, the prescribed course of action, as always, was to form an organization.

In the spring of 1918, Grant approached Charles Benedict Davenport of the Eugenics Record Office and Henry Fairfield Osborn of the American Museum of Natural History (with whom Grant had already created the Bronx Zoo, the American Bison Society, the Half-Moon Club, and the Save-the-Redwoods League) and proposed that they form a new, racially oriented anthropological organization to rival the culture-ridden American Anthropological Association. To distinguish his new organization from the university-based Boasians, Grant resolved that it would be headquartered in Osborn's museum, and he decided to name it the Galton Society in honor of the founder of eugenics.

Grant explained that the Galton Society, like all of his organizations, would be aristocratic in structure and governed by a "self-elected and self-perpetuating" oligarchy of nine charter fellows. In these nine would be vested the power to elect officers and approve the regular members, who would be limited in number to sixteen (later doubled to thirty-two). Grant was quite insistent that he wanted to create an organization "without having a membership which has to be consulted." (A few years later, a semiriot broke out when one bomb thrower made a motion that the regular members be given a say in formulating the agendas of the meetings.) And Grant was adamant that membership in the Galton Society—in contrast to the American Anthropological Association—would be "confined to native Americans, who are anthropologically, socially and politically sound." When it came to the Galton Society, "no Bolsheviki need apply."[17]

After a series of meetings involving Grant, Davenport, and Osborn, the Galton Society for the Study of the Origin and Evolution of Man was formally organized in Madison Grant's office in the Trinity Building on April 2, 1918. The creation of the Galton Society was one of the many reasons why Harry H. Laughlin later stated that "Charles B. Davenport, Madison Grant and Henry Fairfield Osborn constitute the triumvirate of great leaders who built substantially on the biological foundation laid by Galton and Darwin. These three Americans built the main structure of eugenics as a biological science."[18] (And when we add Laughlin himself to the above group, we have the Big Four of scientific racism.)

Once more, Grant paid inordinate attention to the matter of securing properly engraved stationery. With that paramount task accomplished, the three founders selected the six men who would join them as charter fellows: biologist Edwin G. Conklin of Princeton, paleontologist William K. Gregory of the American Museum of Natural History, anatomist George S. Huntington of the College of Physicians and Surgeons, zoologist J. Howard McGregor of Columbia, paleontologist John C. Merriam of the University of California (cofounder with Grant and Osborn of the Save-the-Redwoods League), and psychologist Edward L. Thorndike of Columbia. None of them, we might note, were professional anthropologists. Most of them, we could also observe, were conservationists. Indeed, Grant organized the Galton Society at the exact same time that he was organizing the Save-the-Redwoods League, and in the early years, when Merriam was still at Berkeley, he and Grant synchronized the meetings of the Galton Society with those of the Save-the-Redwoods League so that Merriam would have to make only one trip to New York. In fact, the constitution of the Save-the-Redwoods League was modeled on that of the Galton Society, which in turn was modeled on that of the New York Zoological Society. And why not? All three associations served a common end: one would save the largest mammals of North America, the other would save the tallest trees, and the other the most advanced race.

William K. Gregory, secretary of the Galton Society, wrote articles for a number of scientific journals announcing the formation of the new organization, whose purpose was to promote the study of "racial anthropology."[19] The membership roster soon included the elite of the eugenic establishment, including paleontologist William Diller Matthew, archeologist Nels C. Nelson, geneticist L. C. Dunn, physician Clarence G. Campbell, zoologist Gerrit Smith Miller, journalist Lothrop Stoddard, psychologists Carl C. Brigham and Robert M. Yerkes, anatomists J. Howard McGregor and Charles R. Stockard, biologists Harry H. Laughlin, C. C. Little, Raymond Pearl, Paul Popenoe, and Frederick Adams Woods, and, finally, some actual anthropologists, including E. A. Hooton, George Grant MacCurdy, Herbert J. Spinden, Morris Steggerda, T. Wingate Todd, and Clark Wissler. Madison Grant verified that all of them were "anthropologically, socially and politically sound," and the Boasians watched with concern to see what mischief they might do.

Beginning in 1918, and continuing through the mid-1930s, the Galton Society met once a month (except in the summer) at the American Museum of Natural History. Members were treated to a one o'clock luncheon hosted by Grant and Osborn, and then at 2:30 P.M. they repaired to the Osborn Library for the meeting proper. There, under a portrait of Sir Francis Galton that had been commissioned by Grant, they heard a paper from one of their colleagues, followed by a lengthy (and usually rather sophisticated) discussion. Occasionally a distinguished visitor (G. Elliot Smith, Carl Akeley, Sir Arthur Keith, Dean Inge, Sir Richard Paget) was invited to address the proceedings on some aspect of racial anthropology. (Since this was not a *professional* society of grubby ethnologists, but rather a group of *gentlemen,* the members did not pay dues; the expenses of running the Galton Society were provided primarily by Grant, Osborn, Mrs. E. H. Harriman, and Grant's friend Moses Taylor Pyne.)

At an early meeting of the Galton Society, Madison Grant exhorted the troops by reminding them that only *they* practiced the true anthropology, that is, the study of man as a zoological entity. From now on, authentic anthropology would be confined to "the study of man as a *physical* animal and not confused with an ethnologic study of pottery and blankets." (We can only imagine the disdain with which Grant sputtered out those three words: "pottery and blankets.")[20]

And so the Boasian capture of the American Anthropological Association had been countered by Grant's formation of the Galton Society. At a meeting of the society in 1925, psychologist William McDougall of Harvard summed up the situation neatly. On one side of the nature-nurture debate were the sentimental social workers, egalitarian Bolshevists, and intellectual Jews, all of whom were "biased against racial psychology" and allowed the emotional appeal of humanitarianism to stand "in place of truth." On the other side were the "serious" students of race such as Madison Grant and Lothrop Stoddard, who recognized "the reality" of inequality and stood for "the importance of preserving racial distinctions in their purity." It was clear to McDougall and his auditors which faction had right—and science—on its side.[21]

Still, because of its supposed failure to influence public opinion or to sponsor any large-scale research projects, almost all historians agree with Elazar Barkan that the Galton Society "never achieved the status envisioned by its leaders." But this is to misunderstand the mission of the Galton Society. Research in scientific racism was already being conducted by the Eugenics Record Office and the Eugenics Research Association, while lobbying and propaganda were the purview of the American Eugenics Society, the American Defense Society, and the Immigration Restriction League. The purpose of the Galton Society was not to conduct new research but to provide a *Judenfrei* sanctuary where hereditarian-minded researchers could meet in a noncontentious atmosphere to share their findings with others of their kind. And in this respect the society succeeded admirably. It was within the friendly precincts of the Galton Society, for example, that Robert M. Yerkes and Carl C. Brigham first propounded and refined their racial interpretation of the army intelligence tests that appeared in the seminal *Study of American Intelligence.* It was at the society's meetings that Harry H. Laughlin rehearsed the celebrated testimony he would give before the congressional immigration committees. It was at the Galton Society that Charles Benedict Davenport formulated (at the instigation of Madison Grant) the study that resulted in his influential book *Race Crossing in Jamaica.* And it was after extensive consultations with his fellow Galtonians that anthropologist Clark Wissler produced his magnum opus, *Man and Culture* (1923), in which he thanked "the members of the Galton Society for many illuminating suggestions" and then recommended recent works on race that embodied "the serious thought of eminent men"—a list headed by *The Passing of the Great Race.*[22]

William K. Gregory, the longtime secretary of the Galton Society, found the group's meetings "highly edifying and enjoyable," but on more than one occasion he expressed his concern that having someone like Madison Grant deliver a lecture such as "The Spread of the Nordic" to an audience composed of men like Laughlin, Davenport, and Osborn was akin to the pontiff preaching to the college of cardinals. "Which one of us," asked Gregory rhetorically, "so far has taken up any new line of research, or has modified his old methods, as a result of attending these meetings?" But at a time when their younger and more liberal colleagues over at Columbia were increasingly challenging the pillars of the hereditarian faith, it was salutary for the choir to hear a familiar sermon now and then. The monthly meetings of the Galton Society, which Gregory himself characterized as "inspiring," unified the faithful and bolstered them to venture into the secular world to preach the gospel among the heathen.[23]

"It is important to recognize," writes Steven Selden, "that while its membership was limited, [the Galton Society] was not a marginal organization."[24] To be more specific, a third of the Galtonians were members of the National Academy of Sciences, half were members of the American Philosophical Society, and more than half were members of the American Association for the Advancement of Science (and three served as the association's president). These were, in other words, influential men with important connections, and when, as we shall shortly see, Galtonians like John C. Merriam and Robert M. Yerkes were put in charge of the nation's science establishment in the 1920s, it was upon their fellow hereditarians in the Galton Society that they bestowed government positions and federal grants.

Indeed, Grant, Davenport, Osborn, and the other leaders of the Galton Society understood that their aristocratic little club did not have the resources to influence the course of anthropology all by itself. "Science" in the Progressive Era was becoming the province of the university and the federal government; and since academia was increasingly in the hands of the Boasians, it was toward the government—and more specifically the National Research Council—that the Grantians now directed their efforts. And the effect, as Allan Chase points out, was that "the Galton Society helped retard the development of modern anthropology in America for many years."[25]

The National Research Council

After the German U-boat attack on the *Sussex* in April 1916, astronomer George Ellery Hale understood that the entrance of the United States into the Great War was inevitable, and that American scientists would have to play a leading role if the United States was to have any hope of defeating German militarism. Patriotism aside, the imaginative Hale realized that the war promised to be a boon for science, as public support and federal funding for research were sure to be forthcoming. At the instigation of Hale, the National Academy of Sciences received authorization from President Wilson to create the National Research Council (NRC), whose task was to coordinate the country's scientific resources "in the interests of national security . . . and preparedness."[26]

Hale was named chairman of the National Research Council, and with funding from three sources that Madison Grant knew well (the Carnegie Corporation, the Rockefeller Foundation, and Half-Moon Clubber Cleveland H. Dodge), the NRC created twenty-seven committees that organized the various fields of American science—from agriculture to zoology—for the war effort. Many of Grant's friends from the eugenics movement were involved in running these committees, including Isaiah Bowman, E. G. Conklin, H. H. Donaldson, Edward M. East, Vernon Kellogg, John C. Merriam, Stewart Paton, Michael Pupin, Edward L. Thorndike, William H. Welch, and Robert M. Yerkes.

The Anthropology Committee

There were two National Research Council committees that the Grantians were particularly desirous of controlling: the Psychology Committee and the Anthropology Committee. They were successful in both instances. First, Robert

M. Yerkes was appointed chairman of the Psychology Committee, where his efforts eventually resulted in the program that devised and administered the army intelligence tests. Then, in January of 1917, the NRC tabbed William H. Holmes to chair the Anthropology Committee. It was an obvious choice, as Holmes was a well-respected scholar and former president of the American Anthropological Association who, as head curator of the Department of Anthropology at the Smithsonian, was already located in Washington, D.C. But as Holmes publicly admitted, with more than a little understatement, "The further organization of the Committee presented certain difficulties which were [only] gradually overcome."[27]

In fact, a major struggle broke out behind the scenes over who would comprise the rest of the NRC's Anthropology Committee. William H. Holmes, a descendant of the Puritans and a member of the older, George Bird Grinnell generation of American ethnologists that had explored the Wild West in the 1870s, was a long-standing opponent of Franz Boas. Therefore, when Holmes was selected to be chairman of the Anthropology Committee, Boas demanded that he and his allies be granted slots on the committee so they too would have a voice in its affairs.[28]

For months and months, the factions within anthropology fought over who would serve on the Anthropology Committee. Long after Congress had declared war, the armed forces had mobilized, and the twenty-six other committees of the NRC had begun conducting research, the anthropologists were still bickering over the personnel of the Anthropology Committee. The Boasians pointed out that since they comprised the majority of the American Anthropological Association and the board of *American Anthropologist* it would be absurd to exclude them from the committee. The Grantians scoffed that cultural anthropology was not a true discipline but was rather—in contrast to physical anthropology—an unscientific mishmash.[29]

Furthermore, the Grantians argued that it would be foolish if not dangerous for the National Research Council to include the foreign-born Boas. Not only had he openly opposed U.S. intervention in the war; he had also publicly objected when the Columbia trustees urged students to report any comments by their professors that would "tend to encourage a spirit of disloyalty." Several Grantians wrote to George Ellery Hale (chair of the NRC) charging that the loyalty of Boas himself was therefore suspect. Biologist E. G. Conklin, for instance, reported to Hale that Boas "carried independence of thought to the point of sedition. He ought to be in Germany and not in America, and I should like to help send him there." William H. Holmes fully agreed, declaring: "Self-evidently it is . . . inadvisable to include on [the Anthropology] Committee persons who manifest publicly their pronounced pro-German sympathies."[30]

George Ellery Hale was a descendant of the Puritans, a confirmed eugenicist, and an old friend of Madison Grant's from both the Half-Moon Club and the

Save-the-Redwoods League. Hale agreed that allowing the Boasians onto the Anthropology Committee was unthinkable "on account of [their] pro-German tendencies." And it therefore came as no surprise when the National Research Council finally put a stop to the behind-the-scenes maneuvering and effectively barred the Boasians by announcing that the personnel of the Anthropology Committee would be limited to *physical* anthropologists, whom it deemed the only scholars that could be "of direct assistance in national preparedness."[31]

At the time, it seemed only logical that the NRC should have assessed physical anthropology to be more relevant to the war effort than cultural anthropology. But it seemed so only because of the skillful campaign waged by the Grantians to convince the government that their brand of anthropology was rigorously "practical." The cultural anthropologists could have claimed with equal plausibility that an understanding of the different cultures that the United States was allied with/fighting against was at last as crucial to the war effort (and the peace negotiations) as measuring the length of soldiers' mandibles. But the physical anthropologists won the day, and in July of 1917, when the National Research Council finally revealed the members of the Anthropology Committee, the lineup was dominated by the Grantians. In addition to William H. Holmes as chair, Charles Benedict Davenport of the ERO was tabbed as vice chairman, Ales Hrdlicka (curator of the Division of Physical Anthropology at the Smithsonian) was named secretary, and among the other five members was Madison Grant of the New York Zoological Society.

The newly constituted Anthropology Committee formulated an agenda of three projects that its members felt would be "immediately practical" for the war effort and "of ever increasing value as the centuries pass." First, the committee recommended that the minimum height requirement of the military—which had stood at sixty-four inches for years—be lowered to sixty inches. Publicly, the committee claimed its only concern was that the height regulation "excluded from the ranks many men well fitted to serve," but the true motivation was a hope that more New Immigrants (who were shorter than the Nordics) would perish if they were admitted into the armed forces and sent to the front. The War Department concurred, and revised the minimum height downward to five feet.[32]

The Anthropology Committee's second project, initiated by Grant, was a plan to utilize the draft to compile a comprehensive anatomical record of the U.S. population. Grant was well aware that most of the classic studies of European craniometry had been conducted among military recruits. So now, with the Selective Service System mobilizing and concentrating millions of men, he insisted that American anthropologists seize the opportunity to gather data that would "render a vast service to science." Robert M. Yerkes was masterfully exploiting the war situation to legitimate mental testing, and Grant thought he could likewise use the draft to demonstrate the relevance of physical anthropology. His anthropometric survey would enable anthropologists to ascertain the current physical dimensions of the American "type," and then, by repeating the measurements in future years, scientists would be able to prove that the race was deteriorating as a result of immigration and miscegenation. Such data would be of great interest, Grant felt, to the nation's legislators.[33]

Over in France, Georges Vacher de Lapouge was pleading with his government that he be allowed to make cranial measurements of French recruits, but his repeated requests were routinely denied, an action Grant attributed to "Jewish influence" on the government.[34] But Grant was determined that the Hebraic environmentalists would not stymie the effort in the United States. They had already delayed the project by more than half a year with their failed campaign to gain seats on the Anthropology Committee, and there was no more time to spare.

To oversee the survey, the Anthropology Committee formed a subcommittee on anthropometry in the army, chaired by Charles Benedict Davenport. Davenport was eager to get started. Even before the United States entered the war, he had conducted a Galtonesque study of the pedigrees of historically prominent naval commanders, which he hoped would help the navy determine which characteristics to look for when selecting its officers. Published as *Naval Officers: Their Heredity and Development,* Davenport's investigation concluded that one of the most common traits of great naval officers was "thalassophilia" (love of the sea), a recessive trait inherited in Mendelian fashion that was especially common among Nordics. (This is reminiscent of an earlier study by Davenport, in which he investigated 350 prostitutes and found only one who had become "wayward" for economic reasons; all the other girls chose their profession because of "innate eroticism," a.k.a. "sex-hyperaesthesia.") Surprisingly, the navy expressed little interest in Davenport's findings about sailors, obstinately preferring to rely on its less scientific methods of identifying officer candidates. A disappointed Davenport now saw Grant's army anthropometry project as an opportunity to redeem himself.[35]

Davenport and Grant traveled regularly to Washington, D.C., to consult with the other members of the Anthropology Committee, who were—with one exception—quite excited about the project. The one puzzling exception was Ales Hrdlicka, who seemed strangely and increasingly reluctant to support the work as time went on. He began to complain about "the difficulties" inherent in carrying out such a large-scale endeavor and constantly bemoaned the expense involved.[36]

Hrdlicka's reticence was echoed by the army bureaucracy, which in the middle of a world war did not have much interest in allocating personnel and resources to the measurement of skulls. To gain the army's support, the anthropologists decided to downplay the eugenic aspects of their project and instead emphasize the "practical" benefits to the military of anthropometry. Grant explained to the secretary of war that the army could, for example, use the measurements to sort recruits into companies based on race. It was well known that a platoon composed of soldiers with different leg lengths found it difficult to march in ranks; therefore the army could save itself a lot of grief by employing trained anthropologists to separate, say, the short Russian Jews from the Norwegians. Additionally, given that the army was segregated, it was obviously necessary that anthropologists amass a database of racial measurements so that inductees, especially southerners whose veins might carry hidden amounts of "Negro blood," would not be misplaced in white units. (The latter justification of anthropometry brings to mind the situation in Nazi Germany after the Nuremberg Laws, when anthropologists discovered a new means of livelihood identifying persons with "Jewish blood" who were trying to pass as Aryans.)[37]

The military finally gave its assent to the anthropometry project in 1918. Charles Benedict Davenport and Madison Grant quickly went to work putting together anthropometrical instrument kits, after which Davenport was hastily commissioned as a major in the army, where he trained a team of nineteen anthropologists and sent them out to the various army camps to conduct their measurements, with the exhortation that they were working for the good of "future generations in this country."[38]

So as not to unduly alarm any uncooperative radicals who might object to being measured for eugenic purposes, the soldiers were told that the data would be used to construct clothing mannequins so the army could provide better-fitting uniforms. To extend the ruse, the anthropologists recorded the measurements on an examination blank designed by Grant that bore the heading: "Measurement Card for Clothing Patterns." However, in addition to requiring the usual physical measurements, the card also asked for such seemingly irrelevant information as eye color, mother's native language, and father's religion. Still, no one ever questioned the relationship of eye color to "better-fitting uniforms."

After the measurements were compiled, the information was transferred to Hollerith punch cards in the Surgeon General's Office, and Davenport proceeded to analyze the data according to race. (The Hollerith equipment was later adapted by IBM so the SS could monitor prisoners shipped to Nazi concentration camps.) The results were published by Davenport in a volume titled *Army Anthropology* in 1921, just in time for the congressional debates on immigration restriction. Writing in an official publication of the U.S. Army, Davenport could not be explicit about the eugenic implications of the data—just as Robert M. Yerkes's official report on the army intelligence tests had to be circumspect about declaring Nordic superiority (which is why Carl C. Brigham's *Study of American Intelligence,* which highlighted and popularized the racist implications of Yerkes's work, was so important). Nevertheless, as the reader perused the 635 pages, 333 tables, and 62 plates of *Army Anthropology,* it was difficult not to notice that the Hebrew race ranked last in almost every one of the two dozen anthropometrical measurements. From circumference of chest to height of pubic arch, the Nordics were clearly superior—and the Hebrews were clearly inferior—to all other races. And for the few categories where the Hebrews did not rank last, Davenport often arranged the data in an unusual manner (e.g. by standard deviation, or coefficient of variation) so that the Hebrews still appeared at the bottom of the tables. Furthermore, when the Nordics did not rank first in some measurement, a soothing interpretation was provided to the readers. Thus, for example, when the Nordics ranked in the middle of the pack with regard to shoulder width, the anthropologists explained that since the Nordics were the most "extremely developed of human races, then we may say that evolution has been in the direction of diminished shoulder width." Similar observations were sprinkled throughout the book. What their relevance was to the construction of clothing mannequins—the nominal pretext for the study—was left unexplained.[39]

In addition to lowering the military's height requirement and conducting the anthropometry study, the NRC's Anthropology Committee successfully sponsored a third project. This was Ales Hrdlicka's longtime dream of establishing a journal devoted to physical anthropology. A native of Bohemia, Hrdlicka emigrated to the United States with his family one year before Franz Boas, at the age of thirteen. In 1903 he began a four-decade career at the Smithsonian, where as curator of physical anthropology he became one of the world's most prominent anthropologists and made valuable contributions to the study of early humans. One might expect that Hrdlicka, who, as Ashley Montagu put it, "had come to these shores as a poor immigrant boy," would be a Boasian. But Hrdlicka was, first of all, a physical as opposed to a cultural anthropologist, and he was also rather sympathetic toward Madison Grant's biological determinism, as his original training had been in medicine. Moreover, to gain Grant's approval (and financial backing) for his proposed journal of physical anthropology, Hrdlicka moved ever closer to the hereditarians and became an early member of the American Eugenics Society. In addition to placing Grant on the editorial board of his journal, Hrdlicka solicited articles from members of the Eugenics Research Association and circulated a prospectus for the journal promising that it would assist "in such important coming national movements as . . . the regulation of immigration, eugenics progress, and all other endeavors . . . safeguarding and advancing the physical status of man in this country."[40]

In 1918, the Anthropology Committee of the NRC officially approved the creation of the *American Journal of Physical Anthropology,* and with that the three goals of the committee had been attained. (A fourth project—to include Madison Grant in the U.S. delegation to the Paris Peace Conference so that President Wilson could utilize his racial expertise when redrawing the boundaries of Europe—was rejected by the Democratic administration.)

Ales Hrdlicka now devoted all his time and energy to getting his journal off the ground. In fact, Hrdlicka became monomaniacal about the project, and began neglecting the other work of the Anthropology Committee. This angered Grant, who thought that Hrdlicka's *Journal* could be launched at any time, whereas the army anthropometry study could be conducted only while the draft was still in effect. Grant's cordial relationship with Hrdlicka began to sour, as it became clear that the reason Hrdlicka had tried to put roadblocks in the way of the anthropometry project was because he was afraid it would absorb resources that he coveted for his journal. The two men continued to quarrel over whose project should receive priority, and relations were almost unbearably tense. Grant kept explaining to the National Research Council that the cost of keeping the journal afloat during the war would be prohibitive, while Hrdlicka—deathly afraid that his life's dream was going to be stymied by Grant—kept insisting that his project was entirely feasible.[41]

The important point about these otherwise nugatory machinations is that the rift between Grant and Hrdlicka had almost nothing to do with ideology, principle, or scholarship, and everything to do with personal jealousy and ambition. (We can probably discount Grant's characteristic explanation of Hrdlicka's behavior, which was that he was "unconsciously biased" by the fact that he himself was "an East European round head.")[42] By the middle of 1918, Hrdlicka and Grant were no longer on speaking terms, and as a result Hrdlicka, as we shall see, was edging ever closer to signing a nonaggression pact with the Boasians.

The Division of Anthropology and Psychology

Toward the end of the war, President Wilson ordered the National Research Council to make preparations to continue its work as a permanent peacetime institution. To effect the transition, George Ellery Hale stepped down as chairman of the NRC and was soon succeeded by John C. Merriam, president of the Carnegie Institution, charter member of the Galton Society, member of the Advisory Council of the American Eugenics Society, and cofounder of the Save-the-Redwoods League. (Incidentally the secretary of the NRC during the 1920s was eugenicist Vernon Kellogg, another member of both the AES and the Save-the-Redwoods League.) John C. Merriam was a strong and, in his words, "unequivocal" supporter of eugenics. When Merriam moved to Washington, D.C., to assume command of the nation's science effort, Madison Grant extended the hope that "you and I will be able to cooperate in the interest of science in the future as well as we have in the past." And that they did.[43]

After the war, the National Research Council reorganized its twenty-seven wartime committees into thirteen "divisions," and merged the Anthropology Committee and the Psychology Committee into the Division of Anthropology and Psychology. And, as could have been predicted, once again a battle broke out between the Grantians and the Boasians for control of the anthropology section of the new division.

With the encouragement of Madison Grant, John C. Merriam announced at a meeting of the Galton Society in December of 1918 that American anthropologists should stop frittering away their time observing the cultural practices of American Indians and instead become more "relevant" by devoting themselves to studying "the problems at hand"—that is, the racial traits of the New Immigrants. As with Alan Greenspan and stockbrokers in the 1990s, when John C. Merriam spoke, American scientists listened. Boasian Pliny Goddard fearfully (but correctly) interpreted Merriam's speech as a signal from the National Research Council that "our cultural stuff was getting nowhere, that we aren't scientific anyway, that it is time to take things out of our hands and really get down to business." Goddard's suspicions were confirmed in January 1919 when Franz Boas was pointedly not invited to an organizing meeting for the Division of Anthropology and Psychology (prompting Goddard to chastise John C. Merriam: "the war is over," and "Boas should have been included").[44]

The official response of the Boasians came a few weeks later, when Boas, joined now by none other than Ales Hrdlicka, wrote an open letter to the National Research Council demanding that cultural anthropologists be given a role in the Division of Anthropology and Psychology. "The forces which determine the development of human types," wrote Boas and Hrdlicka, "are to a very large extent cultural forces." The Grantians, of course, sent letters of their own urging the National Research Council to hold firm and exclude the Boasians, and long after the psychologists had elected their representatives (through the medium of the American Psychological Association) and gone about their work, the anthropologists were still fighting over who would represent them in the Division of Anthropology and Psychology.[45]

Anthropologist Clark Wissler observed that the psychologists had never been happy that the NRC had bundled them into the same division with the anthropologists. But pointing to "the lack of harmony in the anthropological camp," he sorrowfully admitted that the differences between the psychologists and the anthropologists were not nearly as great as those between the Boasians and the Grantians. Confusion reigned, and at the end of 1919 the National Academy of Sciences could only report, with great tact, that organizing the Division of Anthropology and Psychology had been somewhat delayed owing to "difficulties experienced by one of the constituent associations in electing its representatives."[46]

Persons on both sides were growing exasperated. William H. Holmes cast himself in the role of an early-day Rodney King and asked Boas: "What influence could a 'house divided against itself' have at the Research Council? Why can we not all be friendly and pull together?" The answer was supplied by Boas's student Robert Lowie, a Viennese-born Jew, who observed that anthropology "is divided by a far-reaching difference as to principles and ideals," and declared that "for myself, I feel that no scientific bond unites me with our opponents." Lowie despised the "arrogant" Grant and the other members of "his clique," and once blasted Grant as a "sadistic Junker" who "gloats over the vision of big blond princes leading little brunet Mediterraneans to be remorselessly butchered." Nonetheless, even Lowie admitted to being "tired of this nonsense . . . . I hardly remember a meeting," he sighed, "when something terrible was not predicted as menacing anthropology unless every mother's son of us who stood for the Right attended, armed to the hilt against the enemy." Lowie dejectedly foresaw that there would be no end to the "disgraceful display," which would go on "from year to year" with the same people saying the same things with "the same bitterness." He told a friend: "It's really beginning to be a positive disgrace to be classed as an anthropologist."[47] Even A. L. Kroeber, as loyal a culturalist as ever filled out a kinship chart, had to admit that he was fatigued by this never-ending anthropological fratricide, and he confided to Elsie Clews Parsons that he was beginning to consider some sort of rapprochement with the enemy. "Don't you think it's nearly time for us to establish positive relations with Madison Grant? He's . . . too great an energy to be allowed to run wild and continue to work mischief." To be sure, Kroeber continued, "[Pliny] Goddard sees only the villain in him, Boas only the perverted scientist," but Kroeber (one of the few gentiles in Boas's circle) held out the hope that Grant "could be turned into good channels" if approached in the proper manner.[48]

After a year of wrangling, a workable compromise was reached when it was decided that the anthropologists would emulate the psychologists and have their professional organization, the American Anthropological Association, elect six representatives to the NRC's Division of Anthropology and Psychology. This represented something of a setback for the Grantians, since the AAA was a stronghold of the Boasians. But Grant could think of no legitimate reason to oppose the plan. What could be fairer than to have the discipline's professional society oversee the selection process? It proceeded to do so, and although Clark Wissler of the Galton Society was chosen, the rest of the slate (consisting of Franz Boas, Roland B. Dixon, J. Walter Fewkes, A. L. Kroeber, and Berthold Laufer) was pretty much in the Boasian camp.

And then, incredibly, Franz Boas himself, just six weeks after being selected to the Division of Anthropology and Psychology, provided the Grantians with all the ammunition they needed to have him removed. On December 10, 1919, the *Nation* published a letter from Boas titled "Scientists as Spies." Boas claimed that he had in his possession "incontrovertible proof" that at least four anthropologists (whom he did not name) had spied abroad for the U.S. government during the war under the guise of conducting research. Not only did Boas accuse the men of "prostitut[ing] science," but he implied as well that President

Wilson was a liar and a hypocrite, and that "our democracy was a fiction." Given that the nation was in the middle of the Red Scare, it was probably not the most auspicious moment for a German-Jewish immigrant such as Boas to level such charges. A shocked A. L. Kroeber remarked to Edward Sapir that "the old man should have had sufficient instinct to realize that he could not write his letter without stirring up almost unlimited trouble."[49]

Trouble indeed. A three-paragraph letter about scholars moonlighting as spies might appear to have little relevance to the nature-nurture debate, but it is the character of a cold war that marginal conflicts take on a symbolic role far larger than the events would otherwise warrant. On December 30, 1919, just ten days after the letter was published, Boas's opponents moved to censure him at a meeting of the American Anthropological Association in Cambridge. After a lengthy and ugly debate, the Council of the AAA reprimanded Boas by a vote of twenty to ten, expelled him from the Council, and forced him to resign from the NRC's Division of Anthropology and Psychology. An infuriated Robert Lowie bitterly referred to the incident as a "degrading spectacle" in which "the foremost representative of our science" had been scorned by a group of men "who are anthropologists only by courtesy."[50]

The censure at Cambridge was the result of numerous schisms within American anthropology, but at its heart it reflected years of friction between aristocratic WASPs (mainly physical anthropologists) anxious to stem the rising tide of color and immigrant Jews (mainly cultural anthropologists) seeking to impede the ascent of scientific racism.[51]

Emotions were running high, and the rout of the Boasians was completed a few weeks later when Pliny Goddard (whose chief concern, he admitted to A. L. Kroeber, had always been "to help build up the 'Boas school'") was stripped of the editorship of *American Anthropologist.* Goddard was well aware that the move was "intended to complete what was done last year at Cambridge," and that the hostility was "directed primarily against Prof. Boas and only incidentally against me." Robert Lowie maintained that the situation was "intolerable" and that he was sick and tired of being "trampled on" by the anti-Boasians, who were "perpetually plotting" against Papa Franz and his supporters.[52] Lowie warned Kroeber, in language reflecting the contentious atmosphere of the era of the Red Scare: "We are confronted once more with Armageddon." And he called on his fellow Boasians to display "gang solidarity" by backing Pliny Goddard and telling the Grantians "to go to hell where the sons of bitches belong." Claiming that he welcomed "an open schism," Lowie suggested that if "the opposition" did not reinstate Goddard, the Boasians ought to walk out of the American Anthropological Association and move en masse to some other organization where they could publish an "independent" journal devoted to real "anthropological science." Pliny Goddard agreed with Lowie and promised that if *American Anthropologist* were placed in the hands of an editor lacking "the proper scientific insight" he would leave the AAA, causing its "complete disruption."[53]

But writing from California, Kroeber chastised Lowie for "inflaming" the situation. (In a letter to Elsie Clews Parsons, Kroeber was more blunt. He admitted that "scraps of anthro-political fur" were "raining even here," and blamed Lowie for "making a bad situation worse by his infantile blustering.") Kroeber and Boas both advised Lowie and Goddard to calm down, and declared that no one was going to abandon the American Anthropological Association, into which they had invested so many years of effort.[54]

On the other side, Henry Fairfield Osborn of the Galton Society was giddy that the Boasians were quarreling among themselves and that Boas himself had finally been put in his place. He triumphantly assured Charles D. Walcott that the *Nation* incident was the last they would ever hear of Franz Boas, who "now occupies a comparatively obscure and uninfluential position." But Walcott had been on unfriendly terms with Boas for years and did not want to let the matter rest there. He angrily told E. G. Conklin that Boas's letter to the *Nation* read as if it had "emanated from the headquarters of the Reds in New York," and he told Osborn that because he was "a great believer in 100 per cent Americans" he had no use for "aliens who are hostile to our institutions and ideals." As head of the Smithsonian, Walcott proceeded to fire Boas as honorary philologist in the Bureau of American Ethnology, and he then wrote to physicist Michael Pupin of Columbia University (Madison Grant's friend from the Half-Moon Club) to get a sense of the faculty's feeling about Boas. Pupin, himself an immigrant (from Hungary), agreed with Walcott. He found it disgraceful that Boas "attacks the United States . . . and yet he is allowed to teach our youth and enjoy the honors of being a member of the National Academy of Science. This thought makes me long for the good old days of absolutism where the means were always at hand for ridding oneself of such a nuisance as Franz Boas."[55]

Satisfied that Boas had no support among the faculty of Columbia University, Walcott contacted William Barclay Parsons, a trustee of Columbia, and urged that Boas be fired posthaste. Parsons admitted that he desperately wanted "to get the scalp of [Boas], whose personality and views I detest," but he regretted that owing to the anonymous nature of Boas's charges in the *Nation,* there was nothing actionable in his letter. A frustrated Walcott, whose son had been killed in the war, contacted President Wilson and demanded that something be done. Wilson instructed Attorney General A. Mitchell Palmer to make "a thorough inquiry into the past and present activities of Boas, in order to ascertain whether or not he has been identified with any of the pernicious radical activities in this country." The investigation of the Justice Department revealed that Boas had broken no laws, and he was allowed to retain his position at Columbia, though the administration punished him by squeezing the Anthropology Department's funding for years to come.[56]

The Committee on Scientific Problems of Human Migration

The main goal—removing Boas from the National Research Council's Division of Anthropology and Psychology—had been achieved, and the Grantians in the division were now free to shelve the cultural projects of the Boasians and sponsor their own studies in scientific racism. They accordingly launched a number of ventures, of which the most important and best financed was the Committee on Scientific Problems of Human Migration, which was charged with distributing funding to investigators who would provide data to support immigration restriction legislation.[57]

Perhaps we should pause here to recap (for those of us whose inheritance does not include the gene—so ubiquitous among the progressives—for organizational aptitude): In 1916, the National Academy of Sciences created the National Research Council, which during the war established twenty-seven committees, including the Anthropology Committee. After the war, the National Research Council reorganized its twenty-seven committees into thirteen operating divisions, one of which was the Division of Anthropology and Psychology. The Division of Anthropology and Psychology then created numerous committees, one of which was the Committee on Scientific Problems of Human Migration.

With $130,000 from the Laura Spelman Rockefeller Foundation and $10,000 from the Russell Sage Foundation, the Committee on Scientific Problems of Human Migration operated from 1922 through 1928. It was a very high-powered committee. Its three most active directors, all members of the Galton Society and the American Eugenics Society, were Robert M. Yerkes, Clark Wissler, and John C. Merriam (who had just finished reading the fourth edition of *The Passing of the Great Race* when the committee began its work, and told Madison Grant that he was "having the pleasure of passing it on to several others to read.")[58] The committee sponsored sixteen research projects, and it need hardly be stated that all of them were congenial to mainstream eugenics and most were headed by longtime associates of Madison Grant who were also members of the Galton Society. Carl C. Brigham, for example, received funding to conduct measurements of mental ability, Henry Pratt Fairchild gained approval for a study of the causes of immigration, Raymond Pearl investigated race and pathology, Clark Wissler researched the results of race intermixture, and so forth. These projects eventually resulted in some fifty scientific publications, which—as intended—cumulatively validated the arguments of the eugenicists and gave even more force to Grant's campaign (covered in chapter 9) to put science in the service of politics. George W. Stocking, Jr., in fact, calls the Committee on Scientific Problems of Human Migration "a *de facto* research arm of the immigration restriction movement."[59]

In sum, by the early 1920s the National Research Council was firmly in the hands of the Grantians, who had managed to gain control of both the wartime Anthropology Committee and the postwar Division of Anthropology and Psychology. Furthermore, the Galton Society had been established as a viable alternative to the American Anthropological Association, and Franz Boas was viewed in many circles as a discredited, quasi-scientific traitor. In Grant's cold war against Boas, containment had been replaced by rollback.

The Revenge of the Roundheads

Back in 1918, when Ales Hrdlicka launched the *American Journal of Physical Anthropology,* it seemed obvious that the Grantians would control the journal. After all, the Boasians (in those pre–letter to the *Nation* days) dominated both the AAA and its organ, the *American Anthropologist,* and it was only logical that since the Grantians already ran the Galton Society (devoted to physical anthropology) they should have the *Journal of Physical Anthropology* as well. The Galton Society, in fact, had been mulling over the idea of launching just such a journal, and now Hrdlicka had done all the work for them. Not only did the inside cover of the *Journal* list eugenics as one of the areas falling within the scope of physical anthropology, but Hrdlicka also chose as associate editors numerous members of the Galton Society and other eugenicists (including Dr. John Harvey Kellogg of the Battle Creek Sanitarium, who was eager to support Hrdlicka because, among other things, he hoped that anthropologists would find evidence that pre-Columbian Native Americans had administered colon-cleansing enemas).[60]

One other Grantian had originally appeared on Hrdlicka's list of associate editors: Madison Grant. And yet by the time volume 1, number 1 of the *Journal of Physical Anthropology* was published in March 1918, Grant's name had disappeared from the masthead. Hrdlicka—a difficult, testy man under the best of circumstances—had developed a grudge against Grant during the war that had nothing to do with Grant's despicable politics or suspect scholarship. Rather, as we have seen, Hrdlicka was upset that the Anthropology Committee of the NRC had given priority to Grant's army anthropometry project rather than to Hrdlicka's launching of the *Journal.* Out of such personal resentments are scientific revolutions sometimes made.

The final break came when Grant, despite a number of entreaties by Hrdlicka, declined to financially underwrite the *Journal.*[61] In a fit of pique, Hrdlicka removed Grant from the list of associate editors and replaced him with Franz Boas and his sympathizers A. L. Kroeber and Henry H. Donaldson (for whom Boas had named his younger son). This radically changed the editorial policy of the journal and the character of physical anthropology in the United States. With Grant removed from the board, the *American Journal of Physical Anthropology* was free to become a respectable, scholarly publication, which in the coming years routinely declined to publish articles by—or carry news of—physical anthropologists friendly to Madison Grant. And thus it was that, contrary to expectations, the *American Journal of Physical Anthropology* never became a mouthpiece for the hereditarians. (The Galton Society was forced to adopt the *Eugenical News,* a much less prestigious journal, as its official organ.) And thus Madison Grant was handed his first defeat in the long-running cold war with the forces of nurture.

In fact, Hrdlicka was so hurt by Grant's snubbing that he pursued a somewhat unseemly vendetta against him. Upon learning that Grant was at the Battle Creek Sanitarium, for example, Hrdlicka told John Harvey Kellogg that he hoped Grant would "be afflicted with everlasting rheumatism of all his writing organs." Robert Lowie (of all people) counseled Hrdlicka that "nothing is gained by a violent controversial manner"—but he then admitted that one probably had to make an exception when dealing with "obvious charlatans like Madison Grant."[62]

Ales Hrdlicka was an ambitious and authoritarian editor who, in his quarter-century reign as editor of the *Journal of Physical Anthropology,* was famous for demanding complete editorial and managerial control. In one of his first acts as editor, Hrdlicka solicited Franz Boas to write a negative review of the second edition of *The Passing of the Great Race,* which had recently appeared. Hrdlicka was well aware that just one year previously Boas had written a scathing review of the first edition of Grant's book for the *New Republic,* but he now pushed for a repeat performance. Boas was taken aback by Hrdlicka's venomous attitude and did not feel wholly comfortable reviewing the book for a second time, but Hrdlicka was a driven man and would not let Boas off the hook. In a letter marked "confidential," Hrdlicka told Boas that Grant's book, "unless promptly shown exactly what it is, may be used to influence men in important positions."[63]

Boas was convinced, and wrote a review of *The Passing of the Great Race* that—per Hrdlicka's request—lambasted Grant. Boas declared that the book "is hardly a subject for a review in a scientific journal" and complained that "the author talks a good deal about inheritance, unit characters, and so on, without, however, approaching anywhere scientific accuracy." (Given that Grant had removed all references to unit characters from the second edition, it was actually quite unfair that Boas used this as an example of Grant's inaccuracy.)[64]

After submitting the review, Boas had second thoughts about its tone, and hoped that Hrdlicka would not find the article "too discourteous." To the contrary, Hrdlicka assured Boas that "the review is no more severe than it deserves to be." But after thinking it over, Hrdlicka did lose his nerve somewhat. Madison Grant, after all, was a man of some influence. So Hrdlicka somewhat bizarrely suggested that Boas write yet *another* review of Grant's book which would run in *Science* concurrently with Boas's review in the *Journal of Physical Anthropology,* so that Grant's wrath would be deflected from the *Journal.* Boas understandably declined to make a career out of writing reviews of *The Passing of the Great Race,* but in the summer of 1918 the *Journal of Physical Anthropology* did run Boas's single review.[65]

The always enterprising Ales Hrdlicka now decided to create a professional alternative to the Galton Society: the American Association of Physical Anthropologists. The AAPA held its first official meeting in 1930; Ales Hrdlicka was elected president of the group, which tabbed the *Journal of Physical Anthropology* to be its official organ. Under Hrdlicka, the AAPA and the *American Journal of Physical Anthropology* provided physical anthropology with the institutional tools it needed to become a reputable field within academia.

In truth, Hrdlicka's creations were important not so much for what they were as what they were *not.* In Germany, for example, the eugenicists succeeded in gaining control of physical anthropology in the 1920s and transformed it into *Rassenkunde* (racial science). Their journals adopted the Grantian version of anthropology, with fatal results for serious scholarship (and for serious scholars). The physical anthropologists in Germany went on to form their own professional organization—the Gesellschaft für Physische Anthropologie—at the same time that Hrdlicka formed the American Association of Physical Anthropologists. The difference was that in the United States the AAPA was headed by Ales Hrdlicka, and its members included Franz Boas, Melville Herskovits, and A. L. Kroeber. In Germany, the Gesellschaft was run by Madison Grant's disciple Eugen Fischer, and its members included Otmar von Verschuer, Fritz Lenz, and Josef Mengele.

The failure of the Grantians to gain control of the *Journal of Physical Anthropology* and the AAPA marked the beginning of a Boasian resurgence in anthropology. Over at Columbia University, Papa Franz may have been relegated, as Henry Fairfield Osborn had put it, to "a comparatively obscure and uninfluential position," but he used that position to churn out a cadre of Ph.D.s who soon comprised the majority of professional anthropologists in the United States. As the Red Scare subsided and hostility to foreigners lessened, Boas's students rapidly moved into, and took over, all the major departments of anthropology in the country, where they in turn trained the succeeding generation of scholars dedicated to the culture idea.[66] When all these Boasians joined the American Anthropological Association they were able to outvote their opponents, and within a few years of the *Nation* incident the AAA was back in the hands of the Boasians, and Robert Lowie ("a very dangerous anthropological Bolshevist" in Madison Grant's estimation)[67] was installed as editor of *American Anthropologist* (a position he held until 1933, when he handed the baton to fellow traveler Leslie Spier).

To a large extent, anthropology had witnessed the law of Lapouge: the lower breeds had simply been more prolific than the older, amateur wing of the profession. As a consequence, academically trained anthropologists hostile to

Madison Grant now set editorial policy at the profession's two journals (the *American Anthropologist* and the *Journal of Physical Anthropology*) and comprised a majority of its two professional organizations (the AAA and the AAPA).

Senator Bilbo of Mississippi, the leading white supremacist in Congress and an ardent disciple of Madison Grant, warned that Franz Boas, via "the tens of thousands of students who came under his influence, [was scattering his] insane and corrupt doctrines of miscegenation, amalgamation, intermarriage, and mongrelization throughout this broad land."[68] But Grant was not fully cognizant of what was happening in the academy. He was distracted in the 1920s and 1930s by other matters, from his numerous preservation activities to the campaigns for immigration restriction, sterilization, and antimiscegenation legislation. Plus, his struggle with arthritis was consuming an enormous amount of his time and energy. The fight to control anthropology simply was not the most important item on his agenda. The Boasians, on the other hand, were battling for their professional lives. So while Grant was purchasing redwood groves in California, or lobbying Georgia to ban intermarriage, or seeking relief at the Battle Creek Sanitarium, Boas and his students were actively infiltrating and taking control of their profession.

The last institutional bastion of the Grantians, other than the Galton Society, was the National Research Council, where the Division of Anthropology and Psychology continued to fund projects on scientific racism and reject anything smacking of culture. The brilliant Melville Herskovits, for example, submitted an application to study cultural diffusion in Africa but was denied for being "irrelevant." In 1922, a discouraged Boas announced that "after all that has happened, I am not taking part in anything that the National Research Council is doing. I consider the matter a complete failure."

But Boas soon thought better of his defeatism. Since it was obvious that the government was going to fund only research involving race, Boas reckoned that he might devise some racial projects of his own and try to milk the hereditarian cow. In 1923, when the National Research Council received $325,000 from the Rockefeller Foundation to establish a Fellowship Program, Boas garnered three of them for his students—a development aided by the fact that eugenicist Robert Yerkes had departed the Division of Anthropology and Psychology to pursue his new interest in primates. Furthermore, the Boasians, having won back control of the AAA, had been able to elect men like A. L. Kroeber and Robert Lowie to represent them at the division. So Melville Herskovits, who had been denied funding to study African culture, now received a fellowship to conduct an anthropometric study of American Negroes. He received the support of the unwitting Clark Wissler and Charles Benedict Davenport, who were only too happy that a cultural anthropologist was devoting himself to a serious topic like anthropometry. Then, in 1925, Boasian Otto Klineberg received a fellowship to study mental differences between the races. And in the same year, after assuring the government that the topic had relevance for contemporary America, Margaret Mead received approval to observe puberty in Samoa. To the old guard, these all seemed to be orthodox, "scientific" investigations. But in fact they were a conscious effort by Boas and his disciples to collect data to bolster the nurture side of the nature-nurture debate.[69]

Secure in the knowledge that his students were finally accumulating the data he needed to refute biological determinism, Boas now decided to take his argument to the public. Starting in the mid–1920s, he managed to publish at least one anti-Nordicist book or article a year in the popular press. His basic thesis had not changed since the 1910s, when he had first attacked Madison Grant. But this time he was armed with a mass of ethnographic research showing that environmental factors were at least as important as genes in determining mental and social traits. In writings like "The Question of Racial Purity" (1924), "What Is a Race?" (1925), "This Nordic Nonsense" (1925), "Fallacies of Racial Inferiority" (1927), and *Anthropology and Modern Life* (1928), Boas accused Madison Grant of being the foremost inciter of racial antagonism in America, and charged him with being "swayed not by scientific arguments but by prejudice." In a campaign that Lothrop Stoddard dismissed as "the desperate attempt of a Jew to pass himself off as 'white,'" Boas relentlessly denounced the immigration laws as being unscientific, attacked the army intelligence tests for being culturally biased, provided biological evidence to show that mongrelization did not lead to deterioration, and cited the findings of cultural anthropologists to show that culture, not biology, accounted for the mental (and a good deal of the physical) nature of man.[70]

By this point, the three recipients of the NRC fellowships had finished their research and were ready to publish their findings (in which they all thanked Boas profusely for his guidance). Margaret Mead's *Coming of Age in Samoa* (1928), with its exotic subject matter, brought cultural determinism to the American reading public by showing that even as biologically determined a condition as puberty varied extensively from culture to culture. (Sinclair Lewis's fictional character George F. Babbitt was certainly impressed, and bragged to his friends about the "racy" anthropological report he was reading about adolescents in the South Seas.) Melville Herskovits's *American Negro* (1928) pointed out: "It now becomes apparent that *social* as well as biological factors are of the utmost importance in the consideration of the American Negro type." And Otto Klineberg's *Negro Intelligence and Selective Migration* (1935) proved "quite definitely that the superiority of the northern over the southern Negroes" was "due to factors in the *environment.*" Additionally, Klineberg's *Race Differences* (1935), which he dedicated to Boas, showed that the army mental tests had ignored the role of cultural bias, and concluded that "there is no scientific proof of racial differences in mentality" and that "those differences which *are* found are in all probability due to culture and the social environment." Klineberg, the grandson of Jewish immigrants from Austria, definitively stated: "There is no reason to make our immigration laws stricter for one people than for another, . . . and there is no reason to pass laws against miscegenation."[71]

These works, along with George Amos Dorsey's "Race and Civilization" (1928), Robert Lowie's *Are We Civilized?* (1929), Ruth Benedict's *Patterns of Culture* (1934), and other influential offerings from the Boasians, cumulatively served to validate cultural anthropology as a viable field and to establish cultural determinism as a legitimate alternative to hereditarianism. An alarmed Grant complained to John Jay Chapman that his theories were the target of "an aggressive attack all along the line." And Henry Fairfield Osborn affirmed that "there is undoubtedly a conspiracy of the radicals against the whole Nordic and racial theory."[72]

After enduring the harsh jeremiads of the hereditarians for over a decade, the American public was more than ready to receive with gratitude the reassuring words of the environmentalists. The fact is that the vast majority of Americans had never felt comfortable with Madison Grant's biological determinism, with its pessimistic and aristocratic emphasis on inequality, degeneration, and the need for the "helots" to fall in line behind their betters. But as long as the Grantians could claim that individual and racial traits were the inevitable products of the immutable germ plasm, thinking men and women had no choice but to accept the cruel findings of science. In the mid–1920s, however, the Boasians started saying that the Grantians were wrong; that the *latest* scientific research showed that we might *not* be prisoners of our genes; that, indeed, almost anybody could rise to the top if provided with a nourishing environment. And once presented with a choice, Americans relievedly rejected cruel nature in favor of benign nurture.

It must be admitted that the evidence of the Boasians was still mainly negative. They did not *prove* that the races were equal; instead, they showed that there was little scientific data to support the assumption that the races were unequal. And this assertion was enough to hold off the Grantians until, as we shall see in the next chapter, the sociologists, psychologists, and geneticists could provide persuasive empirical evidence to discount biological determinism. Americans *wanted* nurture to be right. Environmentalism was so much more amenable than scientific racism to a nation that gazed reverently at the Statue of Liberty. (Of course, it was a lot easier to view that statue as innocuous now that the Immigration Restriction Act of 1924 had cut off the influx of dirty European peasants.) Furthermore, ever since Weismann had disproved the Lamarckian theory of acquired characteristics, those American liberals who believed in peaceful reform had found no place to turn for succor. But in the Boasian scheme of things, *culture* could be improved and passed on from generation to generation, thus providing a mechanism for permanent social reform. With their unquenchable commitment to equal opportunity, Americans spurned the vicious nature of the native-born lawyer in favor of the egalitarian culture of the immigrant anthropologist. Which is one of the reasons why, as Michael Schudson has recently pointed out, a man like Franz Boas is "still read and argued about, while Madison Grant is at most a historical footnote."[73] (Although this biography is certainly a rather lengthy "footnote.")

The ascension of the Boasians had been rapid. As early as 1925, Robert Lowie remarked to Ales Hrdlicka with astonishment that it had been just nine years since *The Passing of the Great Race* had received laudatory notices and Madison Grant himself had been welcomed into the scientific community "as though he were a serious thinker."[74] But those days were all but over. The Boasian point of view, limited to a handful of (primarily Jewish) cultural anthropologists at the end of the First World War, soon began to influence not just other anthropologists but other scholars as well. And as a result, by the beginning of the 1930s the culture idea was becoming the reigning paradigm in American social science. By that time, there was nary a Grantian to be found at the National Research Council—a phenomenon personified by the fact that in 1931 the new chair of the Division of Anthropology and Psychology was Robert Lowie. One year later, the *New York Times* pronounced that "Nordicism" was a "discredited doctrine."[75] And the old order officially passed on May 11, 1936, when Franz Boas was placed on the cover of *Time* magazine, accompanied by a cover story titled, simply, "Environmentalism."

Over at the Galton Society, a frustrated William McDougall asked Lothrop Stoddard: "Are we going to take it lying down, or are we going to react to it?" The answer, apparently, was *Take it lying down.* The Galton Society became increasingly anachronistic and its members thoroughly disheartened. William K. Gregory observed to Grant in 1930 that the group "sadly lacks cohesion and *esprit du corps,*" and Charles Benedict Davenport feared that the society was gaining "the opprobrium of dilettantism."[76]

Grant admitted that "the future looks ominous," but in his book *The Alien in Our Midst* (1930) he gamely tried to rally the partisans by insisting that they had on their side "the increasing force of science, of eugenics, and of an ever-widening acceptance of the fact that heredity and not environment dominates in the evolution and development of man." Anthropologist (and loyal member of the Galton Society) T. Wingate Todd seconded Grant's words, and bravely predicted in 1932 that the Grantian form of anthropology "is going to be more than ever significant in arranging the affairs of the future, and the Galton Society will have a great mission." Grant and Todd, of course, were deluded. The future belonged to the environmentalists. "The anthropological idea of culture," writes George W. Stocking, Jr., "became in time part of the vernacular of a large portion of the American public. . . . By the middle of the twentieth century, it was a commonplace for educated Americans to refer to human differences in cultural terms, and to say that 'modern science has shown that all human races are equal.'" Franz Boas was the individual primarily responsible for introducing culture into the vernacular, which explains why Thomas Gossett was not far off the mark when he stated: "It is possible that Boas did more to combat race prejudice than any other person in history."[77]

13 The Empire Crumbles

In dealing with science we must constantly be prepared to readjust our theories to fit new facts.

Madison Grant

In retrospect, we can see that 1924 was the high point of scientific racism in the United States. That was the year that Congress passed the Immigration Restriction Act and the state of Virginia enacted both the Racial Integrity Act to prevent miscegenation and the Sterilization Law, which was later sanctified by *Buck v. Bell.* Numerous Grantians published important books in 1924, including Henry Pratt Fairchild, Ellsworth Huntington, Vernon Kellogg, Edward A. Ross, Lothrop Stoddard, and A. E. Wiggam. The meetings of the Galton Society and the Half-Moon Club were well attended, and their influential members listened attentively as the leading scientists of the day fervently presented the latest findings about heredity and human nature. American families were eagerly flocking to eugenic lectures, entering Fitter Families contests, and enrolling in courses on eugenics at high schools and colleges throughout the country. The American Museum of Natural History created dioramas to educate the public about the Osborn-Grant view of race and evolution. The Eugenics Record Office, the Eugenics Research Association, and the ECUSA (soon to incorporate as the American Eugenics Society) were active, optimistic, and well connected to other scientific organizations, philanthropic foundations, educational institutions, and government agencies such as the National Research Council. They had also joined with the Immigration Restriction League and the American Defense Society to comprise an interlocking directorate that successfully lobbied for the causes important to Madison Grant.

And yet within ten years of 1924, scientific racism was a discredited doctrine in the United States, and the Grantians were being pushed down the path toward irrelevance. One of the reasons, of course, was the prodigious influence of Franz Boas and the cultural anthropologists. But in addition to Boas, ten other factors contributed to the demise of scientific racism. This chapter will briefly identify those ten factors and then show that the 1930s proved to be a disastrous decade for eugenics in the United States, during which a movement that had once been suffused with reforming zeal and millennial hopes meekly and ignominiously petered out.[1]

The Decline of Scientific Racism

Too Successful

One of the reasons for the decline of scientific racism was that Madison Grant had been too successful for his own good. The Immigration Restriction Act of 1924 had done such a thorough job of shutting out the Alpines and Mediterraneans that most Americans assumed that the threat posed by the inferior breeds had been adequately dealt with, and they now felt free to switch their attention to other issues. Consequently, all further appeals by the Grantians to safeguard the nation's germ plasm fell on ears that wanted to listen only to newly purchased Victrolas and the feverish clicking of ticker tapes. The ending of immigration, and the onset of prosperity, dropped xenophobia from the national agenda, leading a member of the Immigration Restriction League to admit that "the country is somewhat fed up on high brow Nordic superiority stuff."[2]

In 1928, Grant and Charles Stewart Davison jointly edited *The Founders of the Republic on Immigration, Naturalization and Aliens,* a slim volume of excerpts from the writings and speeches of the Founding Fathers. In the foreword, Grant and Davison expressed their fear that owing to the recent "influx of alien races" that were congenitally unfit to uphold American traditions and institutions, the United States would soon degenerate into a "tyranny of the mob, called a Democracy— . . . and, thereupon, will follow chaos." This warning was followed by sixty-nine selectively chosen, out-of-context quotations from John Adams, Benjamin Franklin, Alexander Hamilton, George Washington, and the like, all showing that those wise gentlemen had possessed the foresight to advocate the restriction of immigration and the deportation of dangerous aliens. (Neither Grant nor Davison addressed the inconsistency inherent in the nature of the book, which is that the eighteenth-century immigrants that the Founding Fathers warned about—the Germans, the Irish, the French—were Nordics.)[3]

The *New York Times* was surprisingly complimentary toward the book. Admitting that it had been previously unaware that the Founding Fathers opposed immigration, the *Times* stated that "Messrs. Davison and Grant have performed a real public service in bringing into prominence . . . some of the most vigorous and convincing arguments against unlimited immigration that have ever been penned." Despite the favorable review by the *Times,* Scribner's knew that the book would have a limited sale. It printed just 1,525 copies and was able to sell almost none to the public. The American Defense Society finally bought up most of the copies and distributed them to newspaper editors and members of Congress, whereupon Albert Johnson inserted portions of *The Founders of the Republic* into the *Congressional Record.* But other than that the book was ignored, with the possible exception of Newton B. Drury of the Save-the-Redwoods League, who felt that *The Founders* was both "interesting and enlightening" and that Grant was "entitled to great credit for spreading thought on a subject to which too little regard is being given today." But there were good reasons why "too little regard" was being given to immigration restriction in 1928: Babe Ruth was swatting, Clara Bow was flapping, the automakers' assembly lines were humming, and it was pretty difficult for anybody but the most anxious aristocrat to get worked up about the few immigrants managing to enter the country under the new quota system.[4]

The Great Migration

In addition to rendering Grant superfluous, another unintended result of Madison Grant's immigration restriction legislation was the movement of hundreds of thousands of African Americans from the rural South to the cities of the North, a phenomenon known as the Great Migration (or, as Grant called it, "the Negro invasion of the North"). This process had begun during World War I, but it greatly intensified after Grant and Albert Johnson cut off the supply of cheap European labor. One of the effects of the Great Migration, Grant anxiously told Earnest Sevier Cox in 1930, was that "the enormous Negro population in New York is spreading and ruining real estate values in fairly good sections. In Harlem one constantly sees on the street white and negro boys playing together and white and negro girls walking together."[5]

Grant anticipated that the sight of blacks and whites playing together would stoke the flames of racism in the North. And he was correct. But, paradoxically, the Great Migration also had the opposite effect. For as blacks took advantage of employment and educational opportunities in the North, many sociologists, psycholcgists, anthropologists, and members of the public realized that they had underestimated the capabilities of African Americans, whom they had heretofore encountered mainly as socially and intellectually stunted sharecroppers. In addition, blacks began scoring much higher on intelligence tests once they were exposed to the better schools, health care, and nutrition of the North—a fact seized on by the proponents of nurture. And finally, given the opportunity to enter academia, many blacks themselves began making contributions to antiracist scholarship (of course, the very existence of Negro scholars served to alter perceptions of the innate capabilities of African Americans).[6]

The Jews

African Americans were not the only "inferior" minority enjoying academic success in the 1920s and 1930s. Jews were also taking advantage of higher education, to the point where Grant bewailed that "foreign names are beginning to appear and sometimes predominate in the list of college graduates." In response, institutions such as Harvard, Yale, Columbia, Princeton, Duke, Cornell, and the University of Virginia did establish quotas limiting the number of Jewish students, but Grant cautioned John Jay Chapman that Harvard should never have admitted Jews in the first place and that it was now paying the price for its "foolish liberalism" by having to create "Kykological Tests to save herself from being swamped."[7]

Once they earned their degrees and entered the sciences, many Jews added their voices to the critique of biological determinism. We have already discussed the opposition to Grant by Jewish anthropologists, and they were soon joined by Jewish psychologists (e.g., Gustave Feingold, Sheldon Glueck, and Samuel C. Kohs) and Jewish biologists (e.g., the highly respected Jacques Loeb, who declared that Grant's writings were "beneath contempt").[8]

The ascent of Jewish scholars greatly perturbed Grant, who whined to Ellsworth Huntington as early as 1916 that university professors were hesitant to support him because they were either "saturated with Socialism" or—thanks to the influence of the Jews—"afraid to speak out loud." There had been a time, wrote Grant in the *Journal of Heredity* in 1919, when Jews in America had been justifiably denied "access to social circles and to positions of responsibility." But now that they had been permitted to enter the academy, they were attaining influence wholly out of proportion to their numbers, and working to "deny all discussion of race and raise the cry of 'race prejudice' the moment the subject is opened." Grant's fear of the intellectual sway of the Jews amounted to one of the first critiques of what we call political correctness. In 1921, Grant again protested that "certain religions or races" had become "hysterically sensitive" and created an atmosphere where it was "well-nigh impossible" for racial scientists to publish their findings. And in 1925, the persecutor once more complained of being persecuted, when Grant grumbled in the *Forum* that it was no longer permissible to point out in public that certain races (i.e., the Jews) had a higher incidence of crime, feeblemindedness, and military cowardice. Indeed, when Grant despaired in a 1934 interview with the *Herald Tribune* that the United States had become "a dumping ground for Italians *and others we will not mention,*" he demonstrated that while one could still impugn the Italians publicly, it was no longer acceptable to castigate the Jews by name.[9]

Although it may be difficult for us to respond sympathetically to the plaintive cry of the oppressed WASP, the fact is that by the time the *Encyclopedia of the Social Sciences* was published in 1932, it was not William McDougall who was asked to pen the "Personality" article but the Jewish immigrant Edward Sapir.

The "Anthropometry" article was submitted not by Charles Benedict Davenport but by the son of Jewish immigrants, Melville Herskovits. "Mental Testing" was written neither by Carl C. Brigham nor Robert M. Yerkes but by the grandson of Jewish immigrants, Otto Klineberg. The "Heredity" entry was authored not by Frederick Adams Woods but by yet another Jew: geneticist Alexander Weinstein. And it was not the country's best-known expert (Madison Grant) who was selected to write the "Race" article but rather the Jewish immigrant Franz Boas.

Sociology

Partly as a result of the Great Migration and the entrance of Jews into the professions, increasing numbers of sociologists abandoned the hereditarians and embraced the culture concept. Scholars like George A. Lundberg, William F. Ogburn, and Robert E. Park began to show that human society was not determined mainly by biology but by a complex interaction of nature and nurture.

The Grantians experienced a major setback when respected sociologist Frank Hankins of Smith College defected in 1926. Hankins was a member of the American Eugenics Society and served on the editorial board of *Birth Control Review,* and he still believed it desirable to encourage the breeding of gifted individuals. But Hankins was not a racist, and in *The Racial Basis of Civilization: A Critique of the Nordic Doctrine* he argued that there was no biological justification for racial discrimination and antimiscegenation laws. Hankins did not deny the existence of a Nordic race, nor that the New Immigrants were "in the mass less desirable from the standpoint of general intellectual abilities than the native population." But he rejected the idea that the Nordics possessed a monopoly on excellence, and he even felt that a "well-endowed" Hebrew should be welcomed as a citizen ahead of a "below average" Nordic.[10]

Hankins spent many pages of *The Racial Basis of Civilization* lambasting the works of Carl C. Brigham, Clinton Stoddard Burr, H. J. Eckenrode, Charles W. Gould, William McDougall, William S. Sadler, and Lothrop Stoddard. But he identified Madison Grant as "the chief present-day exponent" of the "pernicious propaganda relating to the Nordic doctrine," and he devoted the greatest space to exposing the contradictions of *The Passing of the Great Race,* which he described as "highly doctrinaire," "hopelessly confused," and "a compound of one part truth and nine parts imagination, suspended in a mixture of exaggeration and suggestion." He accused Grant of "dogmatic simplification" and "anthropological charlatanry," and regretted that Grant's prestige had given his "crudely bald and dangerous" book "an enormous vogue" that resulted in the "irresponsible outpourings of a host of imitators."[11]

It is not surprising that while Harry H. Laughlin in the *Eugenical News* concluded that Hankins's *Racial Basis of Civilization* could "hardly be ranked as a work of science," Ales Hrdlicka in the *American Journal of Physical Anthropology* highly recommended the sociologist's book as an antidote to "all those who have read Grant, Brigham, or Stoddard."[12]

Psychology

Along with the sociologists, many psychologists were forsaking the hereditarians in the early 1930s. Scientists such as William Bagley, Edwin Boring, and Kimball Young began to discredit the claim that there were inherent mental differences between racial groups, arguing that a host of environmental reasons (including poverty, malnutrition, and illiteracy) could cause immigrants to perform below native-born Americans on intelligence tests. These researchers noted—and provided a cultural explanation for the fact—that the IQs of immigrants increased the longer they stayed in the United States, and that the IQs of Negroes increased the longer they resided in the North.[13]

The most famous apostate was the man responsible for popularizing the army intelligence tests in the first place, Carl C. Brigham. In a landmark article in *Psychological Review* in 1930, Brigham stunningly refuted the findings of his own book (*A Study of American Intelligence*), which had done so much to influence Congress to pass the Immigration Restriction Act of 1924. After exposing the flawed methodologies employed by practitioners of mental testing, Brigham humbly admitted that "one of the most pretentious of these comparative racial studies—the writer's own—was without foundation." This unprecedented retraction was a devastating blow to the Grantians. The *New York Times* confessed that "it was a shock when Professor Brigham branded his own studies as valueless," and the *Saturday Review* commented that Brigham's reversal was "as gallant an exhibition of scientific integrity as one is likely to find." Brigham later stated that the claim that intelligence tests measured innate intelligence, without regard to environment, had been "one of the most serious fallacies in the history of science," and he asserted that the only thing intelligence tests measured was how well one performed on intelligence tests.[14]

In addition to the experts on intelligence, two other factions within psychology—the Freudians and the behaviorists—also played a role in undermining the reputation of eugenics. Sigmund Freud taught Americans to look to childhood experiences, not germ plasm, for the origins of human personality and neuroses. And psychologist John B. Watson, who originated behaviorism at the same time as the Boasians were developing the culture idea, claimed that environmental stimuli were what caused humans to act, and explained that "nurture—not nature" was thus responsible "for what the child becomes." "Give me a dozen healthy infants," boasted Watson in his best-selling *Behaviorism* (1925), "and I'll guarantee to take anyone at random and train him to become any type of specialist I might select—doctor, lawyer, artist, merchant-chief, and, yes, even beggar-man and thief, regardless of his talents, penchants, tendencies, abilities, vocations, and *race of his ancestors*." (It is indeed ironic that Watson first formulated his antihereditarian ideas at Madison Grant's Bronx Zoo, after noticing that his children had no innate fear of wild animals.)[15]

Genetics

In the 1920s, geneticists joined the anthropologists, sociologists, and psychologists in undercutting the foundations upon which scientific racism had been built. Researchers such as Herbert S. Jennings, H. J. Muller, and Sewall Wright were revealing the fallacies of much eugenic thinking, and replacing inaccurate concepts like "unit characters" with new ideas about "balanced lethals," "multiple factors," "crossing-over," and "dosage compensation." In 1922, eugenicist Herbert Walter observed: "The biological Rip van Winkle of today who, awaking after a decade of somnolence, gazes again upon the genetic village of Falling Waters, will indeed need to rub his astonished eyes at the changed scene that now spreads out before him. Many old familiar landmarks, such as 'unit characters' and 'dominance,' show signs of dilapidation, while strange children, shouting a medley of outlandish words, 'linkage,' 'tetrapolidy,' and 'non-disjunction,' for example, are playing new games on the village green."[16] It is unfortunate that Madison Grant, Charles Benedict Davenport, Henry Fairfield Osborn, and Harry H. Laughlin went to their graves without ever comprehending those outlandish new terms, and still clinging to the "old familiar landmarks."

Up in Columbia's Schemerhorn Hall, their former friend (and future Nobel laureate) T. H. Morgan and his colleagues were studying *Drosophila melanogaster* and beginning to comprehend that the mechanism of heredity was far more complicated than the Grantians had admitted publicly (or understood privately). "In the fruit fly," explained Herbert S. Jennings in 1924, "at least fifty genes are known to work together to produce so simple a feature as the red color of the eye." How absurd, therefore, for eugenicists to think they could understand and control all the factors that went into creating something as complex as human "intelligence" or as elusive as "feeblemindedness."[17]

Further, the geneticists were now able to demonstrate that, biologically speaking, there is no such thing as a pure "race"—all human population groups are products of the intermixture of many biological strains, and concepts like "Nordic" and "Alpine" are genetically meaningless.

In addition, it finally dawned on thinking persons that since the genes for many undesirable traits are recessive (and hence hidden in carriers who are phenotypically "normal"), the only sure way to prevent the propagation of harmful genes would be to sterilize millions upon millions of healthy adults. Since few Americans were willing to embark on such a program, negative eugenics was exposed as a totally unrealistic scheme, and the last sterilization law in the United States was passed in the state of Georgia in 1937 (fittingly, the year of Madison Grant's death).

Beginning in the early 1920s, and continuing through the 1930s, a steady stream of geneticists abandoned Madison Grant and publicly declared their disapproval of scientific racism. They included such former stalwarts of the eugenics movement as William E. Castle, Edward M. East, Herbert S. Jennings, Vernon Kellogg, Clarence McClung, T. H. Morgan, William H. Welch, and E. G. Conklin (whose defection had been anticipated by Grant because "Dr. Conklin was a Wilsonian and, consequently, is something of a sentimentalist").[18] By 1931, eugenics had sunk so low that geneticist Lancelot Hogben confidently stated, "with full responsibility for my words as a professional biologist," that all the verifiable data eugenicists had accumulated on the inheritance of mental traits could "be written on the back of a postage stamp." (This at a time when the *Eugenical News* was blithely featuring articles such as "The Inheritance of Business Ability" and "The Inheritance of Story Telling.")[19]

For Grant, perhaps the most painful rejection was that of Raymond Pearl, a member of the AES, the ERA, and the Galton Society. In 1921, Pearl had written: "As an animal breeder of some years' experience I have no doubt whatever that almost any breeder of average intelligence, if given omnipotent control over the activities of human beings, could in a couple of generations breed a race of men on the average vastly superior . . . to any race of men now existing in respect of virtually every quality or attribute."[20] Two years later, however, after observing Grant's influence over the U.S. Congress, Pearl confided to Herbert S. Jennings, his colleague at Johns Hopkins: "I have a strong feeling that the reactionary group led by Madison Grant and with Laughlin as its chief spade worker were likely, in their zeal for the Nordic, to do a great deal of real harm." By 1927, genetics had advanced to the point where Pearl felt he should go public with his opposition to eugenics, and he did so in the pages of the influential *American Mercury*. In "The Biology of Superiority," Pearl charged that the Grantians were "full of emotional appeals to class and race prejudices, solemnly put forth as science, and unfortunately accepted as such by the general public. . . . It would seem high time that eugenics cleaned house, and threw away the old-fashioned rubbish which has accumulated in the attic."[21]

Raymond Pearl's choice of venue was not by happenstance. *American Mercury*'s publisher was Alfred A. Knopf, who also published the books of such Boasians as Alexander Goldenweiser, Melville Herskovits, and Wilson D. Wallis. More importantly, Pearl was a drinking partner of the magazine's editor, his fellow Baltimorean H. L. Mencken (who once remarked to Robert Lowie, as only Mencken could: "My impression, though I am blond and Nordic myself, is that the genuine member of that great race, at least in modern times, is often indistinguishable from a cockroach"). While both Pearl and Mencken remained private bigots, Mencken enthusiastically made the pages of his magazine available to the anti-Nordicists. In *American Mercury*'s inaugural issue in 1924, for example, Harry Elmer Barnes had attacked the "preposterous absurdities" and "worthless nature of this Nordic garbage," and declared: "The most offensive nonsense that has been recently loosed in this field is that in Madison Grant's *Passing of a Great Race*."[22]

The Great Depression

The ravages of the Great Depression spared no races, and as a result it became increasingly untenable to claim that poverty was due to inferior germ plasm. As Nordics and non-Nordics shivered together in the same breadlines, they all could perceive with distressing clarity the power of environmental factors to determine one's socioeconomic standing. And many went on to make the appropriate induction that the supposed inferiority of immigrants and blacks may well have been a matter of nurture rather than nature. Which is why historian Donald J. Pickens concludes that "Galtonian eugenics was a victim of unemployment."[23] (Another victim was Representative Albert Johnson who, after being returned to Congress by the good folks of Washington's Third District for twenty-two straight years, was swept out of office in the Democratic landslide of 1932. Johnson's replacement as chair of the House Immigration Committee was Democrat Samuel Dickstein, a Jewish immigrant from Russia and ardent foe of the restrictionists.)

While Madison Grant's personal investments took a hit in the Crash, he and his brother DeForest still owned a nice chunk of West Forty-eighth Street plus the stretch of Broadway just north of Times Square (currently occupied by the Crowne Plaza, the Renaissance New York, and the headquarters of Morgan Stanley Dean Witter). Grant thus avoided going on the dole, and he managed to retain the services of his housekeeper, chauffeur, cook, valet, personal servant, and male nurse.

During the Depression, the Grantians showed remarkable insensitivity by trying to blame the disaster on its victims. They claimed that the unemployed possessed inferior genes, and pointed to the breakdown of the economy as the ultimate proof that the nation's germ plasm had deteriorated to the point of disfunctionality. Grant advised Raymond Moley, leader of FDR's brain trust, that the "obvious" remedy to the unemployment problem was massive sterilization and deportation.[24]

Incredibly, it was in the year 1930—when more people were emigrating than immigrating—that Grant and Charles Stewart Davison decided to publish their second book of anti-immigration essays, *The Alien in Our Midst, or Selling Our Birthright for a Mess of Pottage*. The book contained articles by twenty-eight of the leading Grantians, who all espoused the same hereditarian message they had been promulgating for years. One decade earlier these xenophobic writings would have been seen as thoughtful and persuasive. But in 1930 they were judged to be hackneyed and glaringly out-of-date.

To the eternal shock of Grant and Davison, Scribner's rejected the manuscript of *The Alien in Our Midst.* Maxwell Perkins assured Grant: "It is certainly our earnest wish to cooperate with you, with whom cooperation in the past has been such a great pleasure." But he pulled no punches in explaining, "we cannot believe otherwise than that the sale would be extremely limited, and that the book could not be given the appearance of being other than propaganda." If this were 1920, Scribner's would have been thrilled to publish such a cogently argued and hard-hitting book. (And Perkins would not have derisively referred to it as "propaganda" but would have proudly labeled it "science.") But it was now 1930 and, frankly, Perkins thought it was rather absurd for Grant and his friends to go railing on about the perils of immigration when the number of immigrants to the United States had fallen to the lowest level since the Civil War. And after the fiasco of *The Founders of the Republic,* Scribner's was wary of spending any more money on such efforts. Grant and Davison were forced to print the book at their own expense (under the auspices of the Galton Publishing Company, a subsidiary of the American Eugenics Society).[25]

As Perkins had predicted, *The Alien in Our Midst* was all but ignored by the press and the public, despite the best efforts of the Immigration Restriction League and the American Eugenics Society, which offered the book to their members at a steep discount. In desperation, a publicist suggested that they garner attention by persuading some prominent figure to write a derogatory review of the book, to which a Grantian would then respond with incendiary effect, thus creating a controversy that would stoke book sales. But that proposal, like most copies of *The Alien in Our Midst,* was consigned to the dustbin.[26]

Assimilation

In 1925, Madison Grant claimed (in "America for the Americans") that the melting pot was a failure, as "those who are alien in race and religion have not amalgamated with the native population." He predicted that "it will take centuries before the foreigners now here become Americans."[27] It did not take centuries; it took one generation. The fact is that Americanization, which Grant felt was a biological impossibility, worked.

Again we can only note that Grant was too successful for his own good. The Immigration Restriction Act of 1924 did exactly what he hoped it would do: by cutting off the supply of foreigners it loosened the ties of the New Immigrants to the old country, led to the decline of the foreign-language press, and condemned immigrant communities to death by attrition. The children of the immigrants, embarrassed by the foreign ways of their parents, set about becoming "Americans." They lost their accents, earned their degrees, and assimilated into American society. As with the professors of African American and Jewish descent, the second-generation immigrants proved, by living successful lives, that heredity was less powerful than environment.

In 1940, Boasian Ruth Benedict wrote: "Racism in America turned out to be no more than a spectacle of immigrants of one decade condemning to everlasting inferiority the immigrants of a later decade. . . . A Norman in the time of Ivanhoe could have written of the impossibility of civilizing the Saxons with much better justification than does Madison Grant of non-Nordics in the 1920's in America."[28] Furthermore, as those non-Nordics assimilated, the public stopped viewing Italians, Jews, Poles, and Greeks as members of different races. Within a few years of the Immigration Restriction Act of 1924, all Americans of European heritage were permitted to claim biological membership in "the white race," and their former racial moniker became merely their "ethnicity." This development was partly a consequence of the strategy employed by the various nationalities during the immigration restriction debate, when they all asserted their whiteness in order to distinguish themselves from the Orientals. It was also a result of the fact that "race suicide" had proved to be a bogeyman. It was true that immigrants produced far more children than old-stock Americans, but the *children* of the immigrants did not—the fertility of the second generation of immigrants was exceptionally low. No longer a demographic threat, the offspring of the immigrants were safely granted entry into the camp of the whites (especially when their aid was needed to counteract the growing demands for equal rights from the Negroes who had moved north in the Great Migration).

And as the Jews and the other ethnic groups joined the great coalition known as the white race, Grant's complex taxonomy of *Homo sapiens* featuring scores of different races was rejected in favor of the (equally arbitrary but much more inclusive) pentagonal racial classification (white-black-red-yellow-brown) that dominated the rest of the twentieth century.

The Nazis

A major factor in the demise of scientific racism in the United States was the public's revulsion from the barbarous racism of the Nazis. Long before Americans learned of the Holocaust, the population policies of the Third Reich provoked a powerful antieugenic backlash in the United States. After visiting Germany in 1935, for instance, it dawned on geneticist L. C. Dunn (a member of the Galton Society and the American Eugenics Society) that scientific racism was leading society toward a "very dark future," and he immediately sent a letter to the Carnegie Institution of Washington recommending that it end its patronage of the Eugenics Record Office.[29]

Once the United States officially entered the Second World War, the reputation of eugenics in America hit rock bottom. The Nazi regime was so inextricably associated with eugenics and anti-Semitism that to oppose Germany was to oppose eugenics and antiSemitism. The enemy of the United States was racist, and so it was only patriotic to be antiracist. And thus even those persons who still believed that the Jews were genetically inferior now found it impossible to express that belief publicly.

And after the horrors of the Holocaust were revealed, eugenic practices such as sterilization were universally acknowledged to be the summit of a horrifyingly slippery slope. Following the war, when he was asked why eugenics had died so quickly in the United States, Paul Popenoe immediately replied with one word: "Hitlerism." The irony was that by putting Madison Grant's theories into practice, the Nazis discredited those theories forever.[30]

The Torch Is Not Passed

Finally, the simple fact of human mortality played an inescapable role in the demise of scientific racism. When the eugenics crusade began at the start of the twentieth century, its leaders were young, progressive, and energetic. But the same men who originated the movement were still in charge three decades later, when they were old, reactionary, and tired. And as they began to die off, almost nobody was interested in taking their place. (It did not help that so many eugenicists never got around to producing any children to fill their shoes. Although the president of the American Eugenics Society, Roswell H. Johnson, admonished his colleagues that "the celibacy of superior individuals is a recreancy of serious dysgenic results," almost 10 percent of the members of the Advisory Council of the AES were unmarried, and a full 25 percent were childless.[31]

The Big Four of scientific racism—Madison Grant, Charles Benedict Davenport, Henry Fairfield Osborn, and Harry H. Laughlin—all passed from the scene in the 1930s. In their prime they had been giants; in their dotage they were irrelevant. Yet each of them departed still clinging to the delusion expressed by Davenport in his presidential address to the Third International Eugenics Congress in 1932 that, over the next few decades, eugenics would continue to rise in the public's esteem until it would one day become regarded as "*the* most important influence in human advancement."[32]

The Third International Eugenics Congress

The Third International Eugenics Congress was held at the American Museum of Natural History from August 21 to 23, 1932, under the sponsorship of the ERO, the ERA, and the Galton Society. Eleven years had passed since the Second International Eugenics Congress was held at the same location, but it soon became clear that little had changed in the field of eugenics. With few exceptions, the same persons who had organized the second congress were in charge of the third. Charles Benedict Davenport served as president, the two honorary presidents were Henry Fairfield Osborn and Leonard Darwin, and the Managing Committee consisted of Clarence G. Campbell, Irving Fisher, Harry H. Laughlin, Frederick Osborn, Leon F. Whitney, and Madison Grant.

The congress heard papers by the usual people on the usual subjects: Campbell on race survival, Davenport on anthropometry, Gosney on sterilization,

Holmes on the fecundity of Negroes, Johnson on intelligence tests, Laughlin on immigration restriction, Osborn on birth control, Popenoe on eugenic marriage counseling, and so forth. In other words, despite the theme of the congress—"A Decade of Progress in Eugenics"—there really had been no progress in eugenics. Speakers still referred to *The Kallikak Family* to defend mandatory sterilization, invoked "biological disharmony" to argue against miscegenation, and cited the army intelligence tests to justify immigration restriction (even though Carl C. Brigham had admitted two years earlier that those tests were worthless).

Dr. Lena K. Sadler of Chicago showed that neither the ideas nor the rhetoric of the Grantians had evolved since *The Passing of the Great Race* when she warned the congress that "civilization is doomed if we continue to drift down the stream of a few more generations on the defenseless raft of mistaken brotherly love and blinded sentimentalism." Growing somewhat hysterical, Sadler cried out that the "monster" of racial degeneracy was "a rising tide, constantly growing in volume, because these enemies of society pass on their faulty mental and moral taints to their rapidly increasing progeny. Must we sit supinely by and let all this go on? No! a thousand times, no!" As a first step, Dr. Sadler called for an end to attempts by physicians "to save every weak child that is born into the world."[33]

Five months later, Adolf Hitler would begin to implement in Germany the practical program of eugenics advocated by experts like Sadler at the Third International Eugenics Congress. But at a time when millions of ordinary Americans—not mongrelized foreigners, not degenerate paupers, but (formerly) middle-class, old-stock Americans—were desperately scouring the country for a morsel of bread or a day's wages, it is hard to imagine how the eugenicists could have devised a message that would have more clearly and heartlessly antagonized and alienated the American public.

In addition to their retrograde rhetoric, another indication of the obsolescence of the eugenicists was that only seventy-three delegates showed up at the congress (even though the organizers had sent out more than five thousand invitations), and a paltry sixty-five papers were delivered. Eleven years earlier, by contrast, twice as many papers had been delivered at the Second International Eugenics Congress to four times as many delegates.

As at the earlier congress, the Third Eugenics Congress included an exhibit hall under the direction of Harry H. Laughlin, and again little had changed. The few visitors who showed up were ushered through an entrance guarded by busts of Darwin, Galton, Davenport, and Osborn. They then proceeded to the exhibits, which included sample intelligence tests, a map (by S. J. Holmes of Berkeley) depicting the Great Migration of Negroes, a poster (by W. A. Plecker) explaining the antimiscegenation laws of the several states, an illustration showing how mental superiority was inherited in the family of Theodore Roosevelt, and so forth. One of the more interesting exhibits was a graph showing that immigrants from northern Europe were more inventive than the New Immigrants, who in turn were superior to the Negroes. This was based on a study funded by Madison Grant, in which Laughlin had surveyed the first ten thousand patent applicants in 1927. Unfortunately, the original computations had revealed that the Jews ranked higher than even the English in inventiveness. Laughlin dealt with that embarrassing finding by simply eliminating the Jews as a category before assembling the exhibit.[34]

The most shocking occurrence at the congress was the attack on eugenics by geneticist (and future Nobel Prize winner) H. J. Muller in a paper titled "The Dominance of Economics over Eugenics." Muller (whose father was an immigrant from Germany, and mother a descendant of Sephardic Jews) bravely told the delegates that eugenics had become an unrealistic, ineffective, and anachronistic pseudoscience. Sounding very much like his cousin, anthropologist A. L. Kroeber, Muller admonished eugenicists for ignoring the obvious influence of environment on IQ, and he suggested that slums were more important than heredity in the creation of criminals.[35]

Unfortunately, Muller's cogent critique of scientific racism was greatly negated by his naïve embrace of scientific socialism as the cure for all of society's ills. Claiming that the proletariat revolution was nigh, he argued that there was no point in agonizing over the differential fertility of different classes "when the very basis for the existence of these classes as such will soon be swept away."[36] By converting from eugenics to socialism, Muller (who earlier that year had attempted suicide) simply moved from one millennial movement to another. Still, the fact that he felt he could go to the eugenics congress and confront the Grantians on their home turf reveals the vulnerable state of the movement in the early 1930s.

Before the opening of the Third International Eugenics Congress, the *Eugenical News* had expressed the hope that the congress would serve "to present to the public the real meaning and content of the science of eugenics." Unfortunately, it succeeded all too well in doing just that. The congress received broad coverage in the press, which is precisely what the eugenicists did not need. By continuing to blame the Great Depression on the faulty moral traits of the unemployed, and by demonstrating that they had no intention of recognizing that any advances had been made since 1921 in anthropology, sociology, psychology, or genetics, the Grantians unmasked themselves as a dogmatic group of reactionaries motivated not by science but by bigotry. Eugenics, concluded the *New York Times* at the end of the congress, "seems to have become a disguise for race prejudice, ancestor worship and caste snobbery."[37]

The fatal condition of eugenics was hammered home when the International Congress of Genetics opened at Cornell University, under the presidency of T. H. Morgan, just one day after the eugenics congress closed. More than three times as many papers were delivered at the Genetics Congress as opposed to the Eugenics Congress, and eight times as many delegates (562) were in attendance. In fact, despite the deteriorating economic situation and adverse exchange rate, the number of delegates from foreign countries who managed to make it to upstate New York (103) was larger than the entire attendance at the Eugenics Congress in Manhattan. "In a single week," noted the *New York Times,* "we are thus granted the opportunity of comparing the theories and methods of two schools. On the one hand, much loose talk about sterilizing the feebleminded and restricting marriages of undesirables, on the other earnest experiments [dealing with] the mechanism of heredity." The *Times* concluded: "It is easy to choose between the two schools." The American public agreed, which is why the Third International Eugenics Congress was also the final international eugenics congress.[33]

The Last Hurrah: *The Conquest of a Continent*

Allan Chase has called *The Alien in Our Midst* (the book edited by Grant and Davison in 1930) "the Last Hurrah of the old scientific racism."[39] Actually, Madison Grant had one more hurrah left in him, and in 1933 it appeared as *The Conquest of a Continent, or the Expansion of Races in America.* It was his first major work since *The Passing of the Great Race,* and it would be his last. But what a work it was: a rollicking, sweeping, 393-page saga describing and explaining the history of North America in terms of race. Where *The Passing of the Great Race* had traced the evolution of the Nordics in the rugged environment of northwestern Europe, *The Conquest of a Continent* picked up the story as the most courageous members of the Great Race traversed the Atlantic Ocean in the seventeenth century, possessed the eastern seaboard, crossed the Appalachians, and vigorously swept their way irresistibly toward the Pacific, generation after generation, mercilessly dispossessing the natives and stamping their unique racial inheritance on the landscape and institutions of the United States and Canada. Within two centuries, this manly race of sailors, hunters, explorers, and soldiers had subdued the richest continent in the world.

But then, according to Grant, the country opened its gates to three unassimilable races, and the nation was placed on the road to ruin. First, lazy southern planters imported African slaves to work for them under the hot tropical sun. Then, greedy northern employers induced inferior Mediterranean laborers to man their machinery for inhuman wages. And finally, during the late nineteenth century, the Polish Jews began their "tumultuous and frantic invasion," invited here by "fatuous humanitarians" who took pity on these victims of Old World persecution and offered them refuge in America, resulting in "the draining into our country of that morass of human misery found in the Polish Ghettos." The main problem with the Jews was that they were "essentially a non-

European people," hence incapable of assimilation. And if the presence of these foreign stocks of Negroes, Italians, and Jews was not dangerous enough by itself, the three races were now actually intermarrying with one another in some of our larger cities, resulting in "a racial chaos such as ruined the Roman Empire."[40]

The Conquest of a Continent advanced four remedies to ensure that "civilization itself may be maintained" in the United States: the absolute prohibition of all immigration from all countries; the immediate and wholesale deportation of all undesirable aliens (including any who were unemployed); the sterilization of criminals and "the intellectually unfit"; and the dissemination of birth control to Negroes and other inferior groups. Grant understood that such a program would lead to "a storm of protest . . . from the vociferous and influential foreign blocs and from the radicals and half-breeds," but—writing in the same year that conservationist Aldo Leopold wrote *Game Management*—Grant declared that it was "the duty of all Americans . . . to face the problem boldly and to take all eugenic means to . . . abate drastically the increase of the unfit."[41]

Grant had been meaning to write this sequel to *The Passing of the Great Race* for years, but his debilitating battle with arthritis had prevented him from engaging in any sustained work. It was his comrade Henry Fairfield Osborn, determined that Grant not be allowed to "rust out in peace," who insisted that Grant commence work on the project "as your contribution to history." As word spread in the early 1930s that Grant was finally working on a new epic, his disciples grew giddy with anticipation. Major Frederick Russell Burnham, for example, wrote to say he was "delighted to know you have again picked up your rusty pen. There is no one now living who has given the study of racial problems the time and thought and keen observation which you have given them." And Charles Benedict Davenport admitted to Grant that lately he had been overwhelmed by the difficulty of counteracting the arguments of the egalitarians, but was encouraged that "your book will play an important part in destroying these idols of the environmentalists. I can only trust that it may have as much influence on civilization as has your *Passing of the Great Race*."[42]

Eugenicist Paul Popenoe devoted over four years to doing research for Grant's book, and compiled its bibliography of almost four hundred works in English, French, German, Italian, and Spanish. Harry H. Laughlin also spent a great deal of time and energy looking up various statistics for Grant, and the two men met numerous times over a two-year period to review drafts of the manuscript. Finally, Charles Stewart Davison, William K. Gregory, Clarence G. Campbell, John B. Trevor, and Henry Fairfield Osborn all edited the final draft. This was the last hurrah, and they knew it. And so they had to get it right. If anything was going to reverse the decline of their movement, it was another best seller by Madison Grant.

As he had done for *The Passing of the Great Race*, Henry Fairfield Osborn supplied the introduction, in which he boldly assures the book's readers that *The Conquest of a Continent* "explodes the bubble" of the environmentalists and reveals their beliefs to be "merely myths" by showing "that moral, intellectual, and spiritual traits are just as distinctive and characteristic of different races as head-form." That such a statement could be made in 1933, twenty-two years after Franz Boas had shown that head form was *not* distinctive of different races, is more than a little unsettling.[43]

In 1930, Scribner's had rejected Grant's manuscript of *The Alien in Our Midst,* but in 1933 Grant's associates all assured the publisher that *The Conquest of a Continent* was a worthy endeavor. After thinking it over, Maxwell Perkins agreed that the book was "a very impressive and important one" and that it had a chance "to sell well and long." And so Scribner's renewed its association with Grant and energetically promoted *The Conquest of a Continent* as "the most important work of recent years." The firm provided review copies to educational journals and anthropology departments. It offered discounts to patriotic societies and literary groups. It sent mailings to thousands of high school teachers and librarians, and tried to get the book adopted as a text in colleges. It sent out circulars claiming that just as "Herr Hitler" was "working out his own solution" to the race problem in Germany, so Grant's book offered a solution to "our own problem."[44]

But when copies of *The Conquest of a Continent* were stacked on the shelves of bookstores at the end of 1933, they remained there.

Grant's friends, to be sure, were thrilled by the book. Their positive embrace of Grant's racist message is highly revelatory of the delusional state of eugenics in the early 1930s. California eugenicist C. M. Goethe, for example, admitted to Lothrop Stoddard that he was "so profoundly stirred" by the book that he sat up all night reading it from cover to cover. He told Grant that *The Conquest of a Continent* was "epoch-making" and thanked him for "still inspiring all of us to continue the struggle." Goethe distributed the book to his friends, mentioned it in all his public speeches, and through his involvement with the federated church movement recommended it to thousands of Protestant groups. Nobel Peace Prize winner Elihu Root thanked "my dear Grant" for his "admirable" book, which he hoped would help preserve "the original Nordic stock." Similar sentiments were expressed by all the other Grantians, including Harvard anthropologist E. A. Hooton, who lyrically confided to Grant: "I have a basic sympathy for you in your opposition to the flooding of this country with alien scum."[45]

Some of the reviews of *The Conquest of a Continent* were actually somewhat favorable. Carl Wittke mildly mocked Grant's Nordicism in the *Mississippi Valley Historical Review* ("The book . . . will be welcomed wherever gentlemen continue to prefer blondes"), but overall Wittke was surprisingly accepting of Grant's theses. And Aldous Huxley was fairly laudatory in his review for the *New York American,* declaring that "every reader of Mr. Madison Grant's book must be grateful to him for the way in which he has set forth the facts of American development."[46]

But most of the reviews were overwhelmingly negative and markedly curt. William L. Langer, in *Foreign Affairs,* penned perhaps the shortest book review on record: "Science submerged by opinion." The *New York Times* was slightly less taciturn but lambasted the "pernicious" Grant for following in the footsteps of Gobineau, Chamberlain, and Hitler. *Christian Century* pointed out that Grant's "doomed" political philosophy had "flared into a sudden ascendancy in Germany" and concluded that "Mr. Grant prides himself upon having produced 'the first history of any nation that has been written in terms of race.' It should also be the last."[47]

Grant had predicted that *The Conquest of a Continent* would be denounced "by those who, like the Jews," were "egalitarians." And, indeed, the Boasians were up in arms over the book. In 1930, they had not even bothered to refute *The Alien in Our Midst,* but *The Conquest of a Continent* was a masterpiece of racial propaganda; it threatened to cause trouble unless discredited. Within a few weeks of one another, scathing attacks on the book were produced by Ales Hrdlicka, Ruth Benedict (who scornfully labeled Grant's thesis "a trifle ridiculous" and noted that the only difference between Grant's book and Nazi propaganda was that "in Germany they say Aryan in place of Nordic"), and Melville Herskovits (who, writing in the *Nation*—which sixteen years earlier had praised *The Passing of the Great Race* for being "original," "courageous," and "important"—now expressed surprise that "such works . . . are apparently still being written"). And finally Franz Boas himself, in a review titled "Nordic Propaganda," attacked *Conquest* in the *New Republic.* It was the same venue in which he had criticized *The Passing of the Great Race,* but this time Boas was able to cite a decade's worth of research by anthropologists and geneticists to support his criticisms.[48]

Maxwell Perkins told Grant not to take the attacks of the Boasians personally, since almost all of them, after all, were Jews. Grant admitted that "the Jews" were "roaring against the book" and explained to Elihu Root that the problem with the Jews was that they had rather unrealistic expectations about themselves—they seemed "to think that they should be considered native Americans."[49]

In fact, in December of 1933 the Anti-Defamation League alerted U.S. and British journals that *The Conquest of a Continent* was "even more destructive than Hitler's *Mein Kampf,*" and the ADL therefore appealed to editors to "stifle the sale of this book" by refusing to run reviews. The censorial reaction of the Anti-Defamation League was unfortunate, as it validated for Grant his suspicion that Jewish influence was stifling freedom of speech in the United States. Furthermore, the ADL need not have bothered, as the reading public had no interest in *The Conquest of a Continent,* which in 1933 was hopelessly antiquated. Certainly, Americans were not interested in listening to a wealthy lawyer who claimed, in the middle of the Great Depression, that the nation's most serious problem was the declining demographics of the dolichocephalics. (And there was something a tad pathetic about a bedridden, childless patrician, who spent his winters in Boca Raton, devoting hundreds of pages to espousing the strong, virile barbarians who had evolved in "the fogs and long winters of the north.") In short, Newton B. Drury could not have been more wrong when he said of *Conquest* that "its appearance could not have been more timely."[50]

By the end of the 1930s, just three thousand copies of *The Conquest of a Continent* had been sold, and after the United States declared war on Germany, Scribner's melted down the plates. The negative reception greatly puzzled and disturbed the Grantians. This was the book they had been counting on to revive the eugenics movement and vindicate them against a decade of attacks by the Jewish environmentalists. Harry H. Laughlin had not been alone in predicting that *Conquest* would not only be a best seller now but continue as "a book of permanent reference" well into the future. When that did not happen, the only rationale the eugenicists could come up with was that Grant had been the victim of a Jewish conspiracy. But the real explanation for the book's poor sales was that the eugenic gospel no longer resonated with the public, and not even a work as smoothly argued as *The Conquest of a Continent* was going to prevent the movement's inexorable slide toward oblivion.[51]

Grant was severely affected by the failure of his book. He had spent an enormous amount of money ($8,000—equivalent to almost $100,000 today) preparing the manuscript, "and that," he told Perkins, "is about all I can spare in these times." In return, he earned less than $500 in royalties. It did not help that Grant was aging rapidly (he was sixty-eight years old when *Conquest* was published) and his mind was no longer what it used to be. He howled that he had been the object of a malicious boycott and spent countless hours writing bitter letters to bookstores complaining that they were not properly displaying his book. On numerous occasions he reported to Scribner's that the book was not in stock in some store when a quick check revealed that plenty of copies were on hand.

Everyone at Scribner's was growing increasingly annoyed with Grant's crotchetiness and his episodes of confusion. Even before the book was published, Grant had angered Maxwell Perkins by constantly pestering him about last-minute revisions and arguing over financial matters. Perkins had never had problems with Grant before. (This was in contrast to Grant's stablemate Ernest Hemingway, whose off-color language was a constant worry to his editor. A. Scott Berg tells of the time that Perkins, while reading the manuscript for Hemingway's *Farewell to Arms,* spotted a number of four-letter words that might have to be excised. Perkins jotted the troublesome words—*shit, fuck, piss*—on his desk calendar, without regard to the calendar's heading: "Things to Do Today." When Charles Scribner walked by and saw the list he dryly informed Perkins that he was in great trouble if he actually needed to remind himself to do those things.)[52] But now that Grant's book had garnered nothing but negative reviews and paltry sales, the exasperation of Scribner's employees turned to scorn. Low-level staffers began ridiculing Grant, and even Perkins could barely hide his contempt. Grant did not help matters by continuing to make unrealistic demands, such as that Scribner's print a revised edition so that the word "Scotch" could be replaced by the word "Scottish" in fifteen places. Of course, there never was or would be a revised edition of *The Conquest of a Continent.* And once, when Grant informed Perkins that he was heading to a luxurious resort in the Poconos to seek relief for his arthritis, a typo in Perkins's response may have revealed the editor's true feelings: "I hope Skytop is as good as it wounds."[53]

Disorganization

Toward the end of their lives, Georges Vacher de Lapouge congratulated Madison Grant: "You have done good work for the benefit of your country and that of the entire human species."[54] But the appreciation of friends was paltry comfort. Not only were Grant's books not selling in the 1930s, but the organizations that comprised the interlocking directorate of scientific racism were disintegrating one by one.

The IRL

After Congress passed the Immigration Restriction Act of 1924, contributions to the Immigration Restriction League declined 70 percent, reflecting the fact that the organization no longer had a reason to exist. As with most such groups, however, its leaders could not face the idea of disbanding. The executive committee immediately rushed out a letter to the membership urging them to stand fast, as "Much remains to be done. . . . The enemy is still very much alive." There was some uncertainty as to just who the "enemy" might be, but after lengthy deliberation the league finally decided it would focus on extending the immigration quotas to Mexicans.[55]

I have recounted elsewhere the league's lengthy but unsuccessful battle to ban immigration from Mexico,[56] a battle that was unsuccessful because, like all of Grant's organizations, the IRL could never overcome the public apathy that set in following the triumphs of 1924. Furthermore, the league's leaders failed to address the fact that from 1932 to 1936 more than twice as many immigrants left as were admitted to the United States, and that the platform of the IRL was wildly irrelevant to the concerns of the American public. The league's income continued to drop, and the only reason the organization was able to stay afloat at all was the continuing generosity of millionaire Wickliffe P. Draper. But in 1937, Draper decided that the IRL was a bad investment. He withdrew his support and switched his allegiance (and $50,000) to the newly created Pioneer Fund. The Immigration Restriction League soon faded away, after five decades of restrictionist agitation, leaving executive secretary Theodore G. Holcombe to mutter: "To me it is appalling—where did it all go?"[57]

The ADS

The Immigration Restriction League was not the only Grantian stronghold that was crumbling during this period. The American Defense Society, which had played a large role in fostering the Red Scare of the early 1920s, was also on the ropes. By the end of that decade the country was heartily tired of the American Defense Society's particular brand of xenophobic paranoia, and membership in the group had declined to some two thousand persons.

But in 1935, outrage over the New Deal led Charles Stewart Davison and Madison Grant to revive the society. Not only had the government fallen into the hands of socialists, but Davison (like Father Coughlin and Adolf Hitler) was certain that FDR and his wife were closet Jews. After pondering the evidence, Grant was forced to agree that his former colleague on the Taconic Park Commission was probably of "Ashkenazian descent." (Neither Grant nor Davison followed this train of thought to its logical conclusion, for if both Franklin and Eleanor were Jewish, then the implications for cousin Theodore, the great Nordic hero, were too disturbing to acknowledge consciously.)[58]

As if it were 1919 all over again, the ADS in 1935 called for a total ban on immigration, suppression of "the Red Menace," removal of "disloyal" teachers, and massive deportation of suspicious aliens. The ADS also distributed three hundred thousand life-size portraits of Theodore Roosevelt, for—Jewish or not—the manly visage of the Oyster Bay Roosevelt was a constant reminder of the failings of the Hyde Park version.

But the American Defense Society was not living completely in the past. Its platform in the 1930s did feature a few items added specifically to counter the menace of "The Jew Deal." It called for reduction in the cost of government, withdrawal of recognition from the Soviet Union, and support for "the American System of Government as against Socialism, Communism, Fascism and all 'isms'" (the latter demand presumably not meant to include capitalism or racism). With this agenda, the society attracted a sprinkling of new members, of whom the three most notable were Mary Pickford (who had worked with Madison Grant to save the redwoods), Tallulah Bankhead, and Alice Roosevelt Longworth.[59]

By the time FDR began his second term (in spite of Charles Stewart Davison's assurances that Alf Landon would crush "That Man in the White House"), the ADS was back in business. The society printed pamphlets by J. Edgar Hoover warning about the menace of the Communists. It sent reports on suspected "subversive elements" to Congressman Martin Dies and his House Un-American Activities Committee. And it helped convince the New York legislature to create the infamous Coudert Committee, which in 1940 investigated "radical infiltration" of the New York City school system and fired several teachers on suspicion of being Communists.

The success of the Coudert Committee notwithstanding, the American Defense Society never came close to achieving the size or the influence it had during the Red Scare, and it quietly disappeared in the late 1940s. Its surviving members eventually moved on to kindred causes, particularly anticommunism in the 1950s and then opposition to the civil rights movement in the 1960s.

And so a pattern of anxiety repeated itself. Grant and his fellow patriots had gone from hating Huns during the First World War, to fearing Bolshies in the aftermath of the war, to opposing uppity coloreds in the following decade. And the members of the next generation of right-wingers went from hating Nazis during the Second World War, to fearing pinkos in the aftermath of the war, to opposing militant blacks in the following decade. With the preparedness crowd, plus ça change, plus c'est la même chose.

The ERA

Madison Grant's many eugenic organizations—like his nativist and patriotic groups—were also fading in the 1930s. The Galton Society, for example, quietly dissolved in 1935 (just four months after the Half-Moon Club held its last dinner and disbanded). The Eugenics Research Association managed to hang on a bit longer, but it was definitely in trouble. Revenues from dues were declining because of the Depression, and reputable scientists were abandoning the association in droves. The annual meeting of the ERA was turning into a convocation of amateurs, eccentrics, and fascist sympathizers.

The group's last president was Sacramento realtor C. M. Goethe. The grandson of a clergyman and the son of a Sunday school superintendent, Charles Matthias Goethe was one of the few Grantians who was actively religious (and vehemently anti-Catholic). He was also a tightfisted businessman, who began his career as a child by advancing his friends' allowances and then ruthlessly collecting the interest. As Goethe grew older he moved into banking and real estate, and was responsible for the construction of hundreds of subdivisions in northern California. Though he always protested that he was not a wealthy man, Goethe wound up leaving an estate of $24 million. Neurotic, obsessive, and rigidly methodical, Goethe—like most Grantians—was a puritan who never smoked or drank. Indeed, over the years Goethe waged active campaigns against alcohol, narcotics, venereal disease, tuberculosis, and prostitution. It is probably superfluous to add that Goethe never fathered any children (although, like many of Madison Grant's friends, he did play a large role in founding the Boy Scouts of America).[60]

Goethe's obsession with personal purity was paralleled by a mania for racial purity, and he always testified to the "profound influence" of Madison Grant on his "philosophy of life." (As a good Methodist, Goethe would hold a "Madison

Grant Hour" in his library once a month, where he would reread *The Passing of the Great Race*). In addition to being president of the ERA, Goethe was also the founder of the Eugenics Society of Northern California (ESNC) and the Immigration Study Commission (ISC), which campaigned against the immigration of Mexicans (who were "disease carriers" and "superstitious savages"), Puerto Ricans (who were "largely moronic" and had "a jungle fecundity"), and Filipinos (who were "living museums of intestinal parasites"). The Eugenics Society of Northern California and the Immigration Study Commission were both similar to Earnest Sevier Cox's White America Society, in that they were a one-man operation. The "press office" of the ESNC was Goethe's desk, and when the ISC announced that its "field workers" were making "overseas studies," it meant that Charles and Mrs. Goethe were on vacation.[61]

Goethe was one of the great right-wing propagandists, who spent over a million dollars of his own money to publish scores of racist pamphlets and place hundreds of anti-immigrant articles in newspapers throughout the English-speaking world. During the Second World War, he was the main financial supporter of several preparedness organizations that advocated the mass evacuation of Japanese Americans from the West Coast. The Home Front Commandos, for example, inundated California with racist literature declaring "No Jap is fit to associate with Human Beings" and urging patriotic Americans to "Slap the Jap Rat."[62] After the war, Goethe was a major supporter of Richard Nixon and one of the contributors to the infamous "slush fund" (which led to the Checkers speech).

It is easy to apply to C. M. Goethe a pained admission of Sir Francis Galton: "It may be said, without mincing words . . . that the subject of eugenics is particularly attractive to 'cranks.'"[63] But we should not dismiss Charles Matthias Goethe too quickly. Like so many of Madison Grant's friends, Goethe was active on many fronts. He worked with Grant to preserve the pronghorn antelope and American bison, and he fought for the creation of Everglades National Park and Point Reyes National Seashore. A great friend of John Muir (whom he first met when they bumped into each other while hiking in the Sierra), Goethe was a leading member of the Sierra Club, and it was Goethe who formulated the idea—and provided the funding—for the popular program in which rangers of the National Park Service give campfire talks in the national parks.[64] He was also a major contributor to the Save-the-Redwoods League, and the Goethe Memorial Grove sits next to the Madison Grant Forest and Elk Refuge in northern California. At the age of seventy-one, Goethe turned to philanthropy and doled out funds with such generosity that March 31, 1962, was proclaimed Charles M. Goethe Day in Sacramento, and California's capital is a veritable shrine to the eugenicist, with a Goethe Road, Goethe Park, Goethe Middle School, and Goethe Arboretum on the campus of California State University, Sacramento.

Back in the 1930s, during Goethe's term as president of the Eugenics Research Association, Benito Mussolini's science adviser, Corrado Gini, was made a member of the ERA, as were several Nazi eugenicists such as Eugen Fischer and Ernst Rüdin. It was clear that the ERA no longer had a future in a democratic society, and the twenty-sixth—and last—annual meeting of the ERA was held in June of 1938.

The ERO

Over at the Eugenics Record Office, Charles Benedict Davenport had become obsolete. In 1932, he had castigated his fellow biologists for their innate conservatism: "It is too bad that we reach a stage when new ideas are abhorrent to us." Unfortunately, Davenport had reached that stage in 1911. As the field of genetics progressed, Davenport did not, and he finally retired as director of the ERO in 1934. He spent the rest of his life organizing taxpayers' associations to stave off the depredations of the welfare state, and anxiously searching the skies above Long Island for incoming Axis bombers. He died in 1944 at the age of seventy-eight after catching pneumonia while boiling a whale's skull.[65]

At the strong urging of Madison Grant and Henry Fairfield Osborn, Davenport's loyal lieutenant Harry H. Laughlin was named his successor at the ERO. But it was public knowledge that Laughlin was suffering from increasingly severe attacks of epilepsy. (In 1937 Laughlin had a seizure while driving through Cold Spring Harbor, crashed his car through a retaining wall, and plunged into Long Island Sound.) More importantly, the directors of the Carnegie Institution of Washington, which sponsored the Eugenics Record Office, had been growing more and more disenchanted with—and embarrassed by—eugenics, and wanted to end their support of the ERO. The president of the Carnegie Institution, however, was John C. Merriam, who was, as he himself put it, "one of the vigorous supporters of eugenics" and a leading ally of Madison Grant. Merriam was well aware that Laughlin's work at the ERO was viewed with disdain by serious geneticists, but for a full decade Merriam managed to forestall the inevitable by employing the traditional tactic of those seeking to avoid action: he appointed a succession of committees to "investigate" the situation at Cold Spring Harbor.

But when Merriam stepped down as president of the Carnegie Institution at the end of 1938, his successor, Vannevar Bush, was in no mood for delay. He immediately sent Laughlin packing back to his hometown of Kirksville, Missouri (where he died four years later), and the ERO was shut down on December 31, 1939.[66]

The American Museum of Natural History

For a number of years Henry Fairfield Osborn's notions about evolution had been viewed by his colleagues as flawed and out-of-date. He was still considered by the public—and by his fellow aristocrats in the social world of Madison Grant—to be one of the most eminent scholars in the world, but in his later years Osborn's immense egotism inflated to absurd proportions, and serious scientists (and his own assistants) ridiculed him behind his back as he demanded exorbitant honorariums to speak at conferences and adopted the annoying habit, as the hyperarrogant are wont to do, of referring to himself in the third person ("Osborn believes that . . . "). In the end, his refusal to give credence to the latest findings of genetics rendered him, and his work, obsolete, and in 1933 he was pressured to resign the presidency of the American Museum of Natural History. It was Franz Boas who pointed out the irony that Osborn's dinosaur museum had itself become a dinosaur.[67]

With Osborn gone, the museum would never again be a center of eugenic thought. Two years after being given the boot, Osborn died at the age of seventy-eight at his Castle Rock estate. A devastated Grant, who had visited or spoken with his friend every day since 1895, and had worked with him to create the Bronx Zoo, the New York Aquarium, the American Bison Society, the Half-Moon Club, the Galton Society, the Save-the-Redwoods League, the American Eugenics Society, and so many other clubs and organizations, told his friend John C. Merriam, "We have sustained a terrible loss . . . and I for one feel his death very keenly."[68]

The AES

The outlook for the Grantians—a dwindling group of childless, crippled, epileptic old white men, examining their bowel movements for impurities and scanning the skies for invaders—was bleak. Major Frederick Russell Burnham, noting that they were surrounded by Boasians and beset by financial woes, jested that the beleaguered band of Grantians should escape to some offshore island where they could live off the land. "It occurred to me," he told Grant, "that if our tribe should have to gather and capture Catalina Island we would have to have a Medicine Man, and if you would like to stand as our nominee for that office you are IT. We would all pay tribute to you in shellfish, goat meat, birds' eggs, sea moss, etc., and in return, you would tell us the propitious hour for any foray and also prevent the destruction of the moon during the time of eclipse."[69]

All joking aside, the eugenics movement was in serious distress. The *Eugenical News* was having financial difficulties, and only an infusion of thousands of dollars from Grant and Osborn kept it alive. In the mid-1930s the journal was forced to pare back to a bimonthly, and later it became a quarterly. With the ERO, the ERA, and the Galton Society fading from the scene and the National Research Council in the hands of the Boasians, the American Eugenics Society was left virtually alone in the eugenic wilderness. But membership was declining drastically: from 1,260 members representing every state in the Union in

1930, the AES shriveled to fewer than 500 members in 1933, and then to 282 in 1936. As to the impressive Advisory Council and the many committees of the AES, once so active and full of promise, one member sadly remembered that they rapidly "shrank into complete desuetude" as reputable researchers stopped participating.[70]

Financially, the American Eugenics Society was actually in fairly decent shape, thanks to John D. Rockefeller, Jr., and George Eastman, who had annually contributed to the organization $5,000 and $10,000 respectively. But in 1929, Rockefeller began cutting back his support of the AES after his advisers concluded that the eugenics movement was overly dominated by the "Nordic complex" of "cranks" like Madison Grant. Just a few years earlier, the Rockefeller people had been fully supportive of the AES and of its program to weed out the unfit. But attitudes had changed rapidly, and Rockefeller lowered his contribution to $3,500 in 1929, $2,000 in 1930, and nothing at all after that. Luckily, in February of 1932 George Eastman pledged to make up the difference by increasing his contribution to $15,000 per year—a substantial amount of money in the middle of the Depression. Unluckily, Eastman committed suicide one month later, transforming the AES into a financial orphan. This development reminds us yet again that Madison Grant's eugenic organizations, like most of his conservation societies, were not grassroots movements but narrowly supported clubs that depended almost exclusively on the largesse of one or two philanthropists.[71]

As the Great Depression worsened, the American Eugenics Society went into the red for the first time, and the executive committee began to panic. Its offices in Chicago, Ann Arbor, and New York City were shut down, its publishing arm (the Galton Publishing Company) folded, the journal *Eugenics* ceased publication, and what was left of the office staff in New Haven was asked to serve without salary. Increasingly desperate appeals were sent out to the (shrinking) membership asking for money, but all in vain. The AES managed to limp through the rest of the decade, thanks to occasional infusions of cash from Wickliffe P. Draper's Pioneer Fund. Its activities were minimal, though it did sponsor occasional conferences where supporters such as Will Durant and Margaret Sanger delivered lectures to the remaining faithful. However, like the American Communist Party in the 1960s, the AES—in the words of one member—was turning into an "old folks home."[72]

After lying low during World War II (when no one in the United States wanted to be associated with eugenics), Madison Grant's intellectual heirs shrewdly modernized their postwar agenda and assumed leading roles in the family planning and the environmental movements. Their prewar anxiety over the high birthrate of Mexicans and Negroes in the United States evolved into a concern about the population explosion in the Third World. And their previous worries over the insidious effects of inferior germ plasm turned into apprehension over the possibility that atomic radiation and industrial pollution were causing genetic mutations. The two issues of overpopulation and environmental degradation were interrelated, because as the planet became more crowded it also became more polluted. "Population control" would thus serve to keep the populace of the Third World in check and to prevent any further damage to the environment, and both of these developments would help to protect the human gene pool.[73]

So, for example, the organizers of Planned Parenthood, the Population Reference Bureau, and the Population Association of America were all former eugenicists. Similarly, the first director of the Population Council (the organization funded by John D. Rockefeller III to promote family planning in the Third World) was eugenicist Frederick Osborn (nephew of Henry Fairfield Osborn and president of the American Eugenics Society). It was the work of the Population Council that led to the development of the IUD, and in 1994 the Population Council obtained the U.S. patent for RU-486 (the abortion pill). Linda Gordon is absolutely correct when she writes that "population control was the successor to eugenics in every respect—ideologically, organizationally, and in personnel."[74]

For us progressives who support family planning and the environmental movement (and see the two issues as intertwined), it is unquestionably painful to acknowledge the ties of those concerns to the eugenics movement. Disturbing though it may be, the fact that there is a connection between the "good" causes of today and the "bad" causes of yesteryear at least serves to remind us that Madison Grant and his peers were convinced they were doing what was best for mankind when they advocated what are regarded today as reprehensible practices. Grant was involved with the now-discredited cause of eugenics for the same reason he was involved with the now-favored cause of conservation: because he was just as concerned about the future of his own society and the fate of the planet as any of us.

Nonetheless, it may be difficult to maintain this relativist appreciation for Grant's position when we examine his connection to the horrific events unfolding in Nazi Germany in the 1930s.

14

The Ever-Widening Circle: The Third Reich

Anthropology is the most dangerous of all sciences.

Sir Arthur Keith

In his presidential address to the Eugenics Research Association in 1929, Clarence G. Campbell predicted: "The truth which a few eugenicists see so clearly today will become commonplace opinion in the course of time."[1] As we saw in the previous chapter, Campbell's prediction did not hold true for the United States. But the Old World was a different story. In fact, in the late 1920s, just as sales of *The Passing of the Great Race* began to slacken and Madison Grant's reputation went into decline in the United States, interest in scientific racism began to swell on the continent of Europe. All four editions of *The Passing of the Great Race* had already been published in Britain, and now in Germany at least five publishers expressed their desire to translate the book. Charles Scribner told Grant that he was quite willing to assist with the production of the German edition; after all, noted Scribner with pride, the book had been "a pioneer" that brought "the race question" to the fore and was "continually referred to as an authority." The rights were awarded to the Munich firm of J. F. Lehmann, an enthusiastically pro-Nazi publishing house. The anti-Semitic Lehmann was an early and close associate of Hitler (joining the Nazi Party in 1920), and he published many of the German racists (as well as such American anti-Semites as Henry Ford and Lothrop Stoddard). And so in 1925 Lehmann brought out *Der Untergang der Grossen Rasse.*[2] One year later, the Paris publishing company Payot produced the French edition of Grant's book, *La Declin de la Grande Race,* with a fawning preface by G. Vacher de Lapouge. And two years later, the Oslo firm of Some & Co. graced Norwegian readers with *Den Store Rases Undergang,* translated by Jon Alfred Mjøen (the most famous eugenicist in Scandinavia). The result of all this

European interest in Nordicism would prove flattering to Grant—and fatal to millions.

Der Untergang der Grossen Rasse

It was in Germany that Grant and his book were received with the most enthusiasm. Grant's Nordicism landed on fertile soil in Weimar Germany, where popular anti-Semitism was taking on a harder, more "scientific" edge and theorists were beginning to argue that the laws of biology rendered the Jews unassimilable. When *Der Untergang der Grossen Rasse* appeared in 1925, the *Literarische Wochenschrift* declared that Grant's book, with its dire warnings of the dying off of the great Nordic race, was far more important than even Spengler's *Decline of the West.* Grant's translator, Professor Rudolf Polland of Graz, confessed in the introduction that as soon as he had read Grant's book he resolved to translate it, since "the lessons and warnings of Grant . . . are exceedingly important and relevant to the German Volk." It was fortuitous, wrote Polland, that Grant had come along to offer to the Germans a eugenic remedy to the problem of racial degeneration. And it was Polland's fervent hope that Grant, the groundbreaking pioneer in scientific racism, would inspire patriotic Germans to take up the Nordic cause for the greater glory of the *Volk.*[3]

An excited Lapouge wrote to Grant with the good news of the German reception: "Your book has exercised, from the publication of the German edition, a powerful effect upon the Nordic movement and everybody knows it." Not only did German scientists "have a great veneration" for Grant, but Lapouge was also heartened to see that the young people of Germany were now "imbued with biological ideas" and were finally beginning to "deplore, for example, the incorporation into Germany of foreign elements." German eugenicist Kurt Holler, echoing the observations of Lapouge, told readers of the *Eugenical News* that the books of Grant and his disciple Lothrop Stoddard had found "a large circle of readers" in Germany and helped spark in that country "the movement for the preservation and increase of the Nordic race." By 1928, Grant could not contain his excitement that "Germany has awakened to the importance of increasing its proportion of pure Nordic blood."[4]

German eugenicists eagerly established close ties with the Grantians in the 1920s. They were full of admiration for the success of their American counterparts in restricting immigration, passing antimiscegenation laws, and implementing coercive sterilization acts. German journals provided timely updates on developments in U.S. eugenics and regularly translated the articles of the Americans. In return, the Grantians worked hard after the Great War to reintegrate their German friends into the international scientific community, and they practiced what they preached by attending eugenics conferences in Germany.[5]

Thus, by the time the Nazis rose to power, the party's policy makers were thoroughly familiar with the Grantians' legislative program. For example, taking their cue from Adolf Hitler (who in *Mein Kampf* had pointed to the Immigration Restriction Act as a model for his program of racial purification), numerous Nazi officials praised the United States for its strict immigration policy and expressed their envy that the Americans had already solved their Jewish question by barring the immigration of "unwanted racial elements."[6]

Adolf Hitler's closest scientific advisers were avid fans of Madison Grant and accepted all the major tenets of his scientific racism. Hitler himself, who wrote in *Mein Kampf* that "the highest aim of human existence is . . . the conservation of the race," sent Grant a letter thanking him for writing *The Passing of the Great Race* and telling him that "the book is my Bible."[7] *Mein Kampf* is riddled with passages that seem directly inspired by *The Passing of the Great Race,* in particular the chapters titled "Race and People" and "The State," which encapsulate all the aspects of Grantian thought (including the primacy of race, the worship of modern science, the subordination of the individual to the state, millennialism, the need for positive and negative eugenics, and the connection between eugenics and animal husbandry).[8]

Despite the similarities between the writings of Hitler and Grant, it is a fair assumption that Hitler absorbed Grant secondhand, through the works of German eugenicists influenced by Grant. Certainly, Nazi racial thinkers often pointed to Madison Grant and his disciples as the inspiration for their work. H. H. von Schneidewind, for example, extolled the role of Grant and Stoddard in crafting America's immigration restriction laws. Alfred Rosenberg, in his masterpiece *Der Mythus des 20. Jahrhunderts* (The Myth of the Twentieth Century), which was second only to *Mein Kampf* as a best seller in Nazi Germany, commended Grant and Stoddard as the two "enlightened" Americans who had blazed the way toward an acceptance in the United States of scientific racism. And in 1936, when the Nazi Party published its official recommendation for essential reading in the field of human heredity, it mentioned only two books by non-German authors: Gobineau's *Inequality of Human Races* and Madison Grant's *Passing of the Great Race.*[9]

Fischer and Günther

Madison Grant had long been the conduit between the most important European eugenicists and their American colleagues. Throughout the 1920s and until his death in 1937, he was in constant contact with a number of Nazi researchers. His closest and longest ties were with two of the regime's most influential racial thinkers: Eugen Fischer and Hans F. K. Günther.

Eugen Fischer

Eugen Fischer was the foremost eugenicist in the Third Reich. Known as "the founder of human genetics" in Germany, Fischer began his career studying anatomy at the University of Jena (the same university where Ernst Haeckel had been a professor of zoology), and he was also a student of August Weismann at Freiburg (as were other important eugenicists, such as Fritz Lenz and Wilhelm Schallmayer). After completing his doctoral thesis on the genitalia of female orangutans, Fischer established his reputation in 1913 with *Die Rehobother Bastards und das Bastardierungsproblem beim Menschen* (The Rehoboth Bastards and the Problem of Miscegenation among Humans). As a leading expert on miscegenation, Fischer became a professor of anthropology at Freiburg, and was the editor of several prestigious anthropological journals and the head of the two leading anthropological societies in Germany. In 1921, he coauthored with Erwin Baur and Fritz Lenz the classic text of German genetics, *Menschliche Erblehre und Rassenhygiene* (Human Heredity and Eugenics), universally referred to as "Baur-Fischer-Lenz." The book touted eugenics as the cure for "diseases in the body politic," and cited the U.S. Army intelligence tests to conclude that, intellectually, "the Nordic race marches in the vanguard of mankind." Baur-Fischer-Lenz was published by J. F. Lehmann, the reactionary publisher responsible for the German edition of *The Passing of the Great Race.* Lehmann gave a copy of Baur-Fischer-Lenz to Adolf Hitler while he was imprisoned in Landsberg, and the book greatly influenced and justified National Socialist policies.[10]

Eugen Fischer was an ardent proponent of Nordic supremacy and an early supporter of Adolf Hitler because he realized, as his collaborator Fritz Lenz put it, that Hitler was "the first politician with truly wide influence who . . . recognized that the central mission of all politics [was] eugenics and who [would] actively support this mission." In 1927, Fischer became the founding director of the Kaiser Wilhelm Institute for Anthropology, Human Heredity, and Eugenics in Berlin-Dahlem, which became an important center of Nazi eugenics. In 1933, while retaining his position at the institute, Fischer was also appointed the first Nazi rector of the University of Berlin (the first anthropologist ever to hold that post). In his inaugural address, Fischer triumphantly declared: "The old worship of culture . . . is past." He further announced: "the theory of the heritability of mental as well as physical traits has finally been vindicated," and "genetics has . . . destroyed the theory of the equality of men." A few days later, Fischer began signing his Jewish subordinates' dismissal notices.[11]

As the Nazis solidified their dictatorship, the increasingly anti-Semitic Fischer performed a number of services for the regime. "When a people wants . . . to preserve its own nature," declared Fischer, "it must reject alien racial elements, and when these have already insinuated themselves, it must suppress them *and eliminate them.* The Jew is such an alien . . . . I reject Jewry with every means in my power, and without reserve, in order to preserve the hereditary endowment of my people.[12] Fischer trained hundreds of Nazi officials and SS doctors in racial theory, he provided expert testimony and issued certificates attest-

Madison Grant kept Nazi scientists abreast of the latest developments in American eugenics. Here, Professor Eugen Fischer, an enthusiastic disciple of Grant, peruses a fresh shipment of journals from the United States in 1935. Archiv der Max-Planck-Gesellschaft, Berlin-Dahlem.

ing to the racial purity (or lack thereof) of numerous individuals, and he helped devise Germany's sterilization program. He then aided in the implementation of the sterilization program by serving as a judge on Berlin's Appellate Genetic Health Court and by working with Hans F. K. Günther and the Gestapo to identify the *Rheinlandbastarde* (the offspring of native Germans and colored French occupation troops), who—per Hitler's order—were secretly sterilized. In return, Fischer's eugenic research was very well funded by the government (as well as the Rockefeller Foundation), and Fischer became, in the words of Robert Proctor, "the recognized Führer" of German anthropology.[13]

In 1937, Madison Grant's *Conquest of a Continent* was published in Germany (as *Die Eroberung eines Kontinents*). Henry Fairfield Osborn directed zoologist Ernst Mayr (of all people) to proofread the translation, and Grant asked Eugen Fischer to write the foreword. Fischer complied "with great pleasure." Writing from the Kaiser Wilhelm Institute, Fischer pointed out that Grant, the "well-known champion of racialist thinking in the United States," was "no stranger to German readers," who were already familiar with his "fascinating" *Passing of the Great Race.* Fischer wrote that Grant's latest book was especially relevant and that "no one today should pay more careful attention to Grant's work than the people of Germany, where racialist thinking has become the chief foundation of the population policies of the National Socialist state."[14]

In his foreword, Fischer could not help but allude to the Grant-Boas conflict: "It will surprise no one that this study met with great opposition in Grant's homeland, where politicians and scholars, led by Jewish anthropologist and ethnologist Franz Boas, have dominated all public opinion with the notion that racial differences are environmentally determined, malleable, and insignificant." But Fischer praised Grant for leading the opposition to Boas and the Jewish environmentalists. "No one should forget what courage, what intellectual independence it took for this scholar, now 72 years old, to promulgate in his country the racial idea as the basis of history." Fischer thanked Grant for inspiring Germans to work toward "a better future for our Volk" and ended with a wish that Grant's book, despite the attacks on it by "the Jewish press," would find "a much deserved German audience."[15]

It did indeed find an audience in Germany. Excerpts of *The Conquest of a Continent* appeared in German newspapers, and the book itself was reprinted in Germany in 1940. And why not? As the *New York Times* had observed: "Substitute Aryan for Nordic, and a good deal of Mr. Grant's argument would lend itself without much difficulty to the support of some recent pronouncements in Germany."[16]

During World War II, Eugen Fischer took part in government discussions about the solution to the Jewish problem, and he helped plan the deportation of European Jews to the slave camps. In 1943, Fischer expressed his gratitude, as a scientist, for having been given such an opportunity: "It is a rare and special good fortune for a theoretical science to flourish at a time when the prevailing ideology welcomes it, and its findings can immediately serve the policy of the state."[17]

Hans F. K. Günther

Eugen Fischer's most prominent student was Hans F. K. Günther, a friend and disciple of Madison Grant who became the Nazis' chief Nordicist and a particular favorite of Adolf Hitler. Günther (who claimed he could identify a Jew just from his posture) was the author of seventeen books on racial science and was the period's most widely read anthropologist. His most famous book, *Rassenkunde des deutschen Volkes* (Racial Science of the German People), appeared in 1922 (published by J. F. Lehmann) and was an immediate best seller. As the classic text of the Nordic school in Germany, it warned that unless negative eugenic measures were implemented, miscegenation (especially with the Jews) threatened the future of the Master Race. *Rassenkunde* went through numerous editions and was widely distributed during the Third Reich. Jacques Barzun reported that a knowledge of *Rassenkunde* was obligatory in Hitler's Germany and that even the Dresden School of Dance required its students to demonstrate proficiency in *Rassenkunde* in order to receive their diplomas.[18]

Günther cited Madison Grant frequently in *Rassenkunde* (and in all his other writings as well), and the influence of *The Passing of the Great Race* (along with the works of Gobineau, Chamberlain, and Lapouge) on both the form and the content of Günther's books is unmistakable. *Rassenkunde* even featured a photograph of a bronze bust of Grant (sculpted by Chester Beach), a tribute that went a long way toward convincing Grant that Günther was "one of the most erudite anthropologists in Europe."[19]

In 1927, Günther wrote *Rassenkunde Europas* (issued in English as *The Racial*

Elements of European History), which was yet another recapitulation of *The Passing of the Great Race.* In the book, Günther warned once more against miscegenation with the Jews, and ominously suggested that the "eugenic" practices of the ancient Spartans could serve as a particularly efficacious model for preserving the strength of the Nordic ruling class. Günther again cited Madison Grant prominently (and again placed a photograph of Grant in the text), and noted the "remarkable success" of *The Passing of the Great Race* in preparing the ground "for the reception of racial and eugenic theories" in the enlightened nations. Grant reviewed Günther's book for the *Eugenical News* and, not surprisingly, concluded that it deserved "the careful attention of all Americans who have the welfare of the country at heart."[20]

The books of Hans F. K. Günther sold hundreds of thousands of copies in Germany, and his influence cannot be overestimated. In 1930, when Wilhelm Frick became the first Nazi minister in Thuringia, one of his first acts (in addition to banning jazz music and the antiwar film *All Quiet on the Western Front*) was to appoint Günther to a specially created chair of racial anthropology at the University of Jena. Adolf Hitler attended Günther's inaugural lecture at which, as in his writings, Günther praised Madison Grant for convincing the U.S. Congress to pass the Immigration Restriction Act of 1924. An excited Grant told Charles Benedict Davenport that his friend Günther was "now in a powerful position at Jena," and Grant was "very much pleased." He sent Günther a library of books by the Grantians and had Earnest Sevier Cox send him a copy of *White America* (explaining to Cox that Günther was "one of the most distinguished anthropologists of Germany" and was "in entire accord with all our views"). In return, Günther arranged for the German edition of Lothrop Stoddard's *Revolt against Civilization,* and a few years later, before an audience at the University of Munich, Günther referred to Grant and Stoddard as the "spiritual fathers" of immigration restriction and proposed that their efforts serve as a model for the Third Reich's population policies.[21]

The Sterilization Law

In 1932, *Birth Control Review* produced a special issue devoted to sterilization. On the frontispiece was the opinion of Oliver Wendell Holmes in *Buck v. Bell.* Inside were numerous articles by such Grantians as Harry H. Laughlin, Leon F. Whitney, and E. S. Gosney, all espousing the great importance of sterilization. The issue also featured an article by Ernst Rüdin, professor of psychiatry at the University of Munich and one of the pioneers of Nazi eugenics, titled "Eugenical Sterilization: An Urgent Need." Rüdin assured his American readers that the "fear of abuses" in regard to state-sponsored compulsory sterilization was totally unfounded. The veracity of that statement was revealed just one year later, when the Nazis passed their infamous Sterilization Law.[22]

Adolf Hitler had been following eugenic developments in the United States for years. He knew that Madison Grant, in *The Passing of the Great Race,* had explained that a government could help preserve the Nordics by implementing a widespread sterilization program.[23] Accordingly, in June of 1933, just weeks after taking office, the Nazis promulgated the *Gesetz zur Verhütung Erbkranken Nachwuchses* (The Law for the Prevention of Hereditarily Diseased Offspring), a.k.a. the Sterilization Law.

The measure was quite consciously based on the model sterilization law of Harry H. Laughlin and the American Eugenics Society. It mandated sterilization (whether or not the victim was institutionalized) for persons found guilty of any of the following dysgenic conditions: congenital feeblemindedness, schizophrenia, manic depression, hereditary epilepsy, Huntington's chorea, hereditary blindness, hereditary deafness, serious physical deformity, or chronic alcoholism. Like so many Nazi measures, the Sterilization Law was conceived and justified as a public health measure. Dr. Arthur Gütt (the director of public health in the Ministry of the Interior), who drew up the law with advice from Ernst Rüdin, Fritz Lenz, Hans F. K. Günther, and other eugenicists, explained that their goal was "to prevent . . . *poisoning* the entire bloodstream of the race."[24]

Interior Minister Wilhelm Frick estimated that "hereditary defectives, whose progeny is therefore undesirable," constituted up to 20 percent of the German population. In the face of such a potentially cataclysmic statistic, he could only express his gratitude that eugenic science "gives us the right and the moral obligation to eliminate hereditary defectives from procreation. No misinterpreted charity nor religious scruples, based on the dogmas of past centuries, should prevent us from fulfilling this duty." If the Nazis were grateful to the eugenicists, the feeling was certainly reciprocated. Lapouge told Madison Grant that he was proud that "Frick has been with us since the beginning," and Ernst Rüdin wrote that thanks to Hitler and the Sterilization Law, "the dream we have cherished for more than thirty years of seeing eugenics converted into action has become reality."[25]

The German Sterilization Law required physicians to report all "unfit" persons to one of the 205 Erbgesundheitsgericht (Genetic Health Courts). The Reich established state medical academies to provide training in racial biology so physicians would be prepared for their new duties as "genetic doctors." The first such course was given at Halle University in August 1933. Doctor Heinz Kürten, the Nazi physician in charge, pointed to the eugenics legislation of the United States as the "model" for the new Germany, and stated that he was following "the American pathfinders Madison Grant and Lothrop Stoddard." Kürten was proud that whereas U.S. legislators had been reluctant to endorse some of Grant's more drastic proposals, the Third Reich was actually carrying them out.[26]

The Nazis were aware that compulsory sterilization was viewed by many as a violation of human rights, but they were convinced that just as the survival of the herd outweighs the life of the animal, so the welfare of society overrides the rights of the individual. A common Nazi slogan was *Gemeinnutz geht vor Eigennutz* (General Need Goes before Individual Need), and the *Handbook* of the Hitler Youth, which was required reading for the seven million members of the Hitler Jugend, explained: "it is a natural right of the community to protect itself against the threat of the individual."[27]

"If I can accept a divine Commandment," said Hitler, "it's this one: 'Thou shalt *preserve the species.*'" And the Nazis never doubted that in human society, as in nature, the species could survive only if the weak and the feeble were encouraged to die. "The life of the individual," admonished Hitler, "must not be set at too high a price." Like any conscientious wildlife manager, he averred: "I would prefer not to see anyone suffer, not to do harm to anyone. But when I realize that the species is in danger, then in my case sentiment gives way to the coldest reason." In a meeting with Ribbentrop, Goebbels, and Rosenberg, Hitler instructed them that it was their task to promulgate the concept that the individual's duty was to live and, if need be, to *die* "for the preservation of the species."[28]

Interior Minister Frick, in presenting the draft Sterilization Law to the Reichstag, had said (in a speech that Madison Grant translated and forwarded for immediate publication in the *Eugenical News*) that it was time for Germans to stop concentrating on "personal hygiene and the care of the individual" and start thinking about "race hygiene" and "the nation as a whole." The *Eugenical News* later concurred that "race purification may be accompanied by hardship to the individual, but society follows nature's method in regarding the progress of the race as more important than that of the individual."[29]

Despite the Sterilization Law, the sterilization of "normal" Germans was illegal in the Reich, as it was "a crime against the German body." Indeed, for the "healthy" portion of the *Volk,* the Nazis implemented a wide-ranging program of positive eugenics that included measures eugenicists had been proposing since Galton. Higher taxes, for instance, were imposed on unmarried persons, and all public officials were required to either marry or resign their office. Like Theodore Roosevelt, Hitler decreed that four children should be the normal goal of every married couple. The regime granted loans to eugenically fit couples to encourage breeding, and families were awarded the Honor Cross of German Motherhood in bronze (for four children), silver (six children), and gold (eight children). The *Lebensborn* program of Heinrich Himmler (who, it is not irrelevant to note, was involved in breeding poultry before becoming head of the Schutzstaffel) allowed Nordic women who were impregnated by SS sperm (whether or not they were married) to receive maternity care in special clinics. And, during the war, the Lebensborn organization abducted thousands of "racially valuable" children from occupied areas and sent them to Germany to be raised by Aryan couples.[30]

As to persons of lesser value, in the Sterilization Law's first year the Genetic Health Courts ordered over 56,000 people to be sterilized against their will; in the second year another 73,000; and by 1939 almost 400,000 Germans had been sterilized. Sterilization was so common that the operation became known as a *Hitlerschnitt*—literally "Hitler cut" but actually a word play on *Kaiserschnitt* (a cesarean operation). The size of the Nazi sterilization program caused a jealous Dr. Joseph DeJarnette (the Virginia eugenicist who had testified in the Carrie Buck case) to exclaim: "The Germans are beating us at our own game!" Leon F. Whitney of the AES agreed. "While we were pussyfooting around," he scolded his fellow Americans, the Nazis were accomplishing great things. Harry H. Laughlin told Madison Grant that he was thrilled that the speeches of Nazi leaders such as Wilhelm Frick "sound exactly as though spoken by a perfectly good American eugenist," but he admitted that he was jealous that he and Grant were only humble researchers, whereas Frick was "a powerful Reichsminister in a dictatorial government" that was "getting things done." (After the war, Frick was hanged at Nuremberg for crimes against humanity.)[31]

Envy aside, the Grantians in the United States were enthusiastic about the Nazi Sterilization Law. Leon F. Whitney wrote that "many far-sighted men and women . . . have long been working earnestly toward something very like what Hitler has now made compulsory . . . . And this represents but a small beginning, we are told!" The *Journal of Heredity,* like the *Eugenical News,* published the complete text of the Sterilization Law and praised the Nazi regime for being "the first example in modern times of an administration based frankly and determinedly on the principles of eugenics." The journal's associate editor, Paul Popenoe, quoted extensively and approvingly from *Mein Kampf* to show that "Hitler himself—though a bachelor"—had "long been a convinced advocate of race betterment through eugenic measures." And C. M. Goethe, president of the Eugenics Research Association, told the annual meeting of that group in 1936 that the two greatest achievements of the century so far had been the Immigration Restriction Act of 1924 and the eugenics program of Nazi Germany.[32]

Goethe's admiration for the Germans demonstrates that whereas in the 1920s German eugenicists had looked to the United States as a model, with the Nazi assumption of power it was the Americans who turned to the Germans for inspiration. In Alabama, state health officer J. N. Baker reported to the legislature in 1935 that "with baited breath, the entire civilized world" was "watching the bold experiment in mass sterilization recently launched in Germany."[33] Throughout the decade, the *Eugenical News* kept its readers updated on developments in the Third Reich. It reported on the activities of the Deutsche Gesellschaft für Rassenhygiene, provided monthly summaries of the *Archiv für Rassen- und Gesellschaftsbiologie* (the main journal of the German eugenics movement—published by J. F. Lehmann), and regularly reviewed the books of the German eugenicists, especially Erwin Baur, Fritz Lenz, Eugen Fischer, and Hans F. K. Günther. Fischer and Günther both corresponded regularly with Madison Grant in the 1930s, and Grant happily translated and forwarded to the *Eugenical News* articles by his Nazi friends (although some were rejected by the editors as being too extreme for even the *Eugenical News* to publish).[34]

Charles Benedict Davenport of the ERO participated in the Festschrift for Eugen Fischer in 1934 and in a later volume for Otto Reche (a Nazi racial anthropologist). Davenport was also responsible for the election of Nazi eugenicist Ernst Rüdin to the presidency of the International Federation of Eugenic Organizations. Indeed, the history of the eugenics movement, from its origins with Sir Francis Galton to its flowering under the Big Four in the United States to its apotheosis under the Nazis, was embodied by the presidency of the International Federation of Eugenic Organizations, which passed from England (Leonard Darwin, 1921–27) to the United States (Charles Benedict Davenport, 1927–32) to the Third Reich (Ernst Rüdin, 1932–36).

Madison Grant warned Harry H. Laughlin in 1934 that "while most people of *our* type are in sympathy with the German eugenical measures," political considerations meant that "we will have to proceed cautiously in endorsing them." But the Grantians found it difficult to conceal their ardor for the Nazis. Frederick Osborn expressed the excitement within the American eugenics community when he stated at a conference of the AES in 1937 that the German sterilization program was "an excellent one" and that "taken altogether, recent developments in Germany constitute perhaps the most important social experiment which had ever been tried." Later that year, the ERO, the ERA, and Wickliffe P. Draper's Pioneer Fund cooperated in distributing Nazi propaganda films on sterilization to American churches, clubs, and high school biology classes. Produced in Berlin by the party's Racial Political Office, one of the films, *Erbkrank* (The Hereditary Defective), revealed that Jews are particularly susceptible to moral degeneracy and mental retardation, and explained that caring for the feebleminded cost the state millions of marks every year at a time when hardworking Nordic citizens were doing their best just to make ends meet. The film closed with a shot of a man and a woman tending their field, with the subtitle: "The farmer who prevents the overgrowth of the weed, promotes the valuable."[35]

In 1933, Erwin Baur, coauthor of the Baur-Fischer-Lenz textbook on genetics, asserted: "No one approves of the new sterilization laws more than I, but I must repeat over and over again, that they constitute only a beginning."[36] Two years later, the world discovered what he meant by that.

The Nuremberg Laws

In 1935, the Nazis passed the Nuremberg Laws, which stripped German Jews of all citizenship rights and outlawed marriage and sexual relations between

Nordics and Jews. To forestall "racial pollution," the laws made it obligatory that betrothed couples be certified as racially suitable. Eugen Fischer publicly thanked the Führer for giving eugenicists the opportunity, via the Nuremberg Laws, to make "the results of their researches useful to the general public."[37]

As with the Sterilization Law, the Nuremberg Laws were defended as a "public health" measure. The head of medicine in the Reich, Gerhard Wagner, explained that the taboo on intermarriage was necessary because the Jews were a "diseased race," and the *Deutsches Ärzteblatt* applauded the Nuremberg Laws as measures that would "*cleanse the body* of our Volk" from "foreign racial elements." Sounding like a Teutonic W. A. Plecker, Julius Streicher (publisher of *Der Stürmer*) argued that "a single act of intercourse between a Jew and an Aryan woman is sufficient to pollute her forever. She can never again give birth to pure-blooded Aryan children, even if she marries an Aryan. Their children will be bastards." So infectious were the Jews that German soldiers were forbidden to marry Nordic women who had previously married Jewish husbands.[38]

As German anti-Semitism became more radicalized in the mid–1930s, the American public grew increasingly hostile toward the Nazi regime. (However, American opposition to the Nazis should not be construed as a newfound love for the Jews. A *Fortune* magazine poll in January 1936 asked: "Do you believe that in the long run Germany will be better or worse off if it drives out the Jews?" Only a bare majority of Americans—54 percent—answered "worse off.") The brutality of the Nazis repulsed most Americans, and—with the conspicuous exception of the Grantians—those American scientists who had not yet done so began dissolving their ties to the Germans. The charge was led by Franz Boas. Though seventy-five years old when Hitler came to power, Boas dedicated himself to organizing academic opposition to Nazi racism. Aside from his revulsion toward totalitarianism and anti-Semitism, Boas captured the deeper issues involved when he wrote to Elsie Clews Parsons: "I cannot keep out of the fight between individualism and the attempt to subjugate all reason to an emotionally fermented group consciousness."[39]

Nazi propagandists, like Soviet commentators of the 1950s, blunted Boas's critique by arguing that German policies toward the Jews were benign compared with the treatment of the Negroes in the United States. Other than Nazi Germany, after all, the only other countries with extensive race-based legislation were the two USAs: the Union of South Africa and the United States of America. German Jews may have been excluded from the universities, but American Negroes were excluded from public restrooms. When people such as Boas protested that German Jews had been barred from practicing medicine (except on other Jews), the Nazis pointed out that the American Medical Association still refused to admit black physicians to its membership. To Americans who criticized the German ban on intermarriage between Jews and Nordics, the Germans pointed out that thirty American states prohibited blacks from marrying whites. Indeed, the Germans correctly claimed that America's miscegenation laws were generally much stricter than the Nuremberg Laws. In Nazi Germany a person who was one-eighth Jewish was legally Aryan, but in many American states a person who was just one-sixty-fourth black was legally black.

Franz Boas nonetheless continued to publicly oppose the Nazis. This was in stark contrast to the Grantians, who saw der Führer as their ideal leader. Physician Charles F. Dight, president of the Minnesota Eugenics Society, member of the AES and the ERA (and former Socialist alderman from the Twelfth Ward of Minneapolis), sent Hitler a fawning letter praising his eugenic program. Dight explained to the *Minneapolis Journal* that the "broad and scientific plans" of Hitler "will make him the leader in the greatest national movement for human betterment the world has ever seen." And an excited Harry H. Laughlin made a note to himself that "Hitler should be made honorary member of the Eugenics Research Association."[40]

Incidentally, the Grantians were almost as enamored of Il Duce—despite his Mediterranean germ plasm—as they were of der Führer. Mussolini's science adviser, Corrado Gini, facilitated personal interviews between Mussolini and such Grantians as Charles Benedict Davenport, Kenneth Roberts, Lothrop Stoddard, and Henry Fairfield Osborn. ("If you could see the marvelous transformation of Italy by Mussolini . . . ," wrote Osborn to Grant.) George Eastman thought Mussolini "one of the ablest men in Europe," and Lothrop Stoddard, a great admirer of Il Duce, actually wrote a play (*The Chief*) inspired by his hero (but the lead actor died just before opening night, and the play was never performed). In 1927, Stoddard had a lengthy interview with Mussolini, and upon his return to the United States he wrote an article for *Harper's* explaining to the American public that Il Duce, by scientifically discarding "the equalitarian, environmental notions of past times," had brought order to the Italian nation.[41]

Even after—or perhaps we should say *especially* after—the Nuremberg Laws were put into effect, American eugenicists still maintained their close contacts with their German counterparts. Harry H. Laughlin continued to submit articles to the *Archiv für Rassen- und Gessellschaftsbiologie,* and Charles Benedict Davenport likewise published regularly in German journals. It was in 1935 that Nazi eugenicists Eugen Fischer and Ernst Rüdin were made members of the Eugenics Research Association, and later that year they were added to the advisory board of the *Eugenical News.* In that same year, Davenport joined the editorial boards of two influential German journals (the *Zeitschrift für menschliche Vererbungs- und Konstitutionslehre* and the *Zeitschrift für Rassenkunde und ihrer Nachbargebiete*). When an obtuse Davenport attempted to solicit subscriptions and contributions to the *Zeitschrift für Rassenkunde* among American scientists, Jewish anthropologist Melville Herskovits politely demurred that "in view of the developments of the last two years in Germany," he was more than a little reticent to support such a journal. "I suppose," added Herskovits mockingly, "I might [submit] an article on the fallacies of the Nordic theory, but I suppose that would be too much!"[42]

While most Americans shared Herskovits's contempt for Nazi totalitarianism, eugenicist Ellsworth Huntington chastised the "emotional bias against the Hitler regime" displayed by American sentimentalists. His colleague C. M. Goethe similarly urged Americans to set aside their distaste for Hitler in order to appreciate that sixty million Germans had been "welded into an efficient unit under a dictatorship [and were now determined] to take every possible step to insure Nordic race purity." The *Eugenical News* enthused that "under the dictatorship, Germany is moving more rapidly toward race purification than any other nation," and Frederick Osborn explained that, as a matter of fact, "Germany's rapidity of change with respect to eugenics was possible only under a dictator."[43]

Madison Grant's friends abroad were equally enthusiastic about the Hitler regime. Eugenicist Jon Alfred Mjøen of Norway triumphantly announced in 1935 that "German scientists are today making world history." And G. Vacher de Lapouge told Grant that he admired the Nazis for their "rare audacity" and their "clear understanding of biological laws." Regarding "the Jewish question," Lapouge declared: "I quite agree with Hitler," and he approvingly predicted that Nazi policy would eventually escalate into "the extermination of the Jews." To be sure, Lapouge was a bit miffed that he was not getting credit for the fact that "all the laws which they have just applied are the literal application of my writing." And he was concerned that as a result of all the refugees fleeing German anti-Semitism, it was "raining German Jews" in France. But overall, he had to admit that the Nazis were providing "a splendid example" of a "solution" to the racial problem of Europe, and he declared that the Third Reich was nothing less than "the birth of a new civilization."[44]

The continual fawning over Hitler by the Grantians was becoming untenable in the United States. In 1934, Edwin R. Embree asked the readers of *Atlantic Monthly* if they were not fed up that "Hitler and Madison Grant still cry urgently the Nordic credo." And in the following year, *Opinion* ran an article titled "A Little Hitler in the New York Zoo" by Annie Nathan Meyer, who asserted that it was no longer possible to ignore the fact that "the Leader of the German Reich lifted the whole of his racial creed" from *The Passing of the Great Race*. Calling Grant "a Fascist of the most fanatic type," Meyer wrote: "it is somewhat astonishing that in the city of New York, with its thousands of Jews, of Italians, of Negroes, that no citizen has before this arisen to question how it is that a man of such extreme intolerance" could have remained as president of the Zoological Society and trustee of the American Museum of Natural History.[45]

In 1935, historian Charles Singer complained to Henry Fairfield Osborn that "the Galton Society and, specifically, its organ the *Eugenical News*, is exhibiting anti-Semitic activities which make it almost indistinguishable from an organ of the Nazi party."[46] Paleontologist William K. Gregory, charter member of the Galton Society, agreed with Singer. In May of 1935, Gregory officially informed Madison Grant that he was resigning from the Galton Society to protest the fact that Grant and the Galtonians were "aiding agents of the German government by advice, correspondence and personal contact" and thereby abetting, in the name of science, "the most shocking violations of elementary human rights." Grant and his colleagues were taken aback by Gregory's principled resignation. As biologist Charles R. Stockard of Cornell put it in a letter to Gregory: "It seems that you may be taking the matter too seriously. . . . The behavior of the German government . . . is so far removed from my personal activities that I cannot become very excited over it."[47]

Indeed, the only concern most Grantians had regarding the German government was that its actions threatened to send Europe's Jews scurrying into the United States. As early as April of 1933, Charles Benedict Davenport expressed the fear of many American eugenicists that Hitler's anti-Semitic program would foster sympathy in the United States for the Jews and increase pressure to open America's gates to the "undesirables" of Germany. In that case, said Davenport, "their gain will be our loss." In the following year, the *Eugenical News* noted that tens of thousands of Jews were fleeing the Nazi regime, and stated its apprehension over the negative effect this would have on the germ plasm of the receiving countries.[48]

As the plight of the German Jews worsened, voices were heard in the United States, as Davenport had feared, calling for the immigration quotas to be raised, if only temporarily, so Jews trying to escape German persecution would have somewhere to go. But every time the issue was raised, the Grantians were there to squelch it. Francis H. Kinnicutt of the Allied Patriotic Societies and John Trevor of the American Coalition were especially energetic in fighting against the admission of German Jews, as was Harry H. Laughlin. In May of 1934, at Trevor's request, Laughlin issued a fifty-one-page report for the New York Chamber of Commerce urging Congress not to provide asylum to the Jews of Germany. Indeed, he recommended a 60 percent *reduction* in the immigration quotas. In a variation of the time-honored "Would you let your daughter marry one?" Laughlin asked Americans to look on the German Jews not as political refugees but as potential "sons-in-laws." He was confident that, viewed in this light, Congress would refuse to admit any more Jews. Laughlin's report was endorsed by the Industrial Board of Commerce (controlled by Madison Grant's brother, DeForest) but was denounced by Jewish leaders. Laughlin was not surprised. "The great trouble with the Jew," he told Grant, "is that he tries to be a Jew first and an American second." As far as Laughlin was concerned, this made them "not morally eligible to citizenship." Congressman Martin Dies (D-Texas) of the House Committee on Immigration (and later chairman of the House Un-American Activities Committee) agreed with Laughlin, and urged his colleagues to leave the immigration laws alone. "We must ignore the tears of sobbing sentimentalists and internationalists, and we must permanently close, lock and bar the gates of our country to new immigration waves and then throw the keys away."[49]

The Nazis knew full well that the United States was not prepared to open its gates to more Jews. As foreign condemnations of Nazi anti-Semitism increased, a bemused Goebbels cynically told the foreign press: "If there is any country that believes it has not enough Jews, I shall gladly turn over to it all our Jews." As Arthur D. Morse remarks: "There were no takers."[50]

President Franklin Roosevelt was asked continually to relax the immigration quotas, but—with the backing of the patriotic societies and the American Federation of Labor—he consistently refused. The American public supported him in this. While surveys showed that 94 percent of the American people disapproved of the German treatment of the Jews, a poll in April 1939 showed that only 8 percent were willing to open the doors of the United States to "a larger number of European refugees than now admitted under our immigration quotas." The Grantians had done their job well.[51]

In 1939, the last serious effort to admit Jewish refugees—the Wagner-Rogers bill, which would have admitted ten thousand children in 1939 and another ten thousand in 1940—was introduced in Congress. The measure was supported by the National Council of Churches and the American Friends Service Committee, as well as Eleanor Roosevelt, Herbert Hoover, and a host of other luminaries from across the political spectrum. But the interlocking directorate denounced "the same old gang" of sentimentalists for "exploiting the horrors of Europe to increase admissions wherever possible," and worked to defeat the bill. In the course of their lobbying effort, the patriotic groups devised the unique argument that admitting refugee children would make the U.S. government a party to the breakup of Jewish families that would be better off if they stayed together in Germany. With logic like that, the Wagner-Rogers bill died an ignominious death in the Senate.[52]

One year later, a similar bill to admit *British* children was introduced into the U.S. Congress. It was quickly approved.

Into the Darkness

While German Jews could not get into the United States, a surprising number of American Nordics were visiting Germany in the mid-1930s. Madison Grant was far too feeble by this time to travel overseas, but many of his closest friends made the pilgrimage to meet with high-level Nazi politicians, confer with eugenicist colleagues, and witness the new sterilization and racial laws in action.

In 1934, for example, Henry Fairfield Osborn traveled to Frankfurt to receive an honorary doctorate of science from Johann Wolfgang Goethe University. Osborn was a strong supporter of the racial policies of Mussolini and Hitler, and the Nazis were eager to have his blessing, as the new regime was still insecure enough to care about international opinion. The vainglorious Osborn was inebriated by the snappy salutes of the Frankfurt students and the fawning of the professors (one-third of whose colleagues had already been dismissed as being racially or politically suspect), and upon his return to the United States he enthusiastically praised the Hitler regime and announced that "the metempsychosis of Germany is one of the most extraordinary phenomena of modern times."[53]

Charles Singer was greatly pained by Osborn's endorsement of the Nazis, and warned him that any encouragement of the Germans "may easily lead to further violence and ultimately to a war in which all the world must suffer." But Osborn countered that the world had a false image of the Nazis because "the American press is controlled by the Jews" and "they tell our people all that is bad about Nazism and omit all that is good." "The only way to learn the truth about Germany is to spend a summer there and freely mingle with those wonderful people who have so much to teach us."[54]

Eugenicist Jon Alfred Mjøen followed Osborn's advice and made a tour of the Reich in 1934, and the following year Marie E. Kopp of the Birth Control Clinical Research Bureau traveled to Germany and reported to the AES upon her return that the "extensive eugenics program" of the Nazis was doing wonders "to correct conditions undermining the health of the nation."[55]

In 1935, the Nazis hosted the International Population Congress in Berlin under the presidency of Eugen Fischer. All the German eugenicists attended, and although the Berlin congress was denounced by the *New York Times* as "a staged propaganda assembly," Americans Harry H. Laughlin, Raymond Pearl, and Clarence G. Campbell (all members of the Galton Society) agreed to serve as vice presidents. When the scientists had assembled, Fischer led them in a prayer of thanksgiving to Providence for having given Adolf Hitler to Germany.[56]

The Germans at the International Population Congress eagerly anticipated the speech of American eugenicist Clarence G. Campbell. They had been heartened by his presidential address to the ERA earlier in the year in New York, in which he urged eugenicists to take "active steps" to rid society of those "parasite" races that "sap the life-blood of their host and destroy it." And at Berlin, as the organizers had hoped, Campbell gave a speech warmly praising Hitler's "comprehensive race policy," which, he predicted, "promises to be epochal in racial history." He commended the Germans for following the American lead in banning miscegenation (as "the difference between the Jew and the Aryan is as unsurmountable as that between black and white"), and he applauded them for understanding that "any patriotism worthy of the name carries with it a willingness on the part of individuals [to sacrifice themselves] in the ultimate interest of the group." The American scientist congratulated the Nazis for deciding to base their policies on "the realities of nature" as opposed to the "Rousseauian social philosophy which . . . bases its program . . . upon the patent fallacy of human equality."[57]

The Nazis gratefully received Campbell's praise, and the German press quoted him extensively. Campbell's speech also garnered a great deal of attention back in the United States, where the press found it particularly discomfiting that at the formal dinner closing the conference Campbell had proposed a toast: "To that great leader, Adolf Hitler!"[58]

Campbell kept in touch with the housebound Madison Grant throughout his visit to Germany. And after the Population Congress, Campbell huddled with Eugen Fischer to plan the German edition of Grant's *Conquest of a Continent.* He then spent a week observing the massive *Partietag* (the Nuremberg rally) and breathlessly told Grant that he found the experience not just "enjoyable" but "inspiring." Indeed, after witnessing the gigantic Wagnerian spectacle, a giddy Campbell told Grant: "Nothing is going to stop the Germans in squelching the Jews."[59]

Jacob Landman, a Jewish professor at City College of New York, resigned from the ERA to protest Campbell's "scandalous" behavior. In response, Campbell wrote a lengthy article (Campbell's articles were always lengthy) for the *Eugenical News* denouncing "the anti-Nazi propaganda" flooding the United States. Campbell, who had retired from a career in medicine to devote himself full-time to eugenics, then repeated the usual Grantian theme: "There is no better way of regarding [the programs of the Nazis] than as measures for improving the health and well-being of the body politic." He conceded that "such a policy requires . . . a certain sacrifice of what are regarded by some as inalienable individual rights . . . when they conflict with racial interest," but Campbell assured his readers that the individual German was only too glad to subordinate himself to "the best interest of the Volk and of the Race." Without mentioning Landman by name, Campbell called on his fellow eugenicists to continue to support Adolf Hitler, who was the realization of "the hopes that eugenicists have entertained for many years."[60]

C. M. Goethe, Campbell's successor as president of the ERA, also took the grand tour in 1935 and was likewise thrilled by the "epoch-making" population program of the Nazis. He contrasted Hitler's "sane" and "practical" methods of dealing with Germany's population problems to FDR's socialist schemes that threw billions of dollars' worth of relief programs at "alien morons."[61]

In the following year, 1936, Harry H. Laughlin was "highly honored" to have bestowed on him an honorary doctorate of medicine from Heidelberg University in recognition of his services to eugenics. Laughlin assured the Germans that the honor was "doubly valued because it will come from a nation which for many centuries nurtured the human seed-stock which later founded my own country." It was the 550th anniversary of the university, and the Nazi regime seized the occasion to stage a propaganda extravaganza celebrating the "new" (i.e., *Judenrein*) spirit of academia in Germany. Propaganda Minister Joseph Goebbels and other Nazi functionaries were slated to be at the ceremonies, at which another American scientist, Cornell University neurologist Foster Kennedy, would also be awarded an honorary doctorate. Kennedy, a member of the immigration restriction lobby, was infamous for advocating euthanasia for the mentally handicapped and other examples of "Nature's mistakes" (a position he defended by citing Justice Holmes's decision in *Buck v. Bell*).[62]

The *New York Times* accused those invited to Heidelberg in 1936 of being pawns of Goebbels's propaganda machine, and called on American scholars to resist the "intellectual despotism" of the Nazis by boycotting the event. Professors from England, Norway, Belgium and Sweden did refuse to attend, but several American universities sent representatives, including Yale, Cornell, Columbia, Vassar, Johns Hopkins, Michigan, and Harvard. Laughlin, however, did not travel to Germany (probably because of pressure from his employer, the Carnegie Institution of Washington). Instead, he received his degree at the German consulate in New York, and sent a letter to Heidelberg proclaiming: "I consider the conferring of this high degree upon me . . . as evidence of the common understanding of German and American scientists of the nature of eugenics."[63]

That common understanding was again put into practice when, a few weeks later, Madison Grant's brother DeForest spent two months in Nazi Germany. And the year after that, W. A. Plecker traveled to Germany (with a transatlantic ticket paid for by Prescott Bush, father of the forty-second president of the United States) to read a paper on Virginia's successful effort to halt the "spread of the mongrel races."

And finally, as late as 1939, yet another Grantian, Lothrop Stoddard, entered Germany. The fact that Germany was now at war with Poland, France, and Britain did not dampen his enthusiasm. He spent four months observing the population policies of the Third Reich, which he praised as "the most ambitious and far-reaching experiment in eugenics ever attempted by any nation." While in Germany, Stoddard conferred with many of Madison Grant's old eugenicist friends, including Eugen Fischer, Hans F. K. Günther, Wilhelm Frick, and Fritz Lenz. The Propaganda Ministry provided the famous American racist access to most of the Nazi hierarchy, including Joseph Goebbels, Walther Darré, Robert Ley, Joachim von Ribbentrop, the "dynamic" Heinrich Himmler, and even the Führer himself. No foreigner (with the exception of diplomats on official business) had been received by the chancellor since the outbreak of the war, but Stoddard was granted an interview when he assured the Foreign Office that he was interested in presenting "the human side" of Hitler to the outside world. Stoddard found Hitler to be an extremely pleasant chap, although he had none of the "personal charm" of Il Duce.[64]

Hitler aside, Stoddard was unabashedly impressed by what he heard and saw in Nazi Germany (which he admiringly called the "New Sparta") and published his observations in *Into the Darkness* (Germany was not "dark"—as one might assume—because it was totalitarian, but because it was hidden from Western view during the war). After spending a day at the Genetic Health Court of Appeals in Charlottenburg, where he witnessed numerous trials of dysgenic Germans condemned to be sterilized, Stoddard—who had earned his Ph.D. at Harvard and had written a book (*Scientific Humanism*) extolling the scientific spirit—concluded that the Nazi Sterilization Law was "weeding out the worst strains in the Germanic stock in a scientific and truly humanitarian way."[65]

On a tour of the Public Health Service, Stoddard was instructed by an official that the level of medical care given to patients in Germany was determined by their "social worth." "Valuable" citizens received the best care, but those of inferior worth were treated accordingly, as "we National Socialists are duty bound to foster [only] individuals of social and biological value." "The Nazis," remarked Stoddard, "are robust pragmatists."[66]

Regarding the situation of the Jews, even Stoddard had to confess being taken aback by the Nazis' forthrightness: "I have been stunned at a luncheon or dinner with Nazis, where the Jewish question had not been even mentioned, to have somebody raise his glass and casually give the toast: *Sterben Juden!*— 'May the Jews Die!'" Stoddard inferred for his American readers that the Jewish problem would soon be settled "by the physical elimination of the Jews themselves from the Third Reich."[67]

The Next Logical Step

Lothrop Stoddard did not realize that during his visit to Germany the coercive sterilization program he so admired was actually winding down. This was partly because the number of eligible victims was dwindling, but mainly because the regime had decided that the war provided an opportunity to transform the sterilization program into a euthanasia program that would permanently rid the country of its mentally and physically handicapped citizens.

The handicapped were deviants from the "type," and the Nazis were driven by a eugenic desire to purge them from the body of the *Volk*. The euthanasia program was initiated when German physicians, viewing themselves as "biological soldiers," began killing handicapped infants in hospitals. The victims were either starved, injected with lethal drugs, or—as Galton, Chamberlain, Haeckel, Lapouge, and Günther had long advocated—exposed to the cold as in ancient Sparta. At first, only grossly deformed newborns were killed, but then the circle of victims was extended to include children up to three years old; then eight; then twelve; and finally persons up to sixteen years of age were murdered. As the age limit increased, the definition of a "handicap" widened to include anything from having an incurable disease to being a slow learner or even a chronic bed wetter.[68]

Having rid society of its degenerate children, the next logical step was the extermination of institutionalized adults. Hitler placed Dr. Karl Brandt, his personal physician and *Reichskommissar für Sanitaets—und Gesundheitswesen* (Reich commissioner for health and sanitation) in charge of this program, in which at least seventy thousand elderly, infirm, insane, and incurably ill citizens were murdered (the term the Nazis preferred was "disinfected"). The original method of disinfection was lethal injection, but the program soon began using gas chambers disguised as showers in mental hospitals. Once the victims were dead, nurses removed them from the showers, extracted the gold from their teeth, and then burned the bodies in crematoria. "In Germany," notes Stephen Trombley, "the 'reign of doctors' foreseen by the defence in the *Buck v. Bell* case had come to pass."[69]

Since all these victims had already been sterilized under the Sterilization Law, Henry Friedlander points out that from a eugenic stance such killing was "both unnecessary and senseless." But the Nazis argued that the executions were performed "for wartime economic purposes"—in other words, to save food, medicine, and bed space that could better be used by more valuable citizens. Surely, the killing of a few thousand *unnütze Esser* (useless eaters) was justified at a time when the very existence of the nation was threatened. In *Buck v. Bell,* Justice Holmes had said that "the public welfare may call upon the best citizens for their lives." And so the Nazis now contended that in time of war, when healthy citizens were making the ultimate sacrifice at the front, the handicapped back home should make similar sacrifices for the fatherland.[70]

Karl Brandt's euthanasia program was very efficient. By 1941, the psychiatric institution of Hadamar (one of six official killing centers for mental patients) celebrated the "processing" of its ten-thousandth patient. A special ceremony was conducted, and all the staff members received a bottle of beer for the occasion. By the end of the war, 85 percent of Germany's mental patients had been killed. Indeed, so determined were the eugenicists to cleanse the German *Volk* that, in the state hospital at Kaufbeuren, the extermination of feebleminded children continued even *after* U.S. troops occupied Bavaria in 1945. The chief psychiatrist of the hospital had been arrested as a Nazi, but for two months the remaining doctors and nurses continued to quietly murder their patients, within a few hundred yards of an American MP unit. Acting on a tip, American soldiers finally entered the hospital and shut down the killing ward.[71]

So far, Adolf Hitler was following rather neatly Madison Grant's blueprint for negative eugenics, which, as laid out in *The Passing of the Great Race,* was to be applied to "an ever widening circle of social discards, beginning always with . . . the diseased, and the insane, and extending gradually to types which may be called weaklings rather than defectives, and perhaps ultimately to worthless race types."[72] To be sure, Grant was no exterminationist. He and his ideological forerunners, from Gobineau to Ripley, were peaceful intellectuals. For them, "eliminating the Jews" meant, at worst, removing them or their germ plasm from society via deportation or sterilization. Hitler, however, had the courage of his convictions; for him, "eliminating the Jews" meant just that. After observing the Führer's climb to power, an excited Lapouge told Grant: "This is no longer a question of theoretic desiderata." The Nazis, Lapouge told his American friend, were actually going to *apply* their theories and "rapidly extinguish undesirable stocks."[73]

Adolf Hitler was a man of action—and, not inconsequentially, he had at his disposal the full power of the centralized state. "Perhaps more than any other man of this century," wrote Telford Taylor, "he was able to transmute his own thoughts into events." Eugenicist Ernst Rüdin admiringly said of Hitler in 1943: "it will always remain the undying, historic achievement of Adolf Hitler and his followers that they dared to [put] into practice the theories and advances of Nordic racism [and won] the fight against parasitic alien races such as the Jews and the Gypsies." Or, as Max Wienrich once put it: "There were in the memory of mankind Genghis Khans and Eugen Fischers, but never before had a Genghis Khan joined hands with a Eugen Fischer."[74]

Following the sterilization of hereditary defectives, the euthanasia of handicapped children, and the disinfection of "unproductive" adults, the Nazis implemented *Die Endlösung,* the Final Solution. Pat Shipman points out: "Once the link was made between those who were genetically defective . . . and euthanasia, it was a simple matter to extend the principle to justify elimination of the Jews." And once again, all the old biomedical metaphors were used to defend the elimination of the Jews. According to Hitler, the Jews were "rats," "leeches," and "poisonous mushrooms," and the Nazis were performing a public health service "by exterminating this pest" that "poisons the blood of others." "The Jew," declared Hitler, "is and remains a parasite, a sponger who, like a pernicious bacillus, spreads over wider and wider areas." The Führer always maintained that the chief function of his government was "to protect our people . . . against racial *infection,*" and he explained to Himmler that "the battle in which we are engaged today is of the same sort as the battle waged, during the last century, by Pasteur and Koch." (It was fitting that the Bureau of Human Betterment and Eugenics was located on Robert Koch Platz in Berlin.) "We shall regain our health," said Hitler, "only by eliminating the Jew."[75]

In March 1941, at a conference on the "total solution of the Jewish question" at which Grantians Eugen Fischer and Hans F. K. Günther were the guests of honor, Alfred Rosenberg proclaimed the coming of the "cleansing biological revolution" in which the *Volkstod* (genocide) of the Jews would be the final act. Eight months later, Rosenberg, by then minister of the Occupied Eastern Territories, confirmed to members of the German press that "the Jewish question" could "only be resolved by a biological elimination of the entire Jewish population of Europe."[76]

To that end, the gassing techniques first used in the 1920s to delouse Jewish immigrants bound for America, and then utilized in the 1930s to murder the handicapped, were now employed to exterminate some six million Jews (and thousands of gypsies, Communists, homosexuals, and Slavs). Indeed, the very same equipment and personnel that had been used in the euthanasia program were used in the Holocaust. Since the psychiatric hospitals had liquidated almost all their patients by the fall of 1941, their gas chambers (along with their attending physicians and technicians) were now transferred to Majdanek, Treblinka, Belzec, Sobibor, and Auschwitz. The SS doctors seemed to have few qualms about murdering the Jews, whom they repeatedly characterized as "germs" and "bacilli."[77] The health of the state, they believed, necessitated the destruction of the individuals that threatened it. And the metastasizing Jewish race was a palpable, biological threat to the existence and welfare of the Reich. Fritz Klein, an SS doctor who participated in the exterminations at Auschwitz, explained that just as a physician "would remove a purulent appendix from a diseased body" so he removed degenerates from the "body of Europe." "Our task here is surgical," agreed Goebbels. "Drastic incisions, or some day Europe will perish of the Jewish disease." And so, as Gary Nash has written, "Hitler had done exactly what Madison Grant had suggested: a eugenics program to sterilize 'defectives' [was] followed by the extermination of 'inferior' peoples."[78]

Neither Madison Grant nor any other eugenicist—German or American—can be said to have caused the Holocaust, but they certainly provided the scientific justification for what occurred. And the mass killings presented an unequaled opportunity for eugenicists to conduct racial research. Dr. Josef Mengele, for example, was able to perform a number of eugenic experiments when posted to Auschwitz as an SS doctor. Mengele had two doctorates: one in physical anthropology, the other in medicine. One of his favorite projects was to see if it was possible to change the color of human eyes (in order to increase the incidence of blue eyes among Germans). He selected twins from the railway ramp at Auschwitz (Himmler had decreed that only physicians trained in racial anthropology could carry out the selection of incoming prisoners at the death camps), killed them, and cut out their eyes. He then mailed the harvested tissue (along with the internal organs of murdered children, skeletons of gassed Jews, and blood samples of inmates he had infected with typhoid) in packages stamped "War Material—Urgent" to eugenicist Otmar von Verschuer at the Kaiser Wilhelm Institute of Anthropology for further research.[79]

Another eugenicist who took advantage of the war was Dr. August Hirt, chair of anthropology at Strasbourg, who was anxious to procure the skulls of "Jewish-Bolshevik Commissars, who represent the prototype of the repulsive but characteristic subhuman." At first, he hoped the Wehrmacht might send him decapitated heads from the Russian front in hermetically sealed tin cans. But after conferring with Karl Brandt and Heinrich Himmler, Hirt decided it would be more efficient to harvest full skeletons from the death camps. An assistant at Strasbourg remembered that the bodies arrived from Auschwitz "still warm, with wide open, bloodshot and glazine eyes, bleeding from the mouth and nose and discharging other liquids." The Jews' left testicles were immediately removed and sent to the Anatomical Laboratory, and the rest of the bodies were kept in vats of formaldehyde in the cellar until they could be properly defleshed. (It was fitting that the library at the University of Strasbourg featured a shrine to Gobineau where, under a portrait of the count by the Countess de la Tour, the *Gobinistische Vereinigung* displayed his books, art collection, and personal effects.) After D-day, orders came down that should Strasbourg be threatened by the Allies, the Jewish bodies in the cellar would have to be destroyed. At first, the researchers protested that "this singular collection would be lost to science," but as the enemy closed in the anthropologists frantically began cutting up the bodies and stuffing them in the incinerator (though a number of "partly processed" corpses remained when the city was liberated).[80]

Such behavior is shocking, but it is not incomprehensible. For some persons, it was simply the final step of a long and rigidly deductive stairway, whose first step had been the fairly innocuous decision in 1913 that it was right and proper, for the good of the herd, to kill the "surplus males" of the elk herd in Yellowstone National Park. From wildlife management to human eugenics to the death camps is a tragic but not so absurd progression (or should we say regression?) in logic. The Holocaust was not the irruption of the irrational, but rather the ultimate application of what Hitler called the "ice cold logic" of the rational. Was it inevitable? Of course not. But was it logically coherent? Certainly. "When all is said and done," writes Sheila Weiss, "it is the *logic* of eugenics far more than its racism that proved to be the most unfortunate legacy of the German race hygiene movement for the Third Reich."[81]

A commonly heard slogan during the Third Reich was "National Socialism is nothing but applied biology." By this the Nazis meant that their philosophy was the next link in the intellectual chain connecting Darwin, Galton, Mendel, Weismann, and Madison Grant. As physician Rudolf Ramm said in 1943: "Biology and genetics are the roots from which the National Socialist world-view has grown." But the slogan was also supposed to remind the citizens of the Reich that the new Germany was a part of—and not above—the natural world. Madison Grant had always maintained that man was an animal like any other, and that the laws of nature applied to humans just as they applied to the other mammals. His ideological colleagues, from Ernst Haeckel to Aldo Leopold, similarly preached the interdependence of all living things, and insisted that man was a part of nature. (It was the German eugenicist Haeckel who coined the term *ecology*.)[82]

In this light, it is not a surprise that the Nazis embraced Madison Grant's preservationism as well as his scientific racism. Both movements stemmed from a recognition that biological catastrophe ensues when man interferes with the natural order of things. "Exterminating millions of lives," remarks Simon Schama, "was not at all incompatible with passionate protection for millions of trees." Forest ecology was certainly of keen interest to Nazi leaders such as Hermann Goering, who took quite seriously his duties as *Reichsforstmeister* (the head of German forestry) and made sure that the subject was taught in the schools. And Nazi minister of agriculture Walther Darré's advocacy of ecology has led one historian to call him "Father of the Greens."[83]

Indeed, the conservationist ethos pervaded many aspects of the Nationalist Socialist program. For example, the *Reichsnaturschutzgesetz* (the Nature Protection Law) created an early version of our Environmental Protection Agency and was, according to Thomas Lekan, "the world's most comprehensive and stringent piece of nature protection legislation for its time." Passed in 1935 (the same year as the Nuremberg Laws), the measure was welcomed by German preservationists, one of whom sounded just like Madison Grant in his elation that the law enshrined the principle "that the right to private property only has validity insofar as it does not violate the well-being of the general public." The government also drafted a comprehensive landscape protection plan for the Middle Rhine Gorge that (although never implemented, because of the outbreak of the war) was very similar to Grant's preservation efforts in the Bronx River Valley.[84]

In addition, legislation protecting endangered species was passed early in the Nazi years, and the regime decreed that death was the punishment for killing an eagle (at a time when Alaska was still paying bounties to anyone who did the same). Four months after the Sterilization Law was enacted, the Law for the Protection of Animals banned cruelty to animals (a favorite cause of Himmler's) and specifically prohibited experiments on animals involving the use of cold, heat, or infection. (It need hardly be said that it was precisely such experiments that the SS doctors under Himmler's command conducted on the Jews in the extermination camps.)[85]

The Nazis were also very concerned about pollution, and worked to eliminate such environmental poisons as heavy metals, pesticides, and asbestos. The regime strongly supported research in public health and cancer prevention, and launched temperance and antismoking campaigns (the Nazis were the pioneers in banning smoking in public buildings). Hitler himself, like the anti-Semite Henry Ford and his doctor John Harvey Kellogg, shunned cigarettes, coffee, tea, and alcohol, and did not permit anyone to smoke or drink in his presence. Like Houston Stewart Chamberlain and Richard Wagner, Hitler was also a vegetarian, and his government promoted a diet high in fruit and fiber, required every German bakery to produce whole-grain bread, and banned artificial coloring and preservatives.

There is no question that this absorption with personal purity was also expressed as a preoccupation with racial purity. And the concern about environmental pollution was related to the anxiety about genetic pollution. It was not very difficult to evolve from a fear of racial *poisons* (e.g., alcohol and tobacco) to the fear of *racial* poisons (i.e., the Jews). Eugenics simply extended the Nazis' concern with individual health to the level of racial health, which is why the comparison of degenerate humans with malignant cancer cells was as recurrent in German as in American eugenic literature. (It was to eradicate "malignant" specimens of humanity from the body of the *Volk* that Konrad Lorenz, the 1973 Nobel laureate in Medicine, joined the Nazi Party in 1938).[86]

Preserving trees, protecting animals, ending pollution, fighting cancer, removing Jews: all were manifestations of the Nazi thirst for a more "natural" mode of existence. It is not a coincidence that the regime installed no-smoking cars in trains just about the time it created *Judenfrei* cars (causing German newspapers to announce: "Germ-Carriers Banned from Railways"). It was all part of the effort to cleanse the body and rid society of toxins.

The New Consensus

In 1945, Germany surrendered to the Allies. The Sterilization Law and the Nuremberg Laws were abrogated, and twenty-three German eugenicists were put on trial at Nuremberg for participating in war crimes and crimes against humanity.

Madison Grant's two foremost German devotees—Eugen Fischer and Hans F. K. Günther—were not among them. Fischer retired happily to the town of Freiburg, where he continued to publish and was even made an honorary member of the postwar German Anthropological Association. He produced his memoirs in 1959 (in which he managed to avoid mentioning the Nazi period) and lived on until 1967, visited occasionally by his friend Martin Heidegger (who had been the Nazi rector of Freiburg University).[87] Hans Günther also resided in Freiburg, where he too resumed his ethnological work and continued to warn of the impending demise of civilization, though this time the threat came not from degenerate Jews but from godless communists. In 1958 he was one of the founding members—along with Roger Pearson (the white supremacist supported by Wickliffe P. Draper's Pioneer Fund)—of the Northern League, an international conference of pan-Nordicists (which also received financial support from C. M. Goethe). Günther died one year after Eugen Fischer, in 1968.

Their eugenic associate Fritz Lenz prospered after the war, taking over the chair of genetics at the University of Göttingen. Similarly, Otmar von Verschuer (mentor of Josef Mengele and friend of C. M. Goethe), after denying that he knew anything about the Holocaust, was appointed to the prestigious chair of human genetics at the University of Münster, and was soon elected president of the German Anthropological Association. He also served on the editorial board of Roger Pearson's *Mankind Quarterly* along with such other prewar Grantians as Walter Scheidt, Corrado Gini, and Charles Conant Josey.[88]

Their twenty-three indicted colleagues from the Nazi eugenics program were collectively put on trial in November 1946 in the U.S. military court in Nuremberg's Palace of Justice. The Nuremberg Doctors Trial (known officially as *United States of America v. Karl Brandt et al.*) lasted more than nine months and involved eight-five witnesses, 1,471 pieces of documentary evidence, and 11,538 pages of testimony. In his opening statement, the chief counsel for the prosecution, Brigadier General Telford Taylor, accused the German scientists of experimenting on concentration camp inmates and other living human subjects without their consent, in the course of which the defendants "willfully and without remorse" committed murders, brutalities, cruelties, tortures, atrocities, and other inhuman acts. Taylor correctly concluded that such behavior was "the inevitable outcome of that sinister undercurrent of German philosophy that preaches the supreme importance of the state and the complete subordination of the individual."[89]

The defendants willingly agreed with this assessment. Dr. Gerhard Rose, for example, testified that as a eugenicist he saw nothing wrong in sacrificing the interests of the few in order to benefit the majority. Rose, head of the Koch Institute of Tropical Medicine in Berlin, recounted that he had initially balked when the Wehrmacht asked him to perform lethal experiments on concentration camp inmates in order to develop a vaccine for typhus. But then, employing logic whose metaphorical roots stretched all the way back to the Kaibab Plateau, Rose concluded that it was absurd to care about the lives of a few hundred inmates (who were doomed to die anyway) when one thousand soldiers a day were succumbing to typhus on the eastern front. Surely, thought Rose, the lives of a few individuals were not as important as the survival of the majority. *Gemeinnutz geht vor Eigennutz.*[90]

The lawyer for Dr. Karl Brandt similarly (and repeatedly) argued that the life of the individual was expendable if it helped ensure the continued existence of the community. During the Reich, he explained to the judges, "everything was done in the interests of *humanity* so that the individual person had no meaning whatsoever." He insisted that it was up to the state to decide which individuals were dispensable. And in words that were remarkably similar to those of Oliver Wendell Holmes in *Buck v. Bell,* he justified Nazi eugenics by saying: "In the same way as the state demands the death of its best men as soldiers, it is entitled to order the death of the condemned in its battle against epidemics and diseases."[91]

Karl Brandt had been the plenipotentiary of medicine and health in Nazi Germany. As the supervisor of the euthanasia program he was responsible for the wholesale murder and cruel torture of hundreds of thousands of innocent human beings. Born into a distinguished medical family in Alsace in 1904,

Dr. Karl Brandt striking an unrepentant pose at the Nuremberg trial, confident that the "ice cold logic" of eugenics is on his side. Brandt was sentenced to death, but many other Nazi scientists were never tried for their actions. U.S. Army Photo, Ray D'Addario.

Brandt was an intelligent and gifted surgeon who considered joining his fellow Alsatian Albert Schweitzer in Africa. He joined the Nazi Party instead, and in 1932 was introduced to Adolf Hitler by his fiancée, a German swimming champion. Hitler was drawn to the tall, elegant physician, and by the summer of 1934 the twenty-nine-year-old Brandt was the attending physician and close confidant of the Führer. Hitler had a high degree of personal and professional confidence in the cultured Brandt. Other doctors held Brandt in high regard as well, and spoke of him as a remarkably decent, ethical, and idealistic physician. Indeed, the prosecutors at Nuremberg were caught off guard by the numerous affidavits submitted by the defense testifying to Brandt's personal character, and showing that he had been motivated during the war not by brutality but idealism. A perplexed General Taylor had to admit that "these defendants did not kill in hot blood, nor for personal enrichment. . . . They are not perverts. They are not ignorant men." They were, he conceded, "outstanding men of science . . . with excellent international reputations."[92]

To explain and defend their actions, Brandt and his fellow defendants pointed to the eugenics movement in the United States. They cited the Immigration Restriction Act of 1924, the sterilization program, the antimiscegenation legislation, and the Supreme Court's decision in *Buck v. Bell.* Brandt also entered into the record excerpts from *The Passing of the Great Race* that he claimed justified the elimination of persons not worthy of living. The Allies did not attempt to challenge the contention that the eugenics program of the Nazis had originated in the United States.[93] But in August 1947, Dr. Brandt and fifteen of his codefendants were found guilty, and seven of them, including Brandt, were ordered to be hanged. The prosecutors, interestingly, used the same justification for executing the Germans as the Germans had used regarding the Jews: "They [the Nazis] must not become a spreading cancer in the breast of humanity. They must be cut out."[94]

While awaiting punishment, Brandt was incarcerated in Landsberg fortress, where his most famous patient, Adolf Hitler, had written *Mein Kampf.* As a good scientist, he offered his living body for medical experiments, but the American authorities rejected his offer. On June 2, 1948, after refusing religious solace, he was hanged.

Having killed the messenger, the Allies now killed the message. The United Nations assembled an international panel of anthropologists, sociologists, geneticists, and psychologists to draft a statement officially repudiating scientific racism. The UNESCO Statement on Race, written mainly by anthropologist Ashley Montagu, was made public in 1950. A follow-up statement, written primarily by geneticist L. C. Dunn, was issued in 1951. Together, the statements asserted:

- All men belong to the same species, *Homo sapiens.*
- "Race" is not a biological reality but rather a social myth, and the term should be dropped in favor of "ethnic group."
- There is no proof that the groups of mankind differ in their innate mental characteristics or intellectual capacity, or that there is any connection between the physical and the mental characteristics of human beings.
- The results of intelligence tests are greatly affected by the environmental opportunities of the testees.
- There is no evidence that miscegenation causes biological deterioration.

In the wake of these announcements, the headline in the *New York Times* read: "No Scientific Basis for Race Bias Found by World Panel of Experts." By expressing the consensus of scientists worldwide that all the major tenets of Madison Grant's epistemology were obsolete, the UNESCO statements were, in the words of Robert Proctor, "the triumph of Boasian anthropology on a world-historical scale."[95]

Epilogue

The Passing of the Great Patrician

And the old nobleman, dispossessed and uprooted, went away. . . . We can well see how the superior racial qualities possessed by such fallen families no longer usefully served the state.

Gobineau, *Ottar Jarl*, 1879

Hunting with Goering

Madison Grant, who had been rendered increasingly infirm by arthritis and old age, died in 1937 at the age of seventy-two. "He had been through a most dreadful time," wrote Kermit Roosevelt, "and slipping out could certainly mean nothing but a release." And yet Grant worked right up to the very end. In his final weeks, despite his decrepit condition, Grant was feverishly trying to finish an article on the preservation of the bison. He was pushing to get a bill through Congress to save the groves of Sitka spruce at the entrance to Yellowstone National Park. He was lobbying to make Admiralty Island a sanctuary for the Alaska brown bear (an effort that finally bore fruit in 1978). He was imploring Interior Secretary Harold Ickes to enlarge Yosemite National Park to include the stand of immense sugar pines just west of its boundary. And, having been active in the campaign to create Everglades National Park in 1934, he was now working at the opposite corner of the Lower 48 to create Olympic National Park (which was realized the following year, in 1938).

In addition, having devoted his life to preserving the flora, fauna, and scenery of North America, Grant in his final years had expanded his activities to include preservation efforts all over the globe. In doing so, Grant and the organizations he headed set the precedent for international cooperation in the field of conservation. I have discussed these endeavors in detail elsewhere,[1] and will simply mention that among the many species that Grant helped rescue from the verge of extinction toward the end of his life were the elephants of Africa, the koalas of Australia, the chinchillas of South America, the gorillas of the Congo, the giant tortoises of the Galapagos Islands, the ibex of Spain, the mountain zebras of South Africa, the elephant seals of Mexico, the giant sable antelopes of Angola, the nyalas of Ethiopia, the white rhinos of the Sudan, the wisents of Poland, and several species of whales. The two things these attempts had in common with his domestic campaigns were that (a) they were almost always on behalf of the large, ancient species that today we would call "charismatic megafauna" and (b) they were always successful.

It is simply staggering the amount of work that the severely crippled Grant was able to accomplish in his later years. Whatever else we might think of Madison Grant, we must join with all those who knew him as an old man and admired him for his measureless stamina and uncomplaining courage. As a patrician, there had never been any need for him to work a day in his life, but he was still fighting for the causes he thought important until the day he died. "What a brave fight you have made," said John Harvey Kellogg to Grant, "in refusing to be laid on the shelf notwithstanding the outrageous manner in which the fates have treated you."[2]

In January 1937, Grant presided over his forty-third and last annual meeting of the Zoological Society. A shaken Edmund Seymour told William T. Hornaday (who himself had only weeks to live): "It must have taken a lot of courage for Grant to have done it because he seemed to be in pretty bad shape." Grant grudgingly confessed that "this arthritis of mine has temporarily crippled me so that I cannot walk or stand on my legs, but this" (he added with characteristic optimism) "I think will pass."[3]

Grant's final undertaking was helping to organize the International Hunting Exposition of 1937, brainchild of Hermann Goering. As commander in chief of the Luftwaffe, president of the Reichstag, prime minister of Prussia, and Hitler's designated successor, the aristocratic Goering was a busy man. But he was also a committed conservationist and a compulsive hunter who never let duties of state interfere with his sporting activities. (This was in contrast to Hitler, who thought that hunting was "a dreary sport" and admitted that "personally, I cannot see what possible pleasure can be derived from shooting.")[4]

In 1937, in his capacity as *Reichsjägermeister* (Reich master of the hunt), Goering planned the massive International Hunting Exposition in Berlin, a three-week-long festival to promote "comradeship, goodwill and understanding among the international hunting fraternity."[5] The exposition took place on the vast grounds of the newly built Berlin Exhibition Center and featured shooting competitions, falconry demonstrations, displays of hunting trophies, exhibits of archaic weapons, and seminars on game preserves and nature conservation. The highlight was when selected delegates joined in a hunting party with Reichsmarshal Goering himself. Madison Grant, of course, was one of those invited, and he intended to take with him to Berlin some of his record trophies.

But Grant did not live long enough to go hunting with Goering. Still, it is

Madison Grant speaking at a 1926 ceremony at the Bronx Zoo honoring William T. Hornaday (seated at the left) for thirty years of service to the New York Zoological Society. © Wildlife Conservation Society.

fitting that Grant's final project had combined hunting and the Nazis, for those were the two poles of his intellectual development. He began as a big-game hunter in the 1890s and evolved over the course of his life into conservationism, preservationism, wildlife management, eugenics, and finally National Socialism.

Charon Beckoned

Grant's final illness lasted one week. He passed away in his Park Avenue residence on Sunday, May 30, 1937, as he was packing to leave for the Battle Creek Sanitarium (where John Harvey Kellogg had promised that "we can improve your chances for a 25 or 30 years' longer sojourn in this mundane sphere").[6] The man who assailed his fellow patricians for their low birthrate left no survivors except for DeForest, his beloved (but also childless) brother.

And so it was over. From Alaska to Zarathustra, from Zangwill to Auschwitz—

and all the people, places, and organizations in between—it had been a despicable, and praiseworthy, and complex life.

W. Redmond Cross, Grant's successor as president of the New York Zoological Society, officially announced "the irreparable loss occasioned by the death of Madison Grant." Laudatory and moving tributes to Grant came flooding in from conservationists worldwide. H. E. Anthony, for example, wrote a long obituary for the *Journal of Mammalogy* explaining that the passing of Grant "removed from the ranks of the conservationists one of their staunchest champions."[7] And George Bird Grinnell was inconsolable. He and Grant had been inseparable for decades, and as all of Grinnell's other peers began dying off he clung ever more tightly to Grant (who admitted in 1934 that he feared the annual report of the Boone and Crockett Club was "turning into a mortuary column.")[8] After Grant's death, Grinnell's memory began to fade, he became quite hard of hearing, and he died one year after his friend at the age of eighty-nine.

The eugenicists were as saddened by Grant's death as the conservationists. C. M. Goethe confided to Harry H. Laughlin his fear that with Madison Grant gone there was no one left to defend the nation from the racial menace. Laughlin, who loved to go for afternoon drives with Grant and often sent him flowers in his final years, agreed: "He was a gallant leader who fought a good fight." In Germany, the journals *Volk und Rasse* and *Zeitschrift für Rassenkunde* mourned the passing of "the courageous researcher" who had stood up to international Jewry.[9]

The *Argonaut* was typical of many journals when it wrote that the death of Grant "marked the passing of a great American . . . A gentleman in the full sense of the word, it was a pleasure to know Madison Grant. He used his talents to better his race. There are all too few of his type left."[10] And Frederick Russell Burnham told Kermit Roosevelt: "we have all suffered the loss of a great soul." The obituary that Burnham wrote for the yearbook of the Boone and Crockett Club was so remarkably maudlin that it made the prose of Henry Fairfield Osborn seem terse by comparison, and it is partly reproduced here in hopes it may serve as a warning to aspiring writers. Under the title "Charon Beckoned," Burnham wrote:

> We feel deeply the loss of our beloved President, Madison Grant. . . . No more will his stalwart figure be seen among us. Around our council fires we shall sadly miss his deep voice, his ready wit, his wise decisions, and the inspiration to spur us on to high endeavor. Yet, believing with the poet that they who dwell in our hearts never die, we shall, like the Norsemen of old, carry him with us to the halls of Valhalla. . . .
>
> In the tundras of the north, where the sun dogs are on guard, men will gather around the flame in the soapstone and tell tales of this tall Nordic, . . . this mighty Nimrod. On the shores of the Pacific, whose towering trees point

to the Infinite, thousands of forest worshippers will cry, "Praise be to God for this man's valiant work. . . ."

Now that he is dead, let us pray that some other powerful man can bend the mighty bow and save our land from alien hordes. This nation is fortunate that he lived. Alas, that he is gone. We shall long seek his equal . . . . Let us carve on the tablets of memory the Saga of Madison Grant.[11]

Funeral services were held at St. Bartholomew's chapel on Park Avenue, and then hundreds of persons attended the burial at Sleepy Hollow Cemetery in Tarrytown. (Washington Irving is buried at Sleepy Hollow, as are two famous immigrants that Grant knew well, Andrew Carnegie and Samuel Gompers.) Grant's grave is on the highest spot in the cemetery, facing east, on guard against any influx of aliens from the Old World.

It may or may not be a coincidence that the same week that Grant was buried, Franz Boas, age seventy-nine, finally decided that he could retire from Columbia University. But even after he stepped down from teaching, Boas continued to fight against scientific racism until literally the day he died. On December 19, 1942, at the age of eighty-five, Boas was giving a speech in the Columbia Faculty Club on the need to combat scientific racism when, in midsentence, he fell over backward in his chair, dead.[12]

Dust in the Wind

Madison Grant never sought personal recognition or public accolades. He preferred to labor in the background and was quite content to let others receive the credit. Hence, with the exception of the various memorials to Madison Grant in the California redwoods (the Grant Grove, the Founders Tree, and the Madison Grant Forest and Elk Refuge), there is little physical evidence extant in the world today to indicate that Grant ever lived. No plaque indicates his place of birth in Manhattan—although, if one did not know better, one might conclude that some bureaucrat in city hall actually possessed a sense of irony: the street where Grant was born has been renamed "Sholom Aleichem Place." And, in a development that would have struck Grant with equal horror, the Polish Consulate now stands only three blocks away. On the other hand, Grant's former residence on Park Avenue is currently a modern high-rise whose ground floor is occupied by that ultimate sign of Nordic conquest: a BMW dealership.

One of Grant's greatest creations, the Bronx Zoo, has no memorial to inform visitors that Madison Grant was its founder, benefactor, and chief executive for four decades. Grant's name appears nowhere on the website of the zoo. The library, which is open only to scholars, does have a full-length portrait of Grant, which hangs on the wall opposite a similar portrait of his friend Henry Fairfield Osborn. And a bronze bust of Grant gathers dust on top of a bookshelf, though few people pay attention to it (other than the occasional professor who climbs on top of a stack of chairs, tape measure in hand, to confirm that Grant's skull was indeed dolichocephalic).

Over at the American Museum of Natural History, where Madison Grant was an influential figure for decades, he is equally hard to find. To this day the museum zealously and absurdly attempts to conceal its eugenic past. (Donna Haraway, after being subjected to the museum's stonewalling tactics, asked incredulously: "Is it still so sensitive after 68 years?" I can exasperatedly attest that the answer to Haraway's question is Yes!)[13] However, the Hall of North American Mammals is still spearheaded by a group of Grant's caribou (opposite an alcove of Osborn's caribou). The hall was planned in the nineteenth century to exhibit for future generations specimens of species that the museum assumed would be extinct in a few years.[14] Thanks to Madison Grant, those massive extinctions did not occur, and when the Hall of North American Mammals finally opened in 1942 the *Rangifer granti* were in a strategic location, as they were the first exhibit seen by visitors reaching the museum by the new subway entrance. Despite the museum's defensive determination to protect its reputation, one assumes it will permit the caribou to remain there in the years to come.

For decades, one of the biggest memorials to Grant was the continued existence of the restrictive immigration quotas, which remained in place through prosperity, depression, world war, and Cold War. Congressman Albert Johnson had sworn in 1930 that "the United States will never again adopt a liberal policy of immigration," and for forty years his prediction held true. Not until 1965, when the Nordics were confident that their future as a species was secure, did the country finally abandon the national origins quotas with the Hart-Celler Act of 1965. Yes: Emanuel Celler, who as a freshman congressman in 1924 had called for Madison Grant's "Nordic spell" to be "exorcised" from the body politic, was patiently reelected to the House twenty-four times by the voters of Brooklyn and in 1965 finally performed the role of exorcist. In October, at a ceremony at the Statue of Liberty, Lyndon Johnson signed the Celler Act to correct "a cruel and enduring wrong in the conduct of the American nation."[15]

Since the Celler Act removed the restrictions on Asian as well as European immigration, the face of America was immediately altered. Within fifteen years, Europeans (whether from northern, southern, or eastern Europe) made up only 17 percent of the immigrants to the United States. As early as 1968, Madison Grant's fears were realized when Walter Kerr noted in the *New York Times* that the native WASP had become "the odd ball, the stranger, the fossil. We glance at him, a bit startled and say to ourselves, 'Where did he go?' We remember him: pale, poised, neatly dressed, briskly sure of himself. And we see him as an outsider, an outlander, a reasonably noble breed in the act of vanishing."[16]

And yet, let us not forget the lesson of the redwood tree, which is called *sempervirens* because, as we know, even after being struck by lightning and crashing to the ground dead, it is only a matter of days before it begins to regenerate itself and new saplings pop up through the forest floor. And so it is with the theories of Madison Grant. The universal revulsion at the Holocaust supposedly destroyed his teachings forever, but they keep reappearing. *The Passing of the Great Race,* for instance, is promulgated on the websites of numerous white-power groups, right-wing militias, and neo-Nazi organizations—often alongside Adolf Hitler's *Mein Kampf.* In recent years, Grant's ideas (and his actual words) have been continually regurgitated by the proponents of stricter immigration laws. Indeed, important segments of the modern anti-immigration movement are financed by groups with ties to the Pioneer Fund, Planned Parenthood, and other groups whose origins can be traced back to Grant and the eugenics movement.

Furthermore, Grant would be a strong supporter of the "new eugenics," with its headline-grabbing advances in cloning, stem cell research, genetic screening, and designer babies. In addition, we have seen in recent years the surprising phenomenon that some of Grant's pseudoscience has actually been validated by research in genetics. Scientists now claim to have found significant and identifiable *genetic* causes of such *personality* traits as optimism, shyness, intelligence, criminality, alcoholism, risk taking, homelessness, homosexuality, musical talent, religiosity, schizophrenia, proficiency in child rearing, and many others. These findings admittedly have been simplified and distorted by the press, but the next few years will witness a growing avalanche of similar claims as the pendulum of the never-ending nature-nurture debate swings swiftly back toward nature. Grant, with his patrician self-confidence, would not have doubted it for a minute.

Stephen Jay Gould once said of eugenics: "Some historical arguments are so intrinsically illogical or implausible that, following their fall from grace, we do not anticipate any subsequent resurrection in later times and contexts."[17] And yet—unlike the Vikings of the days of Rollo—the implausible ideas of Madison Grant are continually resurrected and will apparently be with us forever.

Appendix A

Organizations Served by Madison Grant in an Executive Capacity

American Bison Society (cofounder)
American Coalition (honorary vice president; member of the board)
American Committee for International Wild Life Protection (cofounder; advisory committee)
American Defense Society (trustee)
American Eugenics Society (cofounder)
American Geographical Society (councilor)
American Museum of Natural History (trustee)
American Prison Society (vice president)
American Society of Mammalogists (charter and life member)
Aristogenic Association (cofounder)
Belmont Park Race Track (board of governors)
Boone and Crockett Club (president)
Bronx Parkway Commission (president)
Citizens' Committee on Immigration Legislation (vice president)
Conseil International de la Chasse (vice president)
Eugenics Committee of the U.S.A. (board of directors)
Eugenics Research Association (president)
Galton Society for the Study of the Origin and Evolution of Man (cofounder)
Immigration Restriction League (vice president)
Immigration Restriction League, Inc. (vice president)
International Commission on Eugenics (board of directors)
League of American Sportsmen (board of directors)
Military Order of the Loyal Legion (board of directors)
National Conference on Outdoor Recreation (general council)
National Institute of Social Sciences (vice president)
National Parks Association (cofounder)
National Parks Committee (founder)
Natural Parks Association of Washington (cofounder)
New York State Park Plan Committee (chairman)
New York Zoological Society (president)
Save-the-Redwoods League (cofounder)
Society of Colonial Wars (cofounder)
Taconic Park Commission (board of directors)
Westchester County Park Commission (board of directors)
Woodcraft League of America (national council)

Appendix B

The Interlocking Directorate of Wildlife Conservation

	ABS	*AGP*	*AOU*	*ASM*	*BCC*	*CFC*	*MNH*	*NAS*	*NZS*
Albright, Horace M.	ABS	AGP		ASM	BCC			NAS	
Allen, Joel A.		AGP	AOU	ASM	BCC		MNH	NAS	NZS
Andrews, Roy Chapman				ASM	BCC		MNH		NZS
Anthony, H. E.				ASM	BCC		MNH		NZS
Blair, W. Reid	ABS			ASM	BCC				NZS
Burnham, John B.		AGP			BCC	CFC			
Chapman, Frank			AOU		BCC		MNH	NAS	NZS
Coolidge, Harold J.	ABS			ASM	BCC				
Davenport, Charles B.	ABS			ASM					NZS
Davison, Charles Stewart	ABS			ASM	BCC				
Fisher, Albert K.		AGP	AOU	ASM	BCC				
Grant, DeForest					BCC				NZS
Grant, Madison	**ABS**			**ASM**	**BCC**		**MNH**	**NAS**	**NZS**
Gregory, William K.				ASM	BCC		MNH		NZS
Grinnell, George Bird		AGP	AOU	ASM	BCC		MNH	NAS	NZS
Grinnell, Joseph			AOU	ASM	BCC				
Hornaday, William T.	ABS			ASM	BCC	CFC	MNH	NAS	NZS
Merriam, C. Hart		AGP	AOU	ASM	BCC				
Merriam, John C.				ASM	BCC				
Nelson, Edward W.		AGP	AOU	ASM	BCC				
Osborn, Henry Fairfield	ABS		AOU	ASM	BCC		MNH	NAS	NZS
Palmer, Theodore S.	ABS	AGP	AOU	ASM	BCC		MNH	NAS	
Pearson, T. Gilbert					BCC			NAS	
Phillips, John C.	ABS	AGP		ASM	BCC				
Phillips, John M.	ABS	AGP		ASM	BCC	CFC			NZS
Pinchot, Gifford	ABS				BCC		MNH		NZS
Pratt, George D.	ABS	AGP		ASM	BCC		MNH		NZS
Pyne, Percy R.					BCC		MNH		NZS
Roosevelt, Kermit			AOU	ASM	BCC		MNH	NAS	NZS
Roosevelt, Theodore	ABS	AGP			BCC		MNH		
Sage, Dean					BCC		MNH		
Seton, Ernest Thompson	ABS			ASM		CFC			
Seymour, Edmund	ABS					CFC			
Sheldon, Charles		AGP	AOU	ASM	BCC			NAS	NZS

	ABS	*AGP*	*AOU*	*ASM*	*BCC*	*CFC*	*MNH*	*NAS*	*NZS*
Shiras III, George	ABS	AGP		ASM	BCC				
Townsend, Charles H.	ABS		AOU				MNH		NZS
Van Name, Willard	ABS						MNH	NAS	
Wadsworth, W. A.		AGP			BCC				NZS
Walcott, Frederic C.		AGP		ASM	BCC				

ABS: American Bison Society
AGP: American Game Protective Association
AOU: American Ornithologists' Union
ASM: American Society of Mammalogists
BCC: Boone and Crockett Club
CFC: Camp Fire Club
MNH: American Museum of Natural History
NAS: National Audubon Society
NZS: New York Zoological Society

Appendix C

Selected Members of the Advisory Council of the ECUSA

Anthropologists

John M. Cooper (Catholic U.)

E. A. Hooten (Harvard)

Ales Hrdlicka (National Museum)

Robert J. Terry (Washington U.)

Clark Wissler (Am. Mus. Nat. Hist.)

Biologists

Wesley Roswell Coe (Yale)

E. G. Conklin (Princeton)

Arthur Estabrook (ERO)

David Fairchild (Dept. of Ag.)

J. Arthur Harris (Minnesota)

Samuel J. Holmes (California)

Charles R. Stockard (Cornell)

Herbert E. Walter (Brown)

Paleontologists

William K. Gregory (Am. Mus. Nat. Hist.)

John C. Merriam (Carnegie Inst.)

Physicians

Lewellys F. Barker (J. Hopkins)

Herman Biggs

Harold Bowditch (Boston)

Walter B. Cannon

Hugh S. Cumming

Robert L. Dickinson

Haven Emerson (Columbia)

Lucien Howe

Woods Hutchinson

William Freeman Snow

Victor Clarence Vaughan

William H. Welch (J. Hopkins)

Geneticists

William E. Castle (Harvard

Leon J. Cole (Yale)

Edward M. East (Harvard)

Herbert S. Jennings (J. Hopkins)

Aaron Franklin Shull (Michigan)

Sewall Wright (Chicago)

Psychiatrists

Charles W. Burr

Walter E. Fernald

C. Floyd Haviland

Adolf Meyer (J. Hopkins)

Stewart Paton (Princeton)

Aaron J. Rosanoff

Sociologists

Emory Bogardus (USC)

Katherine B. Davis

Henry Pratt Fairchild (NYU)

Franklin H. Giddings (Columbia)

Robert MacIver (Columbia)

Stuart Rice (Pennsylvania)

E. A. Ross (Wisconsin)

Psychologists

Carl C. Brigham (Princeton)

Henry H. Goddard (Ohio State)

William McDougall (Harvard)

Carl E. Seashore (Iowa)

Lewis Terman (Stanford)

Edward L. Thorndike (Columbia)

Robert M. Yerkes (Yale)

Zoologists

Michael Guyer (Wisconsin)

Vernon Kellogg (Stanford)

Helen Dean King (Wistar Inst.)

Charles A. Kofoid (Scripps Inst.)

C. E. McClung (Pennsylvania)

Maynard M. Metcalf (J. Hopkins)

Horatio H. Newman (Chicago)

George Howard Parker (Harvard)

William Wheeler (Harvard)

Appendix D

Selected Members of the Interlocking Directorate of Scientific Racism

	ADS	*AES*	*ERA*	*ERO*	*GS*	*IRL*
Babbott, Frank L.		AES	ERA	ERO		
Barker, Lewellys F.		AES	ERA	ERO		
Brigham, Carl C.		AES	ERA		GS	
Campbell, Clarence G.		AES	ERA		GS	
Conklin, E. G.		AES	ERA		GS	
Davenport, Charles B.		AES	ERA	ERO	GS	
Davison, Charles Stewart	ADS					IRL
Estabrook, Arthur		AES	ERA	ERO		
Fairchild, Henry Pratt		AES				IRL
Fisher, Irving		AES	ERA	ERO	GS	
Goddard, Henry		AES	ERA	ERO		
Goethe, Charles M.		AES	ERA			
Gould, Charles W.	ADS	AES	ERA			
Grant, Madison	**ADS**	**AES**	**ERA**		**GS**	**IRL**
Gregory, William K.		AES			GS	
Hooton, E. A.		AES			GS	
Howe, Lucien		AES	ERA			IRL
Jenks, Jeremiah W.	ADS	AES				
Johnson, Albert		AES	ERA		GS	IRL
Johnson, Roswell		AES	ERA			
Jordan, David Starr		AES	ERA			IRL
Kinnicutt, Francis H.	ADS	AES				IRL
Laughlin, H. H.		AES	ERA	ERO	GS	
Lee, Joseph		AES				IRL
Little, C. C.		AES	ERA		GS	IRL
Merriam, John C.		AES			GS	
Olson, Harry		AES	ERA			
Osborn, Henry Fairfield		AES	ERA		GS	IRL
Paton, Stewart		AES	ERA			
Pearl, Raymond		AES	ERA		GS	
Popenoe, Paul		AES	ERA		GS	
Rumsey, Mary Harriman		AES	ERA	ERO		
Stoddard, Lothrop	ADS	AES	ERA		GS	IRL
Terman, Lewis		AES	ERA			

	ADS	*AES*	*ERA*	*ERO*	*GS*	*IRL*
Thorndike, Edward L.		AES	ERA		GS	
Trevor, John B.	ADS	AES	ERA			IRL
Ward, Robert DeC.		AES	ERA			IRL
Welch, William H.		AES		ERO		
Whitney, Leon		AES	ERA			
Wiggam, A. E.		AES	ERA			
Wissler, Clark		AES	ERA		GS	
Woods, Frederick Adams		AES	ERA		GS	
Yerkes, Robert M.		AES	ERA	ERO	GS	

ADS: American Defense Society
AES: American Eugenics Society (formerly ECUSA)
ERA: Eugenics Research Association
ERO: Eugenics Record Office
GS: Galton Society
IRL: Immigration Restriction League

Key to Archival Collections

ABCL	Records of the American Birth Control League, microfilm, Houghton Library, Harvard University
ABS	American Bison Society papers, Western History/Genealogy Department, Denver Public Library
AC	Andrew Carnegie papers, Library of Congress
ADS	American Defense Society papers, New-York Historical Society
AES	American Eugenics Society papers, American Philosophical Society
AH	Ales Hrdlicka papers, National Anthropological Archives, Smithsonian Institution
AKF	Albert K. Fisher papers, Library of Congress
ALK	A. L. Kroeber papers, the Bancroft Library, University of California, Berkeley
AM	Adolf Meyer papers, Chesney Medical Archives, Johns Hopkins University
AMH	Archer M. Huntington papers, George Arents Research Library for Special Collections, Syracuse University Library
AMNH	Central Archives, American Museum of Natural History
AREP	Amos R. E. Pinchot papers, Library of Congress
ARK	A. Remington Kellogg papers, Smithsonian Institution Archives
BDA	Records of the U.C. Berkeley Department of Anthropology, the Bancroft Library, University of California, Berkeley
BL	Bert Lord papers, Division of Rare and Manuscript Collections, Carl A. Kroch Library, Cornell University
CBD	Charles Benedict Davenport papers, American Philosophical Society
CCL	Clarence C. Little papers, Bentley Historical Library, University of Michigan
CCL-M	Clarence C. Little papers, Special Collections Department, Raymond H. Fogler Library, University of Maine
CDW	Charles D. Walcott papers, Smithsonian Institution Archives
CEA	Carl Ethan Akeley papers, University of Rochester Library
CFD	Charles Fremont Dight papers, Minnesota Historical Society
CHM	C. Hart Merriam papers, the Bancroft Library, University of California, Berkeley
CMG	Charles M. Goethe papers, Special Collections, California State University, Sacramento
CRVH	Charles R. Van Hise papers, microfilm, Library of Congress
CS	Charles Sheldon papers, the Elmer E. Rasmuson Library, University of Alaska Fairbanks
CSS	Charles Scribner's Sons Archives, Department of Rare Books and Special Collections, Princeton University Libraries

CWB	C. William Beebe papers, Department of Rare Books and Special Collections, Princeton University Libraries
DP	Derrydale Press records, Department of Rare Books and Special Collections, Princeton University Libraries
DSJ	David Starr Jordan papers, Hoover Institution Archives, Stanford University
EAH	Earnest A. Hooton papers, Peabody Museum Archives, Harvard University
EAP	Edward Alexander Preble papers, Smithsonian Institution Archives
EGC	Edwin Grant Conklin papers, Department of Rare Books and Special Collections, Princeton University Libraries
EH	Ellsworth Huntington papers, Yale University
ER	Elihu Root papers, Library of Congress
ES	Edmund Seymour collection, #6138, American Heritage Center, University of Wyoming, Laramie
ESC	Earnest Sevier Cox papers, Special Collections Library, Duke University
FB	Franz Boas papers, American Philosophical Society
FCW	Frederick C. Walcott papers, Sterling Library, Yale University
FDR	Section "Taconic State Park Commission," Family, Business and Personal Papers, Franklin D. Roosevelt papers, Franklin D. Roosevelt Library, Hyde Park, New York
FGN	Francis G. Newlands papers, Sterling Library, Yale University
FRB	Frederick Russell Burnham papers, Hoover Institution Archives, Stanford University
GBG	George Bird Grinnell papers, Yale University
GE	George Eastman papers, George Eastman House, Rochester
GEH	George Ellery Hale papers, on microfilm at the Bancroft Library, University of California, Berkeley
GG	Gabriel Grant papers, New-York Historical Society
GP	Gifford Pinchot papers, Library of Congress
GSM	Gerrit Smith Miller, Jr., papers, Smithsonian Institution Archives
HCIN	Records of the House Committee on Immigration and Naturalization, National Archives
HFO	Henry Fairfield Osborn papers, American Museum of Natural History
HFO-NY	Osborn Family papers, New-York Historical Society
HH	Herbert Hoover papers, Herbert Hoover Library, West Branch, Iowa
HHL	Harry H. Laughlin papers, Special Collections Department, Pickler Memorial Library, Truman State University
H-M	Log of the Half-Moon Club, New-York Historical Society
HSJ	Herbert Spencer Jennings Papers, American Philosophical Society
IRL	Immigration Restriction League records, Houghton Library, Harvard University
JB	James Barnes papers, Department of Rare Books and Special Collections, Princeton University Libraries
JBB	John B. Burnham collection, #6797, American Heritage Center, University of Wyoming, Laramie
JCM	John C. Merriam papers, Library of Congress

JCM-B	John C. Merriam papers, the Bancroft Library, University of California, Berkeley
JFL	John Fletcher Lacey papers, State Historical Society of Iowa
JHK	John Harvey Kellogg papers, Bentley Historical Library, University of Michigan
JJC	John Jay Chapman papers, Houghton Library, Harvard University
JL	Jacques Loeb papers, Library of Congress
JMcC	James McKeen Cattell papers, Library of Congress
JoL	Joseph Lee papers, Massachusetts Historical Society
JRS	James R. Sheffield papers, Sterling Library, Yale University
KR	Kermit Roosevelt papers, Library of Congress
LB	Luther Burbank papers, Library of Congress
LCD	L. C. Dunn papers, American Philosophical Society
LF	Livingston Farrand papers, Division of Rare and Manuscript Collections, Carl A. Kroch Library, Cornell University
LH	Lucien Howe papers, Houghton Library, Harvard University
LMT	Lewis M. Terman papers, Department of Special Collections, Stanford University Libraries
MGL	M. G. Lowman papers, Hoover Institution Archives, Stanford University
NMB	Nicholas Murray Butler papers, Rare Book and Manuscript Library, Columbia University
NWC	Records of the United States Nuernberg War Crimes Trials, National Archives
OLM	Ogden L. Mills papers, Library of Congress
PP	Paul Popenoe collection, #4681, American Heritage Center, University of Wyoming, Laramie
RB	Robert Bridges papers, Department of Rare Books and Special Collections, Princeton University Libraries
RBF	Raymond B. Fosdick papers, Mudd Library, Princeton University
RDW	Robert DeCourcy Ward papers, Boston Public Library
REP	Redfield Proctor papers, Proctor Free Library
RF	Rockefeller Family Archives, Rockefeller Archive Center, North Tarryton, New York
RHL	Robert H. Lowie Papers, the Bancroft Library, University of California, Berkeley
RM	Raymond Moley papers, Hoover Institution Archives, Stanford University
RMY	Robert M. Yerkes papers, Sterling Library, Yale University
ROS	Records of the Office of the Secretary, Smithsonian Institution Archives
RP	Raymond Pearl papers, American Philosophical Society
SC	Surrogate's Court, New York County
SJH	Samuel J. Holmes papers, the Bancroft Library, University of California, Berkeley
SRL	Save-the-Redwoods League files, San Francisco
SRL-B	Save-the-Redwoods League papers, the Bancroft Library, University of California, Berkeley

STM	Stephen T. Mather papers, the Bancroft Library, University of California, Berkeley
TB	Thomas Barbour papers, Harvard University Archives
THC	Thomas H. Carter papers, Library of Congress
TR	Theodore Roosevelt papers, Library of Congress
TR-S	Theodore Roosevelt papers, George Arents Research Library for Special Collections, Syracuse University Library
TSP	Theodore Sherman Palmer papers, Library of Congress
TWW	Woodrow Wilson papers, Library of Congress
VQR	Virginia Quarterly Review, Special Collections Dept., University of Virginia Library
WCS-7	"Office of the President (Madison Grant), 1901 and 1923–1929—Zoological Society History Project," Wildlife Conservation Society Archives, New York Zoological Park
WCS-8	"Office of the President (Madison Grant, W. Redmond Cross), 1906–1940—Records," Wildlife Conservation Society Archives, New York Zoological Park
WCS-10	"Office of the Secretary (Madison Grant), 1895–1905—Correspondence," Wildlife Conservation Society Archives, New York Zoological Park
WCS-11	"Office of the Secretary and Chairman of the Executive Committee (Madison Grant), 1901–1910—Correspondence," Wildlife Conservation Society Archives, New York Zoological Park
WCS-12	"Office of the Secretary and Chairman of the Executive Committee (Madison Grant), 1897–1906 and 1909–1912—Outgoing Correspondence," Wildlife Conservation Society Archives, New York Zoological Park
WCS-82	"American Bison Society, Office of the President (William T. Hornaday), 1905–1912—Records," Wildlife Conservation Society Archives, New York Zoological Park
WCS-84	"American Committee for International Wild Life Protection, Office of the Secretary, 1930–1962—Records," Wildlife Conservation Society Archives, New York Zoological Park
WHT	William Howard Taft papers, Library of Congress
WHW	William H. Welch papers, Chesney Medical Archives, Johns Hopkins University
WK	William Kent Family papers, Sterling Library, Yale University
WKG	William King Gregory papers, American Museum of Natural History
WTH	William T. Hornaday papers, Library of Congress
WW	Whitney Warren papers, Houghton Library, Harvard University

Notes

Introduction (pp. xi–xvi)

1. Grant to Osborn, 13 September 1928, quoted in Kennedy, "Philanthropy and Science in New York City," 208. For a brief analysis of the connections between eugenics and conservationism, see Brechin, "Conserving the Race."

2. Grant to Hornaday, 13 December 1927, folder "New York Zoological Society," container 13, WTH.

3. J. B. Trevor to R. D. Ward, 3 November 1930, RDW.

Chapter 1. Big-Game Hunter (pp. 3–30)

1. Roosevelt, "The Boone and Crockett Club," 267.

2. Roosevelt, "The Boone and Crockett Club," 267.

3. Trefethen, *Crusade for Wildlife,* 14–19; Roosevelt, 'The Boone and Crockett Club," 267.

4. "The Boone and Crockett Club," *Forest and Stream* 31, no. 26 (17 January 1889): 513.

5. "The Boone and Crockett Club," *Forest and Stream* 31, no. 26 (17 January 1889): 513.

6. U.S. Senate, Committee on Veterans' Affairs, *Medal of Honor Recipients, 1863–1973.*

7. "Oatlands," 297. Owing to a typo in Grant's obituary in the *New York Times,* almost every scholar since 1937 has erroneously stated that Grant was raised on an estate named Gatlands.

8. "Oatlands," 298, 299.

9. Grant, untitled essay, biography files, Central Archives, AMNH.

10. Anthony, "Madison Grant," 396.

11. I am surmising Grant's train of thought from his discussion of the red deer in "The Vanished Game of Yesterday," in *Hunting Trails on Three Continents,* ed. Grinnell et al., 2–3; and Grant to Courtland E. Durand, 4 February 1931, folder 2, box 2, "Office of the President (Madison Grant, W. Redmond Cross), 1906–1940: Records," WCS-8.

12. Phelps, *Autobiography,* 136; Robert Lewis Woodbury, "William Kent," 20.

13. Santayana quoted in Morison, *Turmoil and Tradition,* 30.

14. William T. Hornaday, draft of unpublished autobiography, container 17, WTH; "Notes on Madison Grant," folder "Grant, Madison," biography files, AMNH.

15. Society of Colonial Wars, *Annual Register of Officers* (1894), 8–9.

16. James, *The American Scene,* 84; Grant, *The Passing of the Great Race,* 2nd ed. (1918), 219.

17. Adams, *The Education of Henry Adams,* 499.

18. Grant, "The Racial Transformation of America," 344; [Lorimer and Bigelow], "America Last," 22.

19. Society of Colonial Wars, *Annual Register of Officers* (1894), 7.

20. Society of Colonial Wars, *Annual Register of Officers* (1894), 7.

21. William T. Hornaday, "Eighty Fascinating Years," chap. 11, p. 2, container 112, WTH.

22. Grant, *Preserve an America Worth Fighting For,* 4. See also Grant, *The Conquest of a Continent,* 221.

23. Grant, "'Preservation' or 'Restoration,'" 338; W. Redmond Cross, "A Tribute to Madison Grant," *Zoological Society Bulletin* 40, no. 4 (July–August 1937), 104. *Zoological Society Bulletin* hereafter abbreviated *ZSB.*

24. "Dr. G. B. Grinnell, Naturalist, Dead," *New York Times,* 12 April 1938, p. 23, col. 1.

25. Grinnell quoted in Trefethen, *Crusade for Wildlife,* 50.

26. Grant, *Early History of Glacier National Park,* 5–6. For Grant's admiration of Grinnell, see Grant, "George Bird Grinnell."

27. Grant to Grinnell, 30 August 1930, reel 37, #524, GBG; Grinnell to Grant, 31 January 1929, reel 30, #224, GBG.

28. Grinnell to Grant, 2 February 1914, #50, reel 18, GBG; 1 April 1916, #927, reel 20, GBG; Niles to Grant, [n.d.], folder "Sinking Fund," box 1, WCS-10.

29. Roosevelt to Grinnell, 13 January 1894 (quoted in Trefethen, *Crusade for Wildlife,* 353).

30. Roosevelt, "My Life as a Naturalist," 325.

31. Alice Roosevelt, quoted in Collier, *The Roosevelts,* 18.

32. "The Boone and Crockett Club," *Forest and Stream* 48, no. 4 (23 January 1897): 61.

33. George Bird Grinnell and Theodore Roosevelt, "The Boone and Crockett Club," in *American Big-Game Hunting,* ed. Grinnell and Roosevelt, 14–15.

34. Yard, *Our Federal Lands,* 330.

35. Quoted in Murphy, *Wild Sanctuaries,* 33; Trefethen, *Wildlife Management and Conservation,* 8.

36. Trefethen, *An American Crusade,* 64.

37. Quoted in *Bulletin of the American Game Protective Association* 3, no. 1 (January 1915): 3.

38. Cameron, *The Bureau of Biological Survey,* 4.

39. Reiger, *American Sportsmen,* 23; Hornaday, *Our Vanishing Wild Life,* 54.

40. Grant, "The Vanishing Moose," 345.

41. Grant, "The Vanishing Moose," 345.

42. Grant, "Condition of Wild Life in Alaska," *Twelfth Annual Report of the New York Zoological Society* (1908), 127. The New York Zoological Society hereafter abbreviated NYZS.

43. Grant, "The Vanishing Moose," 346; Roosevelt to Grinnell, 13 January 1894, quoted in Trefethen, *Crusade for Wildlife,* 76.

44. O'Brien, "Environmentalism as a Mass Movement," 7.

45. See Grant, "Reform in New York Game Laws," 978; "The Problem of the Deer," *New York Times,* 10 February 1896, sec. 2, p. 9, col. 3; "To Preserve the Deer," *New York Times,* 11 February 1896, p. 9, col. 1.

46. Grinnell to Grant, 6 April 1897, folder "1895–1899, G," box 4, WCS-10; Roosevelt to Grant, 27 April 1897, reel 313, series 2, TR.

47. The full story of Grant's efforts to secure the Adirondack Deer Law is in Spiro, "Pa-

trician Racist." See also Tober, *Who Owns the Wildlife?* 194–96; and Kranz, "Pioneering in Conservation," 110, 439.

48. Grant, "Condition of Wild Life in Alaska," *Twelfth Annual Report of the NYZS* (1908), 126–28; "The Vanished Game of Yesterday," in *Hunting Trails on Three Continents,* ed. Grinell et al., 2; "The Society's Expedition to Alaska," *Sixth Annual Report of the NYZS* (1902), 138; "The National Collection of Heads and Horns," *Eleventh Annual Report of the NYZS* (1907), 41.

49. Grant, "The Caribou," *Seventh Annual Report of the NYZS* (1903), 187.

50. Grant, "The Society's Expedition to Alaska," *Sixth Annual Report of the NYZS* (1902), 140; Grant to Lacey, 1 February 1902, box 252, JFL.

51. Grant to Lacey, 1 February 1902 (letter no. 1), box 252, JFL.

52. Grant, "The Condition of Wild Life in Alaska," in *Hunting at High Altitudes,* ed. Grinnell, 386–87; *The National Collection of Heads and Horns,* part 1, 27; Grant, "Condition of Wild Life in Alaska," *Twelfth Annual Report of the NYZS* (1908), 128, 126.

53. Grant, "The Caribou," *Seventh Annual Report of the NYZS* (1903), 196.

54. Grant to W. P. Dillingham, 23 March 1904, folder "1904 Bi-Bo," container 90, GP.

55. The Alaska Game Law was revised in 1908 and again in 1925. See Spiro, "Patrician Racist," chap. 20.

56. Grant, "Condition of Wild Life in Alaska," *Twelfth Annual Report of the NYZS* (1908), 127. The full story of No-Sale-of-Game Laws and the Migratory Bird Law is in Spiro, "Patrician Racist," chaps. 4 and 7.

57. Nelson to Grant, 1 May 1933, folder 4, box 2, WCS-8. The full story of Grant's efforts to save Alaska's wildlife is in Spiro, "Patrician Racist," chaps. 4 and 20.

58. "The Boone and Crockett Club," *Forest and Stream* 62, no. 5 (30 January 1904): 81.

59. Udall, *The Quiet Crisis,* 149.

60. Anthony, "Madison Grant," 396.

Chapter 2. The Bronx Zoo (pp. 31–51)

1. Grant quoted in folder 2, "Office of the President (Madison Grant), 1901 and 1923–1929: Zoological Society History Project," WCS-7.

2. Grant, "The Origin of the New York Zoological Society," 313, 318; *First Annual Report of the NYZS* (1897), 14–16.

3. Kipling quoted in Hofstadter, *The Age of Reform,* 178.

4. Knerr, "The Mayoral Administration of William L. Strong," 4.

5. [Lorimer and Bigelow], "America Last," 22.

6. Parkhurst quoted in Knerr, "The Mayoral Administration of William L. Strong," 10–11.

7. Butler quoted in Knerr, "The Mayoral Administration of William L. Strong," 238.

8. Holli and Jones, *Biographical Dictionary of American Mayors, 1820–1980,* 349.

9. Roosevelt to Grant, 10 October 1894, in Morison, *The Letters of Theodore Roosevelt,* vol. 1, p. 401; "The Boone and Crockett Club," *Forest and Stream* 62, no. 5 (30 January 1904): 81; Roosevelt to Grant, 4 December 1894, folder 2, WCS-7.

10. Grant, "The Origin of the New York Zoological Society," 314; "The Zoological Garden," *New York Times,* 12 March 1897, p. 6, col. 3; Hornaday, draft of unpublished autobiography, container 17, WTH.

11. Grant to A. Newbold Morris, 20 October 1897, book #1, "Office of the Secretary and Chairman of the Executive Committee (Madison Grant), 1897–1906 and 1909–1912: Outgoing Correspondence," WCS-12 see also Grant to A. Newbold Morris, 1 December 1899, book #2, WCS-12; Grant, "History of the Zoological Society," *ZSB* no. 37 (January 1910): 589.

12. Hornaday, draft of unpublished autobiography, container 17, WTH; correspondence, passim, folder 5, WCS-7; Grant, "The Origin of the New York Zoological Society," 320; Roosevelt to Grant, 22 April 1895, folder 2, WCS-7.

13. Hornaday, draft of unpublished autobiography, container 17, WTH.

14. James Andrew Dolph, "Bringing Wildlife to the Millions," vi; Bridges, *Gathering of Animals,* 21.

15. Hornaday, "Report on the Character and Availability of South Bronx Park," in *First Annual Report of the NYZS* (1897), 34.

16. Hornaday, draft of unpublished autobiography, container 17, WTH; Grant to Osborn, 9 November 1899, book #2, WCS-12.

17. Hornaday, draft of unpublished autobiography, container 17, WTH.

18. Hornaday, *Our Vanishing Wildlife,* 203; Fox, *John Muir and His Legacy,* 95, 148.

19. Hornaday, draft of unpublished autobiography, containers 15–17, WTH.

20. Hornaday, unpublished autobiography (chap. 11, p. 9), container 112, WTH.

21. Bridges, *Gathering of Animals,* 71; "Twelve Years' Perspective of the Zoological Park," *Fourteenth Annual Report of the NYZS* (1910), 109.

22. Quoted in Bridges, *Gathering of Animals,* 72.

23. "Opening of the Zoological Park," *Fourth Annual Report of the NYZS* (1900), 76–77.

24. *ZSB* no. 26 (1907): 354–55.

25. Grant, "The Vanishing Moose," 356; Grant, "A Canadian Moose Hunt," in *Hunting in Many Lands,* edited by George Bird Grinnell and Theodore Roosevelt (New York: Forest and Stream Publishing Co., 1895), 97.

26. Grant, "History of the Zoological Society," *ZSB* no. 37 (January 1910): 599.

27. The full story of Grant's involvement with the aquarium is told in Spiro, "Patrician Racist."

28. Hornaday, "Photographic Privileges in the New York Zoological Park," *ZSB* 25, no. 4 (July 1922): 95.

29. Hornaday, "Report of the Director," *Twentieth Annual Report of the NYZS* (1916), 63; "Report of the Director," *Nineteenth Annual Report of the NYZS* (1918), 63 (emphasis added).

30. Osborn, "Twenty-eight Years of Zoological Society Accomplishment," *Twenty-seventh Annual Report of the NYZS* (1923), xiii (emphasis added).

31. Samuel P. Verner, "The Story of Ota Benga, the Pygmy," *ZSB* 19, no. 4 (July 1916), 1377.

32. Samuel P. Verner, "The Story of Ota Benga, the Pygmy," *ZSB* 19, no. 4 (July 1916): 1379. Hornaday's somewhat garbled and inaccurate account of the incident is in "An African Pigmy," *ZSB* no. 23 (October 1906): 301.

33. Hornaday quoted in "Negro Ministers Act to Free the Pygmy," *New York Times,* 11 September 1906, p. 2, col. 2.

34. "Man and Monkey Show Disapproved by Clergy," *New York Times,* 10 September 1906, p. 1, col. 7.

35. “Man and Monkey Show Disapproved by Clergy,” *New York Times,* 10 September 1906, p. 1, col. 7; “Negro Ministers Act to Free the Pygmy,” *New York Times,* 11 September 1906, p. 2, col. 2; *New York Globe,* 10 September 1906, quoted in Bradford and Blume, *Ota,* 260.

36. Blair, *In the Zoo,* 1929, 77; “Negro Ministers Act to Free the Pygmy,” *New York Times* 11 September 1906, p. 2, col. 2; “The Mayor Won't Help to Free Caged Pygmy,” *New York Times,* 12 September 1906, p. 9, col. 5.

37. “African Pygmy's Fate Is Still Undecided,” *New York Times,* 18 September 1906, p. 9, col. 3; Hornaday to Verner; quoted in Bridges, *Gathering of Animals,* 227.

38. “The Mayor Won't Help to Free Caged Pygmy,” *New York Times,* 12 September 1906, p. 9, col. 5.

39. “Negro Ministers Act to Free the Pygmy,” *New York Times,* 11 September 1906, p. 2, col. 2; Gordon quoted in Bradford and Blume, *Ota,* 183.

40. “Negro Ministers Act to Free the Pygmy,” *New York Times,* 11 September 1906, p. 2, col. 2; “The Pigmy Is Not the Point,” *New York Times,* 12 September 1906, p. 8, col. 5; “Ota Benga Having a Fine Time,” *New York Times,* 13 September 1906, p. 6, col. 6.

41. Hornaday, “Suicide of Ota Benga, the African Pygmy,” *ZSB* 19, no. 3 (May 1916) 1356; “Bushman Shares a Cage with Bronx Park Apes,” *New York Times,* 9 September 1906, p. 9, col. 5; “Send Him Back to the Woods,” *New York Times,* 11 September 1906 p. 6, col. 4.

42. Hornaday, “Suicide of Ota Benga, the African Pygmy,” *ZSB* 19, no. 3 (May 1916): 1356

43. For Ishi, see Starn, *Ishi's Brain.*

44. Hellman, *Bankers, Bones and Beetles,* 86–89.

45. Hornaday to John Cromartie (quoted in Bridges, *Gathering of Animals,* 388).

46. Quoted in Henry F. Osborn, “Twenty-eight Years of Zoological Society Accomplishment,” *Twenty-seventh Annual Report of the NYZS* (1923), xvi.

47. Quoted in Hornaday, “Report of the Director,” *Twenty-second Annual Report of the NYZS* (1918), 61.

48. Hornaday, draft of unpublished autobiography, container 17, WTH; Carnegie to Grant, 17 March 1914, reel 37, GBG.

49. Fosdick to Grant, 8 June 1923, folder 1, box 6, series I, RBF.

Chapter 3. From Conservation to Preservation (pp. 52–72)

1. In Seton's *Lives of Game Animals,* which was *the* hunters' encyclopedia, Grant is often cited as an expert. See vol. 1, p. 365 (coyote); vol. 2, p. 459 (fisher); vol. 2, pp. 493 and 499 (marten); vol. 3, p. 202 (bear); vol. 3, p. 526 (bighorn sheep).

2. Grant, *The Passing of the Great Race,* 33. Unless indicated, all references to *The Passing of the Great Race* are from the first (1916) edition.

3. Grant, “Saving the Redwoods: An Account,” 91; Grant, “The Future of Our Fauna,” *ZSB* no. 34 (June 1909): 504.

4. Grant, “Condition of Wild Life in Alaska,” *Twelfth Annual Report of the NYZS* (1908), 127. See also Grant, “The Future of Our Fauna,” *ZSB* no. 34 (June 1909): 504; and Grant, “Notes on Adirondack Mammals,” 334.

5. For the efforts of the Boone and Crockett Club to protect Yellowstone National Park, see Spiro, “Patrician Racist,” chap. 1.

6. Grinnell to John W. Noble, 28 February 1910, GBG; Beard quoted in Cutright, *Theodore Roosevelt: The Making of a Conservationist,* 178.

7. Roosevelt quoted in Cutright, *Theodore Roosevelt, the Naturalist,* 169.

8. Udall, *The Quiet Crisis,* 161.

9. Pinchot, *Breaking New Ground,* 145.

10. Pinchot quoted in Nicholas Roosevelt, *Conservation,* 17.

11. Pinchot quoted in Samuel Hays, *Conservation and the Gospel of Efficiency,* 42.

12. Pinchot, *The Fight for Conservation,* 42; Pinchot quoted in O'Brien, "Environmentalism as a Mass Movement," 9; Pinchot, *Breaking New Ground,* 31.

13. Lacey to Grant, 16 January 1907, box 257, JFL.

14. Pinchot, *Breaking New Ground,* 46–48.

15. "Boone and Crockett Club Meeting," *Forest and Stream* 58, no. 5 (1 February 1902): 88; Roosevelt, "Wilderness Reserves," in *American Big Game in Its Haunts,* ed. Grinnell, 46.

16. *Nineteenth Annual Report of the NYZS* (1915), 59; Grant, "Wild Life Protection," *ZSB* 19, no. 1 (January 1916): 1320; Grant, "The Future of Our Fauna," *ZSB* no. 34 (June 1909): 504. See also Grant, "Condition of Wild Life in Alaska," *Twelfth Annual Report of the NYZS* (1908), 127.

17. Boone and Crockett Club, *Report of Committee on Game Refuges,* 15, 17 (emphasis added).

18. Pinchot quoted in Trefethen, *An American Crusade,* 263.

19. Pinchot to Grant, 4 October 1909, folder "1909 Go-Gu," container 121, GP.

20. Phelan to J. R. Garfield, 21 November 1907, quoted in Fox, *John Muir and His Legacy,* 141; Kent quoted in Graham, *Man's Dominion,* 164.

21. Grant to J. C. Sperry, 16 June 1920, folder "Grant, Madison," SRL; Grant, "Report of the Executive Committee," *Twenty-fifth Annual Report of the NYZS* (1921), 57.

22. This advice was also given to Grant by John F. Lacey (see Lacey to Grant, 25 July 1906, folder 6, box 1, WCS-11).

23. Trefethen, *Crusade for Wildlife,* 88; Grinnell to John F. Lacey, 5 July 1905, box 254, JFL.

24. Conductor in Murphy, *Wild Sanctuaries,* 27.

25. Greeley quoted in Graham, *Man's Dominion,* 18; Sheridan in Garretson, *A Short History of the American Bison,* 128.

26. Fifty/500 rule in Teresi, "Monster of the Tub," 92; Hornaday's census in Tober, *Who Owns the Wildlife?* 102.

27. Grinnell, "In Buffalo Days," in *American Big-Game Hunting,* ed. Grinnell and Roosevelt, 211; Grant to W. P. Dillingham, 23 March 1904, folder "1904 Bi-Bo," container 90, GP.

28. Grant to John F. Lacey, 26 February 1902, box 252, JFL.

29. Grant to Hornaday, 7 February 1906, folder 1, box 1, WCS-82.

30. Elwin R. Sanborn, "The National Bison Herd," *ZSB* no. 28 (January 1908): 408–9.

31. Grant, "Report of the Executive Committee," *Eighteenth Annual Report of the NYZS* (1914), 50.

32. Murphy, *Wild Sanctuaries,* 109.

33. American Bison Society to Dear Sir, 15 January 1906, folder 1, box 1, WCS-82.

34. Hornaday to Osborn, 23 January 1914, folder "New York Zoological Society," container 13, WTH.

35. Grant, "The Vanished Game of Yesterday," in *Hunting Trails on Three Continents,* ed. Grinnell et al., 8; Osborn, "Report of the President for 1915," *Tenth Annual Report of the American Bison Society* (1916), 14. The full story of Grant's efforts to save the bison is in Spiro, "Patrician Racist," chap. 6.

36. Albright, review of *The American Bison,* by Martin Garretson, in *ZSB* 41, no. 3 (May–June 1938): 111.

37. See Spiro, "Patrician Racist," chaps. 6, 8, 9, 20.

38. Trefethen, *Crusade for Wildlife,* 150; A. Brazier Howell to Grant, 14 April 1931, folder 2, box 2, WCS-8.

39. See Spiro, "Patrician Racist," chap. 8.

40. Quoted in Grant, "The Beginnings of Glacier National Park," in *Hunting and Conservation,* ed. Grinnell and Sheldon, 452.

41. Trefethen, *Crusade for Wildlife,* 101.

42. Trefethen, *Crusade for Wildlife,* 105.

43. Grant to Carter, 1 February 1910, folder "G 1910," container 71, ER; Grant to Carter, 3 May 1910, and Carter to Grant, 7 May 1910, both in reel 7, THC.

44. For the story of how the rights of the Blackfeet Indians were ignored in the creation of the park, see Spence, *Dispossessing the Wilderness,* chaps. 5 and 6.

45. See, for example, advertisement in *Collier's Magazine* 67, no. 26 (25 June 1921): 25.

46. Grant to Interior Secretary West, 16 January 1929, folder 9, box 1, WCS-8.

47. W. Redmond Cross, "A Tribute to Madison Grant," *ZSB* 40, no. 4 (July–August, 1937): 103.

48. *New York Herald Tribune,* 13 April 1937.

Chapter 4. Wildlife Management (pp. 73–87)

1. Yard, *Our Federal Lands,* 333.

2. Anthony, "Madison Grant," 397.

3. Hornaday, *Thirty Years War,* 159, 251; Hornaday to Seymour, 18 June 1920, folder 2, box 4, ABS.

4. Blair, "William Temple Hornaday," *ZSB* 40, no. 2 (March–April 1937): 48; Trefethen, *Crusade for Wildlife,* 329; Hornaday, *Thirty Years War,* ix.

5. Hornaday, unpublished autobiography, chap. 20, p. 2, container 112, WTH.

6. Gabrielson, *Wildlife Refuges,* 10–11.

7. Trefethen, *Crusade for Wildlife,* 240. Trefethen covers both the Yellowstone elk situation and the Kaibab deer problem in great detail in *Crusade for Wildlife,* 229–50.

8. Shiras quoted in Trefethen, *An American Crusade,* 197.

9. Cameron, *The Bureau of Biological Survey,* 191; Mather quoted in Shankland, *Steve Mather,* 273.

10. Committee quoted in Cameron, *The Bureau of Biological Survey,* 192.

11. Leopold, *Game Management;* Leopold, "Ten New Developments."

12. Leopold, "Ten New Developments," 7 (emphasis added).

13. Leopold, *Game Management,* 20. The term *ecosystem* was introduced in 1935 by Oxford botanist A. G. Tansley.

14. Leopold quoted in Fox, *John Muir and His Legacy,* 246; Leopold, "Ten New Developments," 7.

15. Herbert C. Hanson, *Dictionary of Ecology,* 121, quoted in Reiger, *American Sportsmen,* 19; Leopold quoted in Alston Chase, *In a Dark Wood,* 45.

16. Leopold, *Sand County Almanac,* 210.

17. Roosevelt, "Our Vanishing Wild Life," 161; Boone and Crockett Club, *Report of the Game Preservation Committee,* 18. See also Roosevelt, "Three Books of the Wilderness," in *Literary Essays,* 379–81.

18. Roosevelt to Grinnell, April 1918, quoted in Cutright, *Theodore Roosevelt, the Naturalist,* 263 (emphasis added).

19. *Annual Report of the American Bison Society* (1931), 53; *Annual Report of the American Bison Society* (1927), 13–14. A good summary of the position of the interlocking directorate on wildlife management is John C. Phillips, "Conservation of Big Game," in Boone and Crockett Club, Committee on Records of North American Big Game, *North American Big Game.*

20. Cameron, *The Bureau of Biological Survey,* 192.

21. Grant to Pierre Lorillard, 29 December 1911, book #20, WCS-12.

22. Biological Survey quoted in Mighetto, *Wild Animals,* 96; Cameron, *The Bureau of Biological Survey,* 51, 52, 178, 57.

23. Hornaday and Burroughs quoted in Mighetto, *Wild Animals,* 75.

24. Cameron, *The Bureau of Biological Survey,* 51, 64; Mighetto, *Wild Animals,* 85.

25. Grant to Walcott, 26 May 1930, folder 10, box 1, WCS-8.

26. "Annual Meeting of the Board of Managers," *ZSB* 32, no. 2 (March–April 1929): 57; "Resolution of 10 January 1929," in Boone and Crockett Club, *Officers, Constitution, Members* (1928–29), 42; Merriam quoted in Sterling, *Last of the Naturalists,* 257. See also "A Protest," in A. Brazier Howell to Paul G. Redington, 4 April 1930, #770, reel 42, GBG.

27. Paul G. Redington to Grant, 8 October 1931, folder "Madison Grant," container 82, JCM; H. E. Anthony to John C. Merriam, 1 December 1931, folder 3, box 2, WCS-8.

Chapter 5. From Mammals to Man (pp. 88–116)

1. Osborn, *Creative Education,* 250. For more on the founding of the Museum, see Kennedy, "Philanthropy and Science in New York City," 2–42, 90–91, 104–5.

2. Rainger, *An Agenda for Antiquity,* 47.

3. Osborn, *The Age of Mammals,* 1.

4. Hornaday, unpublished autobiography, chap. 11, pp. 2–3, container 112, WTH; Osborn, *Fifty-two Years of Research.*

5. Shapiro quoted in Hellman, *Bankers, Bones and Beetles,* 203.

6. Grant to Osborn, 14 November 1910, book #18, WCS-12; Grinnell to Grant, 3 December 1910, #637, reel 14, GBG (emphasis added).

7. Hornaday, draft of unpublished autobiography, container 17, WTH.

8. For a listing of the meetings of the Half-Moon Club, see Spiro, "Patrician Racist," appendix B.

9. Ripley, *The Races of Europe;* Ripley, *A Selected Bibliography.*

10. Ripley, *The Races of Europe,* 600, 32.

11. Ripley, "The European Population of the United States," 221–24, 240. The same article appeared as "Races in the United States," *Atlantic Monthly* (December 1908): 745–59.

12. Ripley, "The European Population of the United States," 224–25, 234.

13. Ripley, *The Races of Europe*, 395, 396.

14. Ripley, *The Races of Europe*,95, 349–382.

15. Ripley, *The Races of Europe*, 372–73.

16. Statistics from Rischin, *The Promised City*, 20, 57, 59, 80; Daniels, *Coming to America*, 226; Lloyd P. Gartner, "Jewish Migrants en Route from Europe to North America," in *The Jews of North America*, ed. Rischin, 32; Burrows and Wallace, *Gotham*, 1117.

17. Ripley, *The Races of Europe*, 383–85; Ripley, "The European Population of the United States," 239.

18. Ripley, "The European Population of the United States," 238.

19. Hornaday to Herbert Kaufman, 3 March 1919, folder "1919-Mar," container 67, WTH; Grant to Taft, 22 November 1910, series 6, case file 77, WHT. Number of synagogues in Daniels, *Coming to America*, 228.

20. The statement is made by David, a young Jewish immigrant, in Zangwill, *The Melting Pot*, 33.

21. Walker, "Restriction of Immigration," 824; Walker, "Immigration and Degradation," 638, 642 (emphasis added).

22. Walker, "Restriction of Immigration," 822; "Immigration and Degradation," 643.

23. Walker, "Restriction of Immigration," 828.

24. Walker, "Immigration and Degradation," 643–44 (see also "Restriction of Immigration," 823).

25. Ross, "The Causes of Race Superiority," 88–89.

26. Roosevelt, "Race Decadence," 763–68.

27. Roosevelt, "Race Decadence," 766; Roosevelt, "Letter on Race Suicide," 551; TR quoted in Wister, *Roosevelt*, 66.

28. Yale fecundity based on statistics gleaned from Hill, *Quarter-Century Record*.

29. Grant, *The Passing of the Great Race*, 3.

30. Ripley, *The Races of Europe*, v.

31. Von Törok in Barzun, *Race*, 161.

32. Ripley, *The Races of Europe*, 108.

33. Gobineau, *The Inequality of Human Races*, 107, 205–6.

34. Gobineau, *The Inequality of Human Races*, 205–7, 85. "Sketch" in Todorov, *On Human Diversity*, 131.

35. Gobineau, *The Inequality of Human Races*, 209.

36. Gobineau, *The Inequality of Human Races*, 207, xv, 210.

37. Gobineau, *The Inequality of Human Races*, xiv.

38. Gobineau, *The Inequality of Human Races*, 25, 38, 37, 852, 173. "Detritus" in Poliakov, *The Aryan Myth*, 237. For Grant's opposition to imperialism, see Grant, *The Conquest of a Continent*, 294, 344.

39. Gobineau, *The Inequality of Human Races*, 98–100, 34.

40. "Apathetic orbits" in Poliakov, *The Aryan Myth*, 237. See also Gobineau, *The Inequality of Human Races*, 102, 208, 870, 872.

41. Tocqueville quoted in Biddiss, *Father of Racist Ideology*, 58; and Todorov, *On Human Diversity*, 127.

42. "We are not descended from the ape, but we are headed in that direction" (quoted in Gossett, *Race*, 345).

43. Stanton, *The Leopard's Spots,* discusses Gobineau's reception in the South (pp. 174–75).

44. Gobineau, *The Inequality of Human Races,* 102, 110; Wittgenstein quoted in Barzun, *Race,* 100.

45. Nietzsche, *Thus Spake Zarathustra,* 100; Cosima quoted in Field, *Evangelist of Race,* 153; Hitler, *Mein Kampf,* 240–43.

46. Chamberlain quoted in Poliakov, *The Aryan Myth,* 315. The best English-language biography of Chamberlain is Geoffrey G. Field's *Evangelist of Race.* It is a marvelous book and serves also as a masterful introduction to the history of anti-Semitism in Wilhelmine Germany.

47. Chamberlain, *Foundations of the Nineteenth Century,* 1: lxxxiii, 457.

48. Wagner quoted in Poliakov, *The Aryan Myth,* 198, and Field, *Evangelist of Race,* 153; Chamberlain, *Foundations,* 1:327, 331.

49. Chamberlain, *Foundations,* 1:262, 327; Kaiser quoted in Field, *Evangelist of Race,* 2; Montagu, *Man's Most Dangerous Myth,* 282.

50. Chamberlain, *Foundations,* 1:490, 577, lxvii, 257.

51. Chamberlain, *Foundations,* 1:542, 578, 257, 338, 329, 11, 482, 480.

52. Chamberlain, *Foundations,* 1: lxxviii, 217, 386; 2:57–65, 246; 1:330, 459, 336.

53. Chamberlain, *Foundations,* 1:390, 286–88, 271, 286, 299, 327.

54. Chamberlain, *Foundations,* 1:328, 291, 260, 261, 277. See also 1:262, 281, 283, 284, 290, 310.

55. Chamberlain, *Foundations,* 1:331, 2:384. See also 1:14, 261.

56. Chamberlain, *Foundations,* 1:537.

57. Schweitzer in Field, *Evangelist of Race,* 18; Shaw in Field, *Evangelist of Race,* 464; the Kaiser in Field, *Evangelist of Race,* 250–52.

58. McWilliams, *A Mask for Privilege,* 58; Grant to Prescott F. Hall, 4 December 1918, folder "Madison Grant," IRL; American reception in Field, *Evangelist of Race,* 465–66; Roosevelt, "The Foundations of the Nineteenth Century."

59. Chamberlain, *Foundations,* 1:315; Chamberlain quoted in Field, *Evangelist of Race,* 384.

60. Lapouge, *L'Aryen,* ix; "Race Studies in Europe," *Eugenical News* 13, no. 6 (June 1928): 82. All translations of Lapouge are mine.

61. Lapouge, *L'Aryen,* 30–31; Osborn, "Address of Welcome," in *Eugenics, Genetics and the Family,* 1; Kaiser quoted in Poliakov, *The Aryan Myth,* 270.

62. Lapouge, "The Fundamental Laws of Anthropo-sociology," 58; Lapouge, *L'Aryen,* 372–73.

63. Lapouge, *L'Aryen,* 22, 481; "From Count de Lapouge," *Eugenical News* 15, no. 8 (August 1930): 117.

64. Lapouge, *L'Aryen,* 464; "Post-War Immigration into France," *Eugenical News* 17, no. 4 (July–August 1932): 94.

65. Ammon quoted in in Barzun, *Race,* 221.

66. Lapouge, *L'Aryen,* vii; Lapouge, "L'Anthropologie et la Science Politique," 150–151.

Chapter 6. The Eugenics Creed (pp. 117–40)

1. Galton, "Hereditary Talent and Character," 157.

2. Galton, "Hereditary Talent and Character," 318, 161, 163. Galton's use of "nature

and nurture" in *English Men of Science: Their Nature and Nurture* (London, Macmillan & Co., 1874), 12.

3. Galton, "Hereditary Talent and Character," 319; *Hereditary Genius,* 64.

4. Galton, "Hereditary Talent and Character," 165.

5. Galton, "Eugenics: Its Definition, Scope and Aims," in *Essays in Eugenics,* 42; "Studies in National Eugenics," in *Essays in Eugenics,* 62; *Memories of My Life,* 311; "The Possible Improvement of the Human Breed under Existing Conditions of Law and Sentiment," in *Essays in Eugenics,* 20; *Hereditary Genius,* 352.

6. Galton, "Hereditary Talent and Character," 326; "Eugenics as a Factor in Religion," in *Essays in Eugenics,* 70; *Hereditary Genius,* 357.

7. Darwin, *The Descent of Man,* 125, 141, 28.

8. Darwin, *The Descent of Man,* 133–43.

9. For a brief discussion of the political effect of Haeckel's ideas on Wilhelmine and Nazi Germany, see Gasman, *The Scientific Origins of National Socialism.*

10. Ernst Haeckel, *The Riddle of the Universe,* 252, 233.

11. Gasman, *The Scientific Origins of National Socialism,* 161. See also Bölsche, *Haeckel.*

12. Haeckel, *Riddle of the Universe,* 239; Gasman, *The Scientific Origins of National Socialism,* 63.

13. Haeckel, *Riddle of the Universe,* 328.

14. Haeckel quoted in Gasman, *The Scientific Origins of National Socialism,* 159; and Shipman, *The Evolution of Racism,* 102.

15. Haeckel quoted in Shipman, *The Evolution of Racism,* 279.

16. Osborn, "The Present Problem of Heredity," 364.

17. Haeckel quoted in Gasman, *The Scientific Origins of National Socialism,* 95, 96.

18. Davenport quoted in Higham, *Strangers in the Land,* 151. For Gertrude Davenport, see MacDowell, "Charles Benedict Davenport," 15.

19. Davenport quoted in Haller, *Eugenics,* 62; Davenport, *Heredity in Relation to Eugenics,* iii.

20. Davenport, *Heredity in Relation to Eugenics,* 1, 4, 249, 7, 1, 6, 7, 2. See Kimmelman, "The American Breeders' Association."

21. Davenport quoted in MacDowell, "Charles Benedict Davenport," 29.

22. Davenport to Galton, 26 October 1910, in Pearson, *Francis Galton,* vol. 3b, 613.

23. The best study of Harry H. Laughlin is Hassencahl, "Harry H. Laughlin."

24. Davenport, *Heredity in Relation to Eugenics,* 267–68.

25. Davenport, *Heredity in Relation to Eugenics,* 256–66; Davenport, "Report of Committee on Eugenics," 129.

26. Davenport, *Heredity in Relation to Eugenics,* 263, 267.

27. Davenport, *Heredity in Relation to Eugenics,* 212–18.

28. Davenport, *Heredity in Relation to Eugenics,* 227.

29. Rosenberg, "Charles Benedict Davenport," 276.

30. Goddard, *The Kallikak Family,* 50, 29.

31. *Independent* review quoted in Smith, *Minds Made Feeble,* 62.

32. Galton, "Probability, the Foundation of Eugenics," in *Essays in Eugenics,* 75; Davenport to Mrs. Harriman, 21 February 1911, folder "Mrs. E. H. Harriman," CBD.

33. "Gonads," *Eugenical News* 10, no. 9 (September 1925): 125; Whitney, *The Case for Sterilization* 291.

34. Huntington, *Tomorrow's Children,* 103, 105; Osborn, *Creative Education,* 265–66; Grant, *The Passing of the Great Race,* 85–86.

35. Wiggam, introduction to *The Next Age of Man;* Wiggam, *The New Decalogue of Science,* 79, 99. For Galton's views on eugenics as a religion, see Galton, *Essays in Eugenics,* 42, 43, 58–59.

36. Grant, *The Passing of the Great Race,* 43–47.

37. Laughlin, *Report of the Committee,* 10; Van Hise quoted in Haller, *Eugenics,* 76.

38. Terman quoted in Samelson, "Putting Psychology on the Map," 116; Davenport, preface, in *Eugenics, Genetics and the Family,* ix.

39. Davenport quoted in Haller, *Eugenics,* 65.

40. Davenport to Prescott F. Hall, 20 May 1911, folder "P. F. Hall," CBD.

Chapter 7. The Passing of the Great Race (pp. 143–66)

1. Grant to Perkins, 19 August 1933, folder "Madison Grant," box 67, Author Files I, CSS.

2. Osborn, preface to Grant, *The Passing of the Great Race,* vii–ix.

3. Charles Scribner's to Roosevelt, 22 November 1916, folder 12, box 125, Author Files, CSS.

4. Grant, *The Passing of the Great Race,* xv, xvii.

5. Grant, *The Passing of the Great Race,* 87.

6. Grant, *The Passing of the Great Race,* 214.

7. Grant, *The Passing of the Great Race,* 150.

8. Grant, *The Passing of the Great Race,* 197.

9. Grant, *The Passing of the Great Race,* 152–59.

10. Grant, *The Passing of the Great Race,* 144, 200, 66, 178, 67. (See also 38, 77, 165–68, 173, 177, and 179.)

11. Grant, *The Passing of the Great Race,* 175.

12. Grant, *The Passing of the Great Race,* 199.

13. Grant, *The Passing of the Great Race,* 147, 199; 4th ed., 375; 1st ed., 191; 2nd ed., 75.

14. Grant, *The Passing of the Great Race,* 79, 77; 2nd ed., 90.

15. Grant, *The Passing of the Great Race,* 79, 77, 80.

16. Grant, *The Passing of the Great Race,* 2nd ed., 89–90, 1st ed., 80–81 (emphasis added).

17. Grant, *The Passing of the Great Race,* 2nd ed., 10–12; 1st ed., 67, 45.

18. Grant, *The Passing of the Great Race,* 96–97, 45–46.

19. Grant, *The Passing of the Great Race,* 69, 44, 45; 2nd ed., 18; 1st ed., 45.

20. Grant, *The Passing of the Great Race,* 14.

21. See Boas, "Changes in Bodily Form." A version of Boas's report was published as "Changes in the Bodily Form of Descendants of Immigrants," *American Anthropologist* 14, no. 3 (July–September 1912): 530–62.

22. Grant, *The Passing of the Great Race,* 14.

23. Grant, *The Passing of the Great Race,* 82, 15–16.

24. Grant, *The Passing of the Great Race,* 200; 2nd ed., 168.

25. Grant, *The Passing of the Great Race,* 34, 40, 190, 35 (emphasis added).

26. Grant, *The Passing of the Great Race,* 39, 185–86, 82.

27. Grant, *The Passing of the Great Race,* 80–81.

28. Grant, *The Passing of the Great Race,* 46.

29. Grant, *The Passing of the Great Race,* 14; 2nd ed., 12; 1st ed., 178; 2nd ed., 227–28; 1st ed., xvi.

30. Grant, *The Passing of the Great Race,* 5; 2nd ed., 8; 1st ed., 7, 69.

31. Grant, *The Passing of the Great Race,* 2nd ed., 79.

32. Grant, *The Passing of the Great Race,* 2nd ed., 79; 1st ed., xx, xvii, 45.

33. Grant, *The Passing of the Great Race,* 2nd ed., 49.

34. Grant, *The Passing of the Great Race,* 69, 55–56.

35. Grant, *The Passing of the Great Race,* 45–46; Grant to Charles Benedict Davenport, 6 July 1914, folder "Madison Grant," CBD; Grant, *The Passing of the Great Race,* 49–50, 98.

36. Grant, *The Passing of the Great Race,* 47 (emphasis added).

37. Grant, *The Passing of the Great Race,* 228.

38. Perry Miller, *The New England Mind,* 153.

39. Osborn, preface to Grant, *The Passing of the Great Race,* viii.

40. Leslie M. Scott to Grant, 27 November 1922, folder "Grant #3," HFO.

41. Roosevelt to Grant, 30 October 1916, reel 386, series 1, TR; Charles Scribner's Sons, *List of Spring Publications—1917,* 1.

42. *New York Herald* quoted in Charles Scribner's Sons, *List of Fall Publications—1920,* 16; review of *The Passing of the Great Race,* by Madison Grant, *Nation;* "A Book to Stir You Up."

43. Balch, review of *The Passing of the Great Race.*

44. Woods, "A Review of Reviews," 95; Kallen, "Racial Mythology."

45. Grant to Prescott F. Hall, 4 December 1918, folder "Madison Grant," IRL.

46. Boas, "Inventing a Great Race," 305.

47. Boas, "Inventing a Great Race," 306, 305.

48. Boas, "Inventing a Great Race," 307, 306, 305.

49. Boas, "Inventing a Great Race," 307; Kroeber, "The Superorganic," 181.

50. Proctor, *Racial Hygiene,* 36.

51. Paul, *Controlling Human Heredity,* 11.

52. Charles Stewart Davison, "A Plea for the Republic," in *The Alien in Our Midst,* ed. Grant and Davison, 68–69; Atherton, "The Alpine School of Fiction," 26; Perkins to Grant, 30 April 1917, folder "Madison Grant," box 67, Author Files I, CSS.

53. Grant, *The Passing of the Great Race,* 75; 2nd ed., 84, 231–32, 187 (emphasis added).

54. Grant to Roosevelt, 3 April 1918, series 1, reel 270, TR; Davenport to Grant, 30 September 1921; Grant-Davenport correspondence, February–November, 1917; Grant to Davenport, 2 April 1917; all in folder "Madison Grant," CBD. See also Davison to John Jay Chapman, 3 October 1918, folder 3, JJC.

55. Grant, *The Passing of the Great Race,* 18; 2nd ed., 20.

56. Grant, *The Passing of the Great Race,* 2nd ed., xx.

57. Grant, *The Passing of the Great Race,* 2nd ed., 12, 219.

58. Chase, *The Legacy of Malthus,* 164.

59. Maxwell Perkins to Grant, 15 March 1920, folder "Madison Grant," box 67, Author Files I, CSS.

60. Masters, "The Great Race Passes," 22.

61. Scribner to Grant, 19 January 1920, folder "Madison Grant," box 67, Author Files I, CSS.

62. Perkins to Grant, 23 November 1920, folder "Madison Grant," box 67, Author Files I, CSS.

63. Osborn to Grant, 6 June 1922, folder "Grant #3," HFO.

64. Grant, *The Passing of the Great Race,* 4th ed., 275.

65. Grant to Perkins, 13 July 1918, folder "Madison Grant," box 67, Author Files I, CSS.

Chapter 8. Grant's Disciples (pp. 167–95)

1. Gregory, review of *The Passing of the Great Race,* 135; Davison, "A Plea for the Republic," in *The Alien in Our Midst,* ed. Grant and Davison, 68–69; "Re-forging America," *Eugenical News* 12, no. 11 (November 1927): 161.

2. See Spiro, "Patrician Racist," chap. 9.

3. Cravens, *The Triumph of Evolution,* 53.

4. Gould, *America,* 125, 164. See Grant, "Failures of the Melting Pot."

5. Wiggam to S. J. Holmes, [n.d.], folder "Wiggam, Albert Edward," SJH; Wiggam, *The Next Age of Man,* 328.

6. Sadler, *Long Heads and Round Heads*, 66–67; Grant, *The Passing of the Great Race,* 15–16.

7. Evans, "The Klan's Fight for Americanism," 52, 49, 43; Evans, *The Menace of Modern Immigration,* 24, 6, 13, 28.

8. Darrow quoted in Bryan and Bryan, *Memoirs,* 546.

9. Bachman, "Theodore Lothrop Stoddard," 81; Stoddard, *The French Revolution in San Domingo,* vii.

10. Ward to Stoddard, 28 April 1927, quoted in Bachman, "Theodore Lothrop Stoddard," 215; see also 184, 280–82.

11. Stoddard, *Re-forging America,* 339.

12. Stoddard, *Rising Tide of Color,* vi; *Racial Realities in Europe,* 17.

13. "The Burbanks of a People," *Saturday Evening Post* 193, no. 44 (30 April 1921): 20; "The Great American Myth," *Saturday Evening Post* 193, no. 45 (7 May 1921): 20.

14. Roberts, *Why Europe Leaves Home,* 48, 22.

15. Roberts, *Why Europe Leaves Home,* 117, 15, 13, 11, 13, 62, 13, 117.

16. Roberts, "The Existence of an Emergency," 90.

17. Conklin to Terman, 6 February 1923, folder 3, box 16, LMT; Bagger, "Ethnology by Ear."

18. Slotkin, *Gunfighter Nation,* 198–202; Atherton, "The Alpine School of Fiction," 28 (Atherton's article was summarized in "Inferior Races in American Fiction," *Literary Digest* 73, no. 5 [29 April 1922]: 27–29); Leider, *California's Daughter,* passim.

19. Kyne, *Never the Twain Shall Meet,* 189–90.

20. Fitzgerald, *The Great Gatsby,* 15–16. For the influence of Grant and Stoddard on Fitzgerald, see Turlish, "*The Rising Tide of Color*"; and Gidley, "Notes on F. Scott Fitzgerald."

21. Wyeth to Dear Mama, 5 October 1919, quoted in Wyeth, *The Wyeths,* 624–25. See also the version quoted in Michaelis, *N. C. Wyeth,* 485.

22. Coolidge, "Whose Country Is This?"

23. "Harding Proposes Immigration Curb," *New York Times,* 15 September 1920, p. 3, col. 1; Harding and Stoddard quoted in Bachman, "Theodore Lothrop Stoddard," 153, 184.

24. "Harding Says Negro Must Have Equality in Political Life," *New York Times,* 27 October 1921, p. 1. col. 1.

25. Davis, "Our Labor Shortage and Immigration," 322–323; Davis quoted in U.S. Congress, Senate, Committee on Immigration, *Selective Immigration Legislation,* 99. See also Davis, "America and Her Immigrants," 291–93; and "A Century of Immigration." For *Rising Tide of Color* and the Harding administration, see Bachman, "Theodore Lothrop Stoddard," 105.

26. Fisher, *Report of the President,* 3.

27. Ad Interim Committee of the U.S.A. to Dear Sir, (n.d.), folder 2, unit 231, series II, AM; also in folder 4, box 2, LMT.

28. Mehler, "American Eugenics Society," 310–449, provides a brief biography of each of the 170 men and women who served on the board or the Advisory Council between 1923 and 1940.

29. Fisher, Davenport, Crampton, Grant, Little, Olson, and Osborn to Dear Sir, [n.d.], folder "1923 D–F," container 16, TSP.

30. F. Osborn, "History of the American Eugenics Society," p. 7, folder "Osborn, Frederick: History of the American Eugenics Society," AES. Dewey in Poliakov, *The History of Anti-Semitism,* 225.

31. Fund-raising letter in folder 22, box 6, CCL.

32. Whitney to J. D. Rockefeller, Jr., 28 January 1925, folder 588; Rockefeller to Woods, 30 January 1925, folder 599; Woods to Rockefeller, 9 July 1925, folder 599; all in box 60, Cultural Interests series, RG III2E, RF.

33. "Ultimate Program to Be Developed by the American Eugenics Society," in Fisher, *Report of the President,* 13–20; Huntington, *Tomorrow's Children,* 98.

34. Whitney, *Pigeon City.*

35. The names and personnel of the committees changed over the years; this list is representative of the lineup in the mid-1920s.

36. Paul, *Controlling Human Heredity,* 13; Mehler, "American Eugenics Society," 87.

37. Davenport quoted in Fisher, *Report of the President,* 5; Watts quoted in Paul, *Controlling Human Heredity,* 11; Phil Eastman to Watts, 22 February 1926, quoted in Fisher, *Report of the President,* 28.

38. Lewis, *Arrowsmith,* 259.

39. AES to Dear Fellow Minister, October 1927, folder 3, unit 40, series II, AM; Mayer and Osgood quoted in Mehler, "American Eugenics Society," 89, 90.

40. American Eugenics Society, *A Eugenics Catechism,* 2–9.

41. Stoddard, *The Revolt against Civilization,* 94; Huntington, *Tomorrow's Children,* 44–45. See also Whitney, "A Hunt for Society's Danger Spot."

42. Grant, foreword, in Grant and Davison, *The Founders of the Republic,* vi; Atherton, "The Alpine School of Fiction," 30; Johnson, "The Opponents of Restricted Immigration," in *The Alien in Our Midst,* ed. Grant and Davison, 10.

43. Goethe, *Seeking to Serve,* 122; Huntington and Whitney, *The Builders of America,* 75; Osborn, *Creative Education,* 265; "Our Disease Inheritance from Slavery," *Eugenical News* 14, no. 12 (December 1929): 176.

44. Hall quoted in Solomon, *Ancestors and Immigrants*, Wister, "Shall We Let the Cuckoos"; Roberts, "Slow Poison," 58; Creel, "Melting Pot or Dumping Ground?" 25; Huntington, *Tomorrow's Children*, 44–45. See also Hassencahl, "Harry H. Laughlin," 140; Rice, *Racial Hygiene*, 308. For a look at eugenicists who supported Prohibition because alcohol was a racial poison, see Jones, "Prohibition and Eugenics."

45. Woodruff, *Expansion of Races*, 382–85; Hall to Grant, 4 February 1913; and Grant to Hall, 3 March 1913; both in folder "Madison Grant," IRL; Roberts, *Why Europe Leaves Home*, 17.

46. Roberts, *Why Europe Leaves Home*, 65, 119, 4; Roberts, "Slow Poison," 58; American Legion quoted in Wang, *Legislating Normalcy*, 47; Burgess quoted in Saveth, *American Historians and European Immigrants*, 49; Strother, "The Immigration Peril," 634.

47. Sadler, *Race Decadence*, 404–8.

48. Holmes, *Studies in Evolution and Eugenics*, 184; Popenoe to Grant, 14 April 1928, folder "Madison Grant," CBD.

49. Giddings quoted in Sanger, *Sixth International Conference*, 1:184.

50. Sanger, *The Pivot of Civilization*, 78, 279; "Intelligent or Unintelligent Birth Control?" 12.

51. Sanger, *The Pivot of Civilization*, 265; Sanger, "The Need for Birth Control," 48, 18; Sanger, "The Reproduction Rate of Genius," 23.

52. Sanger, "Birth Control and Racial Betterment," 11; "The Meaning of Radio Birth Control," 111; *Woman and the New Race*, 32, 229.

53. Ellis, *The Task of Social Hygiene*, 401.

54. Sanger, "The Need for Birth Control," 15; *The Pivot of Civilization*, 116–17; "Address of Welcome," 100; "The Function of Sterilization," 299.

55. Ellis, "The World's Racial Problem," 15.

56. Popenoe, "Birth Control and Eugenics," 6; "Immigration and Birth Control," 219; Laughlin, "Eugenists on the Place of Birth Control," 7.

57. Sanger, *The Pivot of Civilization*, 82, 98; "Selected Reading List on Sterilization," 111.

58. Paul, *Controlling Human Heredity*, 133; Ellis in Trombley, *The Right to Reproduce*, 47; Stopes quoted in Rose, *Marie Stopes*, 135.

59. Kennedy, *Birth Control in America*, 121; Sanger quoted in Kennedy, *Birth Control in America*, 22.

60. Sanger to Grant, 24 September 1923, folder 93, reel 2, ABCL; Sanger to Laughlin, [date illegible], folder "American Birth Control League," box E-1-1, HHL; Davenport to Whitney, 5 April 1928, folder "Leon F. Whitney," CBD; Grant to Whitney, 15 April 1928, folder "Madison Grant," CBD.

61. Grant to Whitney, 15 April 1928, folder "Madison Grant," CBD. See also Whitney, "The American Eugenics Society," 257.

62. Davenport to Grant, 3 May 1920, folder "Madison Grant," CBD.

Chapter 9. Creating the Refuge (pp. 196–233)

1. Cravens, *The Triumph of Evolution*, 53.

2. The most thorough histories of the IRL are in Thomas J. Curran, *Xenophobia and Immigration;* Higham, *Strangers in the Land;* and Solomon, *Ancestors and Immigrants.*

3. Grant, "The Racial Transformation of America," 345.

4. Ross quoted in Gossett, *Race,* 293; Grant to Root, 10 May 1912, folder "G 1912," carton 94, ER.

5. Grant to Hall, 13 February 1907, folder "National Liberal Immigration League," IRL; Patten to J. Lee, 29 September 1912 and 2 May 1911, box 1, JoL.

6. Grant to Lacey, 28 July 1906, box 255, JFL; Hall to Grant, 23 February 1910, folder "Madison Grant," IRL.

7. Commission quoted in Ludmerer, *Genetics and American Society,* 97.

8. Grant to Root, 10 May 1912, folder "G 1912," carton 94, ER.

9. Grant to Taft, 24 November 1908, series 3, WHT.

10. Grant to Roosevelt, 15 November 1915, reel 202, series 1, TR; Wickersham to Grant, 21 December 1912, folder "Madison Grant," IRL.

11. Wilson, *A History of the American People,* 5:212–13.

12. Hall to Shiras, [December 1912], folder "George Shiras," IRL; Immigration Restriction League to The Honorable—, [n.d.], folder "Madison Grant," IRL.

13. Grant to Hall, 13 December 1913, folder "Madison Grant," IRL; Grant to Wilson, 6 November 1913, folder "Woodrow Wilson," IRL.

14. Grant to Roosevelt, 15 November 1915, reel 202, series 1, TR.

15. Grant repeated TR's remark in Grant to Hall, 15 November 1915, folder "Madison Grant," IRL. For the league's reaction to Hughes, see J. H. Patten to J. Lee, 10 August 1916, and Lee to Crane, 30 October 1916; both in box 2, JoL.

16. Grant to Hall, [n.d.], and Hall to Grant, 12 February 1915, folder "Madison Grant," IRL; *Congressional Record,* 64th Cong., 1st sess., 1916, 53, pt. 13: 12763.

17. Grant to Hall, 14 May 1919, folder "Madison Grant," IRL.

18. Grant to Johnson, 23 November 1920, folder "Madison Grant," IRL.

19. U.S. Congress, House Committee on Immigration and Naturalization (hereafter cited as HCIN), *Restriction of Immigration: Hearing,* 475.

20. Trevor to Johnson, 18 February 1927, 69th Congress, HR69A-F20.1, HCIN.

21. HCIN, *Biological Aspects of Immigration,* 3–22.

22. Laughlin quoted in Ludmerer, *Genetics and American Society,* 108. Laughlin's activities as expert eugenics agent are discussed in detail in King, *Making Americans.*

23. Grant to Hall, 20 October 1920 and 15 November 1920, folder "Madison Grant," IRL; F. A. Wallis, quoted in *Congressional Record,* 66th Cong., 3d sess., 1920–21, 60, pt. 4: 4549; Surgeon General in Wang, *Legislating Normalcy,* 16.

24. Grant to Hall, 26 November 1920 and 15 November 1920, folder "Madison Grant," IRL.

25. Grant to Hall, 27 August 1920, folder "Madison Grant," IRL; *Congressional Record,* 66th Cong., 3d sess., 1920–21, 60, pt. 1: 172, 182; pt. 5: 4561–62.

26. HCIN, *Temporary Suspension of Immigration,* 4; *Congressional Record,* 66th Cong., 3d sess., 1920–21, 60, pt. 1: 189–90.

27. *Congressional Record,* 66th Cong., 3d sess., 1920–21, 60, pt. 1: 138; 67th Cong., special session, 1921, 61, pt. 1: 567.

28. *Congressional Record,* 66th Cong., 3d sess., 1920–21, 60, pt. 5: 4563; 66th Cong., 3d sess., 1921, 60, pt. 4: 3965; 66th Cong., 3d sess., 1920–21, 60, pt. 5: 4739–41. See also *Congressional Record,* 67th Cong., special session, 1921, 61, pt. 1: 501 and 513–14.

29. Hendrick quoted in Baltzell, *The Protestant Establishment,* 204; *Congressional Record,* 66th Cong., 3d sess., 1921, 60, pt. 4: 3970.

30. Grant, "Restriction of Immigration," 47–53; *Congressional Record,* 67th Cong., 1st sess., 1921, appendix to pts. 1 to 9: 8937–38; *Congressional Record,* 66th Cong., 3d sess., 1920–21, 60, pt. 1: 137. The editorial was "Cashing In," *Saturday Evening Post* 193, no. 17 (23 October 1920): 26. For another *Post* editorial linking immigration and conservation see *Congressional Record,* 66th Cong., 3d sess., 1920–21, 60, pt. 5: 4715.

31. *Congressional Record,* 66th Cong., 3d sess., 1920–21, 60, pt. 5: 4692; pt. 1: 143; pt. 5: 4563. See also the speech by Taylor (pt. 5: 4564) comparing immigration to pollution.

32. "No Admittance," *Saturday Evening Post* 193, no. 28 (8 January 1921): 24; Roberts, "Plain Remarks," 21; Roberts, "The Existence of an Emergency," 89, 93. See also Roberts's "Beer Worshippers," "Ports of Embarkation," and "Scotland for Scotch." Conklin, "Some Biological Aspects of Immigration," 358, 353; "The Unpopular Branch," *New York Times,* 9 February 1921, p. 8, col. 3 (see also "The Main Immigration Question, *New York Times,* 5 January 1921, p. 12, col. 4).

33. U.S. Congress, Senate, Committee on Immigration, *Emergency Immigration Legislation,* 10–23; 679; Grant to Davenport, 20 December 1920, folder "Madison Grant," CBD.

34. Hall to C. S. Davison, 30 April 1920, box "1919–1921," *ADS;* U.S. Congress, Senate, Committee on Immigration, *Emergency Immigration Legislation,* 652–53.

35. Grant to Hall, 20 October 1920, folder "Madison Grant," IRL; *Congressional Record,* 67th Cong., special session, 1921, 61, pt. 1: 497–99. For Harding and the literacy test, see Wang, *Legislating Normalcy,* 25.

36. Fairchild, *Immigration,* 456.

37. *Congressional Record,* 67th Cong., special session, 1921, 61, pt. 1: 515 (see also pt. 9: 8960).

38. Wiggam, *The New Decalogue of Science,* 110.

39. Grant to Mrs. Wortham James, 12 April 1921, folder "Madison Grant," CBD.

40. "Preliminary Announcement of the Second International Congress of Eugenics," box 15, R445, ROS; Davenport quoted in Haller, *Eugenics,* 156.

41. Osborn, "Address of Welcome," in *Eugenics, Genetics and the Family,* 3–4.

42. Laughlin to J. Walter Fewkes, 9 August 1921, box 15, R445, ROS.

43. Osborn, "Address of Welcome," in *Eugenics, Genetics and the Family,* 1; Lapouge, "La Race Chez les Populations Mélangées," in *Eugenics in Race and State,* 1–6 (translation mine).

44. "Tracing Parentage by Eugenic Tests," *New York Times,* 23 September 1921, p. 8, col. 2; "Eugenists Dread Tainted Aliens," *New York Times,* 25 September 1921, sec. 2, p. 1, col. 8.

45. Grant to Davenport, 17 October 1921 and 29 September 1921, folder "Madison Grant," CBD.

46. "Lillian Russell Would Bar Aliens," *New York Times,* 29 March 1922, p. 4, col. 2; *Annual Report of the American Museum of Natural History* 53 (1921): 31.

47. Gompers in Higham, *Strangers in the Land,* 410; Roberts in HCIN, *Immigration,* 98; "It's Now or Never," *Saturday Evening Post* 194, no. 33 (11 February 1922): 22; Creel, "Melting Pot or Dumping Ground?" 9; and Creel, "Close the Gates!" 9.

48. Osborn to Grant, 30 December 1922, folder "Grant #3," HFO; Lewis M. Terman, "Were We Born that Way?" 658–60.

49. HCIN, *Analysis of America's Modern Melting Pot,* 757.

50. HCIN, *Analysis of America's Modern Melting Pot,* 731; HCIN, *Restriction of Immigration: Hearing,* 1233.

51. Johnson to Whitney, 4 March 1926, quoted in Fisher, *Report of the President,* 6; HCIN, *Analysis of America's Modern Melting Pot,* 738; Roberts, "Lest We Forget," 162.

52. Stocking, *Race, Culture, and Evolution,* 300–301.

53. Gould to Charles B. Davenport, 19 October 1921, folder "C. W. Gould," CBD.

54. Brigham, *A Study of American Intelligence,* 182. For Grant supplying the figures to Brigham, see Brigham to Yerkes, 24 June 1922, folder 115, box 7, RMY.

55. Brigham, *A Study of American Intelligence,* 184, 192.

56. Kamin, *Science and Politics of I.Q.,* 20; Goddard quoted in Smith, *Minds Made Feeble,* 132.

57. Robert M. Yerkes, foreword to Brigham, *A Study of American Intelligence,* vii (emphasis added); Yerkes to Albert Johnson and LeBaron B. Colt, 11 February 1921, cited in Samelson, "World War I Intelligence Testing," 278; Yerkes to Princeton, quoted in Samelson, "Putting Psychology on the Map," 130; Yerkes, "Testing the Human Mind," 359, 364.

58. Pinter, "Intelligence of the American Nation," 184, 185; Terman quoted in Samelson, "Putting Psychology on the Map," 105; Osborn quoted in Chase, *The Legacy of Malthus,* 274. For an attack on the SAT because of its ties to Madison Grant, see Fish, "Affirmative Action and the SAT."

59. U.S. Congress, Senate, Committee on Immigration, *Amendment to Immigration Law: Hearing,* 20 February 1923, 82.

60. Grant, "The Racial Transformation of America," 343.

61. Harding quoted in Wang, *Legislating Normalcy,* 90; "Keep America 'White'!" 399.

62. Grant to Johnson, 22 December 1923, 68th Congress, HR68A-F18.1, HCIN; Lorimer quoted in Tebbel, *George Horace Lorimer,* 94.

63. Coolidge quoted in Garis, *Immigration Restriction,* 170.

64. HCIN, *Restriction of Immigration: Hearing,* 249.

65. Grant, "Restriction of Immigration" 44; HCIN, *Restriction of Immigration: Hearing,* 333, 350, 333–34.

66. HCIN, *Restriction of Immigration: Hearing,* 388–89, 392–94.

67. HCIN, *Restriction of Immigration: Hearing,* 298, 289.

68. HCIN, *Restriction of Immigration: Hearing,* 298. See also "Louis Marshall Warns," *New York Times,* 14 April 1924, p. 4, col. 2.

69. HCIN, *Restriction of Immigration: Hearings,* 387, 316, 334, 373 (emphasis added). For African American support of the immigration restriction law, see Hellwig, "Afro-American Reactions to the Japanese."

70. U.S. Congress, Senate, Committee on Immigration, *Selective Immigration Legislation,* 243–44; HCIN, *Europe as an Emigrant-Exporting Continent.*

71. *Congressional Record,* 68th Cong., 1st sess., 1924, 65, pt. 6: 5656.

72. Boas, "This Nordic Nonsense," 502. Almost all scholars (Divine, Haller, Higham, Ludmerer, Taylor, Wang, etc.) are agreed that the racial factor was overwhelmingly the most important consideration leading to passage of the 1924 act. Kevin MacDonald, on the other hand, claims that "arguments related to Nordic superiority . . . played remark-

ably little role in Congressional debates over immigration" (*The Culture of Critique*, 252ff.).

73. HCIN, *Restriction of Immigration: Hearings*, 570–71.

74. HCIN, *Restriction of Immigration: Report No. 176*, 3.

75. *Congressional Record*, 68th Cong., 1st sess., 1924, 65, pt. 4: 4170; pt. 2: 1329–30; pt. 6: 5914, 5915.

76. *Congressional Record*, 68th Cong., 1st sess., 1924, 65, pt. 6: 5811, 5810.

77. *Congressional Record*, 68th Cong., 1st sess., 1924, 65, pt. 6: 5578–80. See also HCIN, *Restriction of Immigration: Report No. 350, part 2* 4–5.

78. Johnson to Charles W. Gould, 3 March 1924, 68th Congress, HR68A-F18.1, HCIN; "Congress to Tighten Immigration Curb," *New York Times*, 27 January 1924, sec. 2, p. 1, col. 1.

79. Wister, "Shall We Let the Cuckoos," 47; Stoddard, "Racial Realities in Europe," 157; Roberts, "Slow Poison," 8; Roberts, "And West Is West," 13; Roberts, "East Is East," 145; "And West is West," 94. See also Hays, "Immigration and Eugenics," 406.

80. Grant, "The Racial Transformation of America," 343–49; *Congressional Record*, 68th Cong., 1st sess., 1924, 65, pt. 6: 5849.

81. "Minute Men," *Saturday Evening Post* 196, no. 32 (2 February 1924): 24. See also "American Laws for America," *Saturday Evening Post* 196, no. 39 (29 March 1924): 24; and "They Want Unrestricted Immigration," *Saturday Evening Post* 196, no. 40 (5 April 1924): 36.

82. James J. Davis, "One Hundred Years of Immigration," *New York Times*, 17 January 1924, sec. 8, p. 1, col. 1. See also Davis, "A Century of Immigration," 1–19. For scholars who feel that the army intelligence tests played only a limited role in the passage of immigration restriction, see Samelson, "Science and Politics of the IQ"; Snyderman and Herrnstein, "Intelligence Tests"; Degler, *In Search of Human Nature*, 52; MacDonald, *The Culture of Critique*, 32–35, 267.

83. "Eugenics and Immigration," *New York Times*, 16 February 1924, p. 12, col. 6; "Quotas for Immigration," *New York Times*, 1 March 1924, p. 12, col. 7; "New Immigration Quotas," *New York Times*, 2 March 1924, sec. 8, p. 17, col. 1; "Estimates of Population," *New York Times*, 16 March 1924, sec. 8, p. 13, col. 1; "Discrimination in Quotas," *New York Times*, 23 March 1924, sec. 8, p. 12, col. 3.

84. "Lo, the Poor Nordic!" *New York Times*, 8 April 1924, p. 18, col. 7. See the response of Franz Boas in *New York Times*, 13 April 1924, sec. 9, p. 19, col. 1.

85. "Report of the Committee on Selective Immigration of the Eugenics Committee of the United States of America," 2, in folder 2, unit 231, series II, AM.

86. *Congressional Record*, 68th Cong., 1st sess., 1924, 65, pt. 4: 3424; pt. 6: 5911, 5886; 5887; Celler quoted in Hyatt, *Franz Boas*, 136.

87. "American Laws for America," *Saturday Evening Post* 196, no. 39 (29 March 1924): 24; "purify" quoted in Wang, *Legislating Normalcy*, 58; *Congressional Record*, 68th Cong., 1st sess., 1924, 65, pt. 6: 5854, 5872 (emphasis added). See also Ward, "What Next in Immigration Legislation?"

88. "Quotas for Immigration," *New York Times*, 1 March 1924, p. 12, col. 7 (Kinnicutt's letter was reproduced in U.S. Congress, Senate, Committee on Immigration, Selective Immigration Legislation, 246); "Estimates of Population," *New York Times*, 16 March

1924, sec. 8, p. 13, col. 1; "Discrimination in Quotas," *New York Times,* 23 March 1924, sec. 8, p. 12, col. 3; *Congressional Record,* 68th Cong., 1st sess., 1924, 65, pt. 6: 5643, 5921, 5468, 5673. See also the remarks of McReynolds, *Congressional Record,* 68th Cong., 1st sess., 1924, 65, pt. 6: 5854; and see "Discriminating against the American Stock," 14.

89. "Examining the 1890 Census," *New York Times,* 1 March 1924, p. 12, col. 2. (The editorial was reproduced in U.S. Congress, Senate, Committee on Immigration, *Selective Immigration Legislation,* 246.) See also "Attacking the 1890 Quotas," *New York Times,* 26 February 1924, p. 16, col. 4; and "The 1890 Quota up Again," *New York Times,* 19 March 1924, p. 20, col. 3.

90. *Congressional Record,* 68th Cong., 1st sess., 1924, 65, pt. 6: 6263, 6161, 5705, 5693, 6461; Bruce quoted in Wang, *Legislating Normalcy,* 44; *Congressional Record,* 68th Cong., 1st sess., 1924, 65, pt. 6: 5667, 5648 (emphasis added). See also "A Word to the Wise," *Saturday Evening Post* 195, no. 26 (23 December 1922): 20.

91. See the discussion of the national origins concept in Higham, *Strangers in the Land,* chap. 11.

92. Grant to Johnson, 22 December 1923, 68th Congress, HR68A-F18.1, HCIN.

93. *Los Angeles Times* quoted in Higham, *Strangers in the Land,* 300; *Boston Globe* in "Our New 'Nordic' Immigration Policy," 12; *Chicago Tribune* in "End of the 'Melting-Pot' Theory," 14.

94. Statistics in Garis, *Immigration Restriction,* 256–57, 262.

95. Divine, *American Immigration Policy,* 23; Wang, *Legislating Normalcy,* 110, 112.

96. Grant, "America for the Americans," 354; Minutes of the 46th Meeting of the Galton Society, folder "Galton Society," CBD; Grant, "Further Notes on *The Racial Elements of European History,*" *Eugenical News* 13, no. 9 (September 1928): 119; Ward quoted in "Our New Immigration Policy," *Eugenical News* 9, no. 11 (November 1924): 96; Stoddard, *Re-forging America,* 349, 342, 339, 201, 192–93; "President Signs Immigration Bill," *New York Times,* 27 May 1924, p. 2, col. 4; Evans, "The Klan's Fight for Americanism," 43.

97. Osborn to Johnson, 2 May 1924, folder "Johnson, Albert." HFO; "Presidential Address," *Eugenical News* 12, no. 8 (August 1927): 95; "A Great Work of Constructive Legislation," 121.

98. Davis to Coolidge, 29 May 1924, quoted in Wang, *Legislating Normalcy,* 113; Shortridge in "President Signs Immigration Bill," *New York Times,* 27 May 1924, p. 2, col. 4; Ward to Coolidge, 26 May 1924, quoted in Wang, *Legislating Normalcy,* 113; "Notes on Madison Grant," HHL.

99. Grant to Ward, 27 May 1924, RDW.

Chapter 10. Culling the Herd (pp. 234–65)

1. Curran, "The New Immigrant," 25.

2. Gould, *The Mismeasure of Man,* 232; Roberts, "Florida Loafing."

3. Grant to Ward, 27 May 1924, RDW; Hooton quoted in Kevles, *In the Name of Eugenics,* 114.

4. Sanger, "The Function of Sterilization," 299; Sanger quoted in Gordon, *Woman's Body,* 282; Sanger quoted in Ellen Chesler, *Woman of Valor,* 417.

5. Grant, *The Conquest of a Continent,* 351; Huntington to Frederick A. Stokes, 10 March 1934, folder 2851, box 73, series III, EH.

6. See also Rice, *Racial Hygiene,* 323, 313.

7. Paul Popenoe, "Eugenic Sterilization in California"; Laughlin, *The Legal Status of Eugenical Sterilization,* 7; Laughlin, *Report of the Committee,* 59, 58, 64.

8. Laughlin, *Report of the Committee,* 16; Davenport quoted in Pickens, *Eugenics and the Progressives,* 92; Conklin, *The Direction of Human Evolution,* 125–26 (emphasis added); Robinson quoted in Kevles, *In the Name of Eugenics,* 93–94 (emphasis added); Laughlin quoted in Pickens, *Eugenics and the Progressives,* 93.

9. Laughlin, *Eugenical Sterilization,* 446–47.

10. DeJarnette quoted in Dudziak, "Oliver Wendell Holmes," 852; social worker quoted in Smith and Nelson, *The Sterilization of Carrie Buck,* 108, 109, 111; Laughlin and the court in Laughlin, *The Legal Status of Eugenical Sterilization,* 17, 20.

11. *Buck v. Bell,* 274 U.S. 202 (1927), 206–8. See Dudziak, "Oliver Wendell Holmes," 833–67.

12. *National Council of Catholic Women* in Proctor, "Eugenics among Social Sciences," 190; Chesterton, *Eugenics and Other Evils,* vi; Pope Pius XI quoted in Whitney, *The Case for Sterilization,* 191.

13. Whitney in *The Case for Sterilization,* 194–95 (emphasis added). For a Catholic (and Belgian priest) who was pro-eugenics, see Fallon, *Eugenics.*

14. Laski quoted in Chase, *The Legacy of Malthus,* 316.

15. McCormick, "Eugenic Sterilization," 242; Laughlin, "Legal Status," 80.

16. Reilly, *The Surgical Solution,* xiii. See also Gould, "Carrie Buck's Daughter," 18.

17. Holmes quoted in Degler, *In Search of Human Nature,* 47; Laughlin, *The Legal Status of Eugenical Sterilization,* 61, 60; Olson, introduction to Laughlin, *The Legal Status of Eugenical Sterilization,* 5.

18. Shipman, *The Evolution of Racism,* 130; Laughlin, *Report of the Committee,* 6, 47; Whitney, *The Case for Sterilization,* 13; Haller, *Eugenics,* 46; Kevles, *In the Name of Eugenics,* 366.

19. Laughlin to E. S. Gosney, 8 September 1934, folder "Immigration Report-letters in response," box C-4-7, HHL; Whitney, *The Case for Sterilization,* 139. For sterilization in Scandinavia, see Broberg and Roll-Hansen, *Eugenics and the Welfare State.*

20. Grant, *The Conquest of a Continent,* 281.

21. Grant, *The Conquest of a Continent,* 13; *The Passing of the Great Race,* 87.

22. Grant, *The Conquest of a Continent,* 37; *The Passing of the Great Race,* 69, 14, 69; *The Conquest of a Continent,* 12, 345.

23. Grant, *The Passing of the Great Race,* 2nd ed., 13; Grant, *The Conquest of a Continent,* 21; Grant to Davenport, 6 July 1914, folder "Madison Grant," CBD; *The Passing of the Great Race,* 19, 193; Osborn, "The Evolution of the Human Races," 3–5.

24. Grant to Taft, 22 November 1910, series 6, case file 77, WHT; *The Passing of the Great Race,* 16.

25. Grant, *The Conquest of a Continent,* 9.

26. Grant, *The Passing of the Great Race,* 19 (see also p. 56 as well as Grant, "The Racial Transformation of America," 352); *The Conquest of a Continent,* 15; Grant to Cox, 2 December 1920, 14 December 1920, 2 December 1920; Lapouge to Grant, 25 October 1920; all in box 1, ESC.

27. Boas, *Anthropology and Modern Life,* 80. For the Jewish leadership of civil rights

groups, see David Levering-Lewis, "Shortcuts to the Mainstream: Afro-American and Jewish Notables in the 1920s and 1930s," in *Jews in Black Perspective,* ed. Washington, 85. See also Levering-Lewis, "Parallels and Divergences."

28. Popenoe and Johnson, *Applied Eugenics,* 294–97; East, *Mankind at the Crossroads,* 133; "Studies on Races in America," *Eugenical News* 1, no. 2 (February 1916): 5. For the eugenicist position on Negro inferiority see, for example, Charles Benedict Davenport, "As the Twig's Bent," in *Alien in Our Midst,* ed. Grant and Davison, 51–52; Davenport, "Galton Society," *Eugenical News* 13, no. 2 (February 1928): 27; East, *Heredity and Human Affairs,* 188–89; East and Jones, *Inbreeding and Outbreeding,* 253; East, "Population," 603–24; Guyer, *Being Well-Born,* 297; Holmes, "Race Prejudice," *Eugenical News* 19, no. 1 (January–February, 1934): 28; Holmes, *The Negro's Struggle for Survival;* Humphrey, *Mankind,* 223; Laughlin, *Immigration and Conquest,* 22; McDougall, *Is America Safe for Democracy?* vii, 56–57, 67; Sorokin, *Contemporary Sociological Theories,* 298–304; Stoddard, *Re-forging America;* Stoddard, "Impasse at the Color-Line." For a review of the position of geneticists on miscegenation, see Provine, "Geneticists and Race."

29. Davenport and Steggerda, *Race Crossing in Jamaica,* 472, 471, 364.

30. Grant, *The Conquest of a Continent,* 353.

31. Newby, *Jim Crow's Defense,* 61. Cox's reminiscences of his journeys are in Cox, *Black Belt around the World.*

32. Grant, *The Passing of the Great Race,* 69, 79.

33. Cox, *White America,* 23.

34. Cox, *White America,* 246.

35. Ethel Wolfskill Hedlin, "Earnest Cox," 36.

36. Cox to G. P. Putnams Sons, 20 November 1919, box 1; Cox to Grant, 25 February 1921, box 2; both in ESC.

37. Cox to Henry L. West, [n.d.]; Grant to Cox, 2 December 1920 and 14 December 1920, all in box 1; Kinnicutt to Cox, 20 January 1925, box 2; all in ESC.

38. Cox, *White America,* 9–10.

39. Perkins to Grant, 13 November 1924, folder "Madison Grant," box 67, Author Files I, CSS.

40. Cox to Grant, 25 February 1921, box 2, ESC.

41. Grant's review in box 2, ESC; Eckenrode in the *Richmond Times-Dispatch,* 22 November 1923, quoted in flyer for *White America,* folder "Madison Grant," CBD; Stoddard, *Re-forging America,* 324; "White America," *Eugenical News* 9, no. 1 (January 1924): 3. See also Stoddard to Cox, 21 January 1924; Ross to Cox, 22 January and 24 April 1924; McDougall to Cox, 3 May 1924; all in box 2, ESC.

42. Calvin B. Hoover, review of *White America, South Atlantic Quarterly* 27 (January 1927): 104–5; *Salt Lake Tribune,* 28 May 1926; and *Norfolk Journal and Guide,* 22 August 1925; all quoted in Hedlin, "Earnest Cox," 60–61; Herskovits, "Extremes and Means in Racial Interpretation," 551; Newby, *Jim Crow's Defense,* 61. For another southern racist tract directly influenced by Madison Grant, see Sayers, *Can the White Race Survive?*

43. Grant to Merriam, 14 January 1922, container 82, folder "Madison Grant," JCM; Grant to Johnson, 29 December 1923, 68th Congress, HR68A-F18.1, HCIN; Burnham, "Charon Beckoned," folder "Boone & Crockett Club 35–38," container 106, KR.

44. Osborn to Grant, 11 April 1924 and 6 November 1924, folder "Grant #3," HFO; Grant to Roosevelt, 25 September 1925, folder "Correspondence A–Mar," box 34, FDR; Cross, "A Tribute to Madison Grant," *ZSB* 40, no. 4 (July–August, 1937): 104.

45. Caro, *The Power Broker,* 288.

46. Roosevelt to Grant, 5 May 1928; Grant to Roosevelt, 28 April 1928; all in folder "Correspondence A–Mar," box 34, FDR.

47. Kellogg to Gifford Pinchot, 20 December 1936, reel 4, JHK; Kellogg to Burbank, 25 April 1922, folder "Kellogg, John Harvey 1917–1926," container 6, LB; Mrs. Kellogg in Schwarz, *John Harvey Kellogg;* celibacy in Carson, *Cornflake Crusade* (New York: Rinehart & Company, Inc., 1957), 111.

48. Kellogg to Pinchot, 20 December 1936, reel 4, JHK; Burbank to Kellogg, 21 September 1918, folder "Kellogg, John Harvey 1917–1926," container 6, LB; Rockefeller to Kellogg, 29 November 1922, reel 2, JHK.

49. Kellogg, "Needed—A New Human Race," in *Proceedings of the First National Conference on Race Betterment,* 431; Kellogg, "The Eugenics Registry," in *Official Proceedings of the Second National Conference on Race Betterment,* 87; Whitney to C. C. Little, 4 December 1925, folder 25, box 3, CCL.

50. Davenport to Kellogg, 17 August 1923, folder "John Harvey Kellogg," CBD.

51. Kellogg to Burbank, 9 April 1926, folder "Kellogg, John Harvey 1917–1926," container 6, LB.

52. Kellogg to Hornaday, 27 August 1923, folder "Corres 1923 July/Aug," container 72, WTH.

53. Grant to Kermit Roosevelt, 24 August 1929, folder "Madison Grant," container 49, KR; Grant to James Barnes, 8 August 1930, folder 4, box 1, JB.

54. Cox to Grant, 29 November 1921, box 2, ESC.

55. Minutes of the Galton Society, 5 May 1925, folder "Galton Society," CBD.

56. Plecker, *Eugenics in Relation to the New Family,* 14, 28; Plecker to the Editor of the *Buffalo Express,* 3 April 1924, folder "Madison Grant," CBD; Plecker, "Shall We All Be Mulattoes?"

57. Pascoe, "Miscegenation Law," 59.

58. Smith, *Managing White Supremacy,* 81; Powell, blurb on inside back cover of Cox, *Let My People Go;* quoted in *Richmond Times-Dispatch,* 22 July 1923, quoted in Hedlin, "Earnest Cox," 62; quoted in Sherman, "'The Last Stand,'" 87.

59. Letter to the editor, *Survey Graphic* (13 March 1925), quoted in Smith, *The Eugenic Assault on America,* 61; Plecker, *Eugenics in Relation to the New Family.*

60. Grant to Davenport, 8 April 1924, folder "Madison Grant," CBD; *Congressional Record,* 68th Cong., 1st sess., 1924, 65, pt. 2: 1329. Details of the bill's path through the legislature are in Sherman, "'The Last Stand,'" 69–92, and Smith, *Managing White Supremacy,* chap. 3. Text of the law in Plecker, *Eugenics in Relation to the New Family,* 29–32; also "Against Miscegenation," *Eugenical News* 9, no. 5 (May 1924): 48.

61. Davenport to Grant, 14 April 1924, folder "Madison Grant," CBD; Stoddard to Cox, 25 March 1924, box 2, ESC; Cox quoted in Plecker, "Shall We All Be Mulattoes?" 23 (see also Cox to Mrs. Marcus Garvey, 7 July 1925, box 2, ESC); Plecker, "Shall We All Be Mulattoes?" 23.

62. Plecker, "Virginia's Effort to Preserve Racial Integrity," in *A Decade of Progress in*

Eugenics, 111; Plecker to Grant, 18 June 1931, box 3, ESC; Plecker to E. B. Ford, 22 March 1939, folder "W. A. Plecker," box E-2-1, HHL; Stoddard to Cox, 21 October 1924, box 2, ESC.

63. Powell quoted in Sherman, "'The Last Stand,'" 81; Grant, *The Conquest of a Continent,* 99; Cook quoted in Sherman, "'The Last Stand,'" 88.

64. Plecker to R. N. Anderson, 31 July 1924, quoted in Smith, *The Eugenic Assault on America,* 70.

65. Plecker to Mascott Hamilton, 10 October 1930, folder "Racial Integrity," box D-4-3, HHL.

66. Plecker, "Shall We All Be Mulattoes?" 23; Grant to Cox, 14 December 1920, box 1, ESC; Plecker to Laughlin, 24 May 1929, folder "Racial Integrity," box D-4-3, HHL. Trinkle's actions in Sherman, "'The Last Stand,'" 79. Grant's actions in Grant to Cox, 10 June 1925, box 2, ESC; "Meeting of the Eugenics Committee," *Eugenical News* 10, no. 12 (December 1925): 160. For eugenics in the South, see Larson, *Sex, Race, and Science.* For a fictional satire about Plecker and Powell, see Schuyler's *Black No More.*

67. Grant to Stoddard, 28 September 1927, quoted in Bachman, "Theodore Lothrop Stoddard," 231.

68. Cox, *White America,* 339 (emphasis added); Grant to Cox, 18 March 1924, box 2, ESC. (See also Grant to Cox, 12 January 1921 and 14 August 1924, box 2, ESC.)

69. Marcus Garvey, "Miscegenation" (1:17), "Race Assimilation" (1:26), and "Race Purity a Desideratum" (2:62), in *Philosophy and Opinions,* ed. Amy Jacques-Garvey.

70. Garvey, "Who and What Is a Negro," in *Philosophy and Opinions,* ed. Amy Jacques-Garvey, 2:18, 21.

71. Cox to Garvey, 16 August 1925, quoted in Edwards, "Racial Purity," 141.

72. Garvey, *Philosophy and Opinions,* 2:414; St. Louis quoted in *Philosophy and Opinions,* 2:340, 342.

73. Cox, *Let My People Go,* 4, 14–15, 34.

74. M. Garvey to Cox, 2 June 1925; A. J. Garvey to Cox, 26 May 1925; both in box 2, ESC.

75. Powell, "An Answer to the Appeal to White America," in Garvey, *Philosophy and Opinions* 2:341; Tolbert, "Outpost Garveyism," 240–41.

76. Rosenberg, *Der Mythus des 20. Jahrhunderts,* 673; Laughlin to Grant, 19 November 1932, folder "Madison Grant," box C-2-1, HHL; Davenport to Grant, 7 April 1925, folder "Madison Grant," CBD.

77. Cox, *Let My People Go,* 4, 7, 29; A. Garvey quoted in Martin, *Race First,* 348.

78. Du Bois quoted in Cronon, *Black Moses,* 190; Garvey, "The Ideals of Two Races," in *Philosophy and Opinions,* ed. Amy Jacques-Garvey, 2:338.

79. Powell, "An Answer to the Appeal to White America," in Garvey, *Philosophy and Opinions,* 2:339–48.

80. Cox to Davis, 8 November 1926, box 3, ESC; Cox to Coolidge, 23 March 1927, in Edwards, "Racial Purity," 135; Plecker to Coolidge, 19 March 1927, in Smith, *The Eugenic Assault on America,* 34.

81. For a marvelous account of Draper's life, see Blackmon, "Silent Partner."

82. See Spiro, "Patrician Racist," chap. 16.

83. Haller, *Eugenics,* 151 (emphasis added).

84. Ford, review of *White America,* 118. Frank Mintz traces the influence of Madison

Grant and E. S. Cox on the Liberty Lobby (the right-wing group of the 1960s and 1970s) in *The Liberty Lobby and the American Right.*

85. Tucker, *The Funding of Scientific Racism,* 26. Tucker's valuable work provides a detailed history of the Pioneer Fund.

Chapter 11. Saving the Redwoods (pp. 266–93)

1. Letter to Donald W. Hoyle, folder "Founders' Grove," SRL.

2. Grant to John B. Cauldwell, 17 November 1921; Grant to Committee on Admissions, 17 November 1921; both in folder "Madison Grant," container 82, JCM.

3. Merriam, "Forest Windows," 733; Newton B. Drury, quoted in "Save-the-Redwoods League Dinner, Commemorating the Establishment of the Madison Grant Forest and Elk Refuge," folder "Madison Grant Forest and Elk Refuge," SRL.

4. *Annual Report of the Save-the-Redwoods League, 1930–31,* 6.

5. Merriam, "Forest Windows," 734; Grant, *Preserve an America Worth Fighting For,* 5; Grant, "Saving the Redwoods: An Account," 97.

6. Henry Fairfield Osborn, "Sequoia," 601; Grant, "Saving the Redwoods," in *Hunting and Conservation,* ed. Grinnell and Sheldon, 192; Merriam quoted in St. Barbe Baker, *The Redwoods,* 87; Grant, "Saving the Redwoods," in *Hunting and Conservation,* ed. Grinnell and Sheldon, 193; Grant to Merriam, 30 May 1928, folder "Madison Grant," container 82, JCM.

7. Grant, "Saving the Redwoods: An Account," 97.

8. Leydet, *The Last Redwoods,* 65.

9. Grant, "Saving the Redwoods: An Account," 94, 95.

10. Tilden, *The State Parks,* 399; Grant, "Saving the Redwoods," in *Hunting and Conservation,* ed. Grinnell and Sheldon, 190; Grant, *Preserve an America Worth Fighting For,* 2; Grant, "Saving the Redwoods," *National Geographic,* 534; Grant, "Saving the Redwoods: An Account," 94–95.

11. Grant, "Saving the Redwoods," in *Hunting and Conservation,* ed. Grinnell and Sheldon, 191.

12. Joseph D. Grant, *Redwoods and Reminiscences,* 142.

13. Walt Whitman, "Song of the Redwood Tree," in *Complete Poetry,* 191.

14. Muir, "Save the Redwoods," 4.

15. Joseph D. Grant, *Saving California's Redwoods,* 5; Grant to Everett Griggs, 16 September 1919, folder "History of the League—1911–1966," SRL.

16. Grant, "Saving the Redwoods: An Account," 109, 99, 107; Udall quoted in Leydet, *The Last Redwoods,* 11; Osborn, "Preservation of Our Wild Animals," in *American Big Game in Its Haunts,* ed. Grinnell, 351.

17. Grant, "Saving the Redwoods," in *Hunting and Conservation,* ed. Grinnell and Sheldon, 182; Grant to Roosevelt, 22 January 1904, book #7, WCS-12; Roosevelt quoted in Cutright, *Theodore Roosevelt, the Naturalist,* 185; Grant, "Saving the Redwoods," in *Hunting and Conservation,* ed. Grinnell and Sheldon, 188.

18. Grant, *The Passing of the Great Race,* 155, 185; Joseph D. Grant, *Saving California's Redwoods,* 11. See also Count Keyserling, *Travel Diary,* 2:287–90.

19. Muir quoted in Dewitt, *California Redwood Parks,* 10; Grant, "Saving the Red-

woods," *National Geographic,* 534; lumber baron quoted in Joseph D. Grant, *Redwoods and Reminiscences,* 148; Grant, "Saving the Redwoods: An Account," 112.

20. Grant, "Saving the Redwoods: An Account," 112; Grant, "Saving the Redwoods," in *Hunting and Conservation,* ed. Grinnell and Sheldon, 196–97.

21. Muir quoted in Fox, *John Muir and His Legacy,* 134; Tilden, *The State Parks,* 6.

22. Grant to T. Roosevelt, Jr., 2 November 1919, folder "Madison Grant," IRL; Grant to J. C. Sperry, 16 June 1920, folder "Grant, Madison," SRL.

23. Grant to J. C. Sperry, 29 June 1920, folder "Grant, Madison," SRL; "Conference, Madison Grant and Mr. Merriam with Secretary Lane," folder "Archives–Old Letters–Important," SRL.

24. Burnham, "Conservation," 672. For Schrepfer's analysis, see *The Fight to Save the Redwoods,* 14–17, 28.

25. Cravens, *The Triumph of Evolution,* 114 (emphasis added).

26. Grant, "Saving the Redwoods: An Account," 91, 112.

27. Grant, "Saving the Redwoods: An Account," 97, 110–11.

28. Grant to Merriam, 15 January 1919, folder "Grant, Madison," SRL.

29. Grant, "Saving the Redwoods: An Account," 112; Grant to John H. Emmert, 9 October 1919, folder "Grant, Madison," SRL; Grant to Merriam, 13 June 1921, folder "Madison Grant," container 82, JCM; Schrepfer, *The Fight to Save the Redwoods,* 27; Grant to Merriam, 22 May 1924, folder "Madison Grant," container 82, JCM; Grant to Drury, 22 February 1937; and M. Grant to J. D. Grant, 17 July 1935, folder "Grant, Madison," SRL.

30. J. D. Grant to M. Grant, 27 October 1925, folder "Grant, Madison," SRL.

31. Joseph D. Grant, *Redwoods and Reminiscences,* 96; J.D. Grant to M. Grant, 2 December 1921, folder "Grant, Madison," SRL.

32. Joseph D. Grant, *Redwoods and Reminiscences,* 209.

33. Joseph D. Grant, *Redwoods and Reminiscences,* 76, 78.

34. Joseph D. Grant, *Redwoods and Reminiscences,* 208, 209; M. Grant to J. C. Sperry, 13 October 1920, folder "Grant, Madison," SRL.

35. Joseph D. Grant, *Saving California's Redwoods,* 11.

36. Grant, "Saving the Redwoods: An Account," 107; Grant to Drury, 26 May 1921, folder "Grant, Madison," SRL.

37. Heald, "Who Saved the Redwoods?" 44; Grant, "Saving the Redwoods," *National Geographic,* 534.

38. Grant to Phillips, 19 January 1920, folder "Bolling Memorial Grove, 1919–1921," SRL.

39. Clipping from *San Francisco Chronicle,* 6 August 1921, in folder "Bolling Memorial Grove, August 1921–1931," SRL.

40. Grant, *Preserve an America Worth Fighting For,* 1.

41. Grant, *Preserve an America Worth Fighting For,* 2–5.

42. Joseph D. Grant, *Redwoods and Reminiscences,* 143; Drury to M. Grant, 10 August 1921, folder "Grant, Madison," SRL.

43. Drury ("in extreme confidence") to J. D. Grant, 18 May 1931, folder "Joseph D. Grant Grove," SRL.

44. W. Redmond Cross, "A Tribute to Madison Grant," *ZSB* 50, no. 4 (July–August, 1937): 104.

45. E. P. French, quoted in D. Grant to Aubrey Drury, 2 August 1948, folder "Madison

Grant Forest and Elk Refuge," SRL; TR quoted in Cutright, *Theodore Roosevelt, the Naturalist,* 85.

46. Grant to Merriam, 14 November 1924, folder "Madison Grant," container 82, JCM; Grant to Mark M. Jones, 17 September 1923, folder 763, box 81, series "Cultural Interests," RG III2E, RF; Joseph D. Grant, "Twenty-Rings in the Redwoods," 49.

47. Grant, "Saving the Redwoods," in *Hunting and Conservation,* ed. Grinnell and Sheldon, 198; Grant to J. C. Sperry, 25 May 1920, folder "Grant, Madison," SRL; Fletcher A. Cutler to Grant, 23 December 1920, folder "Madison Grant," container 82, JCM.

48. Phillips to Grant, 2 August 1922; Drury to Grant, 24 June 1922; folder "Grant, Madison," SRL.

49. Drury to Grant, 11 August 1934, folder "Grant, Madison," SRL; "A Gratifying Service."

50. Grant to Merriam, 12 September 1932, folder "Madison Grant," container 82, JCM.

51. Joseph D. Grant, *Redwoods and Reminiscences,* 159.

52. Merriam, quoted in Schrepfer, *The Fight to Save the Redwoods,* 58. See also Merriam, *Highest Uses,* 16.

53. See Drury, "Point Lobos Reserve," 308.

54. McDuffie quoted in Engbeck, *State Parks of California,* 47.

55. Hurwitz quoted in Zuckerman, *Saving Our Ancient Forests,* 47.

56. Graves, "Our Most Urgent Public Park," 78; Kent quoted in Engbeck, *State Parks of California,* 48.

57. J. D. Grant to M. Grant, 23 December 1927, folder "Grant, Madison," SRL.

58. Merriam to Grant, 26 October 1928, folder "Madison Grant," container 82, JCM.

59. Nicholas Roosevelt, *Conservation,* 187.

60. Schrepfer, *The Fight to Save the Redwoods,* 33; Drury to Grant, 31 January 1928, folder "Grant, Madison," SRL.

61. Grant, "Saving the Redwoods: An Account," 115.

62. *California Globe,* 17 October 1928, clipping in carton 5, SRL-B. See Olmsted, *Report of the State Park Survey,* 4.

63. M. Grant to J. D. Grant, 30 December 1927, folder "Grant, Madison," SRL.

64. M. Grant to J. D. Grant, 30 December 1927; and M. Grant to J. D. Grant, 7 December 1927; folder "Grant, Madison," SRL; Grant to Merriam, 7 December 1927, folder "Madison Grant," container 82, JCM; M. Grant to J. D. Grant, 7 December 1927, folder "Grant, Madison," SRL; Burnham to Grant, 29 January 1930, folder 9, box 1, WCS-8; M. Grant to J. D. Grant, 7 December 1927; and Burnham to J. D. Grant, 15 December 1927, folder "Grant, Madison," SRL.

65. O'Melveny to J. D. Grant, 10 December 1927, folder "Grant, Madison," SRL.

66. Drury to Grant, 14 December 1927, folder "Grant, Madison," SRL. For Grant and Wing, see Grant to Drury, 2 December 1927, and Drury to Grant, 31 January 1928, folder "Grant, Madison," SRL.

67. Olmsted, *Report of the State Park Survey,* 48.

68. Engbeck, *State Parks of California,* 60.

69. Drury to Grant, 1 February 1929, folder "Grant, Madison," SRL.

70. Reagan quoted in clipping from *Oakland Tribune,* 6 October 1966, in SRL-B.

71. Hornaday, draft of unpublished autobiography, container 17, WTH.

72. Shankland, *Steve Mather,* 194; Linden, "The Prairie Creek Redwoods," 148.

73. Grant, "Saving the Redwoods: An Account," 117; M. Grant to J. D. Grant, 6 December 1920, folder "Madison Grant," container 82, JCM.

74. Grant, "Saving the Redwoods: An Account," 95; Grant to Everett Griggs, 16 September 1919, folder "History of the League—1911–1966," SRL.

75. Commission quoted in Joseph D. Grant, *Redwoods and Reminiscences,* 155.

76. "Dedication, Founders Tree," folder "Joseph D. Grant," SRL; Burnham, "Conservation," 672.

77. Letter from Donald W. Murphy, 27 July 1992, folder "Founders' Grove," SRL.

Chapter 12. Nordic and Anti-Nordic (pp. 297–327)

1. Huntington, "Exploding the Idea of Race," 5; Boas quoted in Elazar Barkan, "Mobilizing Scientists against Nazi Racism, 1933–1939," in *Bones, Bodies, Behavior,* ed. Stocking, 183.

2. A. L. Kroeber, "Preface," in *The Anthropology of Franz Boas,* ed. Goldschmidt, vi.

3. Boas to Hrdlicka, 18 March 1919, folder "Ales Hrdlicka," FB.

4. Boas, "An Anthropologist's Credo," 201; Barkan, *The Retreat of Scientific Racism,* 78; Grant to Perkins, 7 March 1923, folder "Madison Grant," box 67, Author Files I, CSS.

5. Grant to Taft, 22 November 1910, series 6, case file 77, WHT; Grant to Simmons, 5 April 1912, folder "Madison Grant," IRL; M. Grant to P. S. Grant, 8 April 1912, folder "Madison Grant," IRL. See also Grant to Davenport, 12 December 1929, folder "Madison Grant," CBD.

6. Boas, *The Mind of Primitive Man,* 144; Leslie Spier, "Some Central Elements in the Legacy," in *The Anthropology of Franz Boas,* ed. Goldschmidt, 147.

7. Grant to Osborn, 9 January 1912, book #20, WCS-12; "The Great Nordic Race," *Eugenical News* 2, no. 2 (February 1917): 10–11. Book burning in Kardiner and Preble, *They Studied Man,* 141.

8. Boas, "Eugenics," 471–78.

9. Grant, *The Passing of the Great Race,* 4th ed., xxviii–xxix.

10. Sollas to Grant, 17 October 1920, folder "Grant #2," HFO.

11. See Frank, "Jews, Multiculturalism, and Boasian Anthropology," 732.

12. Two caveats: (1) This discussion is not meant to imply that to be Jewish was necessarily to be antihereditarian. Many Jews of the same generation as Boas (e.g., Max Reichler, Redcliffe N. Salaman, and Maurice Fishberg) were confirmed eugenicists. (2) The use of the term "Boasians" does not mean that there was within anthropology an identifiable Boasian "school." But whatever theoretical disagreements or personal enmities may have arisen among the Boasians, they were usually unified in their opposition to biological determinism.

13. See Dinnerstein, *Antisemitism in America,* 87.

14. Osborn to W. B. Scott, 22 May 1908, quoted in Deacon, *Elsie Clews Parsons,* 197 (also in Kennedy, "Philanthropy and Science in New York City," 163).

15. Grant to Osborn, 9 January 1912, book #20, WCS-12. For the Boasians and the AAA, see Cravens, *The Triumph of Evolution,* 103; Stocking, *Race, Culture, and Evolution,* 285.

16. Kroeber quoted in Freeman, *Margaret Mead and Samoa,* 45; Freeman, *Margaret Mead and Samoa,* 33, 35. See also Hyatt, *Franz Boas,* x, xii; Barkan, *The Retreat of Scientific Racism,* 93.

17. Grant to Osborn, 9 March 1918, folder "Madison Grant," CBD.

18. Laughlin to Grant, 11 November 1935, folder "Madison Grant," box C-2-1, HHL.

19. Gregory, "The Galton Society," *Science,* 267 (see also Gregory, "The Galton Society," *American Journal of Physical Anthropology,* 77–79; and "Notes," 405).

20. "Galton Society," *Eugenical News* 16, no. 7 (July 1931): 113.

21. "Galton Society," *Eugenical News* 10, no. 3 (March 1925): 31.

22. Barkan, *The Retreat of Scientific Racism,* 68; Wissler, *Man and Culture,* vi, 286.

23. Gregory to Davenport, 11 January 1919, folder "Galton Society," CBD (also in folder "Galton Society, 1918-1920," HFO); Gregory to Davenport, 6 June 1922, folder "William K. Gregory," CBD.

24. Selden, *Inheriting Shame,* 15.

25. Chase, *The Legacy of Malthus,* 166.

26. *Consolidated Report upon the Activities of the NRC* 5.

27. Holmes, "Organization of the Committee," 77.

28. For the origins of the Boas-Holmes hostility, see Hyatt, *Franz Boas,* 30–32, 56–59; and Stocking, *Race, Culture, and Evolution,* 277–81.

29. Holmes to Hale [n.d.], frame 741, roll 18, GEH.

30. Conklin to Hale, 9 February 1917, roll 10, GEH; Holmes cited in Hrdlicka to Pickering, n.d., folder "Hale, George E.," box 29, AH.

31. Hale to W. H. Holmes, 26 May 1917, folder "Hale, George E.," box 29, AH; Holmes and Hrdlicka, "Report of the Committee on Anthropology," 52. For Hale's attitude toward Boas, see Hale-Holmes correspondence, roll 18, GEH.

32. Holmes, "Organization of the Committee," 90, 82; Davenport et al., "Suggestions Relating to the New National Army," 527. See also *Report of the National Academy of Sciences for the Year 1917,* 51; Grant, "America for the Americans," 352.

33. Grant to G. E. Hale, 23 November 1917, folder "Madison Grant," CBD; Holmes, "Organization of the Committee," 79.

34. Introduction to *The Passing of the Great Race,* 4th ed., xxxii. See also Grant, "Discussion of Article on Democracy and Heredity," 165.

35. Davenport, *Naval Officers,* 25. Formation of the subcommittee in *Report of the National Academy of Sciences for the year 1918,* 108. Prostitution in Castle et al., *Heredity and Eugenics,* 286.

36. See Grant to W. H. Holmes, 7 May 1917, folder "Grant, Madison, Mr. 1916–18" box 27, AH.

37. Holmes and Hrdlicka, "Report of the Committee on Anthropology," 53; Davenport et al., "Suggestions Relating to the New National Army," 526; Davenport and Ireland, *Army Anthropology,* 34, 47, 48.

38. Davenport to Grant, 12 January 1918, folder "Madison Grant," CBD. See also Grant to Holmes, 7 May 1917; and Hrdlicka to Grant, 18 May 1917; both in folder "Grant, Madison, Mr. 1916–18" box 27, AH; *Report of the National Academy of Sciences for the year 1918,* 94; *Eugenical News* 3, no. 8 (August 1918): 61.

39. Davenport and Ireland, *Army Anthropology,* 205, 200.

40. Montagu, *Man's Most Dangerous Myth,* 191; "Data Wanted," *Eugenical News* 3, no. 11 (November 1918): 84; Prospectus in folder "Hrdlicka, Ales A., 1918–1925," box 31, ROS. See also Hrdlicka, "Preface," 2.

41. Grant to Davenport, 19 December 1917, folder "Madison Grant," CBD; Hrdlicka to Davenport, 29 December 1917, folder "Ales Hrdlicka," CBD.

42. Grant to Osborn, 9 January 1912, book #20, WCS-12.

43. Grant to Merriam, 7 June 1920, folder "Grant, Madison," SRL. For Merriam and eugenics, see Merriam to Grant, 8 December 1924 and 7 February 1925; both in folder "Madison Grant," container 82, JCM.

44. Merriam's speech in Gregory, "The Galton Society," *Science,* 268; Goddard to Kroeber, 12 and 18 December 1918, folder "Goddard, P. E.," box 63, BDA.

45. Boas, Tozzer, and Hrdlicka, "Report on Anthropology," 110–11; also in *American Anthropologist* 21, no. 11 (April–June, 1919): 216–18.

46. Wissler, "Opportunities for Coordination," 4; *Report of the National Academy of Sciences for the year 1919,* 101.

47. Holmes to Boas, 9 September 1919, folder "Boas, Franz," box 14, AH; Lowie to Kroeber, 30 October 1920, folder "Lowie, R. H.," box 94, BDA; Lowie, "The Eugenicist's Programme"; Lowie, "Mr. Grant's Apologia"; Lowie to Kroeber, 6 January 1919, and Lowie to Wissler, 18 November 1920, folder "Lowie, Robert H: Outgoing, 1916–1922," box 1, RHL; Lowie to Kroeber, 1 December 1920, folder "Lowie, R. H.," box 94, BDA.

48. Kroeber to Parsons, 19 January 1919, quoted in Deacon, *Elsie Clews Parsons,* 248. See also Kroeber to Pliny Goddard, 17 December 1918, folder "Goddard, P. E.," box 63, BDA.

49. Boas, "Scientists as Spies," 797; Kroeber to Edward Sapir, 28 January 1920, frame 650, reel 5, ALK.

50. Lowie to Wissler, 8 November 1920, folder "Lowie, Robert H: Outgoing, 1916–1922," box 1, RHL. See also *American Anthropologist* 22, no. 1 (January–March 1920): 93; Goldfrank, *Notes on an Undirected Life,* 4–17; Kroeber, "Franz Boas: The Man," 19–20; and Stocking, *The Ethnographer's Magic,* 117–18.

51. For a superb analysis of the vote, see Stocking, *Race, Culture, and Evolution,* 275–76.

52. Goddard to A. Tozzer, 28 October 1920, folder "Goddard, P. E.," box 63, BDA; Lowie to Wissler, 18 November 1920, folder "Lowie, Robert H: Outgoing, 1916–1922," box 1, RHL.

53. Lowie to Kroeber, 30 October 1920, folder "Lowie, R. H.," box 94, BDA; Lowie to Wissler, 8 November 1920, folder "Lowie, Robert H: Outgoing, 1916-1922," box 1, RHL; Goddard to A. Tozzer, 28 October 1920, folder "Goddard, P. E.," box 63, BDA.

54. Kroeber to Lowie, 13 December 1920, folder "Lowie, R. H.," box 94, BDA; Kroeber to Elsie Clews Parsons, 16 November 1919, quoted in Deacon, *Elsie Clews Parsons,* 453.

55. Osborn to Walcott, 6 January 1920, folder "The Nation," HFO; Walcott to Conklin, 29 December 1919, box 22, folder 30, EGC; Walcott to Osborn, 12 January 1920, folder "The Nation," HFO; Pupin to Walcott, 12 January 1920, quoted in Cravens, *The Triumph of Evolution,* 119–20.

56. Walcott to Parsons, 24 December 1919, box 3, CDW; Parsons to Walcott, 29 December 1919, box 3, CDW; Palmer to Walcott, 20 January 1920, box 3, CDW. See also Boas to Kroeber, 1 March 1920, reel 17, ALK; Walcott to Osborn, 22 December 1919; Osborn to Oswald Garrison Villard, 16 January 1920; Villard to Osborn, 22 January 1920; Walcott to Osborn, 24 January 1920; all in folder "The Nation," HFO. For the origin of the Boas-Walcott enmity, see Hyatt, *Franz Boas,* chaps 2–4, passim.

57. Yerkes, *The Work of the Committee,* 1; Mehler, "American Eugenics Society," 205.

58. Merriam to Grant, 10 October 1921, folder "Madison Grant," container 82, JCM.

59. Wissler, "Final Report," 9–10; *Consolidated Report upon the Activities of the National Research Council,* 249; Stocking, *The Ethnographer's Magic,* 185.

60. Kellogg to Hrdlicka, 4 December 1933, folder "Kellogg, John Harvey," box 37, AH.

61. See Grant-Hrdlicka correspondence, 1917–1918, AH.

62. Hrdlicka to J. H. Kellogg, 6 November 1923, folder "Kellogg, John Harvey," box 37, AH; Lowie to Hrdlicka, 17 February 1925, folder "American Anthropological Association," box 7, AH.

63. Hrdlicka to Boas, 6 May 1918; Boas to Hrdlicka, 3 June 1918; folder "Ales Hrdlicka," FB; Boas to Hrdlicka, 4 November 1918, folder "Boas, Franz," box 14, AH. See also Hrdlicka to Boas, 2 May 1918, folder "Boas, Franz," box 14, AH.

64. Boas, review of *The Passing of the Great Race,* 363.

65. Boas to Hrdlicka, 28 May 1918, folder "American Journal of Physical Anthropology," box 8, AH; Hrdlicka to Boas, 29 May 1918, folder "Ales Hrdlicka," FB.

66. See Herskovits, *Franz Boas,* 23; Stocking, *Race, Culture, and Evolution,* 296.

67. Grant to Osborn, 5 June 1922, folder "Grant #3," HFO.

68. Bilbo, *Take Your Choice,* 164.

69. Boas to Hrdlicka, 29 April 1922, folder "Boas, Franz," box 14, AH; *Consolidated Report upon the Activities of the National Research Council* 20; *A History of the National Research Council,* 9; Freeman, *Margaret Mead and Samoa,* 55–61, 75–81. See also Stocking, *Race, Culture, and Evolution,* 300; Freeman, "On Franz Boas."

70. Stoddard quoted in Rogin, *Blackface, White Noise,* 89.

71. Lewis, *Babbitt,* 117; Herskovits, *The American Negro,* 66; Klineberg, *Negro Intelligence,* 59; Klineberg, *Race Differences,* 345, vii, 346 (emphases added).

72. Grant to Chapman, 27 March 1925, letter #702, JJC; Grant to Osborn, 13 March 1925, folder "Grant #3," HFO; Osborn to Grant, 17 October 1927, folder "Grant #4," HFO.

73. Schudson, "Paper Tigers," 55.

74. Lowie to Hrdlicka, 17 February 1925, folder "American Anthropological Association," box 7, AH.

75. "Genes and Eugenics," *New York Times,* 24 August 1932, p. 16, col. 3.

76. McDougall to Stoddard, 15 November 1924, quoted in Bachman, "Theodore Lothrop Stoddard," 223; Gregory to Grant, 20 October 1930, folder "William K. Gregory," CBD; Davenport to Gregory, 14 February 1930, folder "William K. Gregory," CBD.

77. Grant, "Closing the Flood-Gates," in *The Alien in Our Midst,* ed. Grant and Davison, 23; Todd to Gregory, 21 March 1932, folder "Third International Commission of Eugenics, 1922–1934," box 2, WKG; Stocking, *Race, Culture, and Evolution,* 306; Gossett, *Race,* 418.

Chapter 13. The Empire Crumbles (pp. 328–54)

1. For the decline of scientific racism, see Barkan, *The Retreat of Scientific Racism;* Cravens, *The Triumph of Evolution;* Degler, *In Search of Human Nature,* chaps. 3–8; Gossett, *Race,* chaps. 16 and 17; Haller, *Eugenics,* chaps. 8, 11, 12; Kevles, *In the Name of Eugenics,* chaps. 8–11; Ludmerer, *Genetics and American Society,* chaps. 4, 6; Paul, *Controlling Human Heredity,* chap. 7; and Pickens, *Eugenics and the Progressives,* chap. 11. For an interpretation that reveals the role of the legal system in the decline of scientific racism, see Pascoe, "Miscegenation Law," 44–69.

2. IRL quoted in Degler, *In Search of Human Nature,* 52.

3. Foreword to Grant and Davison, *The Founders of the Republic,* v.

4. "Aliens in America," *New York Times,* 18 November 1928, sec. 4, p. 25, col. 1; Drury to Grant, 27 July 1928, folder "Grant, Madison," SRL.

5. Grant, *The Conquest of a Continent,* 250; Grant to Cox, 24 June 1930, box 3, ESC.

6. Not all African Americans opposed eugenics. See, for example, Carter, "Eugenics for the Negro"; Garth, "Eugenics, Euthenics, and Race"; and Beckham, "Applied Eugenics."

7. Grant, *The Conquest of a Continent,* 3; Grant to Chapman, 17 November, 1924, letter #701, JJC.

8. Loeb to East, 7 April 1919, folder "East, E. M.," container 4, JL.

9. Grant to Huntington, 1 December 1916, folder 823, box 35, EH; Grant, "Discussion of Article on Democracy and Heredity," 165; Grant, *The Passing of the Great Race,* 4th ed., xxxi–xxxii; Grant, "America for the Americans," 350; "New U.S. Parks Urged to Save Last Big Game," *New York Herald Tribune,* 18 November 1935, in Grant to Drury, 29 November 1935, folder "Grant, Madison," SRL (emphasis added). See also Degler, *In Search of Human Nature,* 202.

10. Hankins, *The Racial Basis of Civilization,* viii–ix. See also Hankins, "Individual Differences and Democratic Theory," 406.

11. Hankins, *The Racial Basis of Civilization,* 9, vii, 193, 178–205.

12. "A Sociologist on Race," *Eugenical News* 14, no. 4 (April 1929): 51; Hrdlicka, review of *The Racial Basis of Civilization,* 527. See also Ross, *Seventy Years of It,* 24.

13. An early (and exceedingly well-done) critique of Nordicism by a psychologist was Joseph Jastrow's "Why the Germans Have Deemed Themselves Superior." The psychologists were aided by the Boasians in criticizing intelligence tests. See, for example, Kroeber, *Anthropology,* 75–79; Herskovits, "Brains and the Immigrant"; Mead, "The Methodology of Racial Testing"; Tozzer, *Social Origins,* 85; Wallis, "Race and Culture"; Boas, "Fallacies of Racial Inferiority"; Mead, "Group Intelligence Tests."

14. Brigham, "Intelligence Tests of Immigrant Groups," 165, 164; "Science and Race," *New York Times,* 17 April 1933, p. 12, col. 2; Overstreet, "The Mind of the Negro," 8; Brigham quoted in Downey, *Carl Campbell Brigham,* 27. See also Brigham to Davenport, 8 December 1929, folder "Carl C. Brigham," CBD.

15. Watson, "What Is Behaviorism?" 729; Watson, *Behaviorism,* 103–04 (emphasis added).

16. Walter, *Genetics,* vii.

17. Jennings, "Heredity and Environment," 228 (also in Jennings, *Prometheus,* 19).

18. Grant to Cox, 1 December 1923, box 2, ESC. For the abandonment of eugenics, see, for example, William E. Castle, "Eugenics," 1031–32; Castle, "Biological and Social Consequences of Race Crossing," *American Journal of Physical Anthropology* (see also "Biological and Social Consequences of Race-Crossing," *Journal of Heredity;* Edwin G. Conklin, "Some Recent Criticisms of Eugenics," *Eugenical News* 13, no. 5 (May 1928): 61–65; and Conklin to Frederick Osborn, 24 October 1928, folder 30, box 17, EGC; East, *Heredity and Human Affairs,* 160, 183; H. S. Jennings, *Prometheus;* Jennings, "Heredity and Environment," 225–38; Jennings, *The Biological Basis of Human Nature,* especially 223–51, 269–88; Kellogg, "Nordics, and the Rest"; Morgan, *Evolution and Genetics,* especially 200–207.

19. Hogben quoted in Kevles, *In the Name of Eugenics,* 123, and Benedict, *Race,* 96; *Eugenical News* 15, no. 10 (October 1930): 148–51.

20. "Science and Sentiment," *Eugenical News* 6, no. 4 (April 1921): 30; see also Pearl, "Breeding Better Men."

21. Pearl to Jennings, 24 November 1923, folder "Ales Hrdlicka," FB; Pearl, "The Biology of Superiority," 260, 261, 266.

22. Mencken to Lowie, 7 February [n.d.], folder "Mencken, Henry Louis," box 12, RHL; Barnes, "The Drool Method in History," 34. See also Barnes, *History and Social Intelligence* 222; and Barnes, *Sociology and Political Theory,* 61. Another magazine that provided a forum for the environmentalists was *Current History.* See, for example, Smertenko, "The Claim of 'Nordic' Superiority," 16–19; and Calverton, "The Myth of Nordic Superiority."

23. Pickens, *Eugenics and the Progressives,* 215.

24. Grant to Moley, 22 May 1934, folder 34, box 20, RM. See also Laughlin to Isaac F. Marcosson, 5 February 1935, folder "Immigration and Unemployment," box C-4-4, HHL; "Immigration Curb Is Urged in Survey," clipping in folder "Madison Grant correspondence," box D-2-5, HHL; "Fears Moron Types Will People Nation," *New York Times,* 22 March 1930, p. 22, col. 4; and "The Unemployables," *Eugenical News* 19, no. 5 (September–October 1934): 128.

25. Perkins to Grant, 14 May 1929, folder "Madison Grant," box 67, Author Files I, CSS.

26. Davison to Davenport, 2 July 1930, folder "Charles Stewart Davison," CBD; Charles Pyle to Ward, 15 April 1930, RDW.

27. Grant, "America for the Americans," 350–51.

28. Benedict, *Race,* 199.

29. Dunn to John C. Merriam, 3 July 1935, folder "Carnegie Committee, 1929-39," LCD.

30. Popenoe quoted in Pickens, *Eugenics and the Progressives,* 99.

31. "An Ethical Code of Eugenics," folder "American Eugenics Society, 1926–1928," box 1, WKG. See "The Fecundity of Eugenists," folder 75, box 9, series IV, EH; Kevles, *In the Name of Eugenics,* 323.

32. Davenport, "Presidential Address," in *A Decade of Progress in Eugenics,* 22.

33. Sadler, "Is the Abnormal to Become Normal?" in *A Decade of Progress in Eugenics,* 193–99. Sadler was the wife of Grantian William S. Sadler.

34. "The Exhibit," in *A Decade of Progress in Eugenics,* 489; Hassencahl, "Harry H. Laughlin," 346–50.

35. Muller, "The Dominance of Economics over Eugenics," in *A Decade of Progress in Eugenics,* 141.

36. Muller, "The Dominance of Economics over Eugenics," in *A Decade of Progress in Eugenics,* 143. See also Carlson, *Genes, Radiation, and Society.*

37. "The Third International Congress of Eugenics," *Eugenical News* 16, no. 3 (March 1931): 30; "Genes and Eugenics," *New York Times,* 24 August 1932, p. 16, col. 3.

38. "Genes and Eugenics," *New York Times,* 24 August 1932, p. 16, col. 3; "The Week in Science: Eugenists and Geneticists at Odds," *New York Times,* 28 August 1932, sec. 8, p. 4, col. 2.

39. Chase, *The Legacy of Malthus,* 357.

40. Grant, *The Conquest of a Continent,* 1, 224–86.

41. Grant, *The Conquest of a Continent,* 349–51.

42. Osborn to Grant, 15 May 1928, folder "Grant #4," HFO; Burnham to Grant, 18 July

1932, folder 2, box 3, FRB; Davenport to Grant, 6 November 1933, folder "Madison Grant," CBD.

43. Osborn, introduction to Grant, *The Conquest of a Continent,* viii, vii–x.

44. Perkins to Grant, 12 January 1932, folder "Madison Grant," box 67, Author Files I, CSS; R. N. Fuller, "In Madison Grant's New Book"; "The Conquest of a Continent," circular from Charles Scribner's Sons; Charles Scribner's Sons to Those Interested in the Future of America, November 1933; all in folder "Madison Grant," box C-2-1, HHL. For Scribner's promotional campaign, see folder "Madison Grant," box C-2-1, HHL; and box 67, Author Files I, CSS.

45. Goethe to Lothrop Stoddard, 16 January 1934; Goethe to Grant, 27 January 1934; Goethe to Rev. John Smith, 10 January 1934; all in folder "January 31, 1919–December 31, 1953," box 85C, CMG; Goethe to Grant, 20 February 1935, folder "Madison Grant," box 67, Author Files I, CSS; Root to Grant, 11 June 1934, folder "G 1934," container 152, ER; Cox to Grant, 26 October 1933, box 4, ESC; Grinnell to Grant, 26 October 1933, folder "Madison Grant," box 67, Author Files I, CSS; Huntington to Grant, 15 November 1933, folder 2640, box 70, EH; K. Roosevelt to Grant, 21 October 1933, folder "Madison Grant," container 49, KR; Daniel C. Beard to Grant, 10 January 1934, folder 1, box 2, WCS-8; Hooton to Grant, 3 November 1933, folder G, box 9, EAH.

46. Wittke, review of *The Conquest of a Continent,* 589; Huxley, review of *The Conquest of a Continent, New York American,* 7 February 1934, clipping in folder "Review of Madison Grant's Conquest of a Continent " box D-2-5, HHL. See also "The Conquest of a Continent," *Eugenical News* 18, no. 6 (November–December, 1933): 113–15; Little, "The Nordic in America"; Phelps, "As I Like It," 110.

47. William L. Langer, review of *The Conquest of a Continent;* Review of *The Conquest of a Continent, Christian Century;* John Chamberlain, "Books of the Times," review of *The Conquest of a Continent,* by Madison Grant, *New York Times,* 24 October 1933, p. 19, col. 3; William MacDonald, "Mr. Grant's Plea for a Nordic, Protestant America," *New York Times,* 5 November 1933, sec. 5, p 4, col. 2.

48. Grant to Osborn, 23 September 1932, folder "Grant #6," HFO; Hrdlicka, review of *The Conquest of a Continent,* 478–79; Herskovits, "Homo Nordicus-U.S.A.," 49; Ruth Benedict, "The Nordics Once More," *New York Herald Tribune,* 3 December 1933, sec. 7, p. 44, col. 1; Boas, "Nordic Propaganda."

49. Perkins to Grant, 4 November 1933, folder "Madison Grant," box 67, Author Files I, CSS; Grant to Root, 15 June 1934, folder "G 1934," container 152, ER.

50. Richard E. Gustadt to the Publishers of Anglo-Jewish Periodicals, folder "G 1934," container 152, ER; also in folder "Grant #7," HFO; Grant, *The Conquest of a Continent,* 34; Drury to Grant, 21 October 1933, folder "Grant, Madison," SRL.

51. Laughlin to Grant, 23 February 1933, folder "Madison Grant," box C-2-1, HHL.

52. Berg, *Max Perkins,* 5.

53. Grant to Perkins, 20 February 1933; Perkins to Grant, 24 July 1933, folder "Madison Grant," box 67, Author Files I, CSS.

54. "Post-War Immigration into France," *Eugenical News* 17, no. 4 (July–August 1932): 94.

55. Executive Committee to the Members of the Immigration Restriction League, 14 June 1924, folder "Immigration Restriction League," HFO. See also Solomon, *Ancestors and Immigrants,* 206.

56. The full story of the effort to ban immigration from Mexico is in Spiro, "Patrician Racist," chap. 16.

57. Holcombe to Henry D. Ward, 11 December 1937, RDW.

58. See box "1934–1935" and box "1937," ADS.

59. "Report of the American Defense Society," box "1937," ADS. For membership see American Defense Society, *The American Defense Society,* 19.

60. Schauer, "Charles Matthias Goethe," 81, 100.

61. Goethe to Eugenics Record Office, 4 January 1924, folder "C. M. Goethe," CBD; Goethe, "The Influx of Mexican Amerinds," 6; "Puerto Rico as a State," *Eugenical News* 20, no. 4 (July–August 1935): 59; Goethe to Grant, 11 July 1928, box "1927-June 1935," ADS; "C. M. Goethe" *Eugenical News* 20, no. 4 (July–August 1935): 56.

62. Commandos quoted in McWilliams, "Racism on the West Coast," 732–33; and McWilliams, *Prejudice,* 238–39.

63. Galton, "Local Associations for Promoting Eugenics," in *Essays in Eugenics,* 106.

64. Miller, "Founder of the World's Largest Summer School," 55; Schrepfer, *The Fight to Save the Redwoods,* 43–44.

65. MacDowell, "Charles Benedict Davenport," 25, 34.

66. Merriam to F. Osborn, 29 October 1935, folder "Osborn, Fred," container 138, JCM. For the demise of the ERO, see Allen, "The Eugenics Record Office," 250–54; Cravens, *The Triumph of Evolution,* 179–80; and Kevles, *In the Name of Eugenics,* 199. For the campaign to promote Laughlin, see Grant to Merriam, 19 October 1933 and 27 November 1933, folder "Madison Grant," container 82, JCM; and Osborn to Merriam, 7 February 1933, folder "Osborn, Henry Fairfield," container 138, JCM.

67. Franz Boas to R. C. Andrews, 19 November 1937, quoted in Kennedy, "Philanthropy and Science in New York City," 204.

68. Grant to Merriam, 9 November 1935, folder "Madison Grant," container 82, JCM.

69. Burnham to Grant, 17 June 1922, folder 2, box 3, FRB.

70. F. Osborn, "History of the American Eugenics Society," p. 8, in folder "Osborn, Frederick: History of the American Eugenics Society," AES. For Grant's support of *Eugenical News,* see Davenport to Grant, 20 November 1928, folder "Madison Grant," CBD; Laughlin to Grant, 14 November 1931, folder "Madison Grant correspondence," box D-2-5, HHL.

71. T. B. Appleget to R. B. Fosdick, 22 November 1928; and memo of 11 January 1929 "Research on Human Heredity"; both in folder 453, box 46, Cultural Interests series, RG III2E, RF. See also Fosdick to Appleget, 23 November 1928; and Appleget memo of 5 December 1928, "Eugenics Research Association," both in folder 453, box 46; and memo of 7 February 1929, "American Eugenics Society, Inc.," folder 599, box 60; all in Cultural Interests series, RG III2E, RF. For Eastman, see Eastman to Irving Fisher, 17 February 1932, GE.

72. "The American Eugenics Society Round Table Conferences and Annual Dinner," folder 8, unit 40, Series II, AM; M. A. Bigelow to Yerkes, 2 May 1950, folder 1519, box 80, RMY.

73. For Raymond Pearl's move from eugenics to population control see Allen, "Old Wine in New Bottles," or Allen, "From Eugenics to Population Control."

74. Gordon, *Woman's Body,* 395. For the origins of the Population Council, see Shapiro, *Population Control Politics,* chap. 3; see also Hodgson, "Ideological Origins"; and Allen, "The Eugenics Record Office," 254.

Chapter 14. The Ever-Widening Circle: The Third Reich (pp. 355–83)

1. "Presidential Address of the Eugenics Research Association," *Eugenical News* 14, no. 7 (July 1929): 99.

2. Scribner to Grant, 28 June 1922, folder "Madison Grant," box 67, Author Files I, CSS. For Lehmann see Proctor, *Racial Hygiene,* 344; Weindling, *Health, Race, and German Politics,* passim.

3. *Literarische Wochenschrift* quoted in flyleaf of *Der Untergang der Grossen Rasse;* Rudolf Polland, "Vorwort des Übersetzers," *Der Untergang der Grossen Rasse* (Munchen: J. F. Lehmann, 1925), 5–8 (I am grateful to Chad C. Bryant for this translation).

4. "The Nordic Movement in Europe," *Eugenical News* 13, no. 10 (October 1928): 132; K. Holler, "The Nordic Movement in Germany," *Eugenical News* 17, no. 5 (September–October, 1932): 117; Grant, "Further Notes on *The Racial Elements of European History,*" *Eugenical News* 13, no. 9 (September 1928): 118.

5. For the German admiration of American eugenics, see Proctor, *Racial Hygiene,* chap. 4; Field, "Nordic Racism," 524; and Holler to Stoddard, 29 November 1927, in Bachman, "Theodore Lothrop Stoddard," 8.

6. Hitler, *Mein Kampf,* 367; see also Kühl, *The Nazi Connection,* 26; Proctor, *Racial Hygiene,* 100, 173; "The Sterilization Law in Germany," *Eugenical News* 19, no. 6 (November–December 1934): 137.

7. Hitler, *Mein Kampf,* 92; Hitler's letter to Grant quoted in Kühl, *The Nazi Connection,* 85, citing the unpublished autobiography of Leon F. Whitney (pp. 204–5), Whitney Papers, American Philosophical Society.

8. See, for example, Hitler, *Mein Kampf,* 338–40.

9. Schneidewind in Kühl, *The Nazi Connection,* 38; Alfred Rosenberg, *Der Mythus des 20. Jahrhunderts,* 673 (translation Chad C. Bryant); *Rassenkunde: Eine Auswahl des wichtigsten Schrifttums aus dem Gebiet der Rassenkunde, Vererbungslehre, Rassepflege und Bevölkerungspolitik,* ed. Institut für Lese- und Schrifttumskunde (Leipzig, 1936), in Kühl, *The Nazi Connection,* 131.

10. Baur, Fischer, and Lenz, *Human Heredity,* 655.

11. Lenz quoted in Friedlander, *The Origins of Nazi Genocide,* 12; Fischer quoted in Trombley, *The Right to Reproduce,* 114, and Proctor, "From *Anthropologie* to *Rassenkunde* in the German Anthropological Tradition," in Stocking, *Bones, Bodies, Behavior,* 148.

12. Fischer quoted in Muller-Hill, *Murderous Science,* 12 (emphasis added).

13. Proctor, "From *Anthropologie* to *Rassenkunde,*" in Stocking, *Bones, Bodies, Behavior,* 157. For the Rockefeller Foundation, see Schafft, *From Racism to Genocide.*

14. Osborn to Mayr, 14 February 1934, folder "Grant #8," HFO; Eugen Fischer, "Vorwort," *Die Eroberung eines Kontinents oder die Verbreitung der Rassen in Amerika,* by Madison Grant (Berlin: Alfred Metzner, 1937), vii (translation by Chad C. Bryant).

15. Eugen Fischer, "Vorwort," viii (see note 14 above).

16. William MacDonald, "Mr. Grant's Plea for a Nordic, Protestant America," *New York Times,* 5 November 1933, sec. 5, p. 16, col. 2.

17. Fischer quoted in Muller-Hill, *Murderous Science,* 18.

18. Barzun, *Race,* 243. See also Weindling, "Weimar Eugenics," 316.

19. Grant, "Further Notes on *The Racial Elements of European History,*" *Eugenical News* 13, no. 9 (September 1928): 118. See also "Races of Central Europe," review of *Rassen-*

kunde des Deutschen Volkes, by Hans F. K. Günther, *Eugenical News* 9, no. 4 (April 1924): 34; Kühl, *The Nazi Connection,* 130; Walter Kaufmann, *Nietzsche,* 256.

20. *The Racial Elements of European History,* 78, 164, 259; Grant, "Further Notes on *The Racial Elements of European History,*" *Eugenical News* 13, no. 9 (September 1928): 120.

21. Grant to Davenport, 11 October 1930, folder "Madison Grant," CBD; Grant to Cox, 11 June 1930, box 3, ESC; Günther quoted in Kühl, *The Nazi Connection,* 38.

22. Rüdin, "Eugenical Sterilization," 103.

23. For Hitler's familiarity with the American sterilization laws, see Hitler, *Mein Kampf,* 338–39; Kühl, *The Nazi Connection,* 37; Trombley, *The Right to Reproduce,* 115–16.

24. Gütt quoted in Kevles, *In the Name of Eugenics,* 117 (emphasis added).

25. Frick, "German Population and Race Politics," *Eugenical News* 19, no. 2 (March–April 1934): 34; Lapouge to Grant, 27 September 1933, folder "Crime Comm.," box E-1-1, HHL; Rüdin quoted in Lapon, *Mass Murderers in White Coats,* 84.

26. "Reich Opens Race Study," *New York Times,* 2 August 1933, p. 6, col. 7.

27. Brennecke, *The Nazi Primer,* 73.

28. Hitler, *Hitler's Secret Conversations,* 116, 37, 119.

29. Frick, "German Population and Race Politics," *Eugenical News* 19, no. 2 (March–April 1934): 35; "Race Hygiene (Eugenics) in Germany," review of *Erblehre und Rassenhygiene im völkischen Staat,* by Ernst Rüdin, *Eugenical News* 19, no. 5 (September–October 1934): 136.

30. Proctor, "Nazi Doctors," 22; Lifton, *The Nazi Doctors,* 41, 43; Burleigh and Wippermannn, *The Racial State,* 66.

31. DeJarnette quoted in Kevles, *In the Name of Eugenics,* 116; Whitney, *The Case for Sterilization,* 136; Laughlin to Grant, 13 January 1934, folder "Madison Grant correspondence," box D-2-5, HHL. See also Larson, *Sex, Race, and Science,* 147. Statistics in Proctor, "Eugenics among Social Sciences," 188; Proctor, *Racial Hygiene,* 108; Paul, *Controlling Human Heredity,* 89; Mehler, "American Eugenics Society," 251. For *Hitlerschnitt* see Bock, 277.

32. Whitney, The Case for Sterilization, 7, 137–38; Popenoe, "The German Sterilization Law"; Goethe, "Patriotism and Racial Standards," Eugenical News 21, no. 4 (July–August 1936): 66. See also "Eugenical Sterilization in Germany," Eugenical News 18, no. 5 (September–October, 1933): 90; Laughlin to Grant, 13 January 1934, folder "Madison Grant correspondence," box D-2-5, HHL.

33. Baker quoted in Larson, *Sex, Race, and Science,* 146.

34. See Davenport to Grant, 7 October and 24 November 1930, folder "Madison Grant," CBD.

35. Grant to Laughlin, 16 January 1934, folder "Madison Grant correspondence," box D-2-5, HHL; Frederic Osborn, "Summary of the Proceedings" of the Conference on Eugenics in Relation to Nursing, 24 February 1937, quoted in Mehler, "American Eugenics Society," 223; film quoted in Kühl, *The Nazi Connection,* 49. See also "Eugenics in Germany," Eugenical News 22, no. 4 (July–August 1937): 66; and Laughlin to Draper, 9 December 1938, cited in Hassencahl, "Harry H. Laughlin," 355.

36. Baur quoted in Proctor, Racial Hygiene, 131, and Weiss, *Ideology of Death,* 356.

37. Muller-Hill, *Murderous Science,* 33.

38. Wagner and *Ärzteblatt* quoted in Proctor, *Racial Hygiene,* 196, 133; Streicher quoted in Muller-Hill, *Murderous Science,* 81.

39. "The Fortune Quarterly Survey," 157; Boas to Parsons, 11 September 1936, quoted in Deacon, *Elsie Clews Parsons,* 357. Boas's manifold anti-Nazi activities are covered in Barkan, "Mobilizing Scientists," 180–205; Barkan, *The Retreat of Scientific Racism,* chap. 6; Herskovits, "Franz Boas as Physical Anthropologist," 45–46; Herskovits, *Franz Boas,* 117; Kuznick, *Beyond the Laboratory,* chap. 6; Proctor, "From Anthropologie to Rassenkunde," in Stocking, *Bones, Bodies, Behavior,* 157.

40. Dight to Hitler, 1 August 1933, Scrapbooks, CFD; *Minneapolis Journal,* 28 July 1933, clipping in Scrapbooks, CFD; Laughlin quoted in Kühl, *The Nazi Connection,* 48.

41. Osborn to Grant, 4 August 1932, folder "Grant #6," HFO; Eastman to Cousin Mary, 28 June 1928, *GE;* Bachman, "Theodore Lothrop Stoddard," 237–38; Stoddard, "Realism."

42. Herskovits to Davenport, 19 February 1935, folder "Melville J. Herskovits," CBD.

43. Huntington, "Exploding the Idea of Race," 5; "P. G.," folder "D-2-4," HHL; "Race Hygiene (Eugenics) in Germany," review of *Erblehre und Rassenhygiene im völkischen Staat,* by Ernst Rüdin, *Eugenical News* 19, no. 5 (September–October 1934): 136; Osborn quoted in Mehler, "American Eugenics Society," 229.

44. Mjøen quoted in Proctor, *Racial Hygiene,* xi; Lapouge to Grant, 27 September 1933, folder "Crime Comm.," box E-1-1, HHL (see also "A French View," *Eugenical News* 19, no. 2 [March–April 1934]: 39).

45. Embree, "Facing East," 350; Meyer, "A Little Hitler." For other critics who directly mentioned Grant in connection with the rise of fascism, see Steiner, "Fascism in America?" 823; and Goldenweiser, review of *Les trois races de l'Europe et du monde,* 79.

46. Singer to Osborn, 16 May 1935, folder "Galton Society, 1931–1935," HFO.

47. Gregory to C. R. Stockard, 23 May 1935, folder "Galton Society, W. K. Gregory's Resignation, 1934," box 2, WKG; Gregory to Clarence G. Campbell, 6 May 1935, folder "Galton Society," RP; Stockard to Gregory, 22 May 1935; folder "Galton Society, W. K. Gregory's Resignation, 1934," box 2, WKG.

48. Davenport to Warren C. Kincaid, 3 April 1933, folder "Charles Stewart Davison," CBD; "Jewish Refugees from Germany, *Eugenical News* 19, no. 2 (March–April 1934): 44.

49. Laughlin quoted in Hassencahl, "Harry H. Laughlin," 234; DeForest in Grant to Laughlin, 18 April 1934, folder "Madison Grant correspondence," box D-2-5, HHL; Laughlin to Grant, 7 August 1934, folder "Madison Grant," box C-2-1, HHL; Dies quoted in Arthur D. Morse, *While Six Million Died,* 140.

50. Morse, *While Six Million Died,* 228.

51. "The Fortune Survey," 102.

52. Bradley to Goethe, 2 May 1940, RDW; Henry L. Feingold, *The Politics of Rescue,* 150.

53. Osborn to Charles Singer, 28 May 1935, folder "Galton Society, 1931–1935," HFO.

54. Singer to Osborn, 16 May 1935; Osborn to Singer, 28 May 1935; Osborn to Gregory, 25 May 1935, all in folder "Galton Society, 1931–1935," HFO.

55. Kopp quoted in Kühl, *The Nazi Connection,* 56.

56. "American Speakers Shun Berlin Meeting," *New York Times,* 30 August 1935, p. 8, col. 5; "Population Parley Hears Nazi Praise," *New York Times,* 28 August 1935, p. 4, col. 2.

57. "Presidential Address," *Eugenical News* 20, no. 5 (September–October 1935): 68; "Praise for Nazis," 21.

58. "Berlin Congress Ends in Dispute," *New York Times,* 31 August 1935, p. 4, col. 4.

59. Campbell to Grant, 18 September 1935, box "1934–1935," ADS.

60. Landman to Laughlin, 13 September 1935, folder "Correspondence—C. G. Campbell," box D-2-3, HHL; Landman, "Race Betterment by Human Sterilization," 293; Campbell, "The German Racial Policy," *Eugenical News* 21, no. 2 (March–April, 1936): 25–28.

61. Goethe quoted in Trombley, *The Right to Reproduce,* 117; Goethe to Ellsworth Huntington, 26 September 1935, folder 338, box 33, series IV, EH.

62. Kennedy quoted in Mehler, "American Eugenics Society," 248; degree quoted in Hassencahl, "Harry H. Laughlin," 354; Laughlin to Carl Schneider, 28 May 1936, folder "Honorary Degree–1936–University of Heidelberg, Germany," box E-1-3, HHL.

63. "The German Universities," *New York Times* 12 April 1936, sec. 4, p. 8, col. 2; Laughlin to Carl Schneider, 11 August 1936, folder "Honorary Degree–1936–University of Heidelberg, Germany," box E-1-3, HHL. See also Lifton, *The Nazi Doctors,* 38; "Nazis Will Guide Heidelberg Fete," *New York Times,* 28 April 1936, p. 9, col. 2.

64. Stoddard, *Into the Darkness,* 190, 255, 202, 209.

65. Stoddard, *Into the Darkness,* 190, 196, 147.

66. Stoddard, *Into the Darkness,* 179, 272.

67. Stoddard, *Into the Darkness,* 288, 189.

68. "Biological soldiers" in Lifton, *The Nazi Doctors,* 30.

69. Aly, "Medicine against the Useless," 37; Trombley, *The Right to Reproduce,* 112.

70. Friedlander, *The Origins of Nazi Genocide,* 62; Nazis quoted in Aly, "Medicine against the Useless," 22.

71. Hadamar in Proctor, "Nazi Doctors," 25, and Proctor, *Racial Hygiene,* 191; Kaufbeuren in Friedlander, *The Origins of Nazi Genocide,* 162–63; Proctor, *Racial Hygiene,* 193–94.

72. Grant, *The Passing of the Great Race,* 47.

73. Lapouge to Grant, 27 September 1933, folder "Crime Comm.," box E-1-1, HHL; see also "A French View," *Eugenical News* 19, no. 2 (March–April 1934): 39.

74. Telford Taylor, introduction to Hitler, *Hitler's Secret Book,* xxv; Rüdin quoted in Muller-Hill, Murderous Science, 61; Weinrich, *Hitler's Professors,* 240.

75. Shipman, *The Evolution of Racism,* 138; Hitler, *Mein Kampf,* 255–64; Hitler, *Hitler's Secret Conversations,* 65, 393, 457, 269–70.

76. Rosenberg quoted in Muller-Hill, *Murderous Science,* 45, 47.

77. See, for exampke, the physicians quoted in Proctor, *Racial Hygiene,* 176; Shipman, *The Evolution of Racism,* 138; and Weinrich, *Hitler's Professors,* 203.

78. Klein quoted in Mehler, "American Eugenics Society," 250; Goebbels quoted in Lifton, *The Nazi Doctors,* 477; Nash, *Forbidden Love,* 155. For transfer of gas chambers, see Proctor, "Nazi Doctors," 25.

79. See Muller-Hill, *Murderous Science,* 70–74; Friedlander, *The Origins of Nazi Genocide,* 134–35.

80. *Trials of War Criminals,* 1:749; assistant quoted in Mitscherlich and Mielke, *The Death Doctors,* 228; researchers quoted in "Opening Statement of the Prosecution," Trials of War Criminals, 1:55.

81. Sheila Weiss, "The Race Hygiene Movement in Germany, 1904–1945," in *The Well-Born Science,* ed. Adams, 49.

82. Ramm quoted in Proctor, *Racial Hygiene,* 10 (see also Gasman, *The Scientific Origins of National Socialism,* xxii; and Haeckel, *The Riddle of the Universe,* 11ff, 244, and chap. 14).

83. Schama, *Landscape and Memory*, 119.

84. Lekan, "Regionalism," 384; Oskar Karpa quoted in Lekan, "Regionalism," 392; Middle Rhine Gorge in Lekan, "Regionalism," 393–94.

85. Proctor, "Nazi Doctors," 19; "Opening Statement of the Prosecution," *Trials of War Criminals*, 1:71. See also Gröning and Wolschke-Bulmahn, "Politics, Planning, and the Protection of Nature."

86. Lorenz, *Seitschrift für angewandete Psychologie und Characterkunde*, quoted in Mehler, "Eliminating the Inferior," 17 (also in Chase, *The Legacy of Malthus*, 349). For a discussion of eugenical thinking in the work of Lorenz, see Kalikow, "Konrad Lorenz's Ethological Theory."

87. Heidegger in Muller-Hill, *Murderous Science*, 108.

88. Verschuer in Kühl, *The Nazi Connection*, 103.

89. "Records of the United States Nuernberg War Crimes Trials, *United States v. Karl Brandt et al.* (Case 1): November 21, 1946–August 20, 1947," NWC; "Opening Statement of the Prosecution," *Trials of War Criminals*, 1:72.

90. See Annas and Grodin, *The Nazi Doctors and the Nuremberg Code*, 268.

91. *Trials of War Criminals*, 2:30 (emphasis added), 1:989.

92. "Opening Statement of the Prosecution," *Trials of War Criminals*, 1:28, 1:68.

93. "Document Karl Brandt No. 52," "Document Karl Brandt No. 51," NWC. See also Hofmann Defense Exhibits 61 and 62, *Trials of War Criminals*, 4:1159, 1164.

94. "Opening Statement of the Prosecution," *Trials of War Criminals*, 1:28.

95. Proctor, "From *Anthropologie* to *Rassenkunde*," in Stocking, *Bones, Bodies, Behavior*, 174.

Epilogue: The Passing of the Great Patrician (pp. 384–90)

1. The full story of Grant's international conservation efforts is in Spiro, "Patrician Racist," chap. 20.

2. Kellogg to Grant, 29 March 1937, reel 4, JHK.

3. Seymour to Hornaday, 13 January 1937, folder 24, box 9, ABS; Grant to Hornaday, 23 October 1936, folder "Ga-Gu," container 12, WTH.

4. Hitler, *Hitler's Secret Conversations*, 77, 554.

5. "The International Sporting Exhibition of Berlin 1937," folder "Boone & Crockett Club 35–38," container 106, KR.

6. Kellogg to Grant, 25 April 1936, reel 4, JHK.

7. Anthony, "Madison Grant," 396.

8. Grant to Roosevelt, 12 June 1934, folder "Boone & Crockett Club, 1934," container 106, KR.

9. Goethe to Laughlin, 4 January 1939, folder "Correspondence-Goethe," box E-2-1, HHL; Laughlin to DeForest Grant, 8 June 1937, folder "Madison Grant," box C-2-1, HHL; *Volk und Rasse* 12 (1937): 365; *Zeitschrift für Rassenkunde* 6 (1937): 272; both quoted in Kühl, *The Nazi Connection*, 131.

10. "The Passing of a Great American," 3.

11. Burnham to Roosevelt, 22 June 1937, folder "Boone & Crockett Club 35–38," container 106, KR; Burnham, "Charon Beckoned," folder "Boone & Crockett Club 35–38," container 106, KR.

12. Herskovits, *Franz Boas,* 120–21.

13. Haraway, *Primate Visions,* 386.

14. Kennedy, "Philanthropy and Science in New York City," 102.

15. Johnson, "The Opponents of Restricted Immigration," in *The Alien in Our Midst,* ed. Grant and Davison, 12; Bogen, *Immigration in New York,* 4.

16. Walter Kerr, "Skin Deep Is Not Good Enough," *New York Times,* 14 April 1968, sec. 2, p. 1, col. 1.

17. Gould, "Singapore's Patrimony (and Matrimony)," 22.

Works Cited

Adams, Henry. *The Education of Henry Adams.* Boston: Houghton Mifflin Company, 1973.

Adams, Mark B., ed. *The Well-Born Science: Eugenics in Germany, France, Brazil, and Russia.* New York: Oxford University Press, 1990.

Allen, Garland E. "The Eugenics Record Office at Cold Spring Harbor, 1910–1940: An Essay in Institutional History." *Osiris* 2 (1986): 225–64.

——. "From Eugenics to Population Control: The Work of Raymond Pearl." *Science for the People* 12, no. 4 (JulyAugust 1980): 22–28.

——. "Old Wine in New Bottles: From Eugenics to Population Control in the Work of Raymond Pearl." In *The Expansion of American Biology,* edited by Keith R. Benson, Jane Maienschein, and Ronald Rainger, 231–61. New Brunswick, N.J.: Rutgers University Press, 1991.

Aly, Götz. "Medicine against the Useless." In *Cleansing the Fatherland: Nazi Medicine and Racial Hygiene,* by Götz Aly, Peter Chroust, and Christian Pross, translated by Belinda Cooper. Baltimore: The Johns Hopkins University Press, 1994.

American Birth Control League, Inc. *Sixth International Neo-Malthusian and Birth Control Conference.* New York: Marstin Press, 1925.

American Breeders Magazine. 4 vols. Washington, D.C., 1910–13.

American Defense Society. *The American Defense Society, Inc.* New York: American Defense Society, 1941.

American Eugenics Society. *A Eugenics Catechism.* New Haven: n.p., [1926].

Annas, George J., and Michael A. Grodin, eds. *The Nazi Doctors and the Nuremberg Code: Human Rights in Human Experimentation.* New York, Oxford: Oxford University Press, 1992.

Annual Report of the American Bison Society. 14 vols. 1905–30.

Annual Report of the American Museum of Natural History 53 (1921): 31.

Annual Report of the New York Zoological Society. 45 vols. New York, 1897–1941.

Annual Report of the Save-the-Redwoods League. 22 vols. 1920–46.

Anthony, H. E. "Madison Grant." *Journal of Mammalogy* 19, no. 3 (August 1938): 396.

Atherton, Gertrude. "The Alpine School of Fiction." *Bookman* 55, no. 1 (March 1922): 26–33.

Bachman, James Robert. "Theodore Lothrop Stoddard: The Bio-sociological Battle for Civilization." Ph.D. diss., University of Rochester, 1966.

Bagger, Eugene S. "Ethnology by Ear." Review of *Why Europe Leaves Home,* by Kenneth Roberts. *New Republic* 30, no. 389 (17 May 1922): 349.

Balch, Emily Greene. Review of *The Passing of the Great Race,* by Madison Grant. *Survey* 39, no. 9 (1 December 1917): 262–64.

Baltzell, E. Digby. *The Protestant Establishment: Aristocracy and Class in America.* New York: Random House, 1964.

Barkan, Elazar. "Mobilizing Scientists against Nazi Racism, 1933–1939." In *Bones, Bodies, Behavior: Essays on Biological Anthropology,* edited by George W. Stocking. Madison: University of Wisconsin Press, 1988.

——. *The Retreat of Scientific Racism: Changing Concepts of Race in Britain and the United States between the World Wars.* Cambridge: Cambridge University Press, 1992.

Barnes, Harry Elmer. "The Drool Method in History." *American Mercury* 1, no. 1 (January 1924): 31–38.

——. *History and Social Intelligence.* New York: Alfred A. Knopf, 1926.

——. *Sociology and Political Theory: A Consideration of the Sociological Basis of Politics.* New York: Alfred A. Knopf, 1924.

Barzun, Jacques. *Race: A Study in Modern Superstition.* New York: Harcourt, Brace and Company, 1937.

Baur, Erwin, Eugen Fischer, and Fritz Lenz. *Human Heredity.* Translated by Eden and Cedar Paul. New York: Macmillan Company, 1931.

Beckham, Albert Sidney. "Applied Eugenics." *Crisis* 28, no. 4 (August 1924): 177–78.

Benedict, Ruth. *Race: Science and Politics.* Rev. ed. New York: Viking Press, 1943.

Berg, A. Scott. *Max Perkins: Editor of Genius.* New York: E. P. Dutton, 1978.

Biddiss, Michael D. *Father of Racist Ideology: The Social and Political Thought of Count Gobineau.* London: Weidenfeld and Nicolson, 1970.

Bilbo, Theodore G. *Take Your Choice: Separation or Mongrelization.* Poplarville, Miss.: Dream House Publishing Co., 1947.

Blackmon, Douglas A. "Silent Partner: How the South's Fight to Uphold Segregation Was Funded up North." *Wall Street Journal,* 11 June 1999, p. 1, col. 6.

Blair, W. Reid. *In the Zoo: Representing Twenty-seven Years of Observation and Study of the Animals in the New York Zoological Park.* New York: Charles Scribner's Sons, 1929.

Boas, Franz. "An Anthropologist's Credo." *Nation* 147, no. 9 (27 August 1938): 201–4.

——. *Anthropology and Modern Life.* New York: W. W. Norton & Company, 1928.

——. "Changes in Bodily Form of Descendants of Immigrants." In *Reports of the Immigration Commission.* Vol. 38. (61st Cong., 2d sess., document no. 208.) Washington, D.C.: GPO, 1911.

——. "Changes in the Bodily Form of Descendants of Immigrants." *American Anthropologist* 14, no. 3 (July–September 1912): 530–62.

——. "Eugenics." *Scientific Monthly* 3 (November 1916): 471–78.

——. "Fallacies of Racial Inferiority." *Current History* 25, no. 5 (February 1927): 676–82.

——. "Inventing a Great Race." Review of *The Passing of the Great Race,* by Madison Grant. *New Republic* 9 (13 January 1917): 305–7.

——. *The Mind of Primitive Man.* New York: Macmillan Company, 1911.

——. "Nordic Propaganda." Review of *The Conquest of a Continent,* by Madison Grant. *New Republic* 78, no. 1005 (7 March 1934): 107–8.

——. "Report on Anthropology and Its Relations to the National Research Council." *American Journal of Physical Anthropology* 2, no. 1 (January–March 1919): 109–11.

——. Review of *The Passing of the Great Race,* by Madison Grant. *American Journal of Physical Anthropology* 1, no. 3 (July–September 1918): 363.

——. "Scientists as Spies." *Nation* 109, no. 2842 (20 December 1919): 797.
——. "This Nordic Nonsense." *Forum* 74, no. 4 (October 1925): 502–11.
Boas, Franz, A. M. Tozzer, and Ales Hrdlicka. "Report on Anthropology and Its Relations to the National Research Council." *American Journal of Physical Anthropology* 2, no. 1 (January–March 1919): 109–11.
Bock, Gisela. "Racism and Sexism in Nazi Germany: Motherhood, Compulsory Sterilization, and the State." In *When Biology Became Destiny: Women in Weimar and Nazi Germany,* edited by Renate Bridenthal, Atina Grossmann, and Marion Kaplan. New York: Monthly Review Press, 1984.
Bogen, Elizabeth. *Immigration in New York.* New York: Praeger, 1987.
Bölsche, Wilhelm. *Haeckel: His Life and Work.* Translated by Joseph McCabe. Philadelphia: George W. Jacobs & Co., 1906.
"A Book to Stir You Up." Review of *The Passing of the Great Race,* by Madison Grant. *New York Sun,* 7 April 1918, sec. 6, p. 6, col. 1.
Boone and Crockett Club. Committee on Records of North American Big Game. *North American Big Game.* New York: Charles Scribner's Sons, 1939.
——. *Officers, Constitution, Treasurer's Report, and List of Members.* 1921/'22–1938/'39.
——. *Report of Committee on Game Refuges.* 1903.
——. *Report of the Game Preservation Committee.* 1915.
Bradford, Phillips Verner, and Harvey Blume. *Ota: The Pygmy in the Zoo.* New York: St. Martin's Press, 1992.
Brechin, Gray. "Conserving the Race: Natural Aristocracies, Eugenics, and the U.S. Conservation Movement." *Antipode* 28, no. 3 (July 1996): 229–45.
Brennecke, Fritz. *The Nazi Primer: Official Handbook for Schooling the Hitler Youth.* Translated by Harwood L. Childs. New York, London Harper & Brothers Publishers, 1938.
Bridges, William A. *Gathering of Animals: An Unconventional History of the New York Zoological Society.* New York: New York Zoological Society, 1974.
Brigham, Carl C. "Intelligence Tests of Immigrant Groups." *Psychological Review* 37, no. 2 (March 1930): 158–65.
——. *A Study of American Intelligence.* Princeton, N.J.: Princeton University Press, 1923.
Broberg, Gunnar, and Nils Roll-Hansen, eds. *Eugenics and the Welfare State: Sterilization Policy in Denmark, Sweden, Norway and Finland.* East Lansing: Michigan State University Press, 1996.
Bryan, William Jennings, and Mary Baird Bryan. *The Memoirs of William Jennings Bryan.* Chicago: John C. Winston Company, 1925.
Bulletin of the American Game Protective Association. 19 vols. 1912–30.
Burleigh, Michael, and Wolfgang Wippermannn. *The Racial State: Germany, 1933–1945.* Cambridge: Cambridge University Press, 1991.
Burnham, Frederick Russell. "Conservation." *Natural History* 31, no. 6 (November–December 1931): 672.
Burrows, Edwin G., and Mike Wallace. *Gotham: A History of New York City to 1898.* New York, Oxford: Oxford University Press, 1999.
Calverton, V. F. "The Myth of Nordic Superiority." *Current History* 242 (August 1926): 671–77.

Cameron, Jenks. *The Bureau of Biological Survey: Its History, Activities and Organization.* Baltimore: The Johns Hopkins University Press, 1929.
Carlson, Elof Axel. *Genes, Radiation, and Society: The Life and Work of H. J. Muller.* Ithaca, N.Y., and London: Cornell University Press, 1981.
Caro, Robert A. *The Power Broker: Robert Moses and the Fall of New York.* New York: Alfred A. Knopf, 1974.
Carson, Gerald. *Cornflake Crusade.* New York: Rinehart & Company, 1957.
Carter, Elmer. "Eugenics for the Negro." *Birth Control Review* 16 (1932): 169–70.
Castle, William E. "Biological and Social Consequences of Race Crossing." *American Journal of Physical Anthropology* 9, no. 2 (April–June 1926): 145–56.
——. "Biological and Social Consequences of Race-Crossing." *Journal of Heredity* 15, no. 9 (September 1924): 363–69.
——. "Eugenics." In *Encyclopedia Britannica.* 13th ed. New York, 1926.
Castle, William Ernest, John Merle Coulter, Charles Benedict Davenport, Edward Murray East, and William Lawrence Tower. *Heredity and Eugenics.* Chicago: University of Chicago Press, 1912.
Chamberlain, Houston Stewart. *Foundations of the Nineteenth Century.* 2 vols. Translated by John Lees. New York: John Lane Company, 1912.
Charles Scribner's Sons. *List of Fall Publications–1920.*
——. *List of Spring Publications—1917.*
Chase, Allan. *The Legacy of Malthus: The Social Costs of the New Scientific Racism.* New York: Alfred A. Knopf, 1977.
Chase, Alston. *In a Dark Wood: The Fight over Forests and the Rising Tyranny of Ecology.* Boston, New York: Houghton Mifflin Company, 1995.
Chesler, Ellen. *Woman of Valor: Margaret Sanger and the Birth Control Movement in America.* New York: Simon & Schuster, 1992.
Chesterton, G. K. *Eugenics and Other Evils.* New York: Dodd, Mead & Company, 1927.
Collier, Peter. *The Roosevelts: An American Saga.* New York: Simon & Schuster, 1994.
Congressional Record. Washington, D.C. 1916–47.
Conklin, Edwin G. *The Direction of Human Evolution.* New York: Charles Scribner's Sons, 1921.
——. "Some Biological Aspects of Immigration." *Scribner's Magazine* 69, no. 3 (March 1921): 352–59.
Consolidated Report upon the Activities of the National Research Council, 1919–1932. Washington, D,C.: 1932.
Coolidge, Calvin. "Whose Country Is This?" *Good Housekeeping* 72 (February 1921): 13–14, 106, 109.
Cox, Earnest Sevier. *Black Belt around the World at the High Noon of Colonialism.* Richmond, Va., 1963.
——. *Let My People Go.* Richmond, Va.: White America Society, 1925.
——. *White America.* Richmond, Va.: White America Society, 1923.
Cravens, Hamilton. *The Triumph of Evolution: The Heredity-Environment Controversy, 1900–1941.* Baltimore and London: The Johns Hopkins University Press, 1988.
Creel, George. "Close the Gates!" *Collier's* 69, no. 18 (6 May 1922): 9–10, 18.
——. "Melting Pot or Dumping Ground?" *Collier's* 68, no. 10 (3 September 1921): 9–10, 25–26.

Cronon, E. David. *Black Moses: The Story of Marcus Garvey and the Universal Negro Improvement Association.* Madison: University of Wisconsin Press, 1955.

Curran, Henry H. "The New Immigrant." *Saturday Evening Post* 198, no. 7 (15 August 1925): 25, 83–86.

Curran, Thomas J. *Xenophobia and Immigration, 1820–1930.* Boston: Twayne Publishers, 1975.

Cutright, Paul Russell. *Theodore Roosevelt: The Making of a Conservationist.* Urbana: University of Illinois Press, 1985.

——. *Theodore Roosevelt, the Naturalist.* New York: Harper & Brothers, 1956.

Daniels, Roger. *Coming to America: A History of Immigration and Ethnicity in American Life.* New York: HarperCollins Publishers, 1990.

Darwin, Charles. *The Descent of Man and Selection in Relation to Sex.* New York: D. Appleton and Company, 1896.

Davenport, Charles Benedict. *Heredity in Relation to Eugenics.* New York: Henry Holt and Company, 1911.

——. *Naval Officers: Their Heredity and Development.* Washington, D.C.: Carnegie Institution of Washington, 1919.

——. "Report of Committee on Eugenics." *American Breeders Magazine* 1, no. 2 (Second Quarter 1910): 129.

Davenport, C. B., Madison Grant, F. L. Hoffman, W. H. Holmes, E. A. Hooton, Ales Hrdlicka, G. M. Kober, and T. A. Williams. "Suggestions Relating to the New National Army by the Anthropology Committee of the National Research Council." *Proceedings of the National Academy of Sciences of the United States of America* 3, no. 8 (August 1917): 526–28.

Davenport, Charles Benedict, and M. W. Ireland. *Army Anthropology.* Vol. 15, part 1, of *The Medical Department of the United States Army in the World War.* Washington, D.C.: GPO, 1921.

Davenport, Charles Benedict, and Morris Steggerda. *Race Crossing in Jamaica.* Carnegie Institution of Washington, 1929.

Davis, James J. "America and Her Immigrants." *Congressional Digest* 2, nos. 10–11 (July–August 1923): 291–93.

——. "A Century of Immigration." *Monthly Labor Review* 18, no. 1 (January 1924): 1–19.

——. "Our Labor Shortage and Immigration." *Industrial Management* 65 (June 1923): 321–323.

Deacon, Desley. *Elsie Clews Parsons: Inventing Modern Life.* Chicago and London: University of Chicago Press, 1997.

A Decade of Progress in Eugenics: Scientific Papers of the Third International Congress of Eugenics. Baltimore: Williams and Wilkins Company, 1934.

Degler, Carl N. *In Search of Human Nature: The Decline and Revival of Darwinism in American Social Thought.* Oxford: Oxford University Press, 1991.

Dewitt, John B. *California Redwood Parks and Preserves: A Guide to the Redwood Parks and a Brief History of the Efforts to Save the Redwoods.* 3rd ed., rev. San Francisco: Save-the-Redwoods League, 1993.

Dinnerstein, Leonard. *Antisemitism in America.* New York: Oxford University Press, 1994.

"Discriminating against the American Stock." *World's Work* 48, no. 1 (May 1924): 13–15.

Divine, Robert A. *American Immigration Policy, 1924–1952.* New Haven, Conn.: Yale University Press, 1957.

Dolph, James Andrew. "Bringing Wildlife to the Millions: William Temple Hornaday, the Early Years, 1854–1896." Ph.D. diss., University of Massachusetts, 1975.

Downey, Matthew T. *Carl Campbell Brigham: Scientist and Educator.* Princeton, N.J.: Educational Testing Service, 1961.

Drury, Newton Bishop. "Point Lobos Reserve." *American Forests* (July 1938): 303–10.

Dudziak, Mary L. "Oliver Wendell Holmes as a Eugenic Reformer: Rhetoric in the Writing of Constitutional Law." *Iowa Law Review* 71, no. 3 (March 1986): 833–67.

East, Edward M. *Heredity and Human Affairs.* New York: Charles Scribner's Sons, 1927.

——. *Mankind at the Crossroads.* New York: Charles Scribner's Sons, 1923.

——. "Population." *Scientific Monthly* 10 (June 1920): 603–24.

East, Edward M., and Donald F. Jones. *Inbreeding and Outbreeding: Their Genetic and Sociological Significance.* Philadelphia: J. B. Lippincott Company, 1919.

Edwards, William A. "Racial Purity in Black and White: The Case of Marcus Garvey and Earnest Cox." *Journal of Ethnic Studies* 15, no. 1 (Spring 1987): 117–42.

Ellis, Havelock. *The Task of Social Hygiene.* Boston and New York: Houghton Mifflin Company, 1912.

——. "The World's Racial Problem." Review of *The Rising Tide of Color,* by Lothrop Stoddard. *Birth Control Review* 4, no. 10 (October 1920): 14–16.

Embree, Edwin R. "Facing East." *Atlantic Monthly* 154, no. 3 (September 1934): 349–59.

"End of the 'Melting-Pot' Theory." *Literary Digest* 81, no. 10 (7 June 1924): 14–15.

Engbeck, Joseph H. *State Parks of California, from 1864 to the Present.* Portland, Ore.: Graphic Arts Center Publishing Co., 1980.

Eugenical News. 31 vols. 1916–46.

Eugenics, Genetics and the Family: Volume I of the Scientific Papers of the Second International Congress of Eugenics. Baltimore: Williams & Wilkins Company, 1923.

Eugenics in Race and State: Volume II of the Scientific Papers of the Second International Congress of Eugenics. Baltimore: Williams & Wilkins Company, 1923.

Evans, Hiram Wesley. "The Klan's Fight for Americanism." *North American Review* 223, no. 830 (March–May, 1926): 33–63.

——. *The Menace of Modern Immigration: An Address Delivered on the Occasion of Klan Day at the State Fair of Texas at Dallas, October 24, 1923.* Atlanta: Knights of the Ku Klux Klan, 1924.

Fairchild, Henry Pratt. *Immigration: A World Movement and Its American Significance.* Rev. ed. New York: Macmillan Company, 1925.

——. *The Melting-Pot Mistake.* Boston: Little, Brown and Company, 1926.

Fallon, Valère. *Eugenics.* Translated by Ernest C. Messenger. London: Burns Oates & Washbourne, 1923.

Feingold, Henry L. *The Politics of Rescue: The Roosevelt Administration and the Holocaust, 1938–1945.* New Brunswick, N.J.: Rutgers University Press, 1970.

Field, Geoffrey G. *Evangelist of Race: The Germanic Vision of Houston Stewart Chamberlain.* New York: Columbia University Press, 1981.

——. "Nordic Racism." *Journal of the History of Ideas* 38, no. 3 (July–September 1977): 523–40.

Fish, Stanley. "Affirmative Action and the SAT." *Journal of Blacks in Higher Education,* no. 2 (Winter 1993–94): 83.

Fisher, Irving. *Report of the President of the American Eugenics Society, Inc., June 26, 1926.* New Haven, Conn.: American Eugenics Society, 1926.

Fitzgerald, F. Scott. *The Great Gatsby.* New York: Charles Scribner's Sons, 1925.

Ford, Robert. Review of *White America,* by Earnest Sevier Cox. *Mankind Quarterly* 7, no. 2 (October–December 1966): 117–19.

Forest and Stream. 21 vols. 1885–1905.

"The Fortune Quarterly Survey." *Fortune* 13, no. 1 (January 1936): 46–47, 141–57.

"The Fortune Survey." *Fortune* 19, no. 4 (April 1939): 84–85, 102–8.

Fox, Stephen. *John Muir and His Legacy: The American Conservation Movement.* Boston: Little, Brown and Company, 1981.

Frank, Gelya. "Jews, Multiculturalism, and Boasian Anthropology." *American Anthropologist* 99, no. 4 (December 1997): 731–45.

Freeman, Derek. *Margaret Mead and Samoa: The Making and Unmaking of an Anthropological Myth.* Cambridge, Mass.: Harvard University Press, 1983.

——. "On Franz Boas and the Samoan Researches of Margaret Mead." *Current Anthropology* 32 (1991): 322–30.

Friedlander, Henry. *The Origins of Nazi Genocide: From Euthanasia to the Final Solution.* Chapel Hill: University of North Carolina Press, 1985.

Gabrielson, Ira N. *Wildlife Refuges.* New York: Macmillan Company, 1943.

Galton, Francis. *English Men of Science: Their Nature and Nurture.* London: Macmillan & Co., 1874.

——. *Essays in Eugenics.* London: Eugenics Education Society, 1909.

——. *Hereditary Genius: An Inquiry into Its Laws and Consequences.* Rev ed. New York: D. Appleton and Company, 1883.

——. "Hereditary Talent and Character." *Macmillan's Magazine* 12 (June and August 1865): 157–66, 318–27.

——. *Memories of My Life.* New York: E. P. Dutton and Company, 1909.

Garis, Roy L. *Immigration Restriction: A Study of the Opposition to and Regulation of Immigration into the United States.* New York: Macmillan Company, 1927.

——. *A Short History of the American Bison: Distribution, Habits, Trails, Exterminations, etc.* New York: Charter Printing Service, 1934.

Garretson, Martin S. *A Short History of the American Bison: Distribution, Habits, Trails, Exterminations, etc.* New York: Charter Printing Service, 1934.

Garth, Thomas R. "Eugenics, Euthenics, and Race." *Opportunity* 8 (July 1930): 206–7.

Garvey, Amy Jacques, ed. *Philosophy and Opinions of Marcus Garvey.* 2 vols. New York: Universal Publishing House, 1923–25.

Gasman, Daniel. *The Scientific Origins of National Socialism: Social Darwinism in Ernst Haeckel and the German Monist League.* New York: American Elsevier, 1971.

Gidley, M. "Notes on F. Scott Fitzgerald and the Passing of the Great Race." *Journal of American Studies* 7, no. 2 (August 1973): 171–81.

Gobineau, Arthur de. *The Inequality of Human Races.* Translated by Adrian Collins. 1853–55. Reprint, New York: Howard Fertig, 1967.
Goddard, Henry Herbert. *The Kallikak Family: A Study in the Heredity of Feeble-Mindedness.* New York: Macmillan Company, 1912.
Goethe, C. M. "The Influx of Mexican Amerinds." *Eugenics* 2, no. 1 (January 1929): 6–9.
——. *Seeking to Serve.* Sacramento: Keystone Press, 1949.
Goldenweiser, Alexander. Review of *Les trois races de l'Europe et du monde: Leur genie et leur histoire,* by Paul Buyssens. *American Historical Review* 43, no. 1 (October 1937): 79–81.
Goldfrank, Esther S. *Notes on an Undirected Life: As One Anthropologist Tells It.* Flushing, N.Y.: Queens College Press, 1978.
Goldschmidt, Walter, ed. *The Anthropology of Franz Boas.* New York: American Anthropological Association, 1959.
Gordon, Linda. *Woman's Body, Woman's Right: A Social History of Birth Control in America.* New York: Grossman Publishers, 1976.
Gossett, Thomas F. *Race: The History of an Idea in America.* Dallas: Southern Methodist University Press, 1963.
Gould, Charles W. *America: A Family Matter.* New York: Charles Scribner's Sons, 1920.
Gould, Stephen Jay. "Carrie Buck's Daughter." *Natural History* 93, no. 7 (July 1984): 14–18.
——. *The Mismeasure of Man.* New York: W. W. Norton & Company, 1981.
——. "Singapore's Patrimony (and Matrimony)." *Natural History* 93, no. 5 (May 1984): 22–29.
Graham, Frank, Jr. *Man's Dominion: The Story of Conservation in America.* New York: M. Evans and Company, 1971.
Grant, Joseph D. *Redwoods and Reminiscences.* San Francisco: Save-the-Redwoods League, 1973.
——. *Saving California's Redwoods.* Berkeley, Calif.: Save-the-Redwoods League, 1922.
——. "Twenty-Rings in the Redwoods: An Informal History of the Save-the-Redwoods League." Unpublished manuscript in Save-the-Redwoods League files, San Francisco.
Grant, Madison. "America for the Americans." *Forum* 74 (September 1925): 346–55.
——. "The Caribou." In *Records of North American Big Game,* by Prentiss Gray, 57–70. New York: Derrydale Press, 1932.
——. *The Conquest of a Continent, or the Expansion of Races in America.* New York: Charles Scribner's Sons, 1933.
——. "Discussion of Article on Democracy and Heredity." *Journal of Heredity* 10, no. 4 (April 1919): 164–65.
——. *Early History of Glacier National Park, Montana.* Washington, D.C.: GPO, 1919.
——. "Failures of the Melting Pot." Review of *America: A Family Matter,* by Charles W. Gould. *New York Times,* 12 November 1922, sec. 3, p. 1, col. 1.
——. "George Bird Grinnell." *Review of Reviews* 71, no. 6 (June 1925): 600–602.
——. "Notes on Adirondack Mammals, with Special Reference to the Fur-Bearers." *Eighth and Ninth Reports of the Forest, Fish and Game Commission of the State of New York,* 319–34. Albany: J. B. Lyon Company, 1903.
——. "The Origin of the New York Zoological Society." In *Trail and Camp-Fire,* edited by George Bird Grinnell and Theodore Roosevelt, 313–20. New York: Forest and Stream Publishing Co., 1897.

———. *The Passing of the Great Race, or the Racial Basis of European History.* New York: Charles Scribner's Sons, October 1916.

———. "'Preservation' or 'Restoration'" *Natural History* 33, no. 3 (May–June, 1933): 337–38.

———. *Preserve an America Worth Fighting For.* Berkeley, Calif.: Save-the-Redwoods League, 1921.

———. "The Racial Transformation of America." *North American Review* 219, no. 820 (March 1924): 343–52.

———. "Reform in New York Game Laws." *Harper's Weekly,* 3 October 1896, 978.

———. "Restriction of Immigration: Racial Aspects." *Journal of National Institute of Social Sciences* 7 (1 August 1921): 44–54.

———. "Saving the Redwoods." *National Geographic Magazine* 37, no. 6 (June 1920): 519–36.

———. "Saving the Redwoods: An Account of the Movement During 1919 to Preserve the Redwoods of California." *Zoological Society Bulletin* 22, no. 5 (September 1919): 91–118.

———. "The Vanishing Moose, and Their Extermination in the Adirondacks." *Century* 47 (January 1894): 345–356.

———. *Der Untergang der Grossen Rasse.* Translated by Rudolf Polland. Munich: J. F. Lehmann, 1925.

Grant, Madison, and Charles Stewart Davison, eds. *The Alien in Our Midst, or "Selling Our Birthright for a Mess of Pottage."* New York: Galton Publishing Co., 1930.

———. *The Founders of the Republic on Immigration, Naturalization and Aliens.* New York: Charles Scribner's Sons, 1928.

Grant, Madison, with W. H. Holmes, C. B. Davenport, F. L. Hoffman, G. M. Kober, Ales Hrdlicka, E. A. Hooton, and T. A. Williams. "Suggestions Relating to the New National Army by the Anthropology Committee of the National Research Council." *Proceedings of the National Academy of Sciences of the United States of America* 3, no. 8 (August 1917): 526–28.

Grant, Madison, and William T. Hornaday. Preface to *The National Collection of Heads and Horns, Part II.* New York: New York Zoological Society, 1908.

"A Gratifying Service." *Humboldt Standard* 17 February 1933.

Graves, Henry Solon. "Our Most Urgent Public Park." *American Forests and Forest Life* 32 (February 1926): 74–78, 106.

"A Great Work of Constructive Legislation." *World's Work* 48, no. 2 (June 1924): 121–22.

Gregory, William K. "The Galton Society for the Study of the Origin and Evolution of Man." *American Journal of Physical Anthropology* 4, no. 1 (January–March 1921): 77–79.

———. "The Galton Society for the Study of the Origin and Evolution of Man." *Science* 49, no. 1263 (14 March 1919): 267–68.

———. Review of *The Passing of the Great Race,* by Madison Grant. *Natural History* 22, no. 2 (March–April 1922): 135–36.

Grinnell, George Bird, ed. *American Big Game in Its Haunts.* New York: Forest and Stream Publishing Co., 1904

———. *Hunting at High Altitudes.* New York: Harper & Brothers, 1913.

Grinnell, George Bird, Kermit Roosevelt, W. Redmond Cross, and Prentiss N. Gray, eds. *Hunting Trails on Three Continents.* New York: Windward House, 1933.

Grinnell, George Bird, and Theodore Roosevelt, eds. *American Big-Game Hunting.* New York: Forest and Stream Publishing Co., 1893.

——. *Hunting in Many Lands.* New York: Forest and Stream Publishing Co., 1895.

Grinnell, George Bird, and Charles Sheldon, eds. *Hunting and Conservation.* New Haven, Conn.: Yale University Press, 1925.

Gröning, Gert, and Joachim Wolschke-Bulmahn. "Politics, Planning, and the Protection of Nature: Political Abuse of Early Ecological Ideas in Germany, 1933–45." *Planning Perspectives* 2, no. 2 (May 1987): 128–48.

Günther, Hans F. K. *The Racial Elements of European History.* Translated by G. C. Wheeler. London: Methuen & Co., 1927.

Guyer, Michael F. *Being Well-Born: An Introduction to Heredity and Eugenics.* Indianapolis: Bobbs-Merrill Company, 1927.

Haeckel, Ernst. *The Riddle of the Universe, at the Close of the Nineteenth Century.* Translated by Joseph McCabe. New York: Harper & Brothers Publishers, 1900.

Haller, Mark H. *Eugenics: Hereditarian Attitudes in American Thought.* New Brunswick, N.J.: Rutgers University Press, 1963.

Hankins, Frank H. "Individual Differences and Democratic Theory." *Political Science Quarterly* 38, no. 3 (September 1923): 388–412.

——. *The Racial Basis of Civilization: A Critique of the Nordic Doctrine.* New York: Alfred A. Knopf, 1926.

Haraway, Donna. *Primate Visions: Gender, Race, and Nature in the World of Modern Science.* New York: Routledge, 1989.

Hassencahl, Frances J. "Harry H. Laughlin, 'Expert Eugenics Agent' for the House Committee on Immigration and Naturalization, 1921 to 1931." Ph.D. diss., Case Western Reserve University, 1970.

Hays, Samuel P. *Conservation and the Gospel of Efficiency: The Progressive Conservation Movement, 1890–1920.* Cambridge, Mass.: Harvard University Press, 1959.

Hays, Willet M. "Immigration and Eugenics." *Review of Reviews* 69, no. 4 (April 1924): 405–6.

Heald, Weldon F. "Who Saved the Redwoods?" *Natural History* 62, no. 1 (January 1953): 24–31, 44.

Hedlin, Ethel Wolfskill. "Earnest Cox and Colonization: A White Racist's Response to Black Repatration, 1923–1966." Ph.D. diss., Duke University, 1974.

Hellman, Geoffrey. *Bankers, Bones and Beetles: The First Century of the American Museum of Natural History.* Garden City, N.Y.: Natural History Press, 1968.

Hellwig, David. "Afro-American Reactions to the Japanese." *Phylon* 38, no. 1 (March 1977): 93–104.

Herskovits, Melville J. *The American Negro: A Study in Racial Crossing.* New York: Alfred A. Knopf, 1928.

——. "Brains and the Immigrant." *Nation* 120, no. 3110 (11 February 1925): 139–41.

——. "Extremes and Means in Racial Interpretation." *Journal of Social Forces* 2, no. 4 (May 1924): 550–51.

——. "Franz Boas as Physical Anthropologist." In *Franz Boas, 1858–1942,* 39–51. Memoirs of the American Anthropological Association No. 61 (1943).

——. *Franz Boas: The Science of Man in the Making.* New York: Charles Scribner's Sons, 1953.

——. "Homo Nordicus-U.S.A." Review of *The Conquest of a Continent,* by Madison Grant. *Nation* 138, no. 3575 (10 January 1934): 49.

Higham, John. *Strangers in the Land: Patterns of American Nativism, 1860–1925.* New Brunswick, N.J.: Rutgers University Press, 1955.

Hill, George Edwin. *Quarter-Century Record of the Class of 1887 in Yale College.* New Haven, Conn.: Yale University Press, 1915.

A History of the National Research Council, 1919–1933. Reprint and Circular Series of the National Research Council, Number 106. Washington, D.C., 1933.

Hitler, Adolf. *Hitler's Secret Book.* Translated by Salvator Attanasio. New York: Grove Press, 1961.

——. *Hitler's Secret Conversations, 1941–1945.* New York: Octagon Books, 1972.

——. *Mein Kampf.* Translated by James Murphy. London: Hurst and Blackett, 1939.

Hodgson, Dennis. "The Ideological Origins of the Population Association of America." *Population and Development Review* 17, no. 1 (March 1991): 1–34.

Hofstadter, Richard. *The Age of Reform: From Byran to F.D.R.* New York: Vintage Books, 1955.

Holli, Melvin G., and Peter d'A. Jones, eds. *Biographical Dictionary of American Mayors, 1820–1980.* Westport, Conn.: Greenwood Press, 1981.

Holmes, S. J. *The Negro's Struggle for Survival: A Study in Human Ecology.* Berkeley: University of California Press, 1937.

——. *Studies in Evolution and Eugenics.* New York: Harcourt, Brace and Company, 1923.

Holmes, W. H. "Organization of the Committee on Anthropology of the National Research Council and Its Activities for the Year 1917." *American Journal of Physical Anthropology* 1, no. 1 (January–March 1918): 77–89.

Holmes, W. H., and Ales Hrdlicka. "Report of the Committee on Anthropology." *Proceedings of the National Academy of Sciences of the United States of America* 4, no. 2 (February 1918): 52–54.

Hornaday, William T. "Eighty Fascinating Years: An Autobiography." Unpublished manuscript in container 112, WTH.

——. *Our Vanishing Wild Life: Its Extermination and Preservation.* New York: New York Zoological Society, 1913.

——. *Thirty Years War for Wild Life: Gains and Losses in the Thankless Task.* New York: Charles Scribner's Sons, 1931.

Hrdlicka, Ales. "Preface." *American Journal of Physical Anthropology* 1, no. 1 (January–March 1918): 1–23.

——. Review of *The Conquest of a Continent,* by Madison Grant. *American Journal of Physical Anthropology* 18, no. 3 (January–March 1934): 478–80.

——. Review of *The Racial Basis of Civilization,* by Frank Hankins. *American Journal of Physical Anthropology* 11, no. 3 (April–June 1928): 527.

Humphrey, Seth K. *Mankind: Racial Values and the Racial Prospect.* New York: Charles Scriber's, 1917.

Huntington, Ellsworth. "Exploding the Idea of Race." *Saturday Review of Literature* 13, no. 18 (29 February 1936): 5.

——. *Tomorrow's Children: The Goal of Eugenics.* New York: John Wiley & Sons, 1935.

Huntington, Ellsworth, and Leon F. Whitney. *The Builders of America.* New York: William Morrow & Company, 1927.

Hyatt, Marshall. *Franz Boas, Social Activist: The Dynamics of Ethnicity.* New York: Greenwood Press, 1990.
"Immigration and Birth Control." *Birth Control Review* 7, no. 9 (September 1923): 219.
"Inferior Races in American Fiction." *Literary Digest* 73, no. 5 (29 April 1922): 27–29.
"Intelligent or Unintelligent Birth Control?" *Birth Control Review* 3, no. 5 (May 1919): 12–13.
James, Henry. *The American Scene.* Bloomington: Indiana University Press, 1968.
Jastrow, Joseph. "Why the Germans Have Deemed Themselves Superior." *Outlook* 120, no. 11 (20 November 1918): 455–58.
Jennings, Herbert S. *The Biological Basis of Human Nature.* New York: W. W. Norton & Company, 1930.
——. "Heredity and Environment." *Scientific Monthly* 19, no. 3 (September 1924): 228.
——. *Prometheus, or Biology and the Advancement of Man.* New York: E. P. Dutton & Company, 1925.
Jones, Bartlett C. "Prohibition and Eugenics, 1920–1933." *Journal of the History of Medicine and Allied Sciences* 18, no. 2 (April 1963): 158–72.
Kalikow, Theodora J. "Konrad Lorenz's Ethological Theory: Explanation and Ideology, 1938–1943." *Journal of the History of Biology* 16, no. 1 (Spring 1983): 39–73.
Kallen, H. M. "Racial Mythology." Review of *The Passing of the Great Race,* by Madison Grant. *Dial* 62, no. 742 (17 May 1917): 432–33.
Kamin, Leon J. *The Science and Politics of I.Q.* New York: John Wiley & Sons, 1974.
Kardiner, Abram, and Edward Preble. *They Studied Man.* Cleveland and New York: World Publishing Company, 1961.
Kaufmann, Walter. *Nietzsche: Philosopher, Psychologist, Antichrist.* Princeton, N.J.: Princeton University Press, 1950.
"Keep America 'White'!" *Current Opinion* 74 (April 1923): 399–401.
Kellogg, Vernon. "Nordics, and the Rest." *New Republic* 35, no. 453 (8 August 1923): 278–80.
Kennedy, David M. *Birth Control in America: The Career of Margaret Sanger.* New Haven, Conn., and London: Yale University Press, 1970.
Kennedy, John Michael. "Philanthropy and Science in New York City: The American Museum of Natural History, 1868–1968." Ph.D. diss., Yale University, 1968.
Kevles, Daniel J. *In the Name of Eugenics: Genetics and the Uses of Human Heredity.* Cambridge, Mass.: Harvard University Press, 1985.
Keyserling, Hermann. *The Travel Diary of a Philosopher.* Vol. 2. New York: Harcourt, Brace & Company, 1925.
Kimmelman, Barbara. "The American Breeders' Association: Genetics and Eugenics in an Agricultural Context, 1903–1913." *Social Studies of Science* 13, no. 2 (May 1983): 163–204.
King, Desmond. *Making Americans: Immigration, Race, and the Origins of the Diverse Democracy.* Cambridge, Mass.: Harvard University Press, 2000.
Klineberg, Otto. *Negro Intelligence and Selective Migration.* New York: Columbia University Press, 1935.
——. *Race Differences.* New York: Harper & Brothers Publishers, 1935.
Knerr, George Francis. "The Mayoral Administration of William L. Strong: New York City, 1895–1897." Ph.D. diss., New York University, 1957.

Kranz, Marvin Wolf. "Pioneering in Conservation: A History of the Conservation Movement in New York State, 1865–1903." Ph.D. diss., Syracuse University, 1961.
Kroeber, A. L. *Anthropology.* New York: Harcourt, Brace and Company, 1923.
——. "Franz Boas: The Man." In *Franz Boas, 1858–1942,* 5–26. Memoirs of the American Anthropological Association No. 61 (1943).
——. "The Superorganic." *American Anthropologist* 19, no. 2 (April–June 1917): 167–213.
Kühl, Stefan. *The Nazi Connection: Eugenics, American Racism, and German National Socialism.* New York, Oxford: Oxford University Press, 1994.
Kuznick, Peter J. *Beyond the Laboratory: Scientists as Political Activists in 1930s America.* Chicago: University of Chicago Press, 1987.
Kyne, Peter B. *Never the Twain Shall Meet.* New York: Cosmopolitan Book Corporation, 1923.
Landman, J. H. "Race Betterment by Human Sterilization." *Scientific American* 150, no. 6 (June 1934): 292–95.
Langer, William L. Review of *The Conquest of a Continent,* by Madison Grant. *Foreign Affairs* 12, no. 2 (January 1934): 342.
Lapon, Lenny. *Mass Murderers in White Coats: Psychiatric Genocide in Nazi Germany and the United States.* Springfield, Mass: Psychiatric Genocide Research Institute, 1986.
Lapouge, Georges Vacher de. "L'Anthropologie et la Science Politique." *Revue d'Anthropologie* (1887): 136–57.
——. *L'Aryen: son rôle social.* Paris: Albert Fontemoing, 1899.
——. "The Fundamental Laws of Anthropo-sociology." Translated by Carlos C. Closson. *Journal of Political Economy* 6 (December 1897): 54–92.
Larson, Edward J. *Sex, Race, and Science: Eugenics in the Deep South.* Baltimore: The Johns Hopkins University Press, 1995.
Laughlin, Harry H. *Eugenical Sterilization in the United States.* Chicago: Psychopathic Laboratory of the Municipal Court of Chicago, 1922.
——. "Eugenists on the Place of Birth Control." *Birth Control Review* 10, no. 1 (January 1926): 7.
——. *Immigration and Conquest.* New York: New York State Chamber of Commerce, 1939.
——. "Legal Status of Eugenical Sterilization." *Birth Control Review* 12, no. 3 (March 1928): 78–80.
——. *The Legal Status of Eugenical Sterilization.* Chicago: Fred J. Ringley Co., 1930.
——. *Report of the Committee to Study and to Report on the Best Practical Means of Cutting Off the Defective Germ-Plasm in the American Population. I. The Scope of the Committee's Work.* Bulletin no. 10A of the Eugenics Record Office. Cold Spring Harbor, N.Y.: Eugenics Record Office, February 1914.
Leider, Emily Wortis. *California's Daughter: Gertrude Atherton and Her Times.* Stanford, Calif.: Stanford University Press, 1991.
Lekan, Thomas. "Regionalism and the Politics of Landscape Preservation in the Third Reich." *Environmental History* 4, no. 3 (July 1999): 384–404.
Leopold, Aldo. *Game Management.* New York: Charles Scribner's Sons, 1933.
——. *A Sand County Almanac, and Sketches Here and There.* New York: Oxford University Press, 1949.
——. "Ten New Developments in Game Management." *Bulletin of the American Game Protective Association* 14, no. 3 (July 1925): 7–20.

Levering-Lewis, David. "Parallels and Divergences: Assimilationist Strategies of Afro-American and Jewish Elites from 1910 to the Early 1930s." *Journal of American History* 71, no. 3 (December 1984): 543–64.

Lewis, Sinclair. *Arrowsmith.* New York: Harcourt, Brace & World, 1925.

——. *Babbitt.* New York: Harcourt, Brace & World, 1922.

Leydet, François. *The Last Redwoods, and the Parkland of Redwood Creek.* San Francisco: Sierra Club, 1969.

Lifton, Robert J. *The Nazi Doctors: Medical Killing and the Psychology of Genocide.* New York: Basic Books, 1986.

Linden, Grace. "The Prairie Creek Redwoods." *Natural History* 32, no. 2 (March–April 1932): 148–58.

Little, C. C. "The Nordic in America." Review of *The Conquest of a Continent,* by Madison Grant. *Journal of Heredity* 25, no. 12 (December 1934): 491–92.

[Lorimer, George Horace, and Frederick S. Bigelow]. "America Last." *Saturday Evening Post* 194, no. 36 (4 March 1922): 22.

——. "American Laws for America." *Saturday Evening Post* 196, no. 39 (29 March 1924): 24.

——. "The Burbanks of a People." *Saturday Evening Post* 193, no. 44 (30 April 1921): 20.

——. "Cashing In." *Saturday Evening Post* 193, no. 17 (23 October 1920): 26.

——. "The Great American Myth." *Saturday Evening Post* 193, no. 45 (7 May 1921): 20.

——. "It's Now or Never." *Saturday Evening Post* 194, no. 33 (11 February 1922): 22.

——. "Minute Men." *Saturday Evening Post* 196, no. 32 (2 February 1924): 24.

——. "No Admittance." *Saturday Evening Post* 193, no. 28 (8 January 1921): 24.

——. "They Want Unrestricted Immigration." *Saturday Evening Post* 196, no. 40 (5 April 1924): 36.

——. "A Word to the Wise." *Saturday Evening Post* 195, no. 26 (23 December 1922): 20.

Lowie, Robert H. "The Eugenicist's Programme." *Freeman* 4, no. 84 (19 October 1921): 129–30.

——. "Mr. Grant's Apologia." *Freeman* 4, no. 98 (25 January 1922): 476–78.

Ludmerer, Kenneth M. *Genetics and American Society.* Baltimore: The Johns Hopkins University Press, 1972.

MacDonald, Kevin. *The Culture of Critique: An Evolutionary Analysis of Jewish Involvement in Twentieth-Century Intellectual and Political Movements.* Westport, Conn.: Praeger, 1998.

MacDowell, E. Carleton. "Charles Benedict Davenport, 1866–1944: A Study of Conflicting Influences." *Bios* 17, no. 1 (March 1946): 3–50.

Martin, Tony. *Race First: The Ideological and Organizational Struggles of Marcus Garvey and the Universal Negro Improvement Association.* Dover, Mass.: Majority Press, 1976.

Masters, Edgar Lee. "The Great Race Passes." *Reedy's Mirror* 24, no. 2 (8 January 1920), 22.

McCormick, C. O. "Eugenic Sterilization." *Birth Control Review* 16, no. 8 (October 1932): 241–42.

McDougall, William. *Is America Safe for Democracy?* New York: Charles Scribner's Sons, 1921.

McWilliams, Carey. *A Mask for Privilege: Anti-Semitism in America.* Boston: Little, Brown and Company, 1948.

——. *Prejudice: Japanese-Americans, Symbol of Racial Intolerance.* Boston: Little, Brown and Company, 1945.

———. "Racism on the West Coast." *New Republic* 110, no. 22 (29 May 1944): 732–33.
Mead, Margaret. "Group Intelligence Tests and Linguistic Disability among Italian Children." *School and Society* 25, no. 642 (16 April 1927): 465–68.
———. "The Methodology of Racial Testing: Its Significance for Sociology." *American Journal of Sociology* 31, no. 5 (March 1926): 657–67.
Mehler, Barry. "Eliminating the Inferior: American and Nazi Sterilization Programs." *Science for the People* (November–December 1987): 14–18.
Mehler, Barry. "A History of the American Eugenics Society, 1921–1940." Ph.D. diss., University of Illinois at Urbana-Champagne, 1988.
Merriam, John C. "Forest Windows." *Scribner's Magazine* 83, no. 6 (June 1928): 733–37.
———. *The Highest Uses of the Redwoods.* Berkeley, Calif.: Save-the-Redwoods League, 1941.
Meyer, Annie Nathan. "A Little Hitler in the New York Zoo." *Opinion* (May 1935): 33–34.
Michaelis, David. *N. C. Wyeth: A Biography.* New York: Alfred A. Knopf, 1998.
Mighetto, Lisa. *Wild Animals and American Environmental Ethics.* Tucson: University of Arizona Press, 1991.
Miller, Don Carlos. "Founder of the World's Largest Summer School." *American Forests* (September 1961): 12–14, 54–55.
Miller, Perry. *The New England Mind: From Colony to Province.* Cambridge, Mass.: The Belknap Press of Harvard University Press, 1953.
Mintz, Frank. *The Liberty Lobby and the American Right: Race, Conspiracy, and Culture.* Westport, Conn.: Greenwood Press, 1985.
Mitscherlich, A., and F. Mielke. *The Death Doctors.* Translated by James Cleugh. London: Elek Books, 1962.
Montagu, Ashley. *Man's Most Dangerous Myth: The Fallacy of Race.* 5th ed. New York: Oxford University Press, 1974.
Morgan, Thomas Hunt. *Evolution and Genetics.* Princeton, N.J.: Princeton University Press, 1925.
Morison, Elting E. *The Letters of Theodore Roosevelt.* 8 vols. Cambridge, Mass.: Harvard University Press, 1954.
———. *Turmoil and Tradition: A Study of the Life and Times of Henry L. Stimson.* Boston: Houghton Mifflin Company, 1960.
Morse, Arthur D. *While Six Million Died: A Chronicle of American Apathy.* New York: Hart Publishing Company, 1967.
Muir, John. "Save the Redwoods." *Sierra Club Bulletin* 11, no. 1 (1920): 1–4.
Muller-Hill, Benno. *Murderous Science: Elimination by Scientific Selection of Jews, Gypsies, and Others, Germany, 1933–1945.* Oxford: Oxford University Press, 1988.
Murphy, Robert. *Wild Sanctuaries: Our National Wildlife Refuges—a Heritage Restored.* New York: E. P. Dutton & Co., 1968.
Nash, Gary. *Forbidden Love: The Secret History of Mixed-Race America.* New York: Henry Holt and Company, 1999.
The National Collection of Heads and Horns. 2 parts. New York: New York Zoological Society, 1907, 1908.
New York Times. 1865–1952.
Newby, I. A. *Jim Crow's Defense: Anti-Negro Thought in America, 1900–1930.* Baton Rouge: Louisiana State University Press, 1965.

Nietzsche, Friedrich. *Thus Spake Zarathustra: A Book for Every One and No One.* Translated by R. J. Hollingdale. London: Penguin Books, 1961.
"Notes." *The American Museum Journal* 18, no. 5 (May 1918): 405.
"Oatlands." *Architecture* 38, no. 5 (November 1918): 297–300.
O'Brien, Jim. "Environmentalism as a Mass Movement: Historical Notes." *Radical America* 17, nos. 2–3 (March–June 1983): 7–27.
Official Proceedings of the Second National Conference on Race Betterment. Battle Creek, Mich.: Race Betterment Foundation, 1915.
Olmsted, Frederick Law. *Report of the State Park Survey of California.* Sacramento: California State Printing Office, 1929.
Osborn, Henry Fairfield. *The Age of Mammals, in Europe, Asia and North America.* New York: MacMillan Company, 1910.
———. *Creative Education, in School, College, University, and Museum.* New York: Charles Scribner's Sons, 1927.
———. "The Evolution of the Human Races." *Natural History* 26, no. 1 (January–February 1926): 3–13.
———. *Fifty-two Years of Research, Observation and Publication, 1877–1929: A Life Adventure in Breadth and Depth.* New York: Charles Scribner's Sons, 1930.
———. "The Present Problem of Heredity." *Atlantic Monthly* 67, no. 401 (March 1891): 353–64.
———. "Sequoia—the Auld Lang Syne of Trees." *Natural History* 19, no. 6 (December 1919): 598–613.
"Our New 'Nordic' Immigration Policy." *Literary Digest* 81, no. 6 (10 May 1924): 12–13.
Overstreet, H. A. "The Mind of the Negro." *Saturday Review of Literature* 28, no. 36 (8 September 1945): 7–9.
Pascoe, Peggy. "Miscegenation Law, Court Cases, and Ideologies of 'Race' in Twentieth-Century America." *Journal of American History* 83, no.1 (June 1996): 44–69.
"The Passing of a Great American." *Argonaut* (4 June 1937): 3.
Paul, Diane B. *Controlling Human Heredity, 1865 to the Present.* Atlantic Highlands, N.J.: Humanities Press, 1995.
Pearl, Raymond. "The Biology of Superiority." *American Mercury* 12, no. 47 (November 1927): 257–66.
———. "Breeding Better Men: The New Science of Eugenics Which Would Elevate the Race by Producing Higher Types." *World's Work* 15, no. 3 (January 1908): 9818–24.
Pearson, Karl. *The Life, Letters and Labours of Francis Galton.* Cambridge: The University Press, 1930.
Phelps, William Lyon. "As I Like It." *Scribner's Magazine* 96, no. 2 (August 1934): 109–12.
———. *Autobiography.* New York: Oxford University Press, 1939.
Pickens, Donald K. *Eugenics and the Progressives.* Nashville, Tenn.: Vanderbilt University Press, 1968.
Pinchot, Gifford. *Breaking New Ground.* New York: Harcourt, Brace and Company, 1947.
———. *The Fight for Conservation.* New York: Doubleday, Page & Company, 1910.
Pinter, Rudolf. "Is the Intelligence of the American Nation on the Decline." Review of *A Study of American Intelligence,* by C. C. Brigham. *Journal of Educational Psychology* 14, no. 1 (January 1923): 184–85.

Plecker, W. A. *Eugenics in Relation to the New Family.* Richmond, Va.: Bureau of Vital Statistics, 1924.

——. "Shall We All Be Mulattoes?" *Literary Digest* 84, no. 10 (7 March 1925): 23–24.

Poliakov, Leon. *The Aryan Myth: A History of Racist and Nationalist Ideas in Europe.* Translated by Edmund Howard. London: Sussex University Press, 1974.

——. *The History of Anti-Semitism.* Translated by George Klim. New York: Vanguard Press, 1985.

Popenoe, Paul. "Birth Control and Eugenics." *Birth Control Review* 1, no. 3 (April–May 1917): 6.

——. "Eugenic Sterilization in California." *Journal of Heredity* 19, no. 9 (September 1928): 405–11.

——. "The German Sterilization Law." *Journal of Heredity* 25, no. 7 (July 1934): 257–60.

Popenoe, Paul Bowman, and Roswell Hill Johnson. *Applied Eugenics.* New York: Macmillan Company, 1918.

"Praise for Nazis." *Time,* 26, no. 11 (9 September 1935): 20–21.

Proceedings of the First National Conference on Race Betterment. Battle Creek, Mich.: Race Betterment Foundation, 1914.

Proctor, Robert N. "Eugenics among Social Sciences: Hereditarian Thought in Germany and the United States." In *The Estate of Social Knowledge,* edited by JoAnne Brown and David K. van Keuren, 175–208. Baltimore: The Johns Hopkins University Press, 1991.

——. "Nazi Doctors, Racial Medicine, and Human Experimentation." In *The Nazi Doctors and the Nuremberg Code: Human Rights in Human Experimentation,* edited by George J. Annas and Michael A. Grodin, 17–31. New York, Oxford: Oxford University Press, 1992.

——. *Racial Hygiene: Medicine under the Nazis.* Cambridge, Mass., and London: Harvard University Press, 1988.

Provine, W. "Geneticists and Race." *American Zoologist* 26, no. 3 (1986): 857–87.

Rainger, Ronald. *An Agenda for Antiquity: Henry Fairfield Osborn & Vertebrate Paleontology at the American Museum of Natural History, 1890–1935.* Tuscaloosa: University of Alabama Press, 1991.

Reiger, John F. *American Sportsmen and the Origins of Conservation.* Rev. ed. Norman: University of Oklahoma Press, 1975.

Reilly, Philip R. *The Surgical Solution: A History of Involuntary Sterilization in the United States.* Baltimore: The Johns Hopkins University Press, 1991.

Report of the National Academy of Sciences. 3 vols. Washington, D.C.: GPO, 1917–19.

Review of *The Conquest of a Continent,* by Madison Grant. *Christian Century* 50, no. 52 (20 December 1933): 1610.

Review of *The Passing of the Great Race,* by Madison Grant. *Nation* 104, no. 2703 (19 April 1917): 466.

Rice, Thurman. *Racial Hygiene: A Practical Discussion of Eugenics and Race Culture.* New York: Macmillan Company, 1929.

Ripley, William Z. "The European Population of the United States." *Journal of the Royal Anthropological Institute* 38 (July–December 1908): 221–40. Published in the United States as "Races in the United States." *Atlantic Monthly* (December 1908): 745–59.

——. *The Races of Europe: A Sociological Study.* New York: D. Appleton and Company, 1899.

——. *A Selected Bibliography of the Anthropology and Ethnology of Europe.* New York: D. Appleton and Company, 1899.
Rischin, Moses, ed. *The Jews of North America.* Detroit: Wayne State University Press, 1987.
——. *The Promised City: New York's Jews, 1870–1914.* Cambridge, Mass.: Harvard University Press, 1962.
Roberts, Kenneth. "And West Is West." *Saturday Evening Post* 196, no. 37 (15 March 1924): 12–13, 90–94.
——. "Beer Worshippers." *Saturday Evening Post* 193, no. 34 (19 February 1921): 6–7, 73–88.
——. "East Is East." *Saturday Evening Post* 196, no. 34 (23 February 1924): 6–7, 138–46.
——. "The Existence of an Emergency." *Saturday Evening Post* 193, no. 44 (30 April 1921): 3–4, 86–94.
——. "Florida Loafing." *Saturday Evening Post* 196, no. 46 (17 May 1924): 20–21, 125–33.
——. "Lest We Forget." *Saturday Evening Post* 195, no. 44 (28 April 1923): 4, 158–62.
——. "Plain Remarks on Immigration for Plain Americans." *Saturday Evening Post* 193, no. 33 (12 February 1921): 21–22, 44–47.
——. "Ports of Embarkation." *Saturday Evening Post* 193, no. 45 (7 May 1921): 12, 72–78.
——. "Scotland for Scotch." *Saturday Evening Post* 193, no. 36 (5 March 1921): 3–4, 79–92.
——. "Slow Poison." *Saturday Evening Post* 196, no. 32 (2 February 1924): 8–9, 54–58.
——. *Why Europe Leaves Home.* New York: Bobbs-Merrill Company, 1922.
Rogin, Michael. *Blackface, White Noise: Jewish Immigrants in the Hollywood Melting Pot.* Berkeley: University of California Press, 1996.
Roosevelt, Nicholas. *Conservation: Now or Never.* New York: Dodd, Mead & Company, 1970.
Roosevelt, Theodore. "The Boone and Crockett Club." *Harper's Weekly* 37, no. 1891 (18 March 1893): 267.
——. "The Foundations of the Nineteenth Century." *Outlook* 98 (29 July 1911): 728–31.
——. "A Letter from President Roosevelt on Race Suicide." *Review of Reviews* (May 1907): 550–51.
——. *Literary Essays.* Vol. 12 of *The Works of Theodore Roosevelt.* New York: Charles Scribner's Sons, 1926.
——. "My Life as a Naturalist." *American Museum Journal* 18, no. 5 (May 1918): 321–50.
——. "Our Vanishing Wild Life." *Outlook* 103 (25 January 1913): 161–62.
——. "Race Decadence." *Outlook* 97 (8 April 1911): 763–69.
Rose, June. *Marie Stopes and the Sexual Revolution.* London, Boston: Faber and Faber, 1992.
Rosenberg, Alfred. *Der Mythus des 20. Jahrhunderts.* Munich: Hoheneichen-Verlag, 1935.
Rosenberg, Charles E. "Charles Benedict Davenport and the Beginning of Human Genetics." *Bulletin of the History of Medicine* 35, no. 3 (May–June 1961): 266–76.
Ross, Edward A. "The Causes of Race Superiority." *Annals of the American Academy of Political and Social Science* 18 (July 1901): 67–89.
——. *Seventy Years of It: An Autobiography.* New York: D. Appleton-Century Company, 1936.
Rüdin, Ernst. "Eugenical Sterilization: An Urgent Need." *Birth Control Review* 17, no. 4 (April 1932): 102–4.
Sadler, William. *Long Heads and Round Heads, or What's the Matter With Germany.* Chicago: A. C. McClurg & Co., 1918.

——. *Race Decadence: An Examination of the Causes of Racial Degeneracy in the United States.* Chicago: A. C. McClurg & Co., 1922.
Samelson, Franz. "On the Science and Politics of the IQ." *Social Research* 42, no. 3 (Autumn 1975): 467–88.
——. "Putting Psychology on the Map: Ideology and Intelligence Testing." In *Psychology in Social Context,* ed. Allan R. Buss. New York: Irvington Publishers, 1979.
——. "World War I Intelligence Testing and the Development of Psychology." *Journal of the History of the Behavioral Sciences* 13, no. 3 (July 1977): 274–82.
Sanger, Margaret. "Address of Welcome to the Sixth International Neo-Malthusian and Birth Control Conference." *Birth Control Review* 9, no. 4 (April 1925): 99–100.
——. "Birth Control and Racial Betterment." *Birth Control Review* 3, no. 2 (February 1919): 11–12.
——. "The Function of Sterilization." *Birth Control Review* 10, no. 10 (October 1926): 299.
——. "The Meaning of Radio Birth Control." *Birth Control Review* 8, no. 4 (April 1924): 110–11.
——. "The Need for Birth Control in America." In *Birth Control, Facts and Responsibilities,* edited by Adolf Meyer. Baltimore: Williams and Wilkins Company, 1925.
——. *The Pivot of Civilization.* New York: Brentano's, 1922.
——. "The Reproduction Rate of Genius." *Eugenics* 2, no. 3 (March 1929): 23.
——, ed. *The Sixth International Neo-Malthusian and Birth Control Conference.* 4 vols. New York: American Birth Control League, 1926.
——. *Woman and the New Race.* New York: Brentano's, 1920.
Saveth, Edward. *American Historians and European Immigrants, 1875–1925.* New York: Columbia University Press, 1948.
Sayers, James Denson. *Can the White Race Survive?* Washington, D.C.: Independent Publishing Company, 1929.
Schafft, Gretchen E. *From Racism to Genocide: Anthropology in the Third Reich.* Urbana and Chicago: University of Illinois Press, 2004.
Schama, Simon. *Landscape and Memory.* New York: Alfred A. Knopf, 1995.
Schauer, Andrew. "Charles Matthias Goethe." Unpublished, in library of California State University, Sacramento.
Schrepfer, Susan R. *The Fight to Save the Redwoods: A History of Environmental Reform, 1917–1978.* Madison: University of Wisconsin Press, 1983.
Schudson, Michael. "Paper Tigers." *Lingua Franca* 7, no. 6 (August 1997): 49–56.
Schuyler, George. *Black No More.* New York: Macaulay Company, 1931.
Schwarz, Richard W. *John Harvey Kellogg, M. D.* Berrien Springs, Mich.: Andrews University Press, 1981.
Selden, Steven. *Inheriting Shame: The Story of Eugenics and Racism in America.* New York and London: Teachers College Press, 1999.
"Selected Reading List on Sterilization." *Birth Control Review* 17, no. 4 (April 1933): 111.
Seton, Ernest Thompson. *Lives of Game Animals.* 4 vols. New York: Doubleday, Doran & Company, 1909.
Shankland, Robert. *Steve Mather of the National Parks.* New York: Alfred A. Knopf, 1951.

Shapiro, Thomas. *Population Control Politics: Women, Sterilization and Reproductive Choice.* Philadelphia: Temple University Press, 1985.

Sherman, Richard B. "'The Last Stand': The Fight for Racial Integrity in Virginia in the 1920s." *Journal of Southern History* 54, no. 1 (February 1988): 69–92.

Shipman, Pat. *The Evolution of Racism: Human Differences and the Use and Abuse of Science.* New York: Simon & Schuster, 1994.

Slotkin, Richard. *Gunfighter Nation: The Myth of the Frontier in Twentieth-Century America.* New York: Atheneum, 1992.

Smertenko, Johan J. "The Claim of 'Nordic' Superiority." *Current History* 20 (April 1924): 15–23.

Smith, J. David. *The Eugenic Assault on America: Scenes in Red, White, and Black.* Fairfax, Va.: George Mason University Press, 1993.

———. *Managing White Supremacy: Race, Politics, and Citizenship in Jim Crow Virginia.* Chapel Hill: University of North Carolina Press, 2002.

———. *Minds Made Feeble: The Myth and Legacy of the Kallikaks.* Rockville, Md.: Aspen Systems Corporation, 1985.

Smith, J. David, and K. Ray Nelson. *The Sterilization of Carrie Buck.* Far Hills, N.J.: New Horizon Press, 1989.

Snyderman, Mark, and R. J. Herrnstein. "Intelligence Tests and the Immigration Act of 1924." *American Psychologist* 38, no. 9 (September 1983): 986–94.

Society of Colonial Wars. *Annual Register of Officers and Members of the Society of Colonial Wars.* New York: Society of Colonial Wars, 1894–95.

Solomon, Barbara Miller. *Ancestors and Immigrants: A Changing New England Tradition.* Cambridge, Mass.: Harvard University Press, 1956.

Sorokin, Pitirim. *Contemporary Sociological Theories.* New York: Harper & Brothers, 1928.

Spence, Mark David. *Dispossessing the Wilderness: Indian Removal and the Making of the National Parks.* New York, Oxford: Oxford University Press, 1999.

Spiro, Jonathan Peter. "Patrician Racist: The Evolution of Madison Grant." Ph.D. diss., University of California, Berkeley, 2000.

St. Barbe Baker, Richard. *The Redwoods.* London: L. Drummond, 1946.

Stanton, William. *The Leopard's Spots: Scientific Attitudes toward Race in America, 1815–59.* Chicago: University of Chicago Press, 1960.

Starn, Orin. *Ishi's Brain: In Search of America's Last "Wild" Indian.* New York: W. W. Norton & Company, 2004.

Steiner, H. Arthur. "Fascism in America?" *American Political Science Review* 29, no. 5 (October 1935): 821–30.

Sterling, Keir B. *Last of the Naturalists: The Career of C. Hart Merriam.* Rev. ed. New York: Arno Press, 1977.

Stocking, George W. *Bones, Bodies, Behavior: Essays on Biological Anthropology.* Madison: University of Wisconsin Press, 1988.

———. *The Ethnographer's Magic and Other Essays in the History of Anthropology.* Madison: University of Wisconsin Press, 1992.

———. *Race, Culture, and Evolution: Essays in the History of Anthropology.* Chicago and London: University of Chicago Press, 1982.

Stoddard, Lothrop. *The French Revolution in San Domingo.* Boston and New York: Houghton Mifflin Company, 1914.

——. "Impasse at the Color-Line." *Forum* 78, no. 4 (October 1927): 510–19.

——. *Into the Darkness: Nazi Germany Today.* New York: Duell, Sloan & Pearce, 1940.

——. "Racial Realities in Europe." *Saturday Evening Post* 196, no. 38 (22 March 1924): 14–15, 156–58.

——. *Racial Realities in Europe.* New York: Charles Scribner's Sons, 1924.

——. "Realism: The True Challenge of Fascism." *Harper's* (October 1927): 578–83.

——. *Re-forging America: The Story of Our Nationhood.* New York: Charles Scribner's Sons, 1927.

——. *The Revolt against Civilization: The Menace of the Under Man.* New York: Charles Scribner's Sons, 1922.

——. *The Rising Tide of Color against White World-Supremacy.* New York: Charles Scribner's Sons, 1920.

Strother, French. "The Immigration Peril." *World's Work* 46, no. 6 (23 October 1923): 633–37.

Tebbel, John. *George Horace Lorimer and the Saturday Evening Post.* Garden City, N.Y.: Doubleday & Company, 1948.

Teresi, Dick. "Monster of the Tub." *Discover* 19, no. 4 (April 1998): 86–92.

Terman, Lewis M. "Were We Born That Way?" *World's Work* 44, no. 6 (October 1922): 655–60.

Tilden, Freeman. *The State Parks: Their Meaning in American Life.* New York: Alfred A. Knopf, 1962.

Tober, James A. *Who Owns the Wildlife? The Political Economy of Conservation in Nineteenth-Century America.* Westport, Conn.: Greenwood Press, 1981.

Todorov, Tzvetan. *On Human Diversity: Nationalism, Racism, and Exoticism in French Thought.* Translated by Catherine Porter. Cambridge, Mass.: Harvard University Press, 1993.

Tolbert, Emory. "Outpost Garveyism and the UNIA Rank and File." *Journal of Black Studies* 5, no. 3 (March 1975): 233–53.

Tozzer, Alfred M. *Social Origins and Social Continuities.* New York: Macmillan Company, 1925.

Trefethen, James B. *An American Crusade for Wildlife.* New York: Winchester Press, 1975.

——. *Crusade for Wildlife: Highlights in Conservation Progress.* New York: Boone and Crockett Club, 1961.

——. *Wildlife Management and Conservation.* Boston: D. C. Heath and Company, 1964.

Trials of War Criminals before the Nuernberg Military Tribunals under Control Council Law No. 10. Vol. 1, 2, 4. Washington, D.C.: GPO, 1949.

Trombley, Stephen. *The Right to Reproduce: A History of Coercive Sterilization.* London: Weidenfeld and Nicolson, 1988.

Tucker, William H. *The Funding of Scientific Racism: Wickliffe Draper and the Pioneer Fund.* Urbana and Chicago: University of Illinois Press, 2002.

Turlish, Lewis A. "*The Rising Tide of Color*: A Note on the Historicism of *The Great Gatsby.*" *American Literature* 43, no. 3 (November 1971): 442–44.

Udall, Stewart L. *The Quiet Crisis.* New York: Holt, Rinehart and Winston, 1963.

U.S. Congress. House. Committee on Immigration and Naturalization. *Analysis of America's Modern Melting Pot: Hearings.* (Report of Dr. Harry H. Laughlin.) 67th Cong., 3d sess., 21 November 1922.

——. *Biological Aspects of Immigration: Hearings.* (Statement of Dr. Harry H. Laughlin.) 66th Cong., 2d sess., 16–17 April 1920.

——. *Europe as an Emigrant-Exporting Continent and the United States as an Immigrant-Receiving Nation: Hearings.* (Statement of Dr. Harry H. Laughlin.) 68th Cong., 1st sess., 8 March 1924.

——. *Immigration.* (Hearing before the Committee on Immigration and Naturalization.) 67th Cong., 2d sess., 12–20 December 1921.

——. *Restriction of Immigration: Hearing before the Committee on Immigration and Naturalization.* (On H.R. 5, H.R. 101, and H.R. 561.) 68th Cong., 1st sess., 26 December 1923–19 January 1924.

——. *Restriction of Immigration: Report No. 176, part 2.* (Minority Report to Accompany H.R. 6540.) 68th Cong., 1st sess., 15 February 1924.

——. *Restriction of Immigration: Report No. 350.* (To Accompany H.R. 7995.) 68th Cong., 1st sess., 24 March 1924.

——. *Restriction of Immigration: Report No. 350, part 2.* (Minority Report to Accompany H.R. 7995.) 68th Cong., 1st sess., 27 March 1924.

——. *Temporary Suspension of Immigration: Report to Accompany H.R. 14461.* 66th Cong., 3d sess., 1920.

——. Senate. Committee on Immigration. *Amendment to Immigration Law: Hearing before the Committee on Immigration, United States Senate.* 67th Cong., 4th sess., 24 January 1923.

——. *Amendment to Immigration Law: Hearing before the Committee on Immigration, United States Senate.* 67th Cong., 4th sess., 20 February 1923.

——. *Emergency Immigration Legislation: Hearing before Committee on Immigration.* 66th Cong., 3d sess., 3–26 January 1921.

——. *Selective Immigration Legislation: Hearings before the Committee on Immigration, United States Senate.* 68th Cong., 1st sess., 13 February–8 April 1924.

——. Committee on Veterans' Affairs. *Medal of Honor Recipients, 1863–1973.* Washington, D.C.: GPO, 1973.

Walker, Francis A. "Immigration and Degradation." *Forum* 11 (August 1891): 634–44.

——. "Restriction of Immigration." *Atlantic Monthly* 77, no. 464 (June 1896): 822–29.

Wallis, Wilson D. "Race and Culture." *Scientific Monthly* 23 (October 1926): 313–21.

Walter, Herbert Eugene. *Genetics: An Introduction to the Study of Heredity.* Rev. ed. New York: Macmillan Company, 1922.

——. "What Next in Immigration Legislation?" *Scientific Monthly* 15, no. 36 (December 1922): 561–69.

Wang, Peter. *Legislating Normalcy: the Immigration Act of 1924.* San Francisco: R and E Research Associates, 1975.

Ward, Robert DeCourcy. "What Next in Immigration Legislation?" *Scientific Monthly* 15, no. 36 (December 1922): 561–569.

Washington, J. R., ed. *Jews in Black Perspective: A Dialogue.* Rutherford: Fairleigh Dickinson University, 1984.

Watson, John B. *Behaviorism.* New York: W. W. Norton, 1925.
——. "What Is Behaviorism?" *Harper's* 152 (May 1926): 723–29.
Weindling, Paul. *Health, Race, and German Politics between National Unification and Nazism, 1870–1945.* Cambridge: Cambridge University Press, 1989.
——. "Weimar Eugenics: The Kaiser Wilhelm Institute for Anthropology, Human Heredity, and Eugenics in Social Context." *Annals of Science* 42, no. 3 (May 1985): 303–18.
Weinrich, Max. *Hitler's Professors: The Part of Scholarship in Germay's Crimes against the Jewish People.* New York: Yiddish Scientific Institute, 1946.
Weiss, John. *Ideology of Death: Why the Holocaust Happened in Germany.* Chicago: I. R. Dee, 1996.
Whitman, Walt. *Complete Poetry and Selected Prose and Letters.* Edited by Emory Holloway. London: Nonesuch Press, 1963.
Whitney, Leon F. "The American Eugenics Society: A Survey of Its Work." *Eugenics* 3, no. 7 (July 1930): 257.
——. *The Case for Sterilization.* New York: Frederick A. Stokes Company, 1934.
——. "A Hunt for Society's Danger Spot." *Eugenics* 1, no. 1 (October 1928): 25.
——. *Pigeon City.* New York: Robert M. McBride & Company, 1931.
Wiggam, Albert Edward. *The New Decalogue of Science.* Indianapolis: Bobbs-Merrill Company, 1922.
——. *The Next Age of Man.* Indianapolis: Bobbs-Merrill Company, 1927.
Wilson, Woodrow. *A History of the American People.* Vol. 5. New York: Harper & Brothers Publishers, 1908.
Wissler, Clark. "Final Report of the Committee on Scientific Problems of Human Migration." *Reprint and Circular Series of the National Research Council, Number 87.* Washington, D.C., 1929.
——. *Man and Culture.* New York: Thomas Y. Crowell Company, 1923.
——. "Opportunities for Coordination in Anthropological and Psychological Research." *American Anthropologist* 22, no. 1 (January–March 1920): 1–12.
Wister, Owen. *Roosevelt: The Story of a Friendship.* New York: Macmillan Company, 1930.
——. "Shall We Let the Cuckoos Crowd Us Out of Our Nest?" *American Magazine* 91, no. 3 (March 1924): 47.
Wittke, Carl. Review of *The Conquest of a Continent,* by Madison Grant. *Mississippi Valley Historical Review* 20, no. 4 (March 1934): 589–90.
Woodbury, Robert Lewis. "William Kent: Progressive Gadfly, 1864–1928." Ph.D. diss., Yale University, 1967.
Woodruff, Charles Edward. *Expansion of Races.* New York: Rebman Company, 1909.
Woods, Frederick Adams. "A Review of Reviews of Madison Grant's *Passing of the Great Race.*" *Journal of Heredity* 14, no. 2 (May 1923): 93–95.
Wyeth, Betsy James, ed. *The Wyeths: The Letters of N. C. Wyeth, 1901–1945.* Boston: Gambit, 1971.
Yard, Robert Sterling. *Our Federal Lands: A Romance of American Development.* New York: Charles Scribner's Sons, 1928.
Yerkes, Robert. "Testing the Human Mind." *Atlantic Monthly* 131, no. 3 (March 1923): 358–70.

——. *The Work of the Committee on Scientific Problems of Human Migration.* Reprint and Circular Series of the National Research Council, Number 58. Washington, D.C., 1924.

Zangwill, Israel. *The Melting Pot: Drama in Four Acts.* New York: Macmillan Company, 1920.

Zoological Society Bulletin. (Known as *Animal Kingdom* since 1942.) 64 vols. New York, 1898–1961.

Zuckerman, Seth. *Saving Our Ancient Forests.* Los Angeles: Living Planet Press, 1991.

Index